INTRODUCTION TO MICROELECTRONIC DEVICES

DAVID L. PULFREY

University of British Columbia
Vancouver, Canada

N. GARRY TARR

Carleton University
Ottawa, Canada

PRENTICE HALL SERIES IN SOLID STATE PHYSICAL ELECTRONICS

Nick Holonyak, Jr., Editor

Prentice-Hall International, Inc.

A Division of Simon & Schuster
Englewood Cliffs, NJ 07632

Printed in the United States of America

10 9 8 7 6 5 4 3 2 1

ISBN 0-13-477275-X

Prentice-Hall International (UK) Limited, *London*
Prentice-Hall of Australia Pty. Limited, *Sydney*
Prentice-Hall Canada Inc., *Toronto*
Prentice-Hall Hispanoamericana, S.A., *Mexico*
Prentice-Hall of India Private Limited, *New Delhi*
Prentice-Hall of Japan, Inc., *Tokyo*
Simon & Schuster Asia Pte. Ltd., *Singapore*
Editora Prentice-Hall do Brasil, Ltda., *Rio de Janeiro*
Prentice-Hall, Inc., *Englewood Cliffs, New Jersey*

Contents

Preface

GOALS AND CONTENT

The objective of this book is to explain clearly the principles underlying the operation and analysis of those diodes and transistors that form the basis of modern microelectronic circuits. The book differs from many other texts on semiconductor devices in that, rather than attempting an encyclopedic coverage of many different devices, it concentrates on the more fundamental devices and treats these in considerable depth. The devices treated are: ***pn junction*** and ***metal-semiconductor (Schottky barrier) diodes***; ***metal-oxide-semiconductor field-effect transistors (MOSFETs)***, ***metal-semiconductor field-effect transistors (MESFETs)***, ***junction field-effect transistors (JFETs), and bipolar junction transistors (BJTs)***. Emphasis is placed on the integrated circuit versions of these devices. It is the authors' opinion that a detailed study of these foundation diodes and transistors not only makes for confident usage of such devices, both in theoretical studies and practical applications, but also makes for an easy understanding of other semiconductor devices, both integrated and discrete, not treated specifically in this book. "Treatment in depth" refers here to a detailed discussion of ***large-signal transient response*** and ***small-signal ac analysis***, and to the inclusion of ***second-order effects*** (an example might be short- and narrow-channel effects in the MOSFET) in steady-state analyses. For most applications of semiconductor microelectronic devices response speed is a crucial issue, yet transient analysis is treated superficially, if at all, in many texts.

In the 1980s interest in the use of application-specific integrated circuits (ASICS)—custom ICs intended for special-purpose applications and produced in correspondingly small volumes—has grown enormously. The ASIC concept has been made economically feasible by another development, that of the silicon foundry, a processing line that fabricates custom integrated circuit designs for a large number of users. These trends are requiring more and more electrical engineers to become familiar with the principles of integrated circuit design. This is already reflected in university curricula, where courses in the basic aspects of IC design are now commonplace at the senior level. For this reason, transistors and diodes are analyzed in this book in the context of the models used to represent these devices in the ***SPICE circuit simulator***. The SPICE program is used very widely in industry, and almost exclusively in universities, to predict the performance of integrated circuit designs. SPICE is used in many examples in this book for tasks such as generation of a transistor's current–voltage characteristics, and investigation of the transient response of devices in simple circuits. Program input is in the form of SPICE model parameters, the values of which must be calculated from knowledge and understanding of the fundamental material and device properties presented in the book. In working through the examples, the reader should acquire both an appreciation of the origin of the values used for SPICE parameters, and sufficient familiarity with the operation of SPICE to use it with ease in the analysis of more complicated circuits.

Another important trend in microelectronics is the emergence of commerical ***GaAs integrated circuits***. Although it is expected that the majority of ICs will be made on silicon substrates, for at least the rest of this century, GaAs ICs will be used increasingly in special applications where their high speed can offset their higher cost. For this reason, an analysis of the GaAs MESFET, the main device used in GaAs ICs, is included in this book.

To summarize, the novel features of this book are:

1. In-depth treatment of the basic transistor and diode structures
2. Coverage of dc, ac, and transient performance of devices
3. Employment of SPICE simulations to bridge the rapidly narrowing gap between the traditionally distinct areas of device analysis and circuit simulation
4. Treatment of GaAs MESFET devices in addition to the more-established silicon devices (diodes, FETs, and BJTs)

The book attempts to be a self-contained treatment of the fundamentals of diodes and transistors used in integrated circuits. It caters both to readers coming to the subject for the first time, and to readers who have already been introduced to semiconductor devices but who wish to know more about the operation and analysis of modern transistors and diodes. Neophytes would be expected to concentrate on the chapters on semiconductor fundamentals and on the sections in the device chapters that deal with principles of operation and dc properties. Veterans would hopefully enjoy the chapter on quantum mechanics as a refresher on semiconductor basics, and then proceed to the sections in the device chapters which deal with transient response, small-signal ac performance, and secondary phenomena.

AUDIENCE ADDRESSED

The approach described above is intended to provide a detailed, yet uncluttered treatment of the subject, one that will imbue the reader with some confidence in either starting to work with real semiconductor devices or in preparing for further studies of the subject. Thus the intended audience ranges from electrical engineering and physics students in their early years at university, through more senior students in other engineering and science disciplines who are seeking general knowledge, to engineers and technical managers who are anxious to acquire quickly a solid basis on which to form an understanding of modern microelectronic devices.

ORGANIZATION

The first part of this book introduces those aspects of the properties of semiconductor materials which are necessary for an understanding of the operation of diodes and transistors, and also formulates the mathematical equations needed to characterize these devices. This material follows a brief introductory chapter and is covered in Chapters 2 through 4. In Chapter 5 we give a brief quantum mechanical treatment of the fundamental origin of some of the semiconductor properties that are introduced merely as facts in the first four chapters.

Chapters 6 through 11 build on the material of the earlier chapters; the operation, analysis, modeling, and some applications of transistors and diodes are described. The traditional practice of covering bipolar devices before field-effect (unipolar) devices has not been followed. Instead, following a chapter on the seminal device, the *pn*-junction diode, the book moves directly to the metal-oxide-semiconductor field-effect transistor, the MOSFET. This appears to be, pedagogically, the most expedient method of reaching the device that is, arguably, the most important of modern transistors. Continuing with this emphasis on field-effect devices, metal-semiconductor FETs are covered after the necessary prerequisite treatment of metal-semiconductor diodes, and a short chapter on junction field-effect transistors is given. The final chapter, on device operation, treats bipolar junction transistors. Obviously, this chapter can be read directly after the one on *pn* diodes if the course emphasis is on bipolar devices.

Each of the device chapters (6 through 11) opens with a description of the basic structure of the device under consideration. Chapter 12 provides a brief outline of the techniques by which such structures can be fabricated as integrated circuits on silicon or gallium arsenide substrates.

COURSE FORMAT

There is enough material in this book for two courses, each of duration one term, one semester, or two quarters and occupying about 25 fifty-minute lectures.

For a first course on semiconductor devices, the following topics might be covered:

Semiconductor fundamentals	Chapters 1 through 4, and possibly 5
Principles of operation, dc performance of devices, device capacitance	Appropriate sections of Chapters 6 through 11

For a more advanced course on semiconductior devices, the following topics are suggested:

Review of semiconductor fundamentals	Chapter 5
Device processing	Chapter 12
Ac and transient, operation and simulation of devices, second-order phenomena	Appropriate sections of Chapters 6 through 11

ACKNOWLEDGMENTS

Much of the material in this book has been taught at UBC and Carleton in various undergraduate courses on microelectronic devices. We thank the students in these classes who, through their questions, have led us to a deeper understanding of the subject. We are also grateful to our senior colleagues, Lawrence Young at UBC, and Roy Boothroyd at Carleton, for many enlightening conversations over the years. Special thanks are due to Dan Camporese of UBC for his willingness to debate some of the finer points of semiconductors and devices. We thank also Debbie Young of Prentice Hall for her diligent work in producing this book. Finally, D.L.P. acknowledges his children, Simon, Tim, and Louisa, for their cheerful tolerance of his need for quiet, long hours while writing.

To Eileen:
whose constant support and
encouragement made possible
the writing of this book.

1 Introduction

1.1 OVERVIEW

Semiconductors have been defined [1] as "a class of electrical materials that are intermediate between conductors and insulators. Their electrical behavior is often sensitive to temperature and to the degrees of impurities in certain regions. They are the basis for transistors and diodes." We could add that transistors and diodes are the active devices that form the heart of all solid-state electronic circuits. It is the purpose of this book to introduce the reader to the principles underlying the operation and analysis of these important devices.

A diode, as the name implies, has two electrodes. The charge that flows through these electrodes in response to a voltage applied to them constitutes a current which has the form shown in Fig. 1.1a. The essential feature of this current–voltage characteristic is that the magnitude of the current depends markedly on the polarity of the voltage.

To fabricate a semiconductor diode it is necessary to create a structure with a built-in electric field. This is conveniently done by juxtaposing either a metal and a semiconductor (a Schottky diode), or two semiconductor regions doped with different types of impurities (a *pn* diode). These diodes are important devices in themselves but, also, as Fig. 1.2 indicates, they form an integral part of the more complex structures of transistor devices.

Transistors have three principal terminals. Typically, the currents at two of these are controlled by the electrical conditions at the other electrode. All transistors can be connected to external circuitry in such a manner as to produce an I–V characteristic of the form shown in Fig. 1.1b.

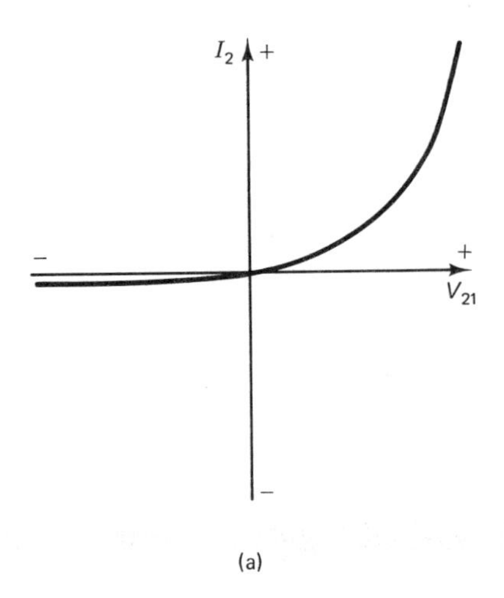

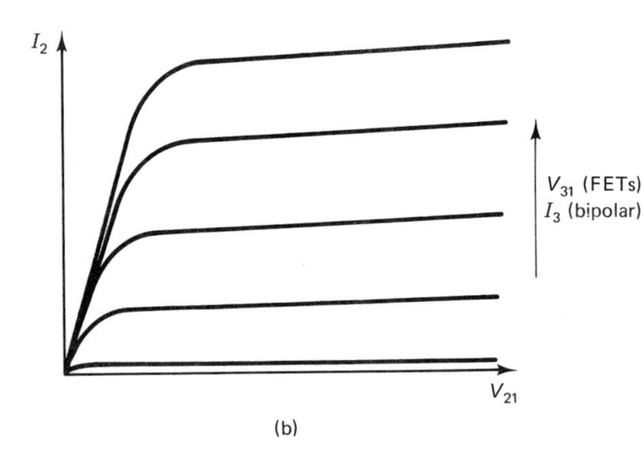

Figure 1.1 Characteristic current–voltage relationships for (a) diodes and (b) transistors. The subscripts refer to the electrode labels given to the devices in Fig. 1.2.

A figure of merit for field-effect transistors (FETs), one that emphasizes the interaction of the three electrodes, is the transconductance. This is defined, using the notation of Figs. 1.1 and 1.2, as

$$g_m = \frac{dI_2}{dV_{31}} \tag{1.1}$$

This parameter is not such a good figure of merit for the bipolar junction transistor (BJT), as it depends, in this case, also on the properties of the circuit in which the transistor is used. However, the term "transconductance" is relevant to how the transistor came to be so-named. The first transistor was a bipolar device. It had metal-semiconductor diode junctions in place of the doped semiconductor junctions of the BJT shown in Fig. 1.2. Apparently, the inventors of the device, J. Bardeen and W. H. Brattain, approached J. R. Pierce, a colleague at Bell

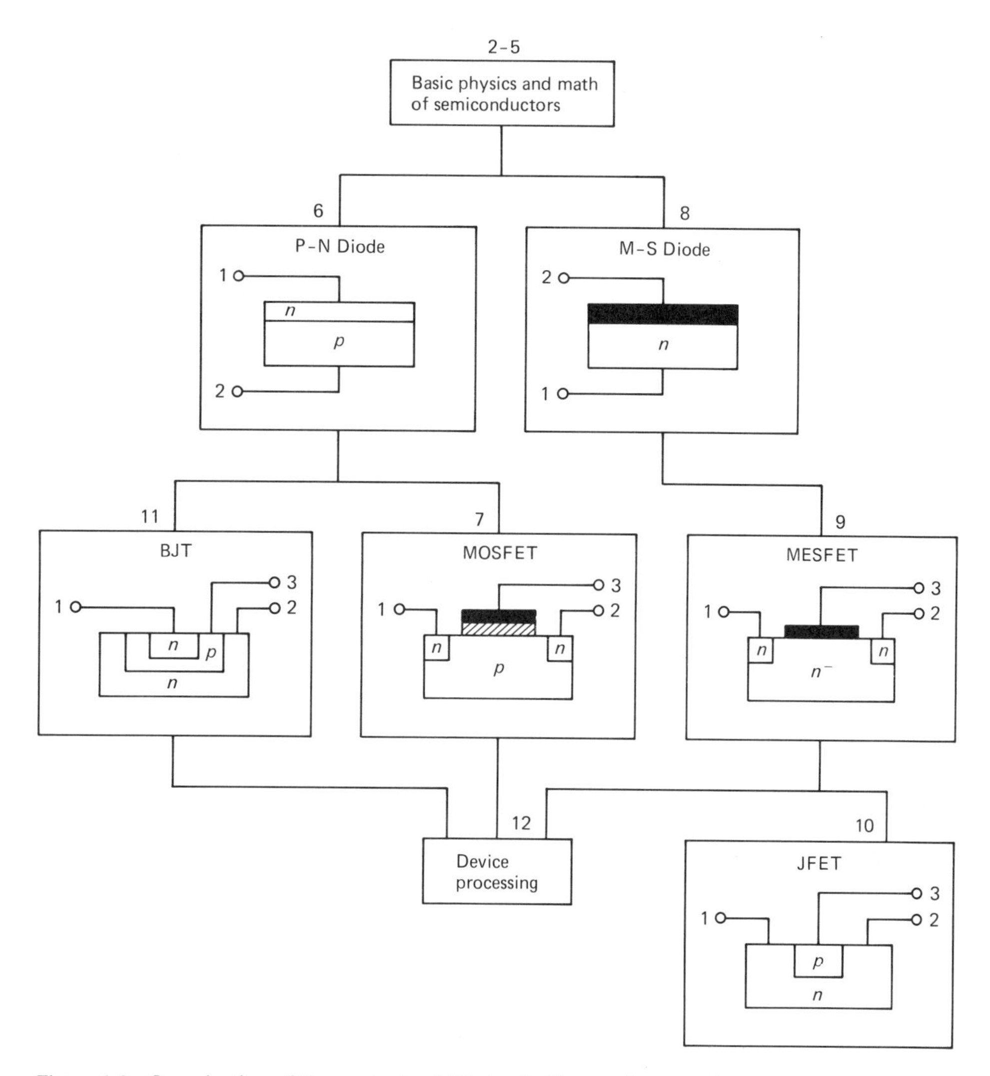

Figure 1.2 Organization of the contents of this book. The numbers on the tops of the boxes indicate the chapters in which the various topics are presented. For the cross-sectional sketches of the devices, n and p refer to differently doped regions of semiconductor material; metal areas are shown in solid black and insulating regions are crosshatched.

Laboratories in New York, seeking an apt name for their new invention [2]. Pierce appreciated that the novel solid-state device was, in the electrical circuit sense of being able to produce amplification and oscillation, the analog, or dual, of the vacuum tube. As transconductance was a significant parameter for vacuum tubes, he mulled over the dual term "transresistance" and, after a few moments' thought, offered the word "transistor." The inventors liked it and the label has stuck.

Transistors now come in a variety of forms, as can be appreciated from Fig. 1.2. In taking an educational route to encompass all of the device types illustrated in the figure, we have chosen to start at the *pn*-junction diode. This structure is an integral part of the metal-oxide-semiconductor field-effect transistor (MOSFET) and of the BJT, the two transistors that presently dominate the electronic marketplace. We visit the MOSFET first in view of its prominence in very large scale integrated (VLSI) circuits. A side trip to metal-semiconductor (Schottky) diodes is then necessary before the journey to the metal-semiconductor field-effect transistor (MESFET) can be completed. This gallium arsenide device is still in its early stages of commercial development, unlike the operationally similar silicon junction field-effect transistor (JFET), which has changed little over the years since it was introduced in the 1950s as the first commercial FET. Our brief stop at the JFET completes the tour of the basic FET devices. Our journey finishes with a detailed study of the silicon bipolar junction transistor. This earliest of commercial transistors presently dominates analog integrated circuit applications and is widely used in high-speed digital switching circuits and in high-power applications.

In Chapter 12 we give an overview of the processing technologies that are used in the fabrication of integrated circuit versions of the three main transistor types: silicon MOSFETs, silicon BJTs, and gallium arsenide MESFETs.

As Fig. 1.2 implies, before the operation of semiconductor devices can be understood, it is necessary to appreciate a number of properties of semiconductor materials and to develop a mathematical framework for device design and analysis. The book begins, therefore, with an exposition of this foundation material.

1.2 REFERENCES

1. D. G. Fink and H. W. Beaty, eds., *Standard Handbook for Electrical Engineers*, 11th ed., p. 2–4, McGraw-Hill Book Company, New York, 1978.
2. W. H. Brattain, "Discovery of the transistor effect," *Adventures in Experimental Physics*, vol. 5, pp. 1–31, 1976.

2

Basic Properties of Semiconductors

2.1 CONDUCTIVITY

The definition given at the beginning of Chapter 1 refers to semiconductors being "intermediate between conductors and insulators" [1]. The "intermediate" property of main relevance to semiconductor devices is the electrical ***conductivity***. It is appropriate, therefore, to open this chapter with a discussion of this material property.

The conductivity σ is related to the more familiar, geometry-dependent property of resistance via the relationship

$$\sigma = \frac{L}{RA} \tag{2.1}$$

where the various parameters are as defined in Fig. 2.1.

Metals have high values of conductivity: for example, for aluminum, $\sigma = 3.5 \times 10^5\ (\Omega\cdot\text{cm})^{-1}$. Insulators have low values of conductivity: for example, for silicon dioxide, $\sigma < 10^{-16}\ (\Omega\cdot\text{cm})^{-1}$. Semiconductors, as we have stated above and as the name implies, have intermediate values of conductivity: for example, for silicon wafers used in electronic devices, typical values span the range $10^{-2}\ (\Omega\cdot\text{cm})^{-1} < \sigma < 10\ (\Omega\cdot\text{cm})^{-1}$. The fact that such a range of conductivity is readily attainable in a controllable manner is one of the main reasons for semiconductors being such an important class of practical materials.

Conductivity in a semiconductor is determined primarily by the concentration of specific types of impurities present. Ideally, these impurities are deliberately introduced into the semiconductor material in a controllable manner. They

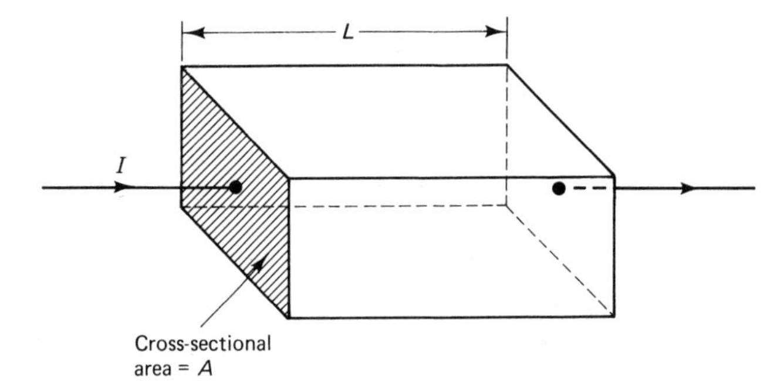

Figure 2.1 Relationship between conductivity and resistance. The resistance of the block to the uniform current shown is given by $R = L/(\sigma A)$, where σ is the conductivity of the material.

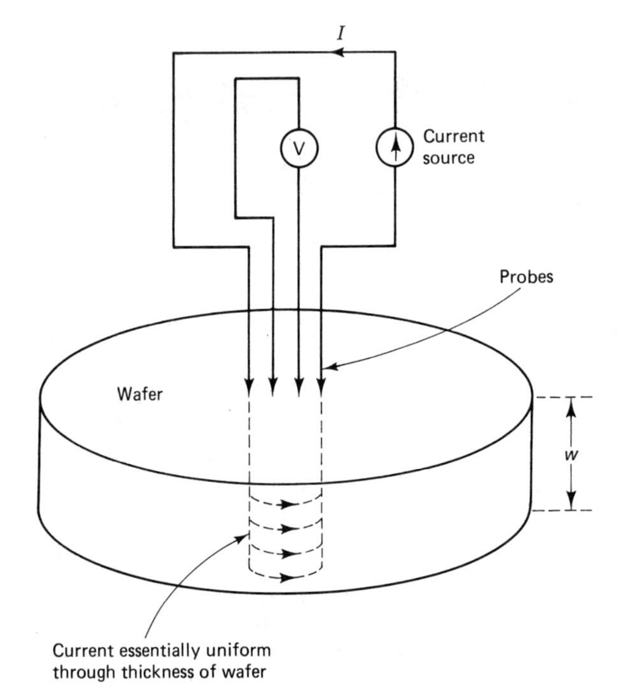

Figure 2.2 Four-point probe method for the measurement of conductivity. This arrangement of driving a current through the wafer, and measuring the voltage via two separate leads which carry virtually no current, eliminates the effect of contact resistance and allows an accurate measurement of the semiconductor conductivity to be made.

are known as ***dopants***. A typical dopant concentration in silicon is about 5×10^{16} cm^{-3}. As the concentration of silicon atoms is 5×10^{22} cm^{-3}, the deliberate doping content is only 1 part in a million! Doping and crystal growth are discussed in Section 2.2, where it is also pointed out that the semiconductor material from which devices are fabricated comes in the form of thin ***wafers***.

The conductivity of wafers is measured by a four-point probe method such as that shown in Fig. 2.2. The separation of the probes in commercial instruments is usually about 0.5 mm; it is certainly considerably less than the wafer diameter, which is typically 100 mm for silicon wafers. If the thickness of the wafer, w, is small enough such that the current can be considered to be uniformly distributed through the thickness of the wafer, the practical arrangement of Fig. 2.2 resembles that of an infinite current sheet. In such a case the conductivity is given by [2]

$$\sigma = \frac{I \ln 2}{V \pi w} \tag{2.2}$$

Semiconductor material specifications and people who make devices (semiconductor process engineers) often refer to the ***resistivity*** of wafers. The resistivity ρ is just the reciprocal of the conductivity and is commonly expressed in units of $\Omega \cdot \text{cm}$.

As we have stated, the conductivity depends on the concentrations of impurities (dopants) in the material. These dopants provide mobile charges, the flow of which constitutes a current. In semiconductors there are two types of mobile charge carrier: ***electrons*** and ***holes***. These entities are described and discussed at length in this chapter, following a section on crystal structure and growth.

2.2 CRYSTAL STRUCTURE AND GROWTH

The electrical properties of silicon and gallium arsenide are directly tied to their crystal structure—the arrangement of atoms within a solid. Solids in general can be grouped into three categories: crystalline, polycrystalline, and amorphous. The atoms in a perfect crystal have perfect long-range order: it is possible to predict precisely the position of the atoms in any part of the crystal based on knowledge of the structure in one localized region a few atomic diameters across. In an amorphous material, there is generally considerable short-range order—the distances and directions of the bonds between neighboring atoms are roughly the same in any area—but there is enough random variation in bond lengths and angles that it is impossible to predict the position of atoms more than a few atomic diameters apart. Polycrystalline material represents something of an "in-between" category, in that it consists of small crystalline regions or ***grains*** separated by disordered zones known as ***grain boundaries***. The differences between the three kinds of solid are illustrated schematically in Fig. 2.3.

It is possible to determine the atomic structure of a solid by observing how x-rays are diffracted as they pass through it. When experiments of this kind are done for silicon crystals, it is found that the atoms are arranged in the ***diamond lattice*** structure shown in Fig. 2.4. (Naturally, this structure receives its name since the diamond form of crystalline carbon also has this pattern.) It is relatively easy to explain the tendency of silicon to crystallize in the diamond structure on chemical grounds. Silicon has four ***valence*** or chemically active electrons, and these prefer to form bonds with equal angular separation, yielding the tetrahedral bonding arrangement shown in Fig. 2.5. The diamond lattice can be constructed by assembling silicon atoms in this tetrahedral bonding configuration.

Gallium arsenide—and for that matter, nearly all other semiconductors—also has a crystal structure based on a tetrahedral bonding arrangement. The GaAs

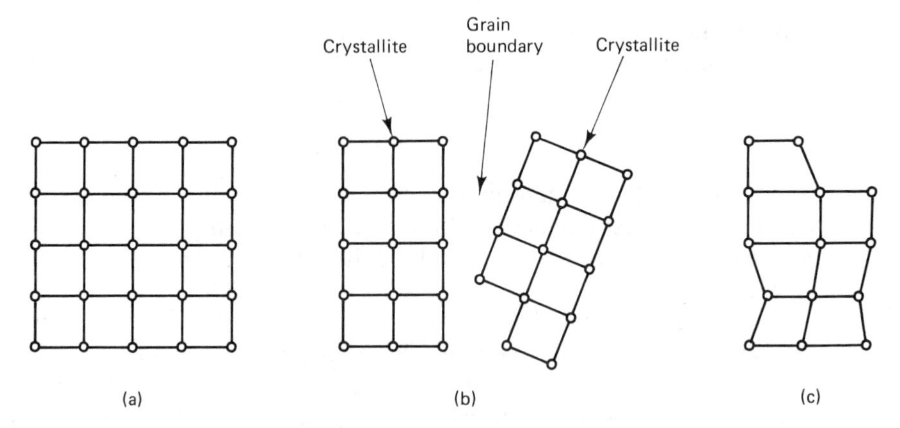

Figure 2.3 Schematic representation of solids: (a) crystalline; (b) polycrystalline; (c) amorphous.

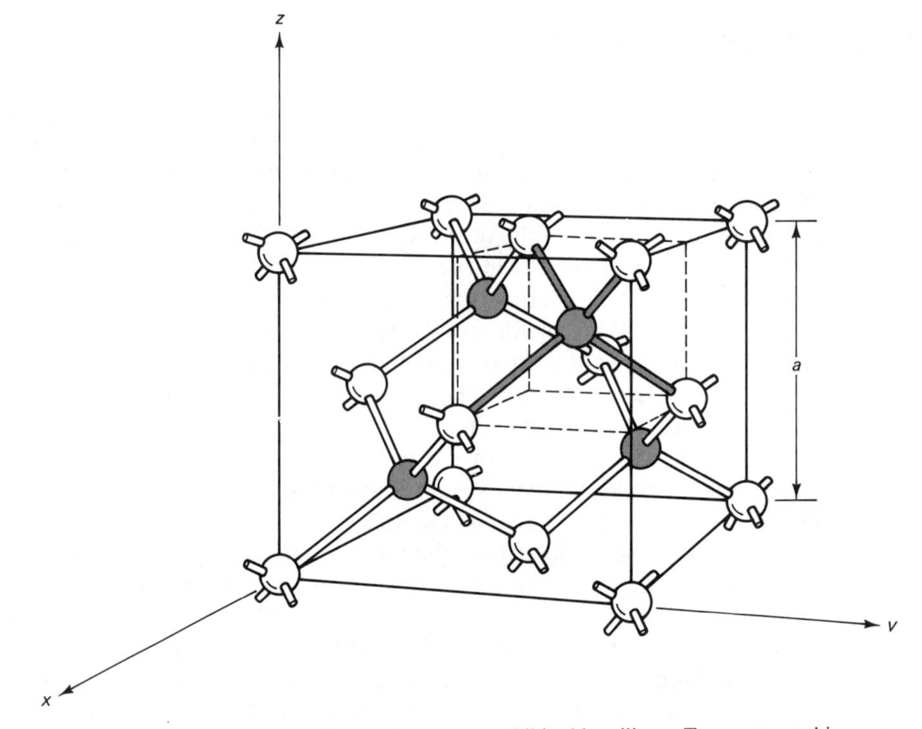

Figure 2.4 The diamond crystal structure exhibited by silicon. To construct this arrangement, draw a face-centered cube (FCC) and add new atoms, displaced by $0.25a(\mathbf{x} + \mathbf{y} + \mathbf{z})$ from each of the atoms of the FCC. (a, the lattice constant, is 0.543 nm for Si; **x**, **y**, and **z** are unit vectors in the x-, y-, and z-directions, respectively.) The four constructed atoms that lie within the original FCC are shaded in the drawing. The dashed lines define the unit cell. (Adapted from Sze [3]. Reprinted with permission of the publisher, John Wiley & Sons, Inc.)

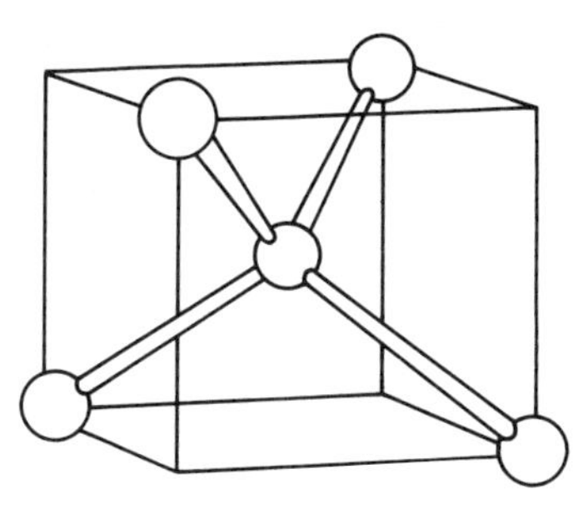

Figure 2.5 Tetrahedral bonding arrangement preferred by silicon. Each silicon atom is bonded to four neighbors. The bonds are of equal length and equal angular separation.

lattice is shown in Fig. 2.6; it can be derived from the silicon lattice simply by replacing half the silicon atoms with gallium and half with arsenic, so that each gallium has arsenic for its nearest neighbors, and vice versa.

2.2.1 Crystal Orientation

The chemical and electrical properties of a silicon surface depend to some extent on the orientation of that surface with respect to the diamond lattice structure. The surface orientation is specified by giving the coordinates of the normal to the surface plane in the natural (x, y, z) Cartesian coordinate system specified by the cubic diamond lattice, as shown in Fig. 2.4. For example, a surface with a normal vector of (1, 0, 0) is referred to as a (100) surface. This notation can equally well be applied to planes of atoms within the crystal that do not form a surface; an example of a (100) plane is shown in Fig. 2.7a. Naturally, surfaces with normals such as $(-1, 0, 0)$ or (0, 1, 0) should have exactly the same properties as surfaces with (1, 0, 0) normals, since the labeling of the x, y, and z axes in Fig. 2.4 is completely arbitrary.

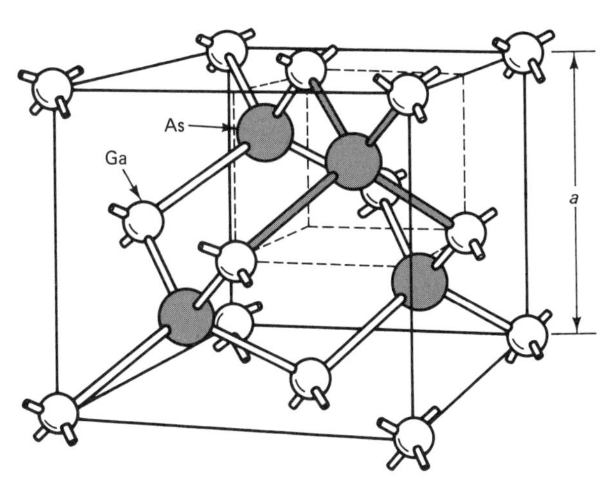

Figure 2.6 Zinc blende crystal structure exhibited by gallium arsenide. The construction is the same as that described in the caption for Fig. 2.4. If the original face-centered cube (FCC) is comprised of Ga atoms, as shown here, the constructed atoms are As. Thus each atom has four nearest neighbors of the other element. The FCC of the Ga atoms interpenetrates the FCC of the As atoms. The lattice constant a is 0.565 nm for GaAs. The dashed lines define the unit cell. (Adapted from Sze [3]. Reprinted with permission of the publisher, John Wiley & Sons, Inc.)

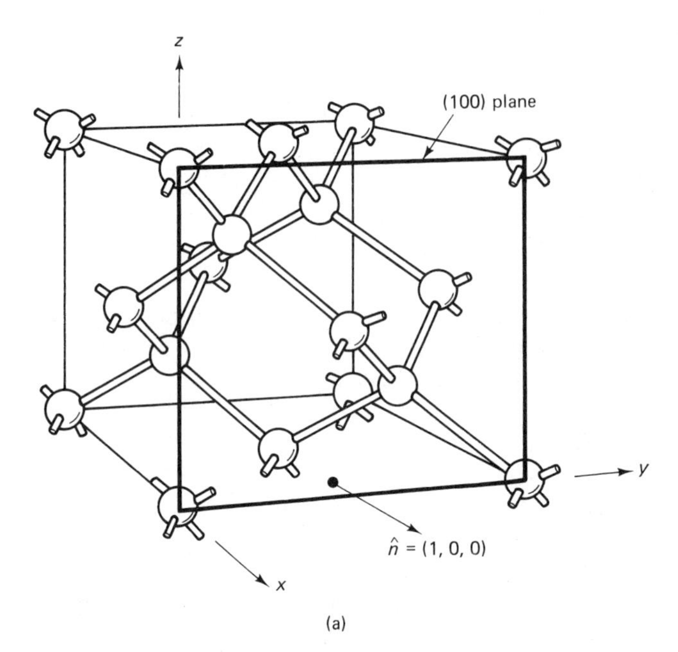

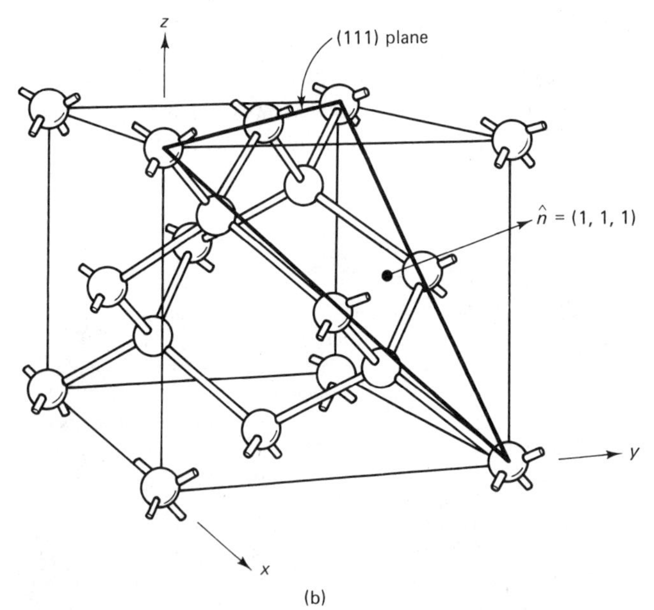

Figure 2.7 (a) A (100) plane, and (b) a (111) plane of atoms in the diamond lattice.

Silicon with a (100)-type surface is most frequently encountered in integrated circuit manufacturing, but (111) surfaces are also used, particularly in the production of bipolar devices. A (111) plane is shown in Fig. 2.7b.

The properties of GaAs surfaces are also dependent on their orientation, but it is not always possible to describe a GaAs surface uniquely by specifying its normal vector. For example, on a (111) GaAs surface the atoms will either be all gallium or all arsenic, each case giving rather different chemical properties.

2.2.2 Crystal Growth

A typical silicon crystal used in integrated circuit manufacturing today is 100 mm in diameter, about 0.5 mm thick, and costs around $20. Commercial-grade silicon ***wafers*** of this kind are almost perfect crystals, in that the lattice structure shown in Fig. 2.4 continues uninterrupted from one side of the wafer to the other. Occasionally, a silicon atom may be missing from the lattice, an impurity atom may replace silicon on a lattice site, or small impurity atoms such as oxygen or carbon may be present in the spaces between the atoms, but otherwise the crystal contains no defects.

Production of such large, near-perfect silicon crystals is an art that has been under development for more than 30 years. Wafer manufacturing begins with the reduction of sand—impure silicon dioxide—with carbon in an arc furnace at a temperature of about 1600°C. The overall reaction is

$$SiO_2 + 2C \longrightarrow Si + 2CO$$

The "metallurgical-grade" silicon produced in this way is about 98% pure, with aluminum and iron being the main contaminants. Most of the world's output of metallurgical-grade silicon is used in the manufacture of silicon lubricants, in the preparation of special steels, and for hardening aluminum alloys. A small fraction is subjected to elaborate chemical purification to produce the semiconductor-grade silicon required by the microelectronics industry. To begin purification, the metallurgical-grade material is ground into a fine powder and then reacted with gaseous hydrogen chloride at a temperature of about 300°C, yielding a variety of ***chlorosilanes***. The most important of these for our purposes is ***trichlorosilane*** ($SiHCl_3$), produced via the reaction

$$Si + 3HCl \rightleftarrows SiHCl_3 + H_2$$

The reason trichlorosilane is of such importance is that it is a liquid with a boiling point of 32°C and so can be easily purified by successive distillation. After purification the trichlorosilane is mixed with hydrogen and heated to about 1100°C, yielding very pure silicon via the reverse reaction given above. A seed rod of high-purity polycrystalline silicon is used to provide a substrate on which the silicon can deposit.

Although the silicon produced by the trichlorosilane process is very pure chemically—the finished product typically contains only a few parts per million

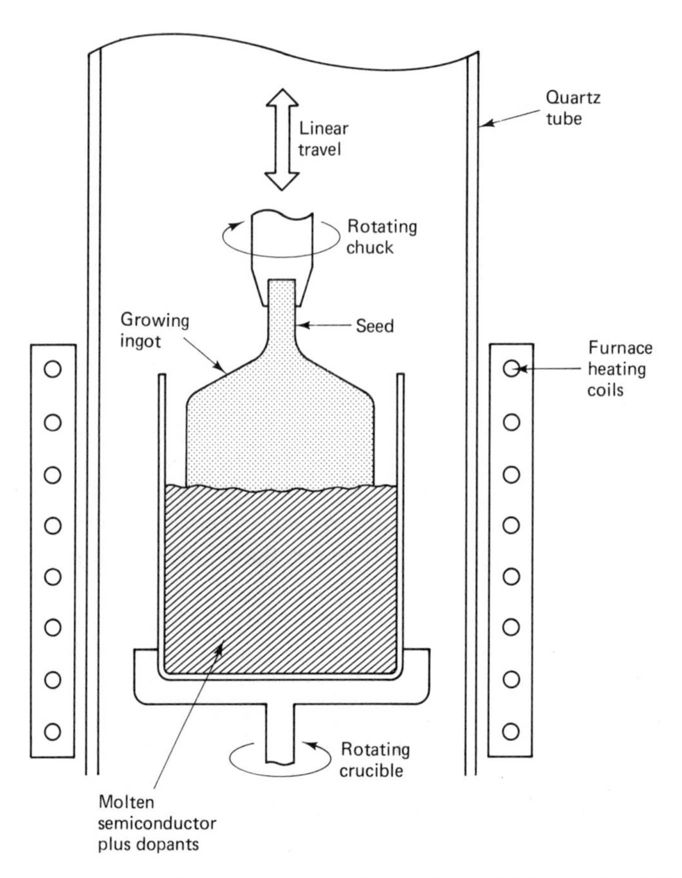

Figure 2.8 Czochralski crystal growth. The seed crystal is lowered until it contacts the melt. The molten, doped semiconductor material adhering to the seed solidifies and assumes the crystal structure and orientation of the seed. The seed is slowly pulled up, drawing with it the growing single crystal. By carefully controlling the melt temperature and the speeds of seed withdrawal, seed rotation, and crucible counterrotation, a long ingot of quite uniform diameter can be obtained.

of impurities—it must still be converted into a single crystal before it can be used to manufacture transistors. This is usually done via the ***Czochralski*** crystal growth technique, which is illustrated in Fig. 2.8. The material in the crucible is molten semiconductor-grade silicon containing dopant impurities which are added in precise amounts (at the parts per million level) so that they can be incorporated in the growing crystal and so determine its conductivity. A seed crystal is dipped into the melt and then slowly removed while being rotated. If the rate at which the seed crystal is withdrawn is carefully controlled, it is possible to "pull" a large, cylindrical crystalline ingot from the melt. Production of defect-free material requires very careful control of the temperature gradients across the ingot during growth, since stress induced by temperature differences can cause plastic deformation of the crystal, creating defects. The growth process must be carried out in an inert atmosphere (usually argon), since the molten silicon is highly reactive, and would immediately form SiO_2 if exposed to air. The high reactivity of molten silicon combined with the high melt temperature (1410°C) severely limits the choice of materials for the crucible containing the melt. Crucibles are almost always made of quartz (fused crystalline SiO_2). Unfortunately, molten silicon tends to erode even quartz crucibles, leading to substantial incorporation of oxygen in the crystal. The presence of large quantities of oxygen in Czochralski silicon has some very important consequences for subsequent processing steps and in the electrical properties of the material, but discussion of these problems is beyond the scope of this book.

Commercial silicon ingots may be up to 200 mm in diameter, weigh as much as 60 kg, and be more than a meter in length. The thin silicon wafers used in integrated circuit manufacture are cut from these ingots using a diamond saw and then polished on one side until optically flat. This last step is needed to enable the fine patterns used in ICs to be transferred onto the wafer surface using a photographic process.

Wafers used in the commercial manufacture of GaAs integrated circuits are normally made using a process paralleling that for silicon. First, high-purity gallium and arsenic are heated in a boron nitride crucible, where they react to form molten GaAs (the melt temperature is 1238°C). Unfortunately, the vapor pressure of arsenic at this temperature is about 60 atm, so steps must be taken to prevent the loss of arsenic from the system. Arsenic evaporation is blocked by overlaying the GaAs with a layer of molten boric oxide, B_2O_3. (The B_2O_3 is added to the GaAs charge in the crucible prior to heating. It has a lower melting point than GaAs and is less dense, and so floats to the surface to provide an encapsulating layer.) To hold the B_2O_3 in place against the high arsenic vapor pressure, it is necessary to pressurize the entire crystal growing apparatus to about 100 atm. In analogy to the silicon growth process, the GaAs crystal is grown by dipping a seed crystal through the B_2O_3 into the melt, and then slowly withdrawing it. This growth technique is known as the ***l***iquid ***e***ncapsulated ***C***zochralski (LEC) method. Ingots are sawn and polished to make circular wafers with optically flat surfaces in much the same manner as for silicon.

From the discussion above it can be appreciated that the growth of GaAs crystals is technically far more complicated than that of silicon and, as a result, the largest commercially available GaAs wafers are only 75 mm in diameter. More important, commercial GaAs wafers generally contain a high density of crystal defects. These defects result largely from plastic deformation of the ingot in response to thermally induced stress during the growth process. The stress required to cause plastic deformation of GaAs is considerably smaller than that for silicon, so GaAs crystals are much more susceptible to this problem. The effect of the high density of crystal defects in GaAs wafers on the performance of transistors and integrated circuits is not entirely clear at present.

2.3 BANDS AND BONDS

The electrons in a silicon atom can be imagined as occupying ***orbitals*** that surround the nucleus, as represented schematically in the upper part of Fig. 2.9. Each of the electrons in the atom has a potential energy by virtue of its proximity to both the positively charged nucleus and the other negatively charged electrons. The allowed values of energy are quantized and are called ***energy levels***. The correspondence between the physical picture of electrons in orbitals and the energy picture of electrons in levels is illustrated in Fig. 2.9. Each energy level can accommodate two electrons in what are called the quantum states or, plainly, the ***states*** of that energy level. The two states in each energy level can be populated, in accordance with Pauli's exclusion principle from quantum mechanics, by two electrons, provided that these electrons have opposite spins. The energy level corresponding to the first orbital is filled by two electrons labeled "1*s*" (see Fig. 2.9). The second orbital can accommodate 8 electrons, two labeled "2*s*" and six labeled "2*p*." Silicon has its full complement of these electrons and they reside in four energy levels. The valence orbital also has four energy levels, corresponding to the eight states of the 3*s* and 3*p* electrons. In an isolated silicon atom at zero kelvin, only the bottom two levels are filled by the four ***valence*** electrons.

In an actual piece of silicon material the bonding situation is one in which the 3*s* and 3*p* orbitals of the silicon atom overlap to form ***$3sp^3$ hybrid molecular orbitals*** containing the four valence electrons. These $3sp^3$ orbitals form four covalent bonds of equal angular separation, as noted in Section 2.2. The bonding arrangement in the silicon crystal can be represented by the two-dimensional model shown in Fig. 2.10. In reality, of course, the semiconductor is a three-dimensional solid and the sharing of valence electrons occurs between nearest-neighbour atoms in three dimensions (see Figs. 2.4 and 2.5).

The bonds between atoms are complete at zero kelvin. The sharing of valence bond electrons means that these electrons can no longer be viewed as belonging to one particular atom. At finite temperatures the valence electrons are in motion throughout the crystal, using the $3sp^3$ hybrid molecular orbitals as interconnecting pathways between atoms. At any instant in time electrons will be present near the atoms to effect the covalent bonding, but no particular electron is associated with any one particular atom all of the time.

Another feature, in addition to the mixing of the 3*s* and 3*p* orbitals, resulting from bringing silicon atoms together to form a solid is the splitting of the $3sp^3$ orbitals into two distinct bands of energy levels. The explanation of how the orbitals interact in this way is impossible to present in a straightforward, qualitative manner. The explanation demands a quantum mechanical representation of electrons in solids. Such a treatment is offered in Section 5.3. For the present, we proceed by noting the result of the interaction, namely: for N interacting silicon atoms, the four valence energy levels of each atom split into two bands each of $2N$ energy levels. Figure 2.11 shows an example for the case of $N = 2$. The $2N$ energy levels of the lower band, the ***valence band***, are completely filled at zero

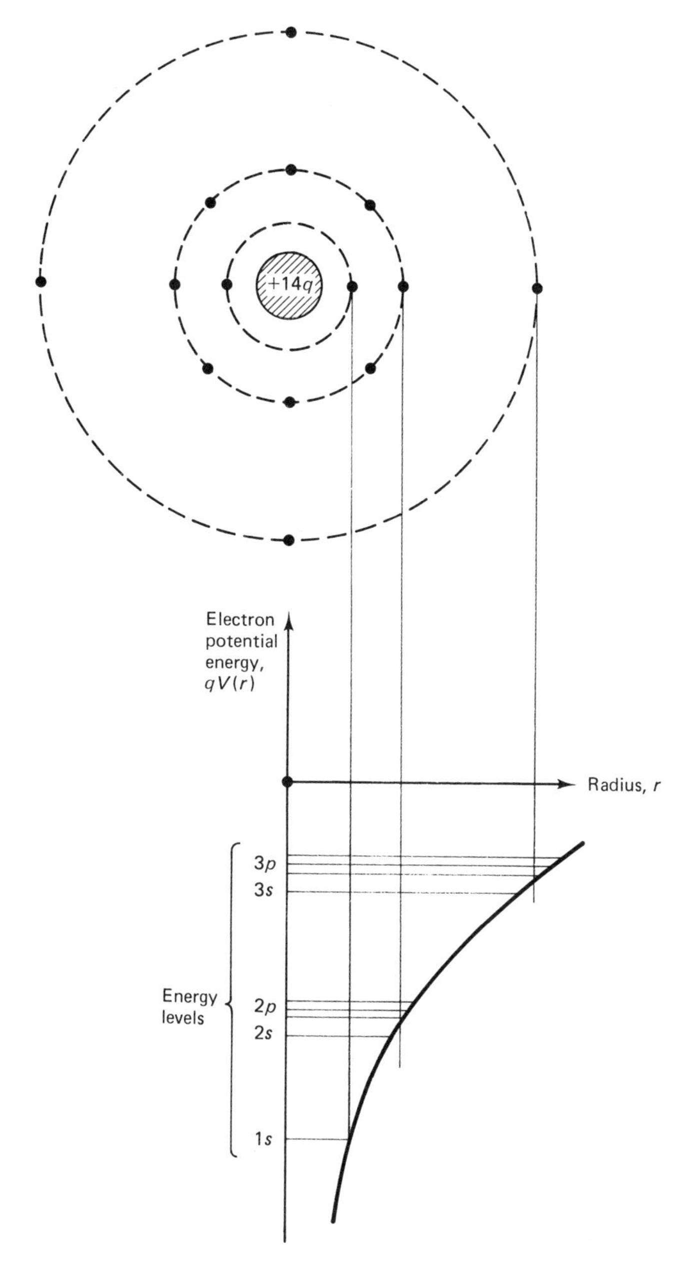

Figure 2.9 Relationship between electron orbitals and electron energy levels. The top part of the figure shows schematically the electrons of an isolated silicon atom at 0 K. The valence electrons reside in the outermost orbital. The potential energy of the electrons in the various orbitals is depicted in the lower part of the figure. In this illustration the potential energy is shown to vary directly as $1/r$, such as would be the case if the potential distribution were Coulombic, arising solely from the attraction of electrons to the nucleus, that is, $-qV(r) = -qQ/4\pi\epsilon r$, where $-q$ is the charge on an electron, Q the charge on the nucleus, r the radius of the orbital (assumed here to be circular), and ϵ the permittivity of silicon.

Figure 2.10 Two-dimensional representation of the silicon lattice for the case of no broken bonds. The circles marked Si represent the nucleus and the electrons of the inner two orbitals. The solid circles represent the valence electrons. The sharing of valence electrons produces covalent bonds between nearest neighbor atoms. The valence electrons are in continual motion "along the orbital pathways."

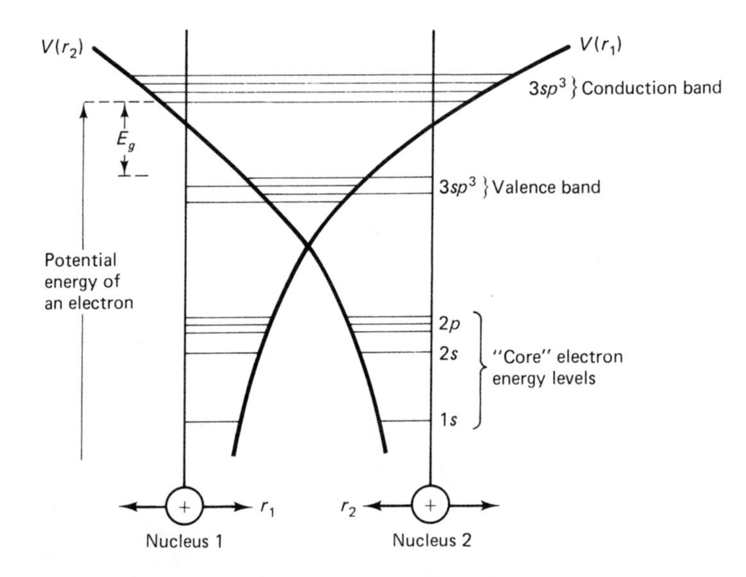

Figure 2.11 Formation of energy bands. Two neighboring Si nuclei are shown, each with its own Coulombic potential distribution $V(r)$. The valence orbitals overlap, leading to an interaction between the $3s$ and $3p$ energy levels. The fact that these energy levels merge is not difficult to appreciate, but why they should split into two separate bands is not at all obvious. For an explanation, see Section 5.3. Each band contains $2N$ energy levels, where N, the number of interacting atoms, is 2 in this example. The top band is called the conduction band and the bottom band is the valence band. Their separation is called the energy bandgap, E_g.

kelvin by the $4N$ valence electrons. The upper band, called the ***conduction band***, is therefore empty at this temperature. The difference in energy between the two bands is called the ***energy bandgap***, E_g.

In a real single crystal of silicon the concentration of interacting atoms is given by

$$C\ (\text{atoms/cm}^3) = \frac{A\ (\text{atoms/mole}) \times \text{density}\ (\text{g/cm}^3)}{\text{GMW}\ (\text{g/mole})} \tag{2.3}$$

where A is Avogadro's number, 6.02×10^{23} atoms/mole; the density of silicon is 2.33 g/cm^3; and silicon's gram molecular weight is 28.09 g/mole. These numbers give a value for C of 5×10^{22} atoms/cm^3 for silicon. For such a large number of atoms, the bands extend throughout the entire crystal, not just over the short distance shown in Fig. 2.11 for the case of $N = 2$. In the valence band of silicon there are thus 10^{23} energy levels/cm^3 and 2×10^{23} electron states/cm^3, all of which are occupied at zero kelvin.

At higher temperatures the finite thermal energy causes the atoms to vibrate, resulting in the ***breaking of some bonds***. From an energy band point of view, this means that the electrons which are freed from bonding duties translate some of this thermal energy into potential energy. If these excited electrons gain potential energy equal to, or in excess of, the energy bandgap, they can be excited out of the valence band of energies representing bonded electrons, to the conduction band of energies representing "free" electrons. Clearly, the number of excited electrons depends strongly on the amount of available thermal energy (i.e., on the temperature). At 300 K in silicon, for example, the concentration of broken bonds is 1.25×10^{10} cm^{-3}. Thus there are 1.25×10^{10} electrons/cm^3 in a conduction band which has space for 2×10^{23} electrons/cm^3. Therefore, this band is nearly empty.

The minimum amount of energy to break a bond (i.e., to excite an electron from the valence band to the conduction band) is the bandgap energy E_g (see Fig. 2.12). E_g is a characteristic of the semiconductor material. For silicon at 300 K, the value, in units of electron-volts, is 1.12. The corresponding figure for gallium arsenide is 1.42 eV. The higher energy for GaAs is indicative of a stronger bond and arises from the ionic nature of the association of a group V atom, arsenic, with a group III atom, gallium. In a perfectly pure and flawless semiconductor there are no allowed energy levels within the energy band gap at which an electron can reside.

Consider now the situation of a full valence band, as illustrated in Fig. 2.13a. This simple figure indicates that electrons in the full band have kinetic energy (i.e., they are not stationary). We have attempted to illustrate this fact in Fig. 2.10 and also show that, at any instant in time, there are eight valence electrons in the vicinity of a particular atom in order to bond it to its four nearest neighbors. The motion of the electrons is random in the sense that the net momentum of the electrons within a full band is zero. As the velocity of the electrons is random, there is no net transfer of charge per unit time across a plane whose normal is

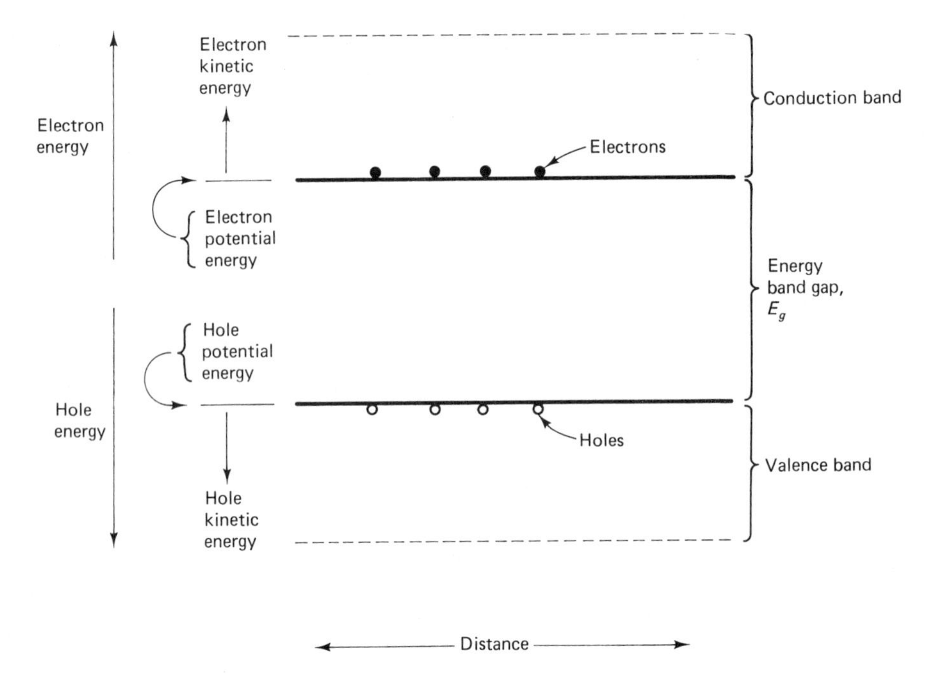

Figure 2.12 Energy band diagram showing electrons and holes resulting from the breaking of bonds. This type of diagram is more than just a simplified representation of the top part of the energy–distance relationship shown in Fig. 2.11. The energy levels corresponding to the top of the valence band and the bottom of the conduction band are the same in both figures and represent the potential energy of a hole and electron, respectively. In Fig. 2.12, however, total energy is represented so that energies above the bottom of the conduction band correspond to electron kinetic energy, and energies below the top of the valence band correspond to hole kinetic energy. The directions of increasing energy for electrons and holes are oppositely directed because of the difference in the sign of the charge associated with these two entities.

parallel to the direction of the velocity. In other words, there is no current due to the motion of electrons in a full band. Mathematically, this can be expressed by

$$\mathbf{J} = \sum_{\text{full band}} - qv_i = 0 \tag{2.4}$$

where J is the current density in A/m^2, $-q$ the charge in coulombs of each electron, and v_i the velocity of an individual electron. This fact about the current can be arrived at in a different way from Fig. 2.13a. If all the states in a band are full, electrons cannot be excited to higher-energy states within the band. Thus if an electric field were externally applied, the electrons could not gain kinetic energy from it in order to create a current.

The situation is quite different when the band is not full (see Fig. 2.13b). In

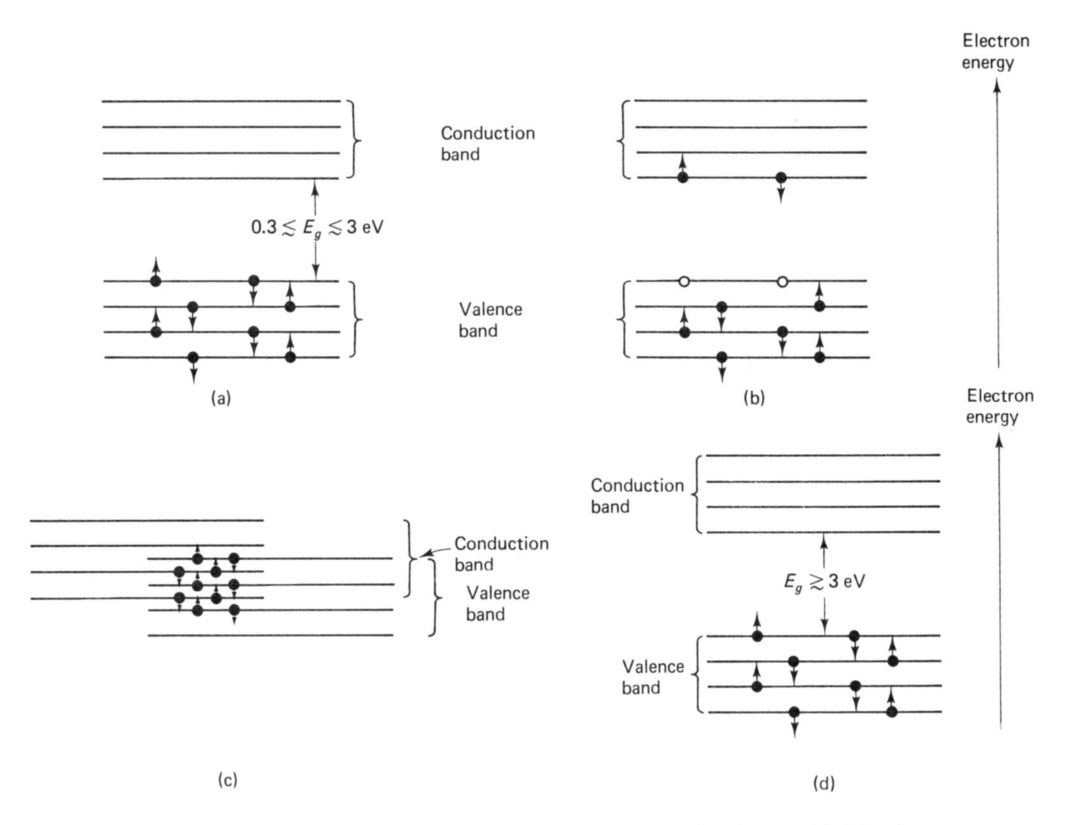

Figure 2.13 Schematic energy band representation of (a) a semiconductor with full valence band, (b) a semiconductor with nearly full valence band, (c) a metal with overlapping bands, and (d) an insulator. Each band is made up of closely spaced energy levels, each of which has two states to accommodate electrons of differing spin.

this circumstance both the electrons in the valence band and the electrons in the conduction band can be excited to higher kinetic energies within their respective bands. If this excitation energy is supplied by an external electric field, there is a tendency for the negatively charged electrons to move toward the positive source of the field. Thus the net motion of the electrons is no longer completely random and there is a current. For the conduction band we have

$$\mathbf{J} = \sum_{\substack{\text{conduction} \\ \text{band}}} -qv_i \neq 0 \tag{2.5}$$

As the electrons in the conduction band transport charge, they are called mobile charge carriers or, simply, ***carriers***.

The situation regarding the motion of electrons in the nearly full valence band is viewed a little differently, as we describe at length in Section 2.3.1.

In concluding this subsection, we note that conduction is easy in a material for which there is no energy bandgap, that is, where the conduction and valence bands overlap (see Fig. 2.13c). Such is the case for metals. In this situation the closely spaced energy levels permit movement to higher kinetic energies in response to even very small excitations. On the other hand, when E_g is large (i.e., greater than about 3 or 4 eV), the valence band remains full up to high temperatures, so that, under normal operating conditions, there is no opportunity for charge conduction. Such materials are classed, therefore, as insulators.

2.3.1 Holes

On excitation of electrons into the conduction band, ***empty states*** appear in the valence band (see Figs. 2.12 and 2.13b). Their concentration, in pure silicon at 300 K, for example, is 1.25×10^{10} cm^{-3}, which is very small compared to the number of electrons remaining in the band (i.e., nearly 2×10^{23} cm^{-3}), and therefore the valence band is nearly full. However, this small number of empty states now allows movement of electrons in the valence band. These empty states play a unique role in the interpretation of electronic conduction in semiconductors and are given the name ***holes***.

The effect of a valence bond electron moving into the state left behind by another electron, which has been excited to the conduction band, is that the hole moves (see Fig. 2.14). The phenomenon is akin to that of all the cars in a nearly full parking lot continually exchanging places. The number of vacant spots remains the same, but the locations of these spaces are forever changing. If all the cars were identical, it would appear as if only the spaces were moving. While this analogy is helpful in visualizing hole movement in physical terms, it does a disservice to the hole by treating it as nothing more than a missing electron. After all, it has not been found necessary to treat the small number of electrons in the conduction band as missing spaces in that band! The subtlety of the difference between a mere empty state where a negatively charged electron might be, and a viable, positively charged particle, needs some explanation.

An electron has both mass and charge. Thus if the removal of an electron accords the resulting space with an effective positive charge, it must also bequeath to the space a mass opposite to that of the electron. Now, the mass of relevance here is not the rest mass, but is the mass that needs to be used in any equations of motion describing the movement of electrons and holes. Why should this mass be any different from the rest, or free, mass? Of course, in the absence of relativistic effects, the particles do not actually change mass when they move. But their motion, due to the force of an external electric field, for example, is different from that of particles in free space subjected to the same field. The mobile electrons and holes in semiconductors experience not only the force of the applied field, but also the forces due to atoms and other electrons in the solid. However, it turns out that the external force on the particle can be written in the classical Newtonian form of the product of a mass and an acceleration, provided some

Figure 2.14 Movement of holes. (a) Empty state (hole) due to a broken bond at A, $t = t_1$. (b) The situation a short time later ($t = t_2 > t_1$) when an electron from B has filled the empty state at A. The hole has effectively moved from A to B.

mass other than the rest mass is used in the equation. This mass is called the ***effective mass*** and the equation defining it is derived in Section 5.4.

The effective mass will depend on the nature of the atoms and their spacing, as these factors help to determine the internal electric field. The effective mass is, thus, material dependent. However, what is not so obvious, and what is needed to appreciate fully the significance of holes, is the fact that the effective mass is also band dependent. The details of this dependency are worked out in Section 5.4. It transpires that electrons near the bottom of the conduction band have positive effective mass, but electrons near the top of the valence band have ***negative effective mass***. The former finding is not unexpected and confirms that electrons in the conduction band can be treated as "regular," positive mass, negatively charged particles. On the other hand, the concept of negative effective mass is not easy to grasp. In pondering on this, it is helpful to recall that conduction in the valence band occurs only when electrons from near the top of the band are missing. The removal of a particle with negative effective mass is akin to creating, in its place, a particle of positive effective mass. Thus the hole can be viewed as just another subatomic particle, one that has positive effective mass and positive charge. Direct evidence for the existence of holes is furnished by the results of Hall effect measurements on doped semiconductors (see Section 2.4.5).

Thus in picturing charge movement in semiconductors, it is convenient, and correct, to think of negative charge as being carried by electrons in the conduction band, and of positive charge as being transported by holes in the valence band. With this interpretation, there is no confusion between electrons in the two bands. Therefore, henceforth, conduction band electrons can be described merely as electrons. Similarly, the term "hole" implies an empty state in the valence band only.

2.4 ELECTRON AND HOLE CONCENTRATIONS IN THERMAL EQUILIBRIUM

2.4.1 Intrinsic Case

The properties of a pure material are intrinsic to that material. Thus the concentrations of electrons in the conduction band and holes in the valence band of a pure semiconductor take on their intrinsic values. From the discussion in Section 2.3, and as illustrated in Fig. 2.12, it follows that the ***intrinsic concentration*** of holes in the valence band equals the intrinsic concentration of electrons in the conduction band, that is,

$$n_i = p_i \tag{2.6}$$

where the subscript i refers to "intrinsic," and n and p refer to the electron and hole concentrations, respectively. It is to be understood, in the light of the discussion in the preceding section, that the holes reside only in the valence band

and that the electrons under consideration here reside only in the conduction band.

The concentration of intrinsic electrons and holes depends on the amount of energy needed to break a bond (i.e., the energy band gap, E_g) and on the amount of energy available (e.g., the thermal energy as characterized by the temperature, T). The exact mathematical relationship between these parameters is derived in Section 4.4. Here, to illustrate the form of the dependencies, we merely quote the general form of the equation:

$$n_i = B(T) \exp\left(\frac{-E_g}{2kT}\right) \tag{2.7}$$

where $B(T)$ is a material- and temperature-dependent parameter and k is Boltzmann's constant, which has the value of 8.62×10^{-5} eV/K. Note that the presence of T in the exponential term gives n_i a very strong temperature dependence.

Measured values of n_i at 300 K for Si and GaAs are 1.25×10^{10} cm^{-3} and 2×10^6 cm^{-3}, respectively. Such low concentrations, relative to those in extrinsic material (see Section 2.4.2), make for near-insulating material. This property is exploited in some GaAs integrated circuits to provide good electrical isolation between neighboring devices. It should be noted, though, that it is very difficult to purify a semiconductor to the extent necessary to realise intrinsic conduction at room temperature. In the case of GaAs, obtaining high-resistivity wafers is something of an art and requires the presence of particular impurities and defects in the material.

2.4.2 Extrinsic Case

Semiconductors doped with a few parts per million of certain impurities can have electron and hole concentrations which are markedly different from those in intrinsic material at the same temperature. In such cases the material is said to be ***extrinsic***. Effective dopant atoms are those which substitute for semiconductor atoms at lattice sites. Phosphorus is an example of such a substitutional impurity in silicon (see Fig. 2.15). Phosphorus is a group V element with atomic mass very close to that of silicon. Thus it can be accommodated in the silicon lattice without causing much physical disturbance. Four of the five valence electrons of phosphorus satisfy the bonding requirements. The remaining valence electron is not needed for bond formation and, consequently, is not strongly bound to the phosphorus atom. Thus only a small amount of thermal energy is required to elevate this electron to the conduction band and render it "free." The phosphorus atom donates the electron to the pool of current carriers, leaving the region of physical space around the dopant with a net positive charge. Thus once the ***donor*** impurity has donated its electron, it becomes an ***ion***.

The situation can be represented on an energy band diagram as shown in Fig. 2.16. The physical positions of phosphorus atoms are marked with short lines to emphasize the ***localized*** nature of the doping. For example, a moderately high

Figure 2.15 Incorporation of a phosphorus atom in the silicon lattice. Only four of the five valence electrons of the group V element are required for bonding to the four nearest-neighbor silicon atoms. The surplus electron (not shown) is easily excited into the conduction band. Note that this loss of an electron does not create an empty state (hole) in the valence band. Instead, it creates a singly charged positive phosphorus ion, that is, the net charge of the central core charge of $+5q$ (nucleus plus core electrons) of the phosphorus and the four valence electrons.

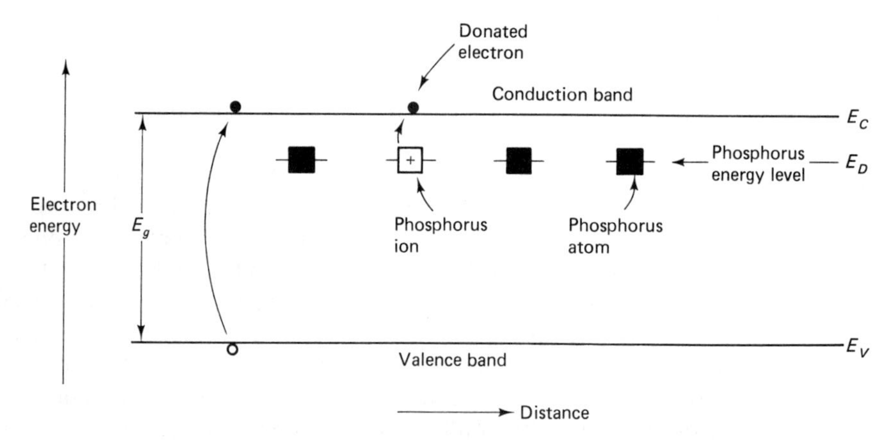

Figure 2.16 Energy band diagram for phosphorus-doped silicon. The incorporation of phosphorus in the silicon lattice introduces an energy level E_D at 0.044 eV below the conduction band. Excitation of an electron (the surplus electron referred to in the caption of Fig. 2.15) from this level results in an extra electron in the conduction band and a localized, positively charged phosphorus ion. The electron and hole shown on the left side of the figure represent current carriers formed by the breaking of silicon bonds.

doping density of 10^{17} atoms/cm^3 in silicon would result in a dopant atom being in only one of about every 100 lattice points in any direction. Because the ions are locked in position at lattice sites they are immobile. Thus even though they are charged, they do not contribute to any current. However, the electron that the phosphorus atom donates to the conduction band can give rise to a current. It increases the concentration of electrons, which thus become, in this instance, the ***majority carriers***. In recognition of the fact that the predominant carrier is negatively charged, semiconductor material doped in this fashion is said to be ***n-type***.

Phosphorus and arsenic are the two group V impurities most commonly employed as donors in silicon. In gallium arsenide, useful donors are silicon (group IV) as a substitutional impurity for gallium and selenium (group VI) in place of arsenic.

Another common dopant in silicon is the group III element boron. The three valence electrons of this small atom are insufficient to satisfy the full bonding requirements of the four neighboring silicon atoms that surround it (see Fig. 2.17). This deficiency in bonding can be repaired by the acceptance of a valence electron from elsewhere in the valence band. This action demands a small amount of energy because the accepted electron is surplus to the needs of the boron atom, and therefore a small repulsive force due to the existing three valence electrons has to be overcome. However, once the electron is accepted, the boron atom becomes a negative ion and the empty state left behind by the electron elsewhere in the valence band becomes a hole. The situation is depicted on an energy band diagram as shown in Fig. 2.18. The boron ion remains fixed in space, and therefore does not contribute to any current, but the hole is free to move in the valence band. In this instance, then, the holes are the majority carriers of charge. This predominance of positively charged carriers is responsible for the designation of boron-doped silicon as ***p-type***.

Boron and, to a lesser extent, aluminum are the group III elements most commonly used as ***acceptor*** impurities in silicon. In gallium arsenide, zinc (group II) is an acceptor impurity when it is incorporated in the lattice in place of gallium atoms. Substitution of arsenic atoms with silicon (group IV) also produces p-type GaAs.

2.4.3 Charge Neutrality

Gauss's famous law in electromagnetics states that the divergence of the electric field over a volume is proportional to the net charge density within the volume. It follows that if there is no electric field in the volume, the charge density therein is zero. Thus field-free semiconductor material, whether it be intrinsic or extrinsic, is ***electrically neutral***. In the intrinsic case, the concentration of negatively charged conduction electrons equals the concentration of positively charged holes. In extrinsic n-type material, the concentration of electrons is balanced by the sum of the concentrations of sources of positive charge, namely the holes and the ionized

Figure 2.17 The incorporation of a boron atom in the silicon lattice. (a) The three valence electrons of the group III element are insufficient to make four covalent bonds with the four nearest-neighbor silicon atoms. (b) On the acceptance of an electron from elsewhere (e.g., point A) in the lattice, to satisfy the bonding requirements, the immediate locality of the boron acquires a net negative charge of $-q$, that is, the boron becomes a singly charged negative ion. The accepted electron creates an empty state (hole) in the valence band.

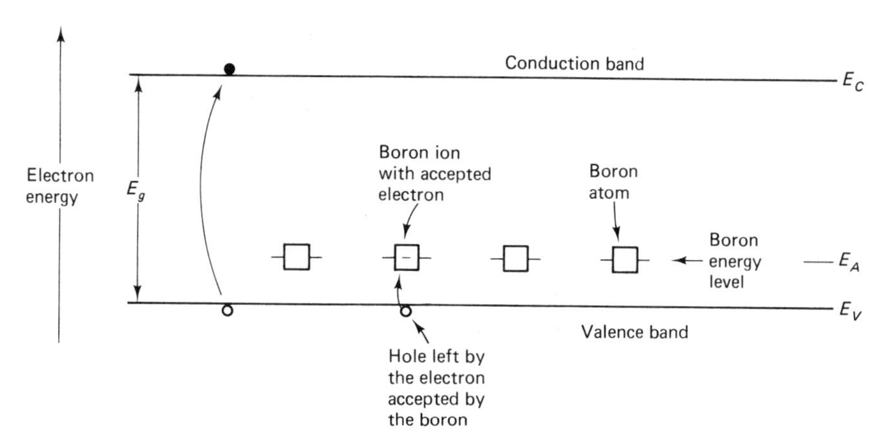

Figure 2.18 Energy band diagram for boron-doped silicon. The incorporation of boron in the silicon lattice introduces an energy level E_A at 0.045 eV above the valence band. Acceptance of an electron at this energy level results in an extra hole in the valence band and a localized, negatively charged boron ion. The regular breaking of silicon bonds results in electron–hole pairs, one of which is shown on the left side of the figure.

donors. Similarly, in field-free p-type material the hole concentration equals that of the sum of the concentrations of electrons and ionized acceptors. The general situation is summarized by

$$p + N_D^+ = n + N_A^- \tag{2.8}$$

Note that N_D^+ does not necessarily equal N_D, the concentration of donor atom impurities. Ionization of the atoms depends on how much energy is needed to donate an electron ($E_C - E_D$ in Fig. 2.16) and on how much energy is available. E_D depends on the particular atom/semiconductor combination; the available energy, if it is thermal in nature, depends on the temperature. Similarly, N_A^- is not necessarily equal to N_A.

Note also that if $N_D^+ = N_A^-$, the electron and hole concentrations are equal and the material is essentially intrinsic. Such a material is said to be exactly ***compensated***.

2.4.4 Dynamic Equilibrium

When the only source of external energy to the semiconductor is heat and the material reaches a steady temperature T, the situation is said to be one of thermal equilibrium. As far as the electrons and holes are concerned, this equilibrium is dynamic. Absorbed energy results in bonds continually being broken, leading to the continual ***generation*** of electrons and holes. Also, energy release accompanies the continual repair of bonds on the ***recombination*** of electrons and holes. The balancing of these two phenomena brings about ***thermal equilibrium***. To illustrate the situation, we consider here the case of ***direct***, or ***band-to-band***, recombination and generation (see Fig. 2.19).

The rate of direct generation of electron–hole pairs depends solely on the temperature. It is not influenced by the presence of impurities, as these do not produce pairs of electrons and holes. Thus

$$G_{\mathrm{th}}(T) = G_i(T) \tag{2.9}$$

that is, the generation rate in units of electron–hole pairs per unit volume per second is that exhibited by intrinsic material.

The rate of direct recombination of electron–hole pairs depends not only on the temperature but also on the concentrations of electrons and holes present. The greater the number of individual species, the greater the opportunity for pairing. This statement is as true for carriers in semiconductors as it is for people at dances! Thus

$$R(n_0, p_0, T) = n_0 p_0 r(T) \tag{2.10}$$

where R is the recombination rate in electron–hole pairs per unit volume per second, r is a rate constant, and the subscript zero for the carrier concentrations denotes explicitly the condition of thermal equilibrium.

At thermal equilibrium the two rates are equal; thus

$$R(n_0, p_0, T) = G_{\mathrm{th}}(T) \tag{2.11}$$

For intrinsic material this equation can be written as

$$R(n_i, p_i, T) = G_i(T) \tag{2.12}$$

Combining (2.9), (2.11), and (2.12) leads to the realization that in thermal equilibrium, the direct recombination rate in extrinsic material equals that in intrinsic material. Using this fact in (2.10) leads to

$$n_0 p_0 = n_i p_i \tag{2.13}$$

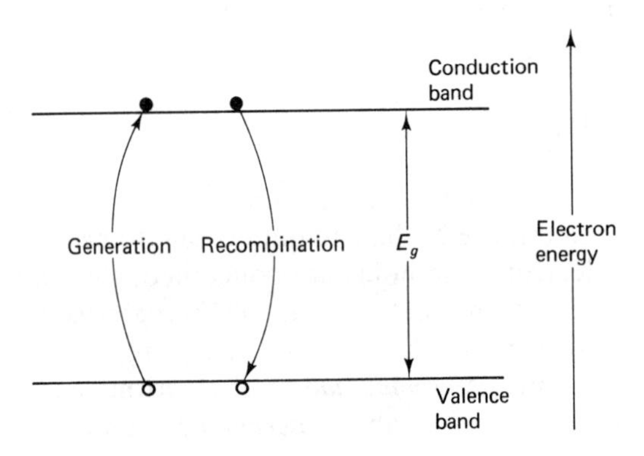

Figure 2.19 Direct, or band-to-band, generation and recombination of electrons and holes.

Making use of (2.6) gives

$$n_0 p_0 = n_i^2 \tag{2.14}$$

This important equation implies that in n-type material, for example, where the electron concentration is greater than n_i, the hole concentration is less than n_i. This is not surprising since a unilateral increase in electron concentration provides more opportunity for electron–hole recombination. The hole concentration diminishes accordingly until the recombination rate is lowered to its equilibrium value. Although our proof of this equation has been developed for the case of direct recombination, we show in Section 4.4.1 the relationship is true in general, provided that the material is not very heavily doped.

2.4.5 Measurement of Majority Carrier Concentration

Provided that the ionized dopant concentrations and the intrinsic carrier concentration are known, (2.8) and (2.14) can be solved simultaneously to compute n_0 and p_0. Details of the calculation are given in Section 4.5. The equilibrium carrier concentrations can also be determined by measurement by making use of the experimental arrangement shown in Fig. 2.20. Consider the case of a p-type sample. The holes that constitute the current due to the applied voltage are deflected

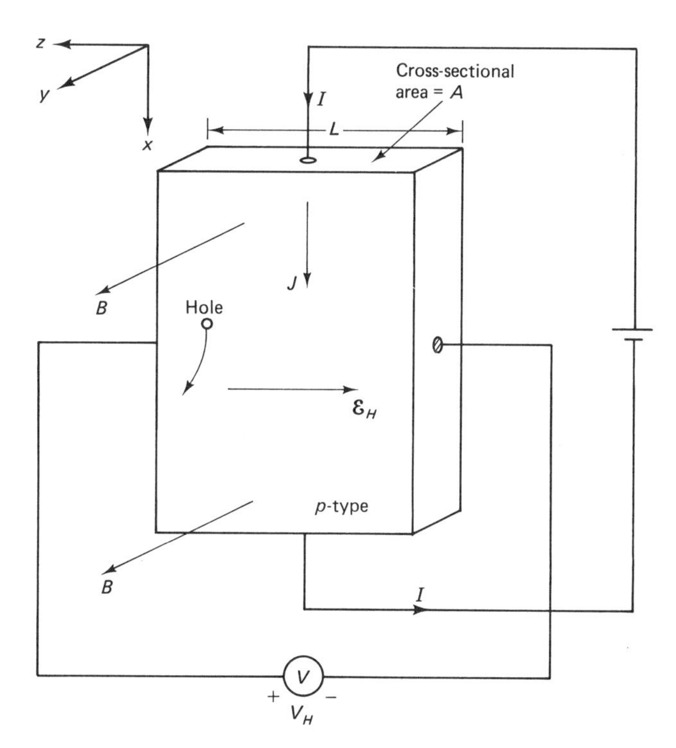

Figure 2.20 Hall-effect arrangement for measuring majority carrier concentration. In the p-type sample shown, the majority carrier holes are deflected to the left side of the sample by the interaction of the current and the magnetic field. The left side of the sample becomes positively charged, as can be detected by a voltmeter.

by the applied magnetic field according to Lorentz's equation [4], that is,

$$\text{Lorentz force} = J\mathbf{x} \times B\mathbf{y} \tag{2.15}$$

where B is the magnitude of the magnetic field intensity in tesla in the y-direction and J, the current density in A/m^2, is given by (I/A), where I is the current due to holes flowing through the sample of cross-sectional area A in the x-direction. The Lorentz force is in the z-direction, so the holes are pushed to the left-hand side of the sample. This accumulation of positive charge can be detected by measuring the potential difference between the left and right sides of the sample. The resultant voltage is called the ***Hall voltage***. The electric field $\mathscr{E}_H$ in the negative z-direction caused by the Hall voltage V_H exerts an electrostatic force on the holes tending to push them to the right; thus

$$\text{electrostatic force} = qp_0\mathscr{E}_H(-\mathbf{z}) \tag{2.16}$$

where (qp_0) is the charge per unit volume of the deflected holes. q has a positive sign and a magnitude equal to that of the electronic charge (i.e., 1.6×10^{-19} C). The concentration of holes is given by the thermal equilibrium value even though there are sources of energy present in addition to thermal energy. The electric and magnetic fields serve only to influence the direction of movement of the carriers and do not affect the carrier concentration.

The two opposing forces given by (2.15) and (2.16) produce a state of physical equilibrium when

$$J\mathbf{x} \times B\mathbf{y} + qp_0\mathscr{E}_H(-\mathbf{z}) = 0 \tag{2.17}$$

Rearranging and substituting for J and $\mathscr{E}_H$ leads to

$$V_H = \frac{IBL}{p_0qA} \tag{2.18}$$

The thermal equilibrium concentration of the majority carrier holes can be calculated from this equation as all the other parameters can be measured. The ***minority carrier*** concentration, electrons in this case, can then be computed from (2.14).

If an n-type sample is used in the Hall-effect measurement arrangement shown in Fig. 2.20, the majority carriers pushed to the left-hand side will be electrons; thus the direction of $\mathscr{E}_H$ will be reversed. This gives for V_H,

$$V_H(n\text{-type}) = \frac{-IBL}{n_0qA} \tag{2.19}$$

where n_0 is the equilibrium concentration of the majority carrier electrons. The polarity of the Hall voltage thus establishes whether the material is n- or p-type.

If it is desired to determine only the type of some unknown wafer and not the magnitude of the majority carrier concentration, a simpler measurement than the Hall-effect measurement can be performed. The setup is shown in Fig. 2.21

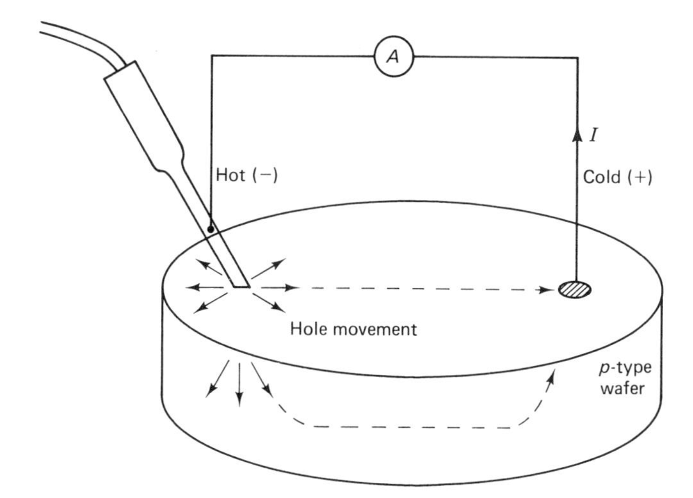

Figure 2.21 Hot-point probe arrangement for determining the doping type of a semiconductor wafer. In the p-type sample shown, the majority carrier holes diffuse away from the hot region. Some of the holes reach the cold probe and cause a measurable current in the external circuit.

for the case of a p-type sample. A hot probe, such as the tip of a soldering iron, is used to raise the temperature of a localized region of the sample. The carriers under the probe tend to move away to the cooler regions of the sample and some will flow through the contact shown in the figure, giving rise to a detectable current. The direction of this current gives the polarity of the majority carrier and hence tells whether the material under test is n- or p-type.

In both the methods described above, the minority carriers move as well as the majority carriers. However, neither the magnitude and the sign of the Hall voltage nor the direction of the "hot-probe" current is much affected by the minority carriers as their concentration is so low. To appreciate this, consider a p-type silicon sample with a hole concentration of 10^{16} cm^{-3}. Such a value is typical of many wafers used as starting material for semiconductor devices. From (2.14) it follows that at 300 K, the electron concentration is only 1.5×10^{-10}% of the hole concentration.

2.5 EXCESS CARRIERS

2.5.1 Excess Carrier Concentrations

If sources of energy other than just thermal energy are present, it is possible for the electron and hole concentrations to be elevated above their thermal equilibrium concentrations. Thus, using the case of electrons, for example,

$$n = n_0 + \hat{n} \tag{2.20}$$

where n is the ***total electron concentration***, n_0 is the thermal equilibrium concentration, and $\hat{n}$ is called the ***excess electron concentration***. Excess carriers are commonly created in semiconductor devices by optical or electrical energy sources. The latter case is usually concerned with the injection of majority carriers from one type of semiconductor into oppositely doped material; for example, holes injected from p-type material into n-type material in response to an applied voltage. This situation gives rise to excess minority carrier holes in the n-type material and is fundamental to the operation of bipolar devices, as treated in Chapters 6 and 11.

In the case of optical generation of excess carriers, bonds are broken by the absorption of photons. It follows that the photons must have an energy at least equal to the energy bandgap, and that this type of excitation produces equal numbers of excess electrons and holes. The situation is illustrated in Fig. 2.22. The total carrier concentrations are given by

$$n = n_0 + \hat{n}$$
$$p = p_0 + \hat{p} = p_0 + \hat{n} \tag{2.21}$$

from which it follows that

$$np \text{ (out of thermal equilibrium)} \neq n_i^2 \tag{2.22}$$

This emphasizes that the earlier expression for the np product, (2.14), must be used only when thermal equilibrium obtains.

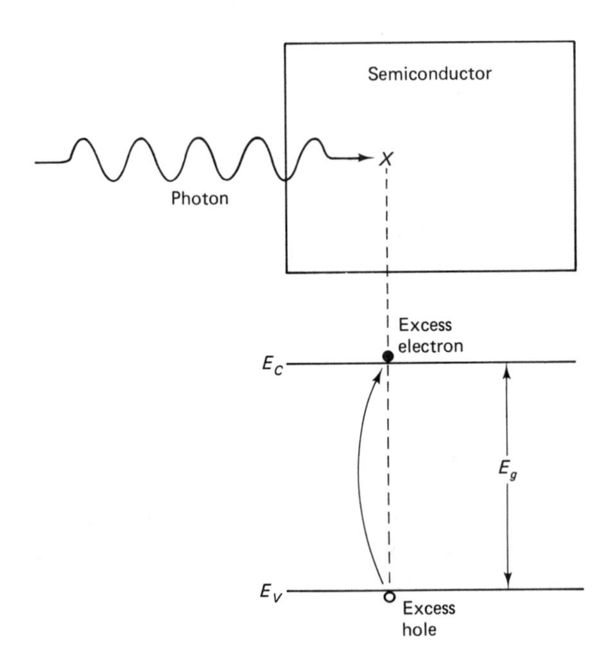

Figure 2.22 Generation of excess carriers by optical absorption. Photons with energy greater than or equal to the energy bandgap can be absorbed, so breaking bonds and generating excess carriers. The figure depicts absorption at point X. Photon energy is given by hf, where h is Planck's constant and f is the optical frequency. Photons with $hf < E_g$ pass right through the material. Thus silicon is transparent to light with frequencies below about 270 THz. The bandgap energy of 1.12 eV is on the infrared side of the visible spectrum (1.7 to 3 eV).

2.5.2 Recombination of Excess Carriers

2.5.2.1 Direct recombination. Out of thermal equilibrium, the total recombination rate of electron–hole pairs can be written as

$$R = \hat{R} + R_0(T) \tag{2.23}$$

where $R_0(T)$ is the thermal equilibrium recombination rate given, for the case of direct or band-to-band recombination (Figs. 2.19 and 2.23) by (2.10), and $\hat{R}$ is the recombination rate of excess carriers. Rearranging (2.23) and using (2.10) and (2.11) leads to

$$\hat{R} = R - R_0(T) = R - G_{\text{th}}(T) = r(T)np - \text{r}(T)n_0p_0 \tag{2.24}$$

Substituting for np from (2.21) gives

$$\hat{R} = r(T)(\hat{n}\hat{p} + \hat{n}p_0 + \hat{p}n_0) \tag{2.25}$$

Often, the excess carrier concentrations will be much less than the equilibrium concentration of majority carriers. For example, bright sunlight is absorbed by silicon and can give rise to excess carrier concentrations around 10^{12} cm^{-3}. Large

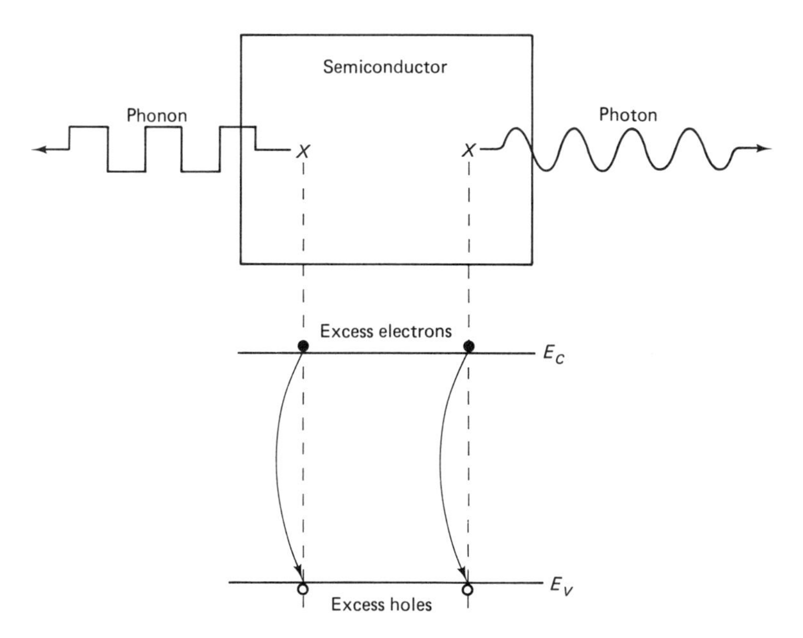

Figure 2.23 Recombination of excess carriers. This example depicts excess electrons recombining directly with holes in a band-to-band transition. The potential energy of the electrons is converted into either phonon energy (heat) or photon energy (light), depending on the nature of the semiconductor material.

as this value is, it is only significant as regards the minority carrier concentration because the equilibrium concentration of majority carriers might typically be 10^{17} cm^{-3} in moderately doped material. Conditions such as these are referred to as ***low-level injection conditions***. Thus, in low-level injection,

$$\begin{aligned} \hat{n} &< n_0 \quad \text{in } n\text{-type material} \\ \hat{p} &< p_0 \quad \text{in } p\text{-type material} \end{aligned} \tag{2.26}$$

In low-level injection, therefore, the ***direct recombination*** rates for excess minority carriers are well approximated by

$$\begin{aligned} \hat{R}(\text{holes}) &= r(T)\hat{p}n_0 \quad \text{for } n\text{-type material} \\ \hat{R}(\text{electrons}) &= r(T)\hat{n}p_0 \quad \text{for } p\text{-type material} \end{aligned} \tag{2.27}$$

2.5.2.2 Indirect recombination. If impurities or flaws are present in a crystalline semiconductor, the perfect periodicity of the lattice is disturbed in the vicinity of these foreign entities. This perturbation leads to localized energy levels within the energy bandgap of the semiconductor. These energy levels act as "stepping stones" for the electrons to use in their transitions between the bands. The situation is illustrated in Fig. 2.24.

The left-hand side of the figure shows an electron falling into a state at the energy level E_t. The electron may be ***trapped*** in this state for a while, before falling into the valence band and annihilating a hole. From the trap's point of view, the latter step (Fig. 2.24b) is akin to the capture of a hole. The net result of (a) + (b) is that one electron–hole pair is destroyed; that is, recombination has taken place indirectly through the trap. This type of recombination is often called ***Shockley–Read–Hall*** recombination, after the three men who first worked out the dynamics of the process. It is possible that after event (a) the electron, instead of recombining with a hole, may be excited back into the conduction band. This is very

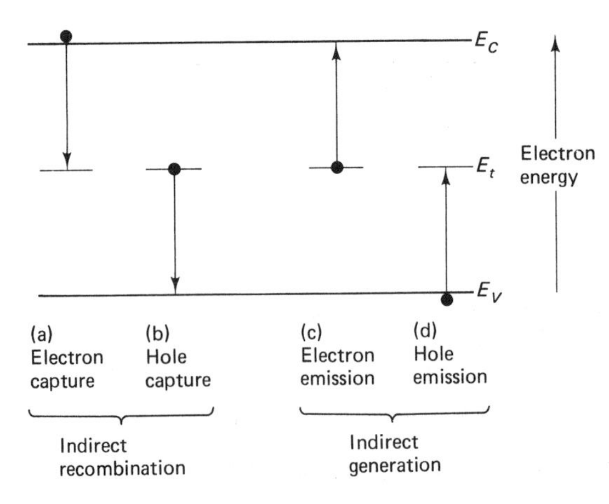

Figure 2.24 Indirect, or Shockley–Read–Hall generation and recombination of electrons and holes. The localized energy levels at E_t are due to impurities and defects in crystallinity.

probable if the energy level E_t is close to the conduction band. Thus, donor energy levels, such as illustrated in Fig. 2.16, are not good recombination centers. They merely serve as temporary "resting places" for electrons dropping in and out of the conduction band.

Returning to the case of near-midgap energy levels, Fig. 2.24d shows the capture of an electron from the valence band. To the trap, this event is the same as the emission of a hole from the trap. If such a step were followed by event (c), one electron–hole pair would be created. This is called ***indirect generation*** via traps.

To analyze fully the dynamics of indirect recombination–generation events, it is necessary to consider the likelihood of each of the four processes in Fig. 2.24 occurring. This depends on the concentrations of the carriers in the conduction and valence bands, and on the concentration and state of occupancy of the traps. The result of such an analysis [5] indicates that, in low-level injection, the indirect recombination rates for excess minority carriers are well approximated by

$$\begin{aligned} \hat{R}(\text{holes}) &= c_p N_t \hat{p} \qquad \text{for } n\text{-type material} \\ \hat{R}(\text{electrons}) &= c_n N_t \hat{n} \qquad \text{for } p\text{-type material} \end{aligned} \tag{2.28}$$

where c_p and c_n are constants involving the effective size, or capture cross section, of the trap. These expressions are similar to those for direct recombination, (2.27), the essential difference being the replacement of the majority carrier concentration in the direct case by the trap concentration N_t in the indirect case. Such a result is easy to appreciate. In indirect recombination, we can expect the trap to have no difficulty in capturing a majority carrier, as there are so many of them. Thus the number of majority carriers available for recombination by the indirect process is limited not by their concentration in the conduction band, but by the number of traps in which they may temporarily reside.

2.5.3 Excess Carrier Lifetime

The generation of excess carriers by absorbed radiation, for example, and the recombination of excess carriers, as described by either (2.27) or (2.28), coact to produce a ***steady-state*** excess carrier concentration ($\hat{n}$ or $\hat{p}$). On removal of the nonthermal energy source, the recombination of excess carriers continues until there are no excess carriers left (i.e., the total carrier concentration becomes the thermal equilibrium concentration). Mathematically, the situation in n-type material following the turning off of ionizing radiation, for example, is written as

$$\hat{R}(t) = -\frac{d}{dt}(\hat{p}(t)) \tag{2.29}$$

Substituting for $\hat{R}$ from either (2.27) or (2.28) and integrating gives

$$\hat{p}(t) = \hat{p}(0) \exp\left(\frac{-t}{\tau_h}\right) \tag{2.30}$$

where $\hat{p}(0)$ is the excess hole concentration at the instant of removing the nonthermal energy souce and τ_h, the hole ***minority carrier lifetime***, is defined as either

$$\tau_h = \frac{1}{r(T)n_0(T, N_D)} \qquad \text{[direct recombination]}$$

or (2.31)

$$\tau_h = \frac{1}{c_p N_t} \qquad \text{[indirect recombination]}$$

Similarly, the electron minority carrier lifetime is defined by either

$$\tau_e = \frac{1}{r(T)p_0(T, N_A)} \qquad \text{[direct recombination]}$$

or (2.32)

$$\tau_e = \frac{1}{c_n N_t} \qquad \text{[indirect recombination]}$$

These equations show that the minority carrier lifetimes depend on the doping density for the case of direct recombination, and on the trap density for the case of indirect recombination. The higher the doping of the semiconductor, the higher the concentration of majority carriers available to recombine directly with the excess minority carriers and thus the shorter the lifetime of the excess carriers in the direct case. However, in the indirect case, the majority carriers communicate with the minority carriers via the traps. The more traps there are, the more majority carriers there are available for indirect recombination, and therefore, the shorter the lifetime of the soon-to-be-annihilated excess minority carriers. Equation (2.30) indicates that on removal of the generating source of the excess carriers, after the elapse of a time equal to the minority carrier lifetime, the excess carrier concentration will be reduced by 63%. Typical hole minority carrier lifetimes for n-type material doped with 10^{17} cm^{-3} donors are about 10^{-9} s for GaAs and about 10^{-7} s for Si. The origin of this difference in minority carrier lifetimes is discussed in Section 5.5.

In some of the examples and problems later in this book we need an equation to describe the doping density dependence of the minority carrier lifetimes in silicon. Expressions of the form given in (2.31) and (2.32) for indirect recombination would be suitable, provided that the doping density dependencies of the capture cross sections and the actual trap densities were accurately known. As this is rarely the case, models for semiconductor device analysis purposes often employ an empirical relationship of the form

$$\tau = \frac{\tau_0}{1 + N/N_0} \tag{2.33}$$

where N is the total doping density in atoms/cm^3. τ_0 and N_0 are empirical factors used to obtain a good fit to experimental data. The values adopted in

SEDAN, the popular device analysis program from Stanford University [6], are $\tau_0 = 5 \times 10^{-7}$ s and $N_0 = 5 \times 10^{16}$ cm^{-3}.

2.6 CHAPTER SUMMARY

In this chapter those properties of semiconductor materials which are fundamental to the understanding of transistors and diodes have been introduced. The reader should now have an appreciation of the following topics:

1. ***Crystal structure and growth.*** Silicon and gallium arsenide have crystal structures based on a ***tetrahedral*** bonding arrangement. Single-crystal ***ingots*** of these semiconductors are grown by the ***Czochralski*** method. The ingots are sawn into thin ***wafers***, which form the substrates of semiconductor devices.

2. ***Conductivity.*** Conductivity is a material property dependent on temperature, ***energy bandgap***, and ***doping density***. Typically, it spans the range of 0.01 to 10 $(\Omega\cdot\text{cm})^{-1}$ in semiconductors and is measured by four-point probe methods.

3. ***Dopants.*** Dopants are the ***donors*** and ***acceptors*** that are incorporated in the semiconductor lattice to produce ***n-type*** and ***p-type*** material, respectively.

4. ***Charge carriers.*** Charge carriers are the ***electrons*** in the ***conduction band*** and ***holes*** in the ***valence band*** responsible for the conduction of electricity in the semiconductor. The more plentiful carrier is called the ***majority*** carrier, the less numerous carrier is the ***minority*** carrier. The majority carrier ***concentration*** can be measured via the ***Hall effect***.

5. ***Charge neutrality.*** The positive charge of the holes and the ***ionized*** donors equals the negative charge of the electrons and the ionized acceptors in uniformly doped material with no applied field.

6. ***Thermal equilibrium.*** Thermal equilibrium is a ***dynamic*** equilibrium in which the rate of thermal ***generation*** of electrons and holes due to the ***breaking of bonds*** exactly balances the ***recombination*** of electrons and holes on restoration of the bonds. For a given temperature, the product of electron and hole concentrations is constant.

7. ***Excess carriers.*** Thermal equilibrium can be disturbed by the presence of nonthermal sources of energy, notably optical energy from absorbed radiation and electrical energy from a voltage supply. Under these conditions excess carriers are introduced into the semiconductor, raising the ***total concentration*** of one or both of the carriers above the equilibrium value. As long as the excess carrier concentrations are less than the thermal equilibrium majority carrier concentration, the situation is referred to as one of ***low-level injection***. Increasing the concentrations of carriers increases the rate of recombination. The ***minority carrier lifetime*** is a measure of the time taken for an excess minority carrier to recombine with a majority carrier. Recombination (and generation) can occur ***directly*** in ***band-to-band*** transitions and ***indirectly*** in ***Shockley–Read–Hall*** processes involving energy levels in the ***forbidden region*** due to impurities and crystalline defects.

2.7 REFERENCES

1. D. G. Fink and H. W. Beaty, eds., *Standard Handbook for Electrical Engineers*, 11th ed., p. 2–4, McGraw-Hill Book Company, New York, 1978.
2. F. M. Smits, "Measurement of sheet resistivities with the four-point probe," *Bell System Technical Journal*, vol. 37, pp. 711–718, 1958.
3. S. M. Sze, *Semiconductor Devices, Physics and Technology*, p. 5, John Wiley & Sons, Inc., New York, 1985.
4. R. P. Feynmann, R. B. Leighton, and M. Sands, *The Feynmann Lectures on Physics*, vol. 2, p. 13–2, Addison-Wesley Publishing Co., Inc., Reading, Mass., 1966.
5. A. S. Grove, *Physics and Technology of Semiconductor Devices*, p. 134, John Wiley & Sons, Inc., New York, 1967.
6. Z. Yu and R. W. Dutton, "SEDAN III—a generalized electronic material device analysis program," Technical Report, Integrated Circuits Laboratory, Stanford University, Stanford, Calif., July 1985.

PROBLEMS

2.1. A four-point probe measurement on a silicon wafer gives $V = 0.5$ V when $I = 5$ mA. The probe spacing, wafer diameter, and wafer thickness (500 μm) are such that the current in the wafer can be considered as being uniformly distributed throughout the thickness of the wafer. Estimate the resistivity of the wafer.

2.2. Figure 2P.1 shows how a resistor may be implemented in an integrated circuit using n-type silicon. The n^+ regions are heavily doped and act as essentially zero-resis-

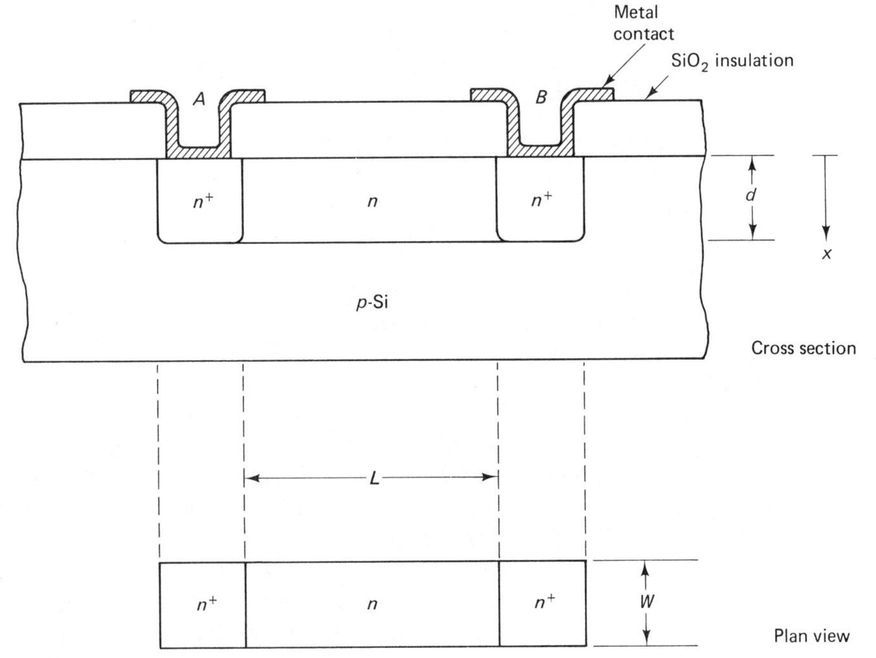

Figure 2P.1

tance connections between the n region and the metallic contacts. To compute the resistance between contacts A and B, use is made of the ***sheet resistance***, i.e.,

$$R_{\text{sheet}} = \frac{\rho}{d} \qquad \Omega/\text{square} \tag{2P.1}$$

This expression follows from (2.1) for the case of $L = W$ (i.e., for a square of material). Given that $R_{\text{sheet}} = 30\ \Omega$/square for the n-type material in Fig. 2P.1, compute R_{AB} when **(a)** $L = 2W$; **(b)** $W = 2L$.

2.3. Construct a projection of the zinc blende structure of GaAs onto the basal plane such as to show that the lattice actually comprises two interpenetrating face-centered cubic sublattices, one composed of Ga atoms and the other of As atoms. (*Note:* The xy plane in Fig. 2.4 is an example of a basal plane.)

2.4. The GaAs lattice may be constructed by stacking planes of Ga and As atoms alternately along the (1, 1, 1) direction (see Fig. 2P.2). The sequence alternates between widely spaced and closely spaced planes, in the pattern . . . Ga–As—Ga–As—Ga–As. . . . Compute the two possible separations between the Ga and As planes.

2.5. In single-crystalline Si, what is the distance from the center of one Si atom to the center of its nearest neighbor?

2.6. Compute the number of atoms per unit area on the (100) and (111) surface planes in the silicon lattice.

2.7. In device fabrication it is sometimes useful to be able to see through a silicon wafer in order to align a pattern on the front of the wafer with one on the back. This requires the use of an infrared microscope, since silicon is opaque to visible light. Given that the energy bandgap of silicon is 1.12 eV, estimate the minimum wavelength of light that could be used in such a microscope.

2.8. Light-emitting diodes (LEDs) can be made from GaAs and are often used in intrusion alarm systems. Photons are emitted as a consequence of the direct recombination illustrated in Fig. 2.23. Confirm that the light emitted from GaAs is outside the visible range.

2.9. Calculate the majority and minority carrier concentrations in extrinsic silicon doped with the following impurities:
(a) $N_D = 10^{17}\ \text{cm}^{-3}$, $N_A = 0$
(b) $N_D = 10^{17}\ \text{cm}^{-3}$, $N_A = 10^{15}\ \text{cm}^{-3}$
(c) $N_D = 10^{17}\ \text{cm}^{-3}$, $N_A = 10^{17}\ \text{cm}^{-3}$
(d) $N_D = 10^{17}\ \text{cm}^{-3}$, $N_A = 10^{19}\ \text{cm}^{-3}$
Assume that all impurities are ionized and that $T = 300$ K.

2.10. Table 2P.1 gives experimental data for the intrinsic carrier concentrations of Si and

TABLE 2P.1

n_i (cm^{-3}) Si	n_i (cm^{-3}) GaAs	T (°C)
$4.4. \times 10^{9}$	$7.3. \times 10^{5}$	0
$1.8. \times 10^{12}$	$1.9. \times 10^{9}$	100
$1.0. \times 10^{14}$	$2.5. \times 10^{11}$	200
$2.1. \times 10^{16}$	$2.6. \times 10^{14}$	500
$4.3. \times 10^{17}$	$1.9. \times 10^{16}$	1000

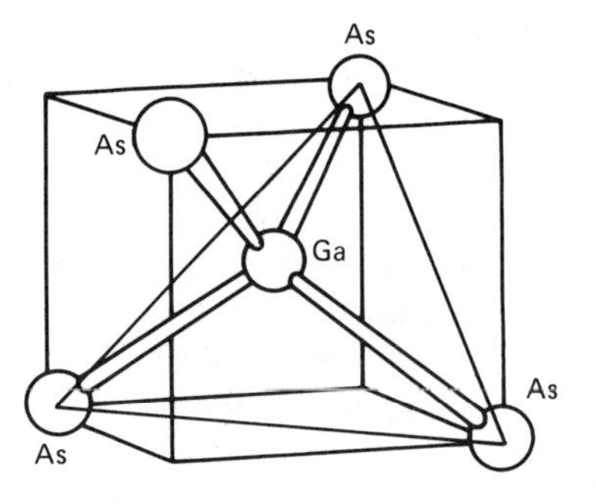

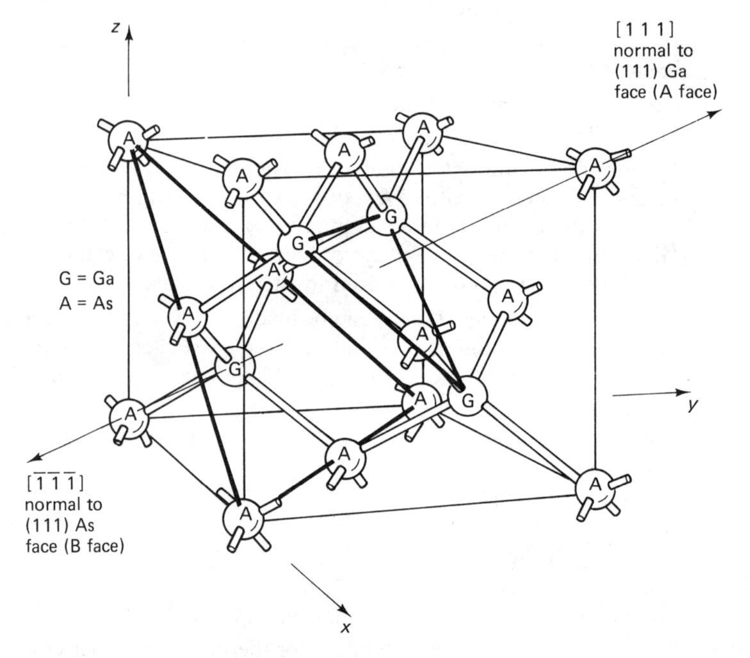

Figure 2P.2

GaAs at various temperatures. By plotting a suitable graph, obtain values for the energy bandgap of the two semiconductors. Use (2.7) but ignore the temperature dependence of the preexponential factor B.

2.11. **(a)** By considering the magnetic and electric forces operating on electrons in an n-type sample in a Hall-effect arrangement, derive (2.19).

(b) Why is the "missing electron" model of a hole inconsistent with Hall-effect results from p-type material?

2.12. Using the parameter values for modeling minority carrier lifetime as suggested at the end of Chapter 2, generate a plot of electron minority carrier lifetime versus doping density for silicon. Use the range 10^{14} to 10^{19} cm^{-3} for the impurity concentration. Describe why the curve has its particular shape.

2.13. Consider a *p*-type silicon photosensor irradiated by light which is uniformly absorbed and which generates a steady-state excess electron concentration of 10^{12} cm^{-3}. If the light source is suddenly removed, what will be the electron concentration 1 μs later? Take the acceptor doping density to be 10^{15} cm^{-3}.

3 Current in Semiconductors

3.1 PARTICLE MOVEMENT

The electrons in the conduction band of a semiconductor, while being free from participating in bonding, are not completely free. They are constrained to move within the solid semiconductor material in an environment of considerable electrostatic complexity. The potential distribution in the material has an underlying periodicity due to the regular spacing of the atoms in the lattice (see Fig. 2.11 for example), with perturbations caused by impurities and crystal defects. In describing the motion of conduction electrons therefore, it follows that a mass other than the true free mass needs to be invoked. This point was stressed in Section 2.3.1, where it was suggested that the mass needed is called the ***effective mass***, m^*. Similarly, the holes in the valence band have an effective mass. Due to the differences between the valence and conduction bands, the effective masses of the holes and electrons are different. This point is elaborated upon in Section 5.4. Finally, because of the different types and arrangements of atoms in different materials, effective masses are a material property.

The thermal energy in a material imparts kinetic energy to the electrons and holes in accordance with the principle of equipartition of energy [1]. The kinetic energies are distributed over a wide range of values with a mean value given by

$$\tfrac{1}{2}m^* v_{\text{th}}^2 = \tfrac{3}{2}kT \tag{3.1}$$

where k is Boltzmann's constant and v_{th} is the ***thermal velocity***. v_{th} has a very high value in semiconductor materials (e.g., about 10^7 cm/s for electrons in silicon).

In thermal equilibrium, the electrons and holes in a semiconductor move about in a random fashion with a mean velocity v_{th}. On average, as many electrons (or holes) pass from the right through a plane in the semiconductor as pass through the same plane from the left (see Fig. 3.1). Thus in thermal equilibrium there can be no electric current. The direction of motion of a charged particle is randomized by its interaction with atoms and ions at lattice sites in the semiconductor material. The charged particle either collides with the atoms or is deflected by the ions (see Fig. 3.2). Both types of event are termed ***collisions***, and in each case the direction of motion of the electron or hole is changed. In the example of Fig. 3.2, the electron is shown as returning, eventually, to its original position. This is merely to emphasize that the random motion produces no net current.

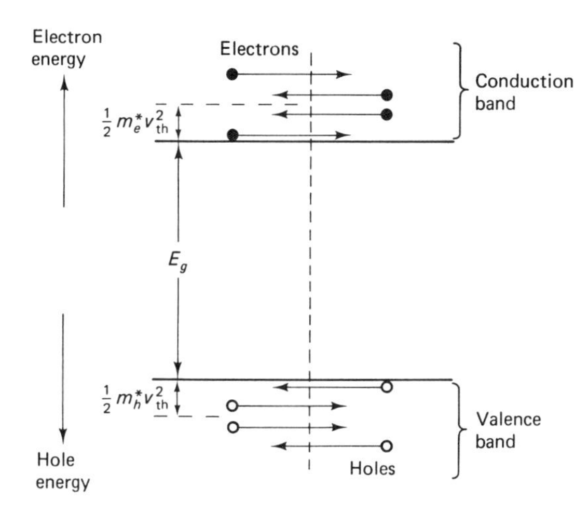

Figure 3.1 Motion of electrons and holes in the energy bands at thermal equilibrium. The carriers are distributed in energy about the mean kinetic energy, $(1/2)m^*v_{th}^2$. In thermal equilibrium, there is no net transfer of charge across any plane perpendicular to the direction of motion, such as the one depicted by the dashed line in the diagram. Thus there is no current in thermal equilibrium.

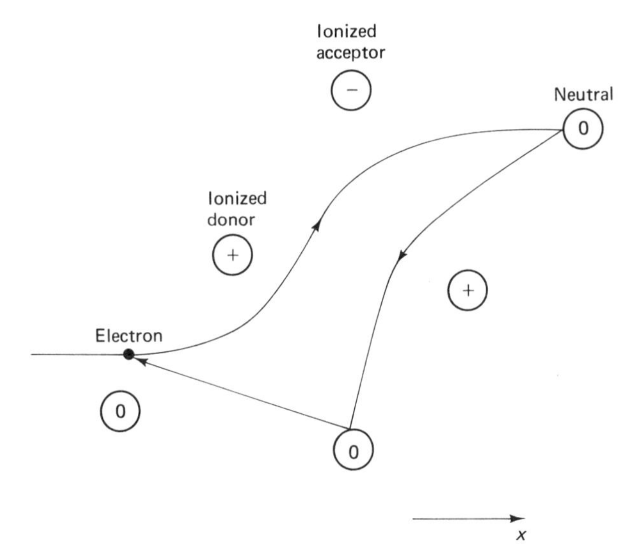

Figure 3.2 Collisions between an electron and ions and atoms at lattice sites, for the case of thermal equilibrium. The electron, moving with a mean velocity v_{th}, is deflected electrostatically by the ions (donors or acceptors) and is deflected physically by the atoms (semiconductor atoms or nonionized impurities). The electron's motion is randomized by these collisions. The electron is shown here as returning to its original position. This is just a diagrammatic way of emphasizing that the random motion results in no current.

Note that in a perfectly periodic lattice, one of the results of the quantum mechanical treatment of solids (see Section 5.3) is that an electron moves unimpeded throughout the crystal. However, when the atoms of the material absorb thermal energy and vibrate about their lattice sites, the perfect periodicity of the atomic spacings is perturbed. The quanta of lattice vibrational energy are called ***phonons***. Thus a collision between an electron and an atom of the host crystal can be viewed abstractly as an interaction of electron and phonon waves. This viewpoint is complementary to the more concrete, "billiard ball" approach that we have adopted in Fig. 3.2.

If some order can be impressed on the random motion of electrons and holes by disturbing the thermal equlibrium, a net movement of charge in a particular direction can result and be manifest as an electric current. Electric fields and gradients of carrier concentrations are two such "ordering agents" and they result in ***drift*** and ***diffusion*** currents, respectively. These types of current are the most prevalent in semiconductor devices and they are described in detail in the following sections. In some special types of diode, other currents, such as tunnel currents or thermionic emission currents, arise. These are described briefly in Section 6.5.3 (Zener diode) and Chapter 8 (Schottky diode).

3.2 DRIFT

3.2.1 Drift Velocity and Mobility

Figure 3.2 shows the random motion of an electron for the case of thermal equilibrium. If this equilibrium is now disturbed by the presence of an electric field, a net displacement of the electron can occur, resulting in a current. The situation is illustrated in Fig. 3.3 for the case of an electric field applied in the negative x-direction. Between collisions, the electron moves with a velocity determined by the vector sum of the random postcollision thermal velocity, v_{th}, and the velocity in the x-direction due to the electric field. The latter velocity is termed the ***drift velocity***, v_d.

To obtain an expression for the drift velocity, we note that during the time between collisions, t_{col}, the impulse due to the force of the electric field causes the electron to gain momentum. The electron accelerates until it suffers another collision. The time between collisions will not be the same for all electrons, nor for any electron all of the time. Thus the velocities attained between collisions will vary. We label the drift velocity, therefore, as the mean of the drift velocities for all electrons. Similarly, t_{col} is the mean of all times between collisions. If all the momentum gained by an electron is lost to the lattice in collisions between the two particles, it follows that

$$-qt_{col}\mathscr{E}(-\mathbf{x}) = m_e^* v_d \mathbf{x} \tag{3.2}$$

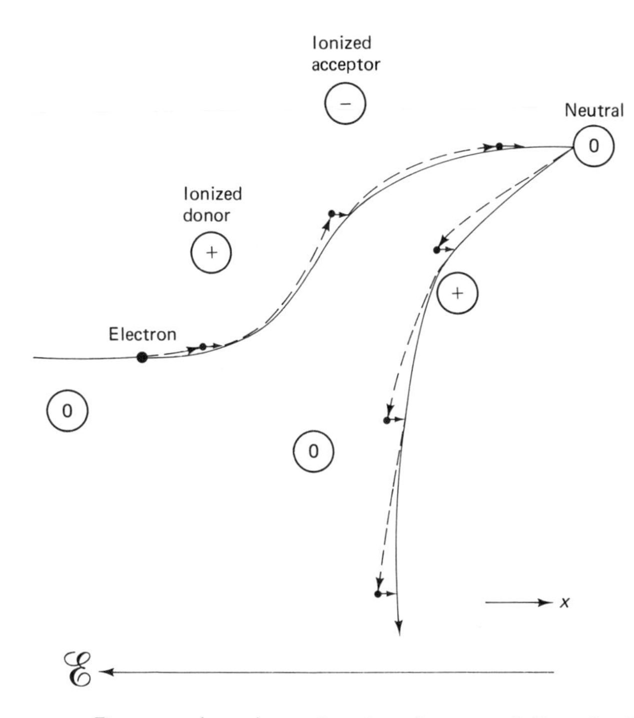

Figure 3.3 Electron drift. The path of the electron in thermal equilibrium (from Fig. 3.2) is shown here, broken into six segments, by the dashed lines. These segments are displaced to the right, in a "vectorial" fashion, by a small component, denoted by the solid arrow, representing the drift of the electron due to the applied field $\mathscr{E}$. As a result of this, the electron does not return to its original position, as it did in the thermal equilibrium case. Instead, there is a net movement to the right (i.e., an electric current).

Rearranging gives, for the electron drift velocity,

$$v_d\mathbf{x} = \frac{qt_{\text{col}}}{m_e^*}\mathscr{E}\mathbf{x} \tag{3.3}$$

Equation (3.3) brings out the important fact that provided that t_{col} is independent of $\mathscr{E}$, the drift velocity is linearly dependent on the electric field. The constant of proportionality is defined as the ***mobility***, that is,

$$v_d\mathbf{x} = \mu_e\mathscr{E}\mathbf{x} \tag{3.4}$$

from which it can be seen that the dimensions of mobility are (length)^2/(voltage·time). The usual units are cm^2/(V·s).

The experimental data for drift velocity displayed in Fig. 3.4 indicate that the linear relationship between v_d and $\mathscr{E}$ holds up to quite high values of field strength. When $\mathscr{E}$ is above about 3×10^3 V/cm in n-type silicon, for example, the time between collisions can no longer be deemed to be independent of $\mathscr{E}$, so the linearity is lost. At fields around 10^5 V/cm the drift velocity saturates at its ***scattering-limited*** value, which is close to that of the thermal velocity v_{th}. Briefly, the situation is as follows. At the low fields we have been considering so far, the drift velocity is so small compared to the thermal velocity that the electric field has very little effect on the electron's overall velocity or energy. However, at high fields when the drift velocity becomes comparable to the thermal velocity, the total energy of an electron increases significantly. As the kinetic energy is

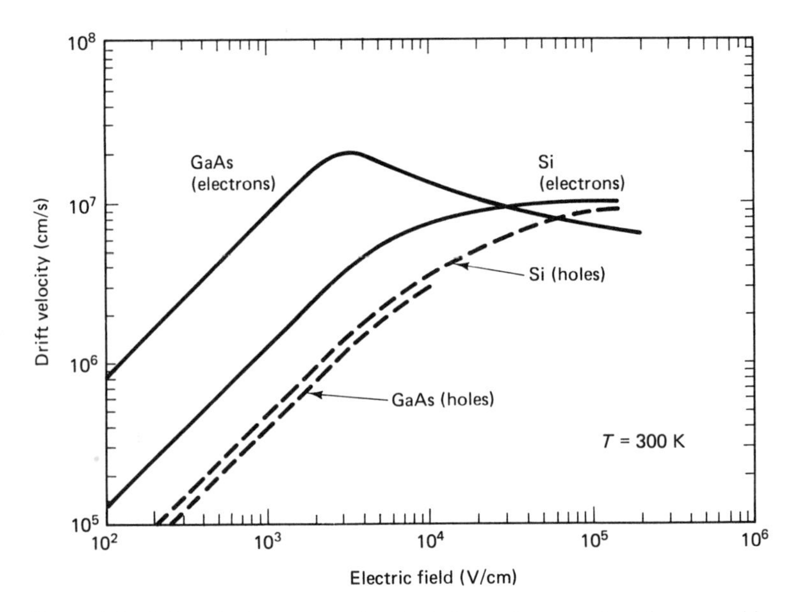

Figure 3.4 Drift velocity versus electric field in silicon and gallium arsenide. From Sze [2]. Reprinted with permission of the publisher, John Wiley & Sons, Inc.

related to the temperature [see (3.1)], such electrons are termed ***hot electrons***. When they collide with lattice atoms they excite the lattice into a new vibrational mode which has a large cross section for intercepting more electrons. Thus the lattice atoms become very effective at scattering mobile electrons, causing the drift velocity to saturate.

Velocity saturation also occurs in gallium arsenide, but in the case of n-type material, not until after a region where $dv_d/d\mathscr{E}$ is negative. This region of ***negative differential conductivity*** is exploited in GaAs microwave oscillator devices (see Ref. 3 for details). The origin of the phenomenon is discussed in Section 5.6.

Mobility is an important parameter for devices in which the current is due to drift. The higher the mobility, the faster the carriers will move. Thus mobility is important in field-effect devices intended for operation at high frequencies or fast switching speeds (see Chapters 7 and 9). In this regard, it follows from Fig. 3.4 that n-type GaAs devices are good candidates for high-speed applications, provided that the devices are designed to operate at fields within the linear regime. For ultrahigh-speed applications, however, the practice is to employ high fields but also to make the devices so small that the transit time of carriers across the device is insufficient for equilibrium conditions to be reached. In such cases this can lead to velocities that ***overshoot*** the equilibrium values (see Section 9.3.2.3).

Figure 3.4 also indicates that mobility is material dependent. This arises through the effective mass term in (3.3). Mobility also depends on temperature and doping density (see Figs. 3.5 and 3.6). Low temperatures mean low kinetic energies for the carriers and hence low velocities. Their time of passage past ionized impurities (see Fig. 3.3) is, therefore, longer than at high temperatures, so the carriers are deflected to a greater extent. As v_d is the component of velocity collinear with the direction of the electric field, this means that the mobility falls as the temperature is reduced. On increasing T, the collisions due to ***ionized impurity scattering*** become less important than the collisions with the neutral atoms of the lattice (***lattice scattering***). This is because the lattice atoms vibrate about their mean position with an amplitude that depends on temperature. Thus the higher the temperature, the greater the "target area," or, less colloquially, the cross section, presented by the atom to the mobile carriers. The frequency of collision is given by the sum of the frequencies of each of the individual scattering mechanisms, thus:

$$\frac{1}{t_{\text{col}}} = \frac{1}{t_{\text{ion}}} + \frac{1}{t_{\text{lat}}} \tag{3.5a}$$

which gives

$$\frac{1}{\mu} = \frac{1}{\mu_I} + \frac{1}{\mu_L} \tag{3.5b}$$

In the analysis of semiconductor devices it is often necessary to know the magnitude of the low-field mobility for different doping densities and temperatures. Although this information is available graphically (Figs. 3.5 and 3.6), it is more convenient when performing a computer-aided analysis to have the data

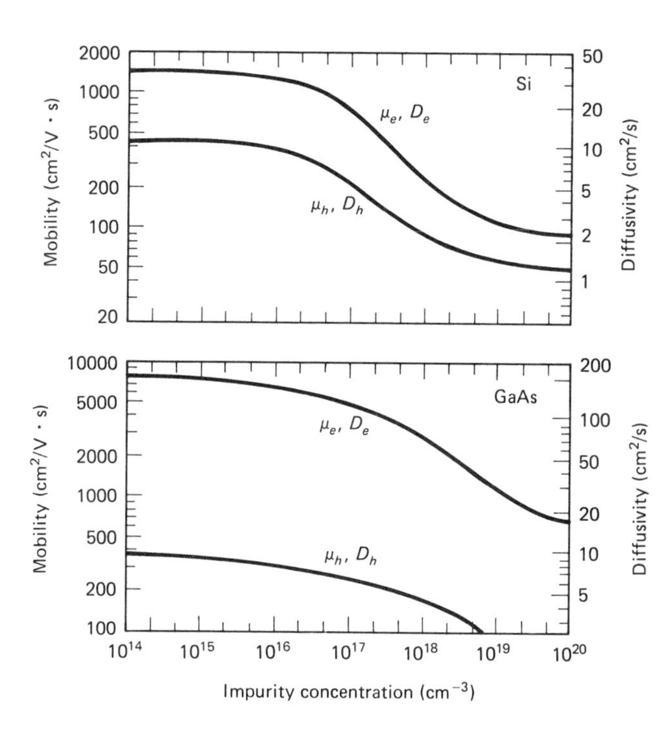

Figure 3.5 Dependence of mobility on impurity concentration at 300 K. The impurity concentration is the sum of all ionized impurities (donors and acceptors) present. Also shown in this figure is the diffusivity, as discussed in Section 3.3. (From Sze [4]. Reprinted with permission of the publisher, John Wiley & Sons, Inc.)

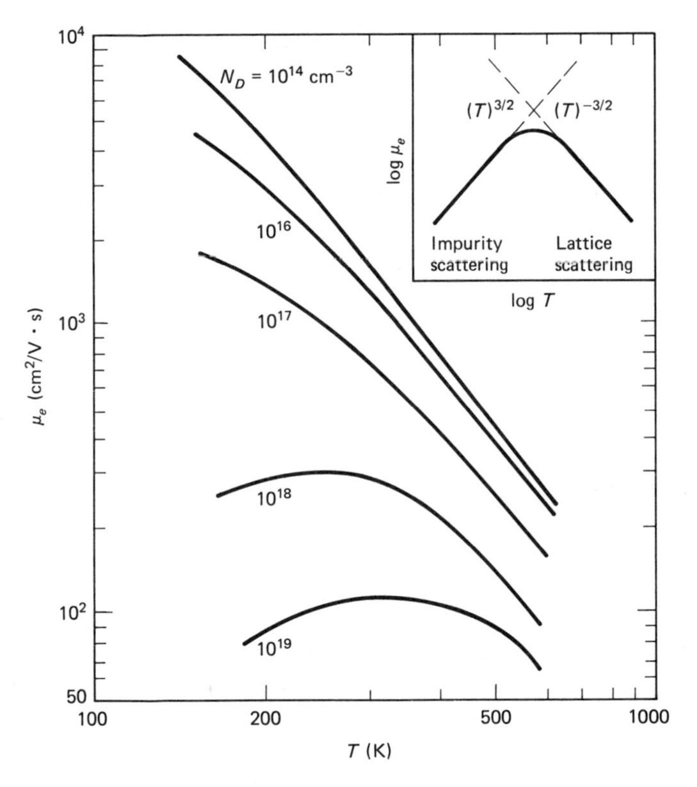

Figure 3.6 Dependence of mobility on temperature and doping density for n-type silicon. The inset shows the theoretical temperature dependence of electron mobility for impurity scattering and lattice scattering. (From Sze [5]. Reprinted with permission of the publisher, John Wiley & Sons, Inc.)

embodied in an equation. Such equations, which give a good fit to the experimental data for the mobility of electrons and holes in silicon, have been developed [6] and are reproduced here:

$$\mu_e = 88T_n^{-0.57} + \frac{7.4 \times 10^8 \times T^{-2.33}}{1 + [N/(1.26 \times 10^{17}T_n^{2.4})]0.88 \times T_n^{-0.146}} \tag{3.6}$$

$$\mu_h = 54.3T_n^{-0.57} + \frac{1.36 \times 10^8 \times T^{-2.23}}{1 + [N/(2.35 \times 10^{17}T_n^{2.4})]0.88 \times T_n^{-0.146}} \tag{3.7}$$

where N is the total impurity concentration in cm^{-3}, T is in kelvin, and $T_n = T/300$. At 300 K the equations reduce to

$$\mu_e = 88 + \frac{1252}{1 + 6.984 \times 10^{-18}N} \tag{3.8}$$

$$\mu_h = 54.3 + \frac{407}{1 + 3.745 \times 10^{-18}N} \tag{3.9}$$

3.2.2 Drift Current

Consider a sample of semiconductor material to which a dc voltage source is attached as shown in Fig. 3.7. The electric field $\mathscr{E}\mathbf{x}$ due to the applied voltage causes electrons to move to the left and holes to move to the right with their respective drift velocities. The lower part of Fig. 3.7 illustrates this situation from

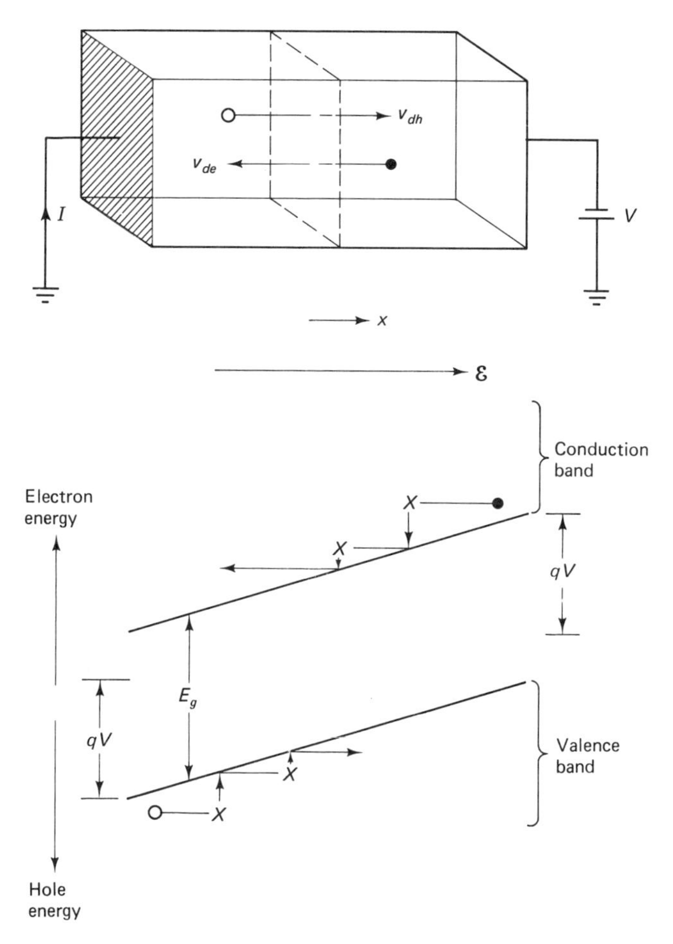

Figure 3.7 Drift of electrons and holes in a field $\mathscr{E}$ caused by the applied voltage, V. As depicted on the energy band diagram, the carriers gain kinetic energy from the field $\mathscr{E}$ between collisions, and lose some or all of it on collision. The bottom of the conduction band, which represents the potential energy of an electron, is raised at the right-hand side by an amount $(-q) \times (-V)$ eV, where $-q$ is the charge on an electron and $-V$ is the voltage on the right-hand side of the sample. Thus electrons can be viewed as having an inclination to fall down a potential energy gradient. For holes a similar situation exists, but note that the potential energy of a hole increases downward. It is often convenient to view holes as having an inclination to float up a gradient in *electron* potential energy.

the perspective of an energy band diagram. The passage of charge across a plane perpendicular to the x-direction is given by

$$J\mathbf{x} = -qnv_{de}(-\mathbf{x}) + qpv_{dh}(\mathbf{x}) \tag{3.10}$$

where n and p are the electron and hole concentrations, respectively, and the dimensions of J are coulombs/(area × time) (i.e., amperes/area). Thus J is a current density. Substituting for the drift velocities from (3.4) gives

$$J\mathbf{x} = q(n\mu_e + p\mu_h)\mathscr{E}\mathbf{x} \tag{3.11}$$

where the subscripts e and h identify the electron and hole mobilities, respectively. J is the magnitude of the current density due to the drift of electrons and holes under the influence of the electric field. $J\mathbf{x}$ is thus the drift current density. The drift current is simply JA, where A is the cross-sectional area of the sample.

3.2.3 Conductivity

The vector form of Ohm's law is

$$J\mathbf{x} = \sigma\mathscr{E}\mathbf{x} \tag{3.12}$$

where σ is the conductivity. Comparing (3.11) and (3.12) reveals that

$$\sigma = q(n\mu_e + p\mu_h) \tag{3.13}$$

σ was introduced in Section 2.1 as a material property. Equation (3.13) indicates that the material parameters determining σ are the carrier concentrations and the mobilities. Both of these parameters depend on the doping densities and the temperature.

Some special cases of importance are:

1. *Intrinsic material*

$$\sigma_i = qn_i(\mu_e + \mu_h) \tag{3.14}$$

Here the mobilities have their largest values (see Fig. 3.5), but the room-temperature carrier concentrations are orders of magnitude lower than in the doped material usually used in devices. Intrinsic silicon and gallium arsenide are, in fact, reasonable insulators, with room-temperature values of σ_i being about 4×10^{-6} and 4×10^{-9} $(\Omega\cdot\text{cm})^{-1}$, respectively.

2. *n-type material*

$$\sigma_n = qN_D^+\mu_e \tag{3.15}$$

Here the equation is valid provided that the ionized doping density is much greater than the intrinsic carrier concentration. This can be checked by substituting appropriate numbers into (2.8) and (2.14) (see Section 4.5 for the resulting equation for n). The decrease of mobility with doping density is not sufficiently strong to prevent the conductivity increasing with the ionized donor concentration. For example, in the range 10^{15} to 10^{19} ions/cm^3, the conductivity of n-type silicon increases from 0.2 to 170 $(\Omega\cdot\text{cm})^{-1}$. The increase for n-type GaAs over the same range of doping is 0.7 to 2600 $(\Omega\cdot\text{cm})^{-1}$.

3. *p-type material*

$$\sigma_p = qN_A^- \mu_h \tag{3.16}$$

For the same conditions as just mentioned, the conductivities for p-type silicon and gallium arsenide range from 0.07 to 110 $(\Omega\cdot\text{cm})^{-1}$ and from 0.05 to 125 $(\Omega\cdot\text{cm})^{-1}$, respectively.

Besides depending on the doping density, the conductivity is also affected by temperature. The dependence of the mobility on temperature is illustrated for the case of silicon in Fig. 3.6. The general dependence of carrier concentration on temperature for n-type material is sketched in Fig. 3.8. Combining the two

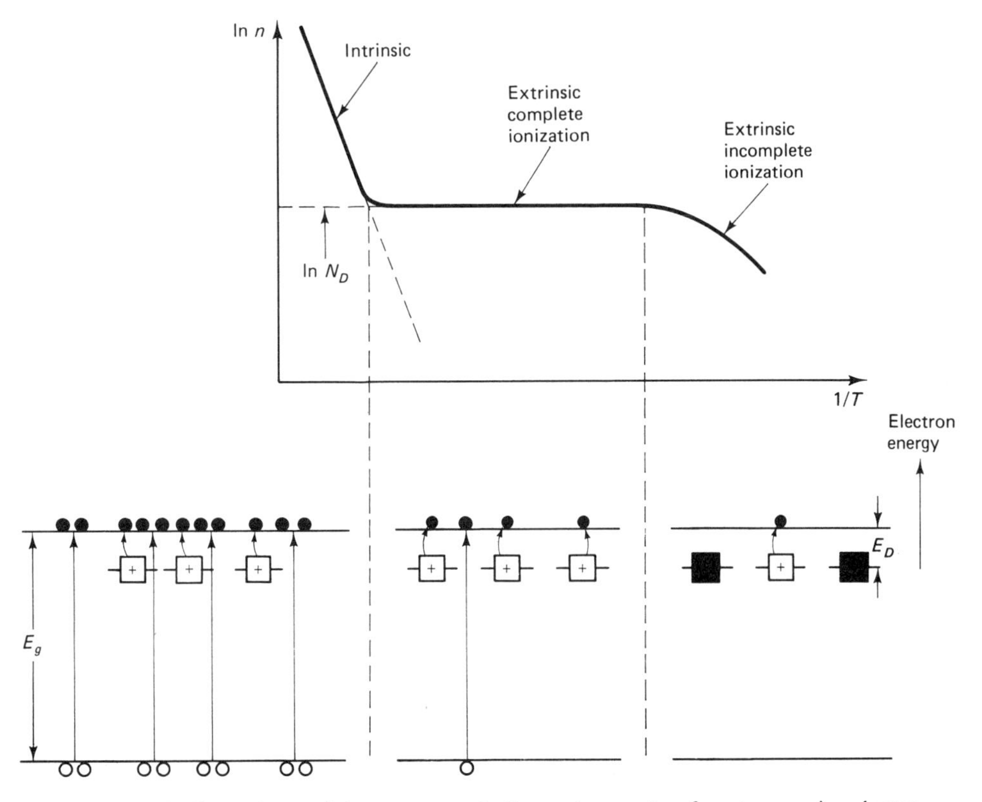

Figure 3.8 Dependence of electron concentration on temperature for n-type semiconductor material. At low temperatures there is not enough energy to break any bonds, and not enough to ionize all the donors (see the energy band diagram at the bottom right). As the temperature increases, the donors ionize and some bonds are broken (see center energy band diagram). At very high temperatures, so many bonds are broken that the intrinsic electron concentration exceeds that of the donated electrons (see left-hand energy band diagram).

effects leads to a general dependence of conductivity on temperature as shown in Fig. 3.9. At high temperatures the large intrinsic carrier concentration [see (2.7)] causes a logarithmic dependence of σ on $1/T$. At lower temperatures the carrier concentration is dominated by the doping density, and the structure of the σ versus $1/T$ plot in this regime reveals the different temperature dependencies of μ_L and μ_I (see Fig. 3.6).

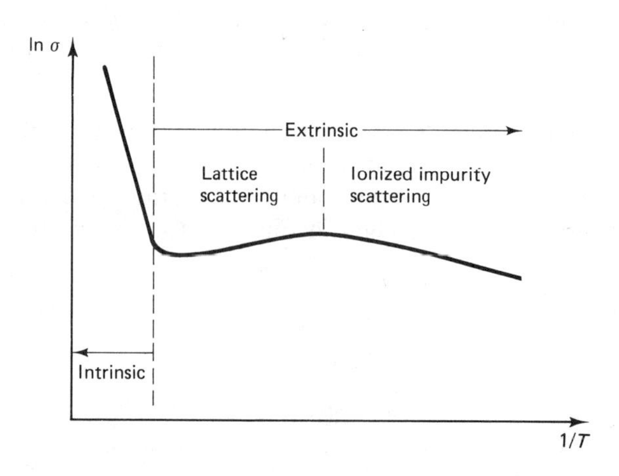

Figure 3.9 Dependence of conductivity on temperature. In the intrinsic region $\sigma(T)$ is determined mainly by $n_i(T)$ and is almost proportional to exp $(-1/T)$ [see (2.7)]. In the extrinsic region $\sigma(T)$ is determined mainly by $\mu(T)$ and is thus almost proportional to $T^{-3/2}$ when lattice scattering dominates, and to $T^{+3/2}$ when ionized impurity scattering is more important (see Fig. 3.6).

3.3 DIFFUSION

Whereas the current in field-effect transistors is principally due to drift, the current in the bipolar diodes and transistors described in Chapters 6 and 11 is due mainly to the diffusion of excess minority carriers. These are introduced into p-type material, for example, by the injection of electrons from contiguous n-type material. The situation is illustrated in Fig. 3.10. There is no electric field in the p-type material to move the carriers by drift; they move instead by ***diffusion***.

The concept of particle diffusion is a familiar one. An everyday example is the permeation through a room of the fragrant molecules emanating from their concentrated source on the body of a perfumed person. The driving force for the movement of the particles is the concentration gradient of the particles. This is because the particles are in random thermal motion, and there will be, therefore, more particles crossing from a region of high concentration to a region of lower concentration than vice versa. In the example of Fig. 3.10, the driving force for the movement of the excess electrons is $d\hat{n}/dx$. This gradient is also given by dn/dx, as the equilibrium electron concentration is spatially uniform in the absence of any doping density variations. The number of electrons flowing per second across a unit area to the right in Fig. 3.10 is given by

$$\text{flux} = -D_e \frac{dn}{dx} \tag{3.17}$$

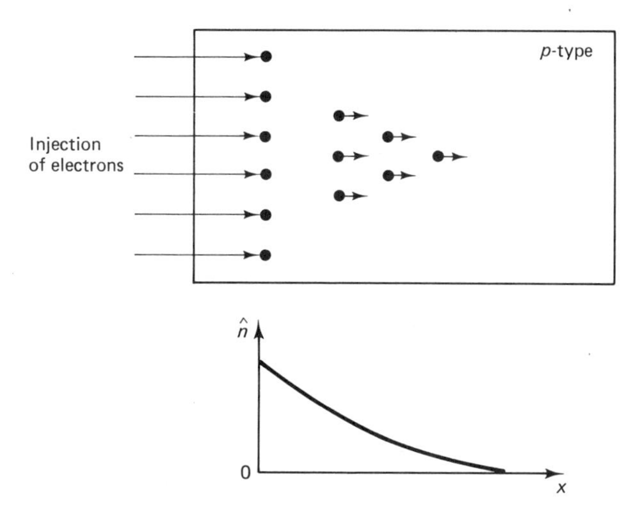

Figure 3.10 Example of excess minority carrier diffusion. A steady flow of electrons into the field-free, p-type material results in a diffusion of the electrons away from their source. The concentration of electrons decreases as the electrons diffuse deeper into the material due to recombination with majority carrier holes.

where D_e is a constant of proportionality called the ***diffusivity*** or ***diffusion coefficient*** for electrons. It has the dimensions of (length)^2/time and the usual units are cm^2/s. Because the electron concentration in Fig. 3.10 is decreasing to the right (i.e., $d\hat{n}/dx$ is negative), the negative sign in (3.17) is needed to describe properly the positive flux of electrons to the right.

As each electron carries a negative charge, the diffusive flow to the right in Fig. 3.10 constitutes a current density in the negative x-direction which is given by

$$J_e(-\mathbf{x}) = -\ D_e(-q)\frac{dn}{dx}(-\mathbf{x}) \tag{3.18}$$

If, in Fig. 3.10, the situation were one of excess holes being injected into n-type material, the hole diffusion current density would be in the x-direction and be given by

$$J_h\mathbf{x} = -D_h q\frac{dp}{dx}(\mathbf{x}) \tag{3.19}$$

The temperature and doping density dependencies of the diffusion currents are governed by the dependence of the diffusivity on these two parameters. D is given by ***Einstein's relation***, namely:

$$D = \frac{kT}{q}\mu \tag{3.20}$$

Thus D and μ have the same dependence on doping density (see Fig. 3.5). The temperature dependence of D is $T^{-1/2}$ when lattice scattering dominates, and is $T^{+5/2}$ when ionized impurity scattering prevails (see the inset to Fig. 3.6).

3.4 TRANSPORT EQUATIONS

It frequently happens in semiconductor devices that drift and diffusion currents are operative at the same time. Figure 3.11 shows an example of such a case and is based on the devices often used to measure minority carrier properties (see Example 3.2 for more details). Minority carrier holes are injected into the n-type region and diffuse away from their p-type source. Simultaneously, the holes drift to the right under the influence of the electric field caused by the power supply V_A. The total hole current density in the x-direction is given by the sum of the drift and diffusion terms, that is, by the appropriate forms of (3.11) and (3.19), namely,

$$J_h\mathbf{x} = \left(q\mu_h p\mathscr{E} - qD_h\frac{dp}{dx}\right)\mathbf{x} \tag{3.21}$$

In its general, three-dimensional form, this equation is known as the ***transport equation*** for holes, that is,

$$\mathbf{J}_h = q\mu_h p\mathscr{E} - qD_h\nabla p \tag{3.22}$$

The corresponding equation for electrons is

$$\mathbf{J}_e = q\mu_e n\mathscr{E} + qD_e\nabla n \tag{3.23}$$

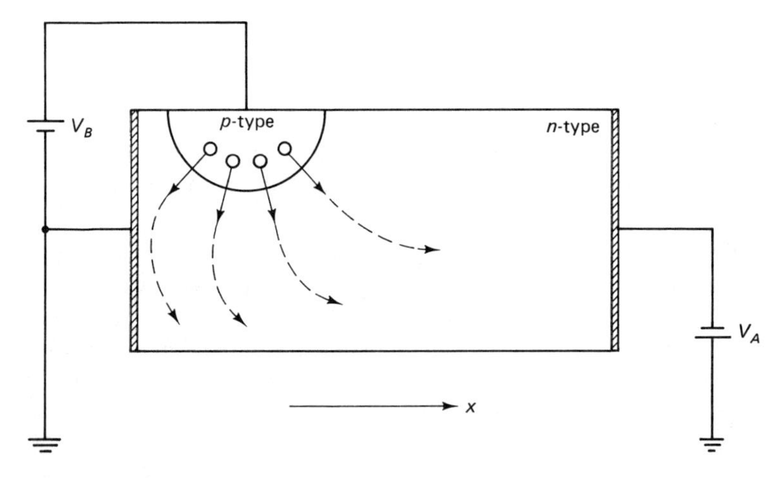

Figure 3.11 Example of diffusion and drift. Excess minority carrier holes are injected from the p-type region into the n-type material. The holes diffuse (solid arrows) away from their source and drift (dashed arrows) in the field set up by the applied voltage V_A.

The signs of the terms used in the transport equations follow from the derivations of the drift and diffusion currents in Sections 3.2 and 3.3. Figure 3.12 emphasizes their significance.

In situations where both electrons and holes contribute to the conduction current, the ***total current density*** is given by the sum of the transport equations, namely,

$$\mathbf{J}_{\text{total}} = \mathbf{J}_h + \mathbf{J}_e \tag{3.24}$$

Often in devices, especially bipolar devices, the electron and hole currents vary with distance. Figure 3.13 shows schematically the situation that might arise in the base region of a long diode (see Section 6.3.1 for more details). When evaluating the total current in such a case, Fig. 3.13 indicates that (3.24) can be evaluated at any plane, provided that the same plane is used for both electron and hole currents.

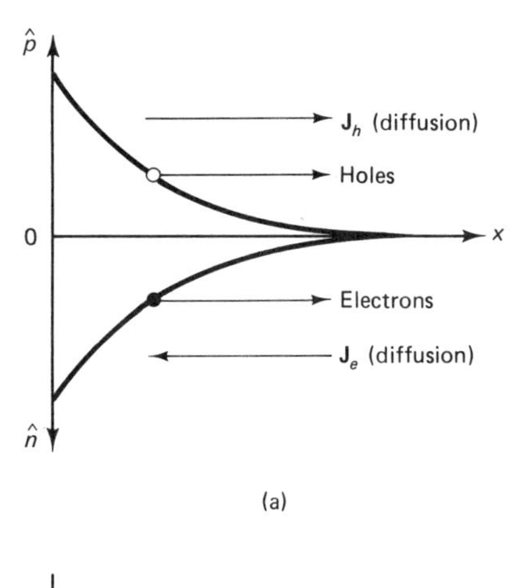

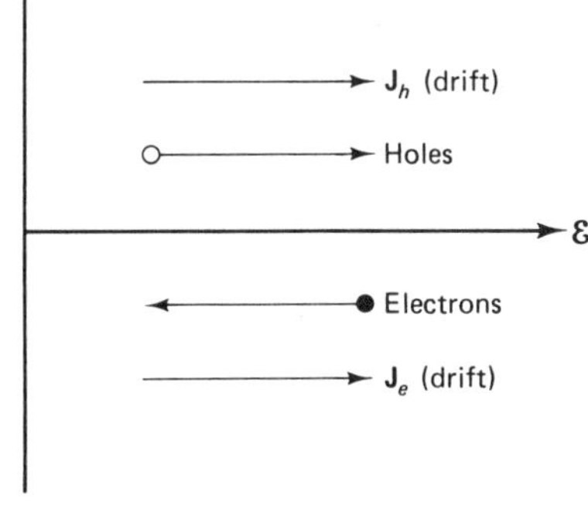

Figure 3.12 Relationship between particle flow and current for carrier movement by (a) diffusion and (b) drift.

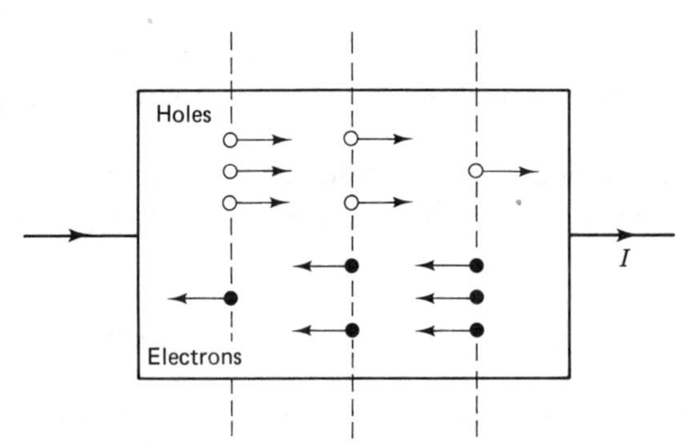

Figure 3.13 Current transport by both electrons and holes. The total current, I, can be evaluated at any plane perpendicular to the direction of particle flow. However, each current component must be computed at the same plane. In this illustrative example, the electrons and holes are considered to have the same mobility (or diffusivity), that is, each particle contributes equally to the current.

3.5 CONTINUITY EQUATION

In designing a semiconductor device one needs to be able to predict the current that will be present in the device under specified operating conditions. The transport equations (3.22) and (3.23) allow this to be done provided the distributions of the concentrations of mobile charge carriers are known. These distributions can be calculated by solving the ***continuity equation***.

To derive this important equation, consider the hole concentration within the small interior volume of semiconductor material shown in Fig. 3.14. The hole concentration will change in time if there is a difference between the hole currents at opposite faces of the elemental volume. Such a divergence of current could occur, for example, if the current were due to diffusion and dp/dx varied over the

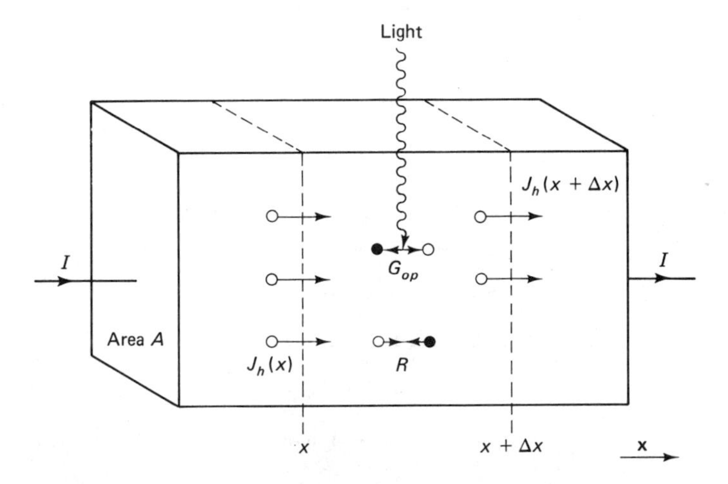

Figure 3.14 Changes in the hole concentration within the volume $A\ \Delta x$ due to variation in the hole current, generation by absorption of ionizing radiation, and recombination. The physical situation depicted here is described mathematically in (3.25), the continuity equation.

distance Δx. The hole concentration will also change if there is a difference between the rates at which holes are generated and recombine within the volume. Mathematically, this state of affairs can be written as

$$\begin{array}{c}\text{rate of change of}\\ \text{number of holes}\end{array} = \begin{array}{c}\text{change due to spatial}\\ \text{difference in current}\end{array} + \begin{array}{l}\text{change due to}\\ \text{difference in rates of}\\ \text{generation and recombination}\end{array}$$

$$\frac{\partial p}{\partial t}(x, t)[\text{volume}] = \frac{I_h}{q}(x) - \frac{I_h}{q}(x + \Delta x) + (G - R)_{\text{holes}}[\text{volume}]$$

Dividing by the volume $A\ \Delta x$ gives

$$\frac{\partial p}{\partial t}(x, t) = \frac{1}{q}[J_h(x) - J_h(x + x)]\frac{A}{\Delta x\, A} + [G - R]_{\text{holes}} \tag{3.25}$$

In the limit that $\Delta x \rightarrow dx$, a Taylor series expansion of $J_h(x + dx)$ can be used to write this equation more succinctly, namely

$$J_h(x + dx) = J_h(x) + \frac{\partial J_h}{\partial x}dx \tag{3.26}$$

and thus

$$\frac{\partial p}{\partial t}(x, t) = \frac{-1}{q}\frac{\partial J_h}{\partial x} + [G - R]_{\text{holes}} \tag{3.27}$$

This equation is the one-dimensional form of the continuity equation for holes. The corresponding equation for electrons is

$$\frac{\partial n}{\partial t}(x, t) = \frac{1}{q}\frac{\partial J_e}{\partial x} + [G - R]_{\text{electrons}} \tag{3.28}$$

These equations are completely general, as no assumptions (other than the validity of a Taylor series truncation) have been made or implied in deriving them. This generality often makes the equations rather difficult to solve exactly. For example, in many instances of semiconductor device, the doping density varies with distance (often in three dimensions), which leads to the diffusion current, and sometimes the generation–recombination rates, being position dependent. Also, the current may include a drift term due to a field that depends on the charge density through ***Poisson's equation***, that is, in one dimension:

$$-\frac{d^2\psi}{dx^2} = \frac{d\mathcal{E}}{dx} = \frac{\rho}{\epsilon_s} = \frac{q}{\epsilon_s}\left(N_D^+ - N_A^- + p - n\right) \tag{3.29}$$

where ψ is the electrostatic potential, ρ the volumetric charge density, and ϵ_s the permittivity of the semiconductor.

It can be appreciated that the combined solution of the continuity equations and Poisson's equation is often best achieved by using numerical methods on a computer. However, in this book we do not need exact solutions to the complete equations. We can afford to make some simplifying assumptions that permit straightforward analytical solutions to be obtained.

3.5.1 Simplified Continuity Equation

We begin our simplifications by recalling the following equations from Chapter 2:

$$R = \hat{R} + R_0 \qquad (2.23)$$

$$G_{\text{th}} = R_0 \qquad (2.11)$$

$$\hat{R} = \frac{\hat{p}}{\tau_h} \qquad (2.27),\ (2.28),\ (2.31),\ \text{and}\ (2.32)$$

and by appreciating that generation can occur by the absorption of both thermal and optical energy (see Section 2.5.1), that is,

$$G = G_{\text{th}} + G_{\text{op}} \qquad (3.30)$$

Amalgamating these relationships into (3.27) yields

$$\frac{\partial \hat{p}}{\partial t}(x, t) = \frac{-1}{q}\frac{\partial J_h}{\partial x} + G_{\text{op}} - \frac{\hat{p}}{\tau_h} \qquad (3.31)$$

where we have made the assumption that neither the temperature nor the doping density can change so rapidly with time that $\partial p_0/\partial t$ could be anything other than zero over the time intervals in which changes in excess carrier concentrations can occur, that is,

$$\frac{\partial p}{\partial t} = \frac{\partial \hat{p}}{\partial t} \quad \text{and} \quad \frac{\partial n}{\partial t} = \frac{\partial \hat{n}}{\partial t} \qquad (3.32)$$

The corresponding simplified continuity equation for electrons is

$$\frac{\partial \hat{n}}{\partial t}(x, t) = \frac{1}{q}\frac{\partial J_e}{\partial x} + G_{\text{op}} - \frac{\hat{n}}{\tau_e} \qquad (3.33)$$

The only serious assumption made in reducing (3.27) and (3.28) to (3.31) and (3.33) is that ***low-level injection*** conditions, as discussed in Section 2.5.2 and embodied in (2.27) and (2.28), apply. This assumption is valid for all the instances where the continuity equation is used in this book.

The continuity equations are second-order, partial differential equations describing the positional and temporal dependencies of, in the simplified case, the excess mobile carrier concentrations. In solving these equations there are many instances in semiconductor device calculations where some of the terms in the equations can be given zero magnitude. These instances should be recognized, as they can often lead to tractable solutions of these otherwise formidable equations. For example:

1. *Steady state.* In this situation dc conditions prevail and there is no temporal change of the mobile carrier concentrations, thus:

$$\frac{\partial p}{\partial t} = \frac{\partial n}{\partial t} = 0 \tag{3.34}$$

2. *Dark operation.* In this case $G_{\text{op}} \ll G_{\text{th}}$, so it is permissible to write

$$G_{\text{op}} = 0 \tag{3.35}$$

3. *Uniform particle currents.* In this instance the particle currents are constant throughout the region of the device under consideration; thus

$$\frac{\partial J_h}{\partial x} = \frac{\partial J_e}{\partial x} = 0 \tag{3.36}$$

4. *Diffusion current.* In this case there is no drift current. This situation would arise in the region of a device where there is no electric field and leads to

$$\frac{-1}{q}\frac{\partial J_h}{\partial x} = D_h \frac{\partial^2 p}{\partial x^2}$$

and (3.37)

$$\frac{1}{q}\frac{\partial J_e}{\partial x} = D_e \frac{\partial^2 n}{\partial x^2}$$

5. *Uniform doping density, field-free region, no light, steady state.* This seemingly restrictive situation is, in fact, quite commonly a good approximation to the conditions in certain regions of semiconductor devices, most notably the neutral base in a bipolar junction diode or transistor (see Sections 6.3 and 11.3). The uniform doping density condition implies that there is no positional variation of the thermal equilibrium carrier concentrations, so the diffusion current, which is the only component present in a field-free region, can be expressed in terms of the excess carrier concentration. Combining this result with the implications of the other conditions specified for this case leads to the compact equations

$$D_h \frac{d^2 \hat{p}}{dx^2} = \frac{\hat{p}}{\tau_h}$$

$$D_e \frac{d^2 \hat{n}}{dx^2} = \frac{\hat{n}}{\tau_e} \tag{3.38}$$

These equations arise so frequently in semiconductor device calculations that it is helpful to state their solution. Taking the hole case as an example, the solution is

$$\hat{p}(x) = \text{A} \exp\left(\frac{-x}{L_h}\right) + \text{B} \exp\left(\frac{x}{L_h}\right) \tag{3.39}$$

where A and B are constants that need to be evaluated from the boundary conditions appropriate to the problem, and L_h, the ***minority carrier diffusion length***, is a measure of how far the minority carrier holes diffuse before they recombine. By deriving (3.39) from (3.38), it will be appreciated that

$$L_h = (D_h \tau_h)^{1/2} \tag{3.40}$$

Situations in which the form of the solution to the continuity equation is represented by (3.39) are covered in Sections 6.3 and 11.3, where the current–voltage characteristics of bipolar diodes and transistors, respectively, are calculated. We close this present chapter with two quite different examples of the use of the continuity equations.

Example 3.1 Use of the Continuity Equations: A Silicon Light Meter

A bar of silicon, when deployed in the same manner shown in Fig. 3E.1, could be used as a device for measuring light intensity. The photon flux, ϕ photons/(cm^2·s), irradiating the sample generates excess current carriers $\hat{n}$ and $\hat{p}$ at a rate G_{op} cm^{-3} s^{-1}. These carriers move under the influence of the applied voltage V to produce a detectable excess current density $\hat{J}$. Thus the current I displayed on the ammeter is related to the incident light intensity through the chain of proportionalities

$$I \propto \hat{J};\ \hat{J} \propto (\hat{p} \text{ and } \hat{n});\ (\hat{p} \text{ and } \hat{n}) \propto G_{op};\ G_{op} \propto \phi$$

The role played by the continuity equation in establishing the link between I and ϕ is in determining the relationship between G_{op} and $\hat{p}$ and $\hat{n}$. In fact, only one of the two types of excess carrier need be considered because optical generation implies that $\hat{n} = \hat{p}$ (see Fig. 2.22). We will consider only the hole continuity equation and begin by examining whether any of the simplifying cases discussed in the preceding section apply. This should be the first step in any problem involving the continuity equation.

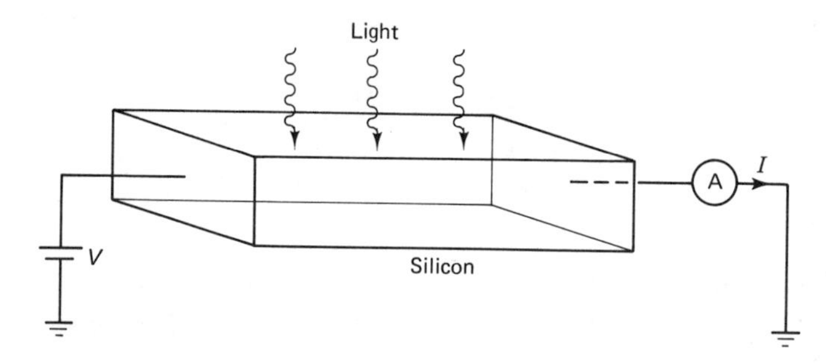

Figure 3E.1 Possible experimental arrangement for a simple lightmeter. Photogenerated electron–hole pairs are separated by the electric field due to V and result in a current I which is proportional to the light intensity.

A meaningful measurement with the simple apparatus shown in Fig. 3E.1 demands that the light intensity not be changing very rapidly, thus steady-state conditions can be assumed to apply. Further, if the silicon sensor is made thin enough, it can be assumed that the light is absorbed uniformly throughout the bulk of the device; thus the condition of uniform particle currents applies. With these simplifications, the hole continuity equation (3.31) reduces to

$$0 = 0 + G_{\text{op}} - \frac{\hat{p}}{\tau_h} \tag{3E.1}$$

which is the desired relationship. Substituting for $\hat{p}$ (and $\hat{n}$) in the transport equations [(3.22) and (3.23)] allows the total current to be calculated via (3.24). The result is

$$J = q\hat{p}\mathscr{E}(\mu_e + \mu_h) + q\mathscr{E}(n_0\mu_e + p_0\mu_h) \tag{3E.2}$$

Note that in this example all the carrier concentrations are independent of position, so there will be no current due to diffusion.

For silicon illuminated by bright sunlight, for example, G_{op} is about 10^{17} electron–hole pairs $\text{cm}^{-3}\,\text{s}^{-1}$. Assuming lightly doped material, (2.33) gives a value of 5×10^{-7} s for the hole minority carrier lifetime which, from (3E.1), yields $\hat{p} = 5 \times 10^{10}\ \text{cm}^{-3}$. The photogenerated current density—given by the first term in (3E.2)—is, for this example, about 14 mA/cm^2 for a field of 1 kV/cm.

Example 3.2 Use of the Continuity Equations: Measurement of Minority Carrier Lifetime

In 1951, shortly after the successful demonstration of transistor action at Bell Labs in the United States, researchers at the same laboratories reported a classic experiment which enabled the minority carrier properties of mobility, diffusivity, and lifetime to be measured. The experiment has since become known as the ***Haynes–Shockley*** experiment, in honor of the two men who conceived and performed it [7]. To extract the minority carrier lifetime from the measured data requires a comprehensive solution to the continuity equation, as is demonstrated here.

The essential features of the experimental arrangement are depicted in Fig. 3E.2a. At point A a pulse of holes can be injected into the n-type semiconductor. This could be done either by irradiating with a short light pulse or by suitably biasing a pn junction (see Fig. 3.11). At point B the holes can be collected and the resulting current displayed on an oscilloscope. At point C a voltage pulse can be applied to create a field in the x-direction to sweep the holes injected at A down to contact B. Figure 3E.2b shows how the hole profile changes due to drift, diffusion, and recombination. Clearly, the hole

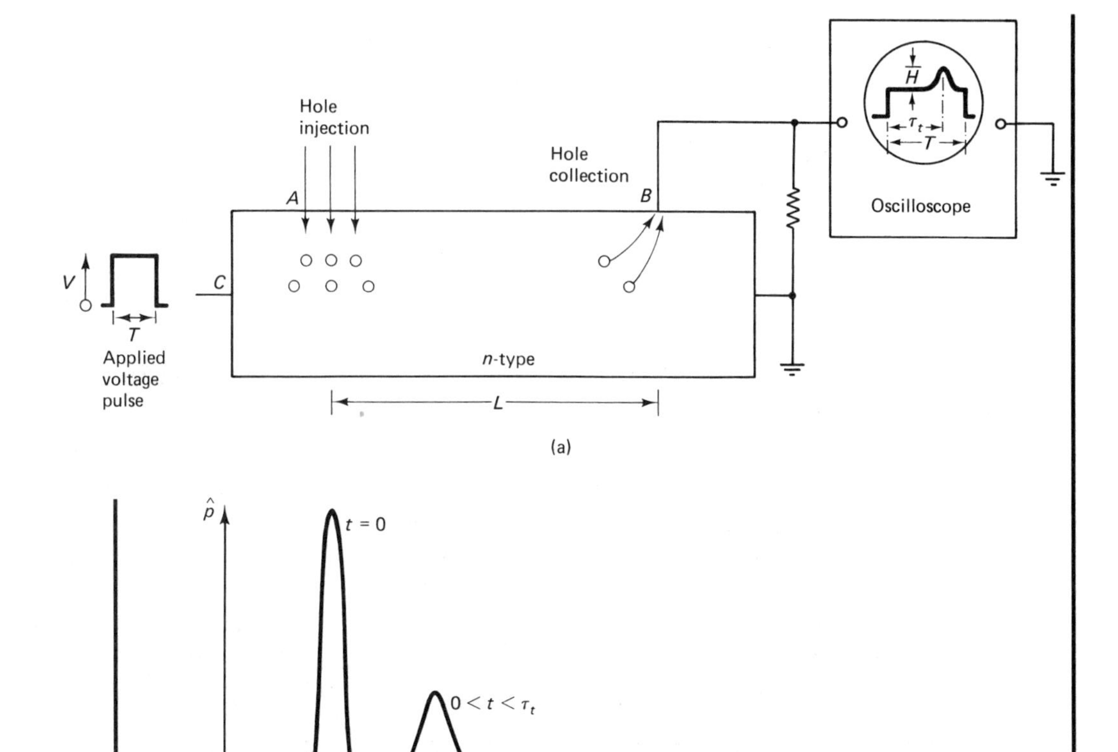

Figure 3E.2 Haynes–Shockley experiment. (a) Experimental arrangement. Minority carrier hole injection at contact A is synchronized with the voltage pulse at C and with the oscilloscope time base. The holes collected at B after elapse of the transit time τ_t give a blip of height H on the oscilloscope. (b) Hole concentration moving to the right due to drift, spreading out due to diffusion, and diminishing in magnitude due to recombination.

distribution changes in both time and space, so little simplification of the continuity equation can be made. The only term that can be omitted is G_{op}. This is true, even for the case of optical creation of the hole pulse, because the region of the sample through which the holes move is not illuminated. This illustrates another general rule to be borne in mind when preparing to solve the continuity equation, namely: try and choose for analysis the region of the device that gives the simplest form of the equation. The relevant form of the hole continuity equation (3.31) is thus

$$\frac{\partial \hat{p}(x, t)}{\partial t} = \left(-\frac{1}{q}\right) \frac{\partial}{\partial x} \left[q\mu_h \hat{p} \mathscr{E} - qD_h \frac{\partial \hat{p}}{\partial x} \right] + 0 - \frac{\hat{p}}{\tau_h} \tag{3E.3}$$

Solution of this example of the continuity equation is by no means easy. However, it can be done with the result that for an initial distribution of excess carriers in the form of a delta function [8],

$$\hat{p}(x, t) = \frac{P}{(4\pi D_h t)^{1/2}} \exp\left[-\frac{(x - \mu_h \mathscr{E} t)^2}{4D_h t}\right] \exp\left(-\frac{t}{\tau_h}\right) \tag{3E.4}$$

where P is the number of holes generated (or injected) per unit area per second.

To obtain an estimate for τ_h, consider the hole distribution under the contact B. Holes will reach B after elapse of the transit time τ_t, which is well approximated by

$$\tau_t = \frac{L}{v_{\text{dr}}} \tag{3E.5}$$

where v_{dr} ($= \mu_h \mathscr{E}$) is the hole drift velocity and L is the separation of contacts A and B. Evaluating (3E.4) at B gives

$$\hat{p}(L, \tau_t) = \frac{P}{(4\pi D_h \tau_t)^{1/2}} \exp\left(-\frac{\tau_t}{\tau_h}\right) \tag{3E.6}$$

The height H of the small pulse on the oscilloscope (Fig. 3E.2a) due to the collection of holes at B will be proportional to $\hat{p}(L, \tau_t)$. Designating the constant of proportionality as K, substituting in (3E.6), rearranging, and taking logarithms yields

$$\ln H\tau_t^{1/2} = \ln \frac{P}{K(4\pi D_h)^{1/2}} - \frac{\tau_t}{\tau_h} \tag{3E.7}$$

This equation is that of a straight line, the slope of which gives τ_h without the need to know the values of K, P, or D_h. A suitable range of experimental points for a graph of $\ln (H\tau_t^{1/2})$ versus τ_t can be obtained by varying the field $\mathscr{E}$ via the pulse height at point C (Fig. 3E.2a), and measuring H for the resulting different values of τ_t.

There are more direct and convenient methods of measuring minority carrier lifetime, but the Haynes–Shockley experiment is an instructive one to perform, as it illustrates so explicitly the effects of minority carrier drift, diffusion, and recombination.

3.6 CHAPTER SUMMARY

In this chapter drift and diffusion, the two most common mechanisms of charge flow in semiconductors, have been introduced. The manner in which the continuity of current due to these driving forces is linked to recombination and generation of excess carriers, to determine the rate of change of carrier concentrations, has also been treated. The reader should now have an appreciation of the following topics:

1. *Particle movement.* In thermal equilibrium the electrons and holes move in a ***random*** fashion with a ***thermal velocity*** that depends on the temperature; there is no current (i.e., net charge flow) in thermal equilibrium.

2. *Drift.* The disturbance of thermal equilibrium by the application of an electric field can cause carriers to acquire a ***drift velocity*** in a direction parallel to the electric field. This net movement constitutes a current. Carriers drift between ***scattering*** events involving interactions with dopant ions and ***vibrating*** lattice atoms. At high electric fields the velocity saturates at its ***scattering limited*** value. At low and moderate fields the drift velocity is proportional to the electric field; the constant of proportionality is called the ***mobility***. Mobility depends on the doping density and the temperature.

3. *Conductivity.* Conductivity is a material property linking current density and electric field in the vector form of Ohm's law. The conductivity depends on the carrier concentrations and the carrier mobilities. Conduction is ***intrinsic*** when the thermal breaking of bonds is the dominant mechanism of carrier production. The conductivity in this regime is essentially exponentially dependent on temperature. Conduction is ***extrinsic*** when donor and acceptor ionization is the dominant mechanism of carrier production. The conductivity in this regime is near-constant at moderate temperatures when all dopants are ionized, but decreases at low temperatures when not all dopants are ionized.

4. *Diffusion.* Diffusion is the net movement of carriers due to a ***concentration gradient*** of carriers. The constant of proportionality between the flux of carriers and the concentration gradient is called the ***diffusivity*** or the ***diffusion constant***. Diffusivity is related to mobility via ***Einstein's relation***. As minority carriers diffuse they travel a distance characterized by the ***minority carrier diffusion length*** before they recombine with majority carriers. The minority carrier diffusion length is related to the minority carrier lifetime via the diffusivity.

5. *Current equations.* The ***transport equations*** express the electron and hole currents in terms of their drift and diffusion components. The ***total current*** is the sum of the electron and hole currents. Summation of the components must be performed at the same plane in the material.

6. *Continuity equation.* This equation relates the rate of change of carrier concentration to spatial differences (divergence) in current, due to spatial differences in carrier concentration, for example, and to temporal changes in carrier concentration, due to differences in the rates of excess carrier recombination and generation. Its prime use is in finding the carrier concentration profiles, from which charge densities and currents can be computed.

3.7 REFERENCES

1. R. P. Feynmann, R. B. Leighton, and M. Sands, *The Feynmann Lectures on Physics*, vol. 1, pp. 41–1, 39–10, Addison-Wesley Publishing Co., Inc., Reading, Mass., 1966.
2. S. M. Sze, *Semiconductor Devices, Physics and Technology*, p. 63, John Wiley & Sons, Inc., New York, 1985.
3. A. G. Milnes, *Semiconductor Devices and Integrated Electronics*, pp. 674–692, Van Nostrand Reinhold Company, Inc., New York, 1980.
4. S. M. Sze, see Ref. 2, p. 34.
5. S. M. Sze, see Ref. 2, p. 33.
6. N. D. Arora, J. R. Hauser, and D. J. Roulston, "Electron and hole mobilities in Si as a function of concentration and temperature," *IEEE Transactions on Electron Devices*, vol. ED-29, pp. 292–295, 1982.
7. J. R. Haynes and W. Shockley, "The mobility and life of injected holes and electrons in germanium," *Physical Review*, vol. 81, pp. 835–843, 1951.
8. A. Bar-Lev, *Semiconductors and Electronic Devices*, 2nd ed., pp. 51–53, Prentice-Hall, Inc., Englewood Cliffs, N.J., 1984.

PROBLEMS

3.1. Calculate the resistance between the contacts of the silicon chip shown in Fig. 3P.1 for the following cases:
(**a**) Intrinsic Si.
(**b**) Si doped with 10^{16}cm^{-3} donors.
(**c**) Si doped with 10^{16}cm^{-3} acceptors.
Assume that all the impurities are ionized and that the temperature is 300 K.

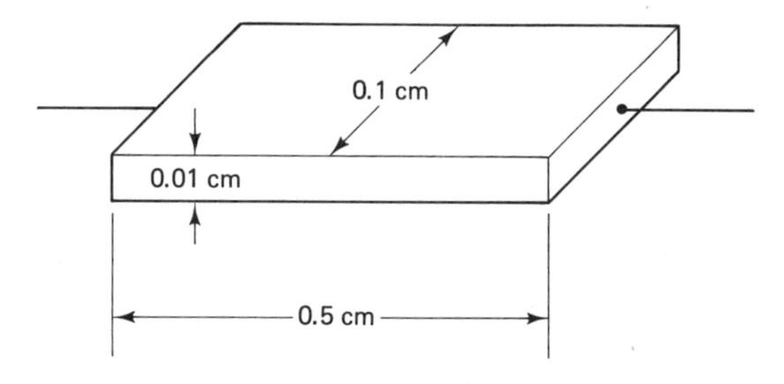

Figure 3P.1

3.2. Show that the conductivity of doped silicon reaches a minimum when the semiconductor is doped very slightly *p*-type. Find the acceptor doping density needed to achieve this minimum. (Note that, in practice, it is not possible to control the doping of silicon to this accuracy.)

3.3. Consider the resistor of Fig. 2P.1 and take the donor doping density to vary with depth. Show that the resistance R_{AB} is given by

$$R_{AB} = \frac{1}{\int_0^d q\mu n(x)\,dx}\,\frac{L}{W} \tag{3P.1}$$

3.4. For the situation in Problem 3.1(b), calculate the ratio of the hole drift current density to the total drift current density if a voltage is applied between the electrodes.

3.5. The donor doping density in a piece of silicon varies as $N_D(x) = N_0 \exp(-ax)$.

(a) Find an expression for the electric field at equilibrium over the range for which $N_D \gg n_i$.

(b) Sketch an energy band diagram for this case, and indicate the direction of the electric field. Explain qualitatively why the electric field is in the direction you have shown.

3.6. An ***exactly compensated*** semiconductor is one in which the mobile carrier concentrations are equal. Assuming complete ionization of all dopants, compute the conductivity for exactly compensated Si at 300 K in which **(a)** $N_A = N_D = 10^{13}$ cm^{-3}; **(b)** $N_A = N_D = 10^{20}$ cm^{-3}.

3.7. A diffused junction diode is made by diffusing phosphorus into p-type silicon of acceptor doping density $N_A = 10^3$ cm^{-3}. The n-type layer has a thickness of 1 μm, a sheet resistivity of 10^{15} Ω/square, and can be considered to be uniformly doped. Assuming complete ionization of all donors and acceptors, compute the donor doping density in the diffused layer.

3.8. A piece of silicon is uniformly doped with 10^{15}-cm^{-3} acceptor atoms. E_A is such that 10% of the acceptors are ionized. The silicon is then uniformly counterdoped with donors. E_D is such that, after counterdoping, 20% of the donors are ionized. A Hall-effect measurement, using the setup of Fig. 3P.2, gives a majority carrier concentration of 10^{13} cm^{-3}. Calculate the donor atom concentration N_D.

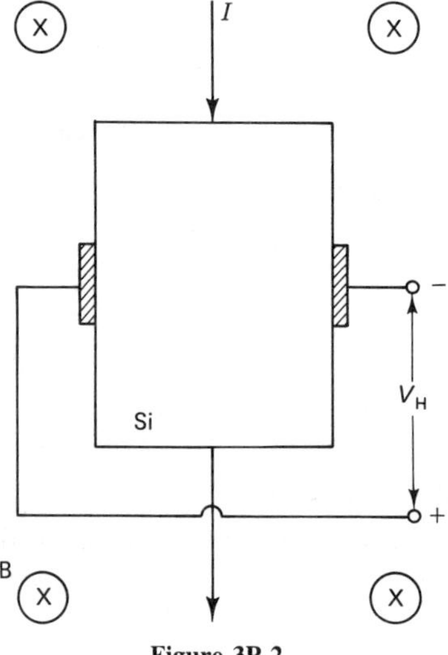

Figure 3P.2

3.9. For silicon at 300 K, plot the electron minority carrier diffusion length versus doping density over the range $N_A = 10^{14}$ to 10^{19} cm^{-3}. Use (3.20) and (3.40), along with data from Fig. 3.5 [or (3.8)] and Problem 2.12. Describe the physical phenomena that give the curve its particular shape.

3.10. Equations (3.6) and (3.7) describe the dependence of the mobility of electrons and holes in silicon on doping density and temperature. Use equation (3.6) to plot the electron mobility versus doping density over the range $N_A = 10^{14}$ to 10^{19} cm^{-3} for Si at 200 K. Compare the plot with the one in Fig. 3.5 and comment on any differences. What can be inferred from this result regarding the desirability of operating devices at low temperatures?

3.11. A silicon rod, used as a photodetector, is 2 mm wide, 1 mm thick, and 5 mm long and has contacts at each end to which a potential difference of 10 V is applied. The rod is irradiated with photons of energy greater than the silicon bandgap. The radiation is absorbed uniformly throughout the rod and leads to a steady-state photocurrent of 4.8 mA. When the radiation is abruptly cut off, the current decreases initially at a rate of 200 A sec^{-1}. If the electron and hole mobilities are 1000 and 500 cm^2/(V·s), respectively, find:

(a) The steady-state concentration of photogenerated electron–hole pairs.
(b) The excess carrier lifetime.
(c) The concentration of excess electrons and holes remaining 0.5 ms after the light is cut off.

3.12. A voltage is applied to the ends of a silicon rod photodetector. A pulse of photons with energy greater than the silicon bandgap is uniformly absorbed within the semiconductor. Show that the rate of increase of current density at the instant the light is pulsed on is equal to the rate of decrease of current density at the instant the light is pulsed off.

3.13. A hole current density of 10^{-5} A cm^{-2} is injected into one end of a long piece of n-type silicon. (This situation could arise in practice in a forward-biased diode, as described in Section 6.3.1.) Assuming that the holes flow by diffusion:

(a) Calculate the steady-state excess hole concentration at the injecting surface.
(b) Calculate the steady-state excess hole concentration at a depth of 1 mm.
(c) Calculate the number of excess holes per cm^2 of contact which are stored in the semiconductor in the steady state.

Take the hole mobility to be 500 cm^2/(V·s) and the minority carrier lifetime to be 25 μs.

4
The Fermi Level

4.1 INTRODUCTION

The thrust of this book is toward an understanding of the principles underlying the operation and analysis of semiconductor devices. This goal is sought in the chapters in this book on actual devices (Chapters 6 through 11) by first characterizing a device in terms of its dc current–voltage relationship. As current in semiconductors is due to the flow of electrons and holes, it follows that knowledge of the concentrations of these carriers is of importance. As discussed in Section 3.5, computation of $n(x, t)$ and $p(x, t)$ is possible by using the continuity equations and Poisson's equation. Recall that the latter, (3.29), relates the potential inside the semiconductor to the charge density. However, the potential inside the semiconductor is also affected by externally applied voltages (see Fig. 3.7). It would be convenient, therefore, to establish a link between the potential inside the semiconductor and the applied voltage. Such a link would relate voltage to current via the charge density, and so provide a means of deriving a device's I–V characteristic. The reference energy level, called the ***Fermi level***, helps in providing this link. The Fermi level is also very helpful in energy band diagrams, where it provides a means of visualizing both voltage drops and the magnitudes of the carrier concentrations.

We demonstrate these attributes of the Fermi level in Chapters 6 through 11. In the present chapter we introduce the Fermi-level concept and show how the Fermi level is related explicitly to the equilibrium concentrations of charge, both mobile and fixed, in the semiconductor. We end the chapter by extending the Fermi-level concept to nonequilibrium conditions and deriving an important, general equation for current in semiconductors.

4.2 FERMI–DIRAC DISTRIBUTION FUNCTION

The Fermi level is an energy level E_F defined via the ***Fermi–Dirac distribution function***:

$$f(E) = \frac{1}{1 + \exp[(E - E_F)/kT]} \tag{4.1}$$

where k is Boltzmann's constant, T the absolute temperature, and E represents energy. The distribution function is named after Fermi and Dirac, two physicists who independently derived the function in the 1920s. Specifically, $f(E)$ is the probability that a particular quantum state at the energy level E is occupied by an electron. The distribution function correctly takes into account the occupancy of the two states in an energy level in accordance with Pauli's exclusion principle; that is, if one state is filled by an electron, the other state can only be occupied by an electron of opposite spin. Equation (4.1) was derived using the methods of statistical mechanics. This branch of mathematical physics is applicable to this problem, as the number of available states in a semiconductor is very large. For example, in Section 2.3 it was shown that there are about 10^{23} states/cm^3 in the conduction and valence bands of silicon. These states are distributed in energy throughout the energy bands and, as was also discussed in Section 2.3, the states near the bottom of the conduction band and near the top of the valence band are the ones of main relevance to semiconductor devices. The actual distributions in energy of these important states near the band edges are derived in Section 5.7; the results are

$$g_c(E)\,dE = \frac{8\sqrt{2}\,\pi}{h^3} m_e^{*3/2}(E - E_C)^{1/2}\,dE \qquad \text{for } E > E_C$$

$$g_v(E)\,dE = \frac{8\sqrt{2}\,\pi}{h^3} m_h^{*3/2}(E_V - E)^{1/2}\,dE \qquad \text{for } E < E_V \tag{4.2}$$

where h is Planck's constant. The dimensions of $g(E)\,dE$ are number (of states) per volume and the term is to be interpreted as the density of quantum states available in the energy interval between E and $E + dE$. A graphical representation of (4.2) is shown in Fig. 4.1a. This figure emphasizes the parabolic nature of the relationship between $g(E)$ and E within the bands. The density of states functions in the conduction and valence bands are not quite identical because of the differences in effective mass of electrons in the conduction band and holes in the valence band. The other point brought out in Fig. 4.1a is that there are no available states in the energy bandgap of a pure semiconductor. On doping the material, however, electron states do become available in this region. These states are represented in Fig. 4.1b as being localized in energy around the ground-state values that were designated E_D for donors and E_A for acceptors in the energy band diagrams of Figs. 2.16 and 2.18.

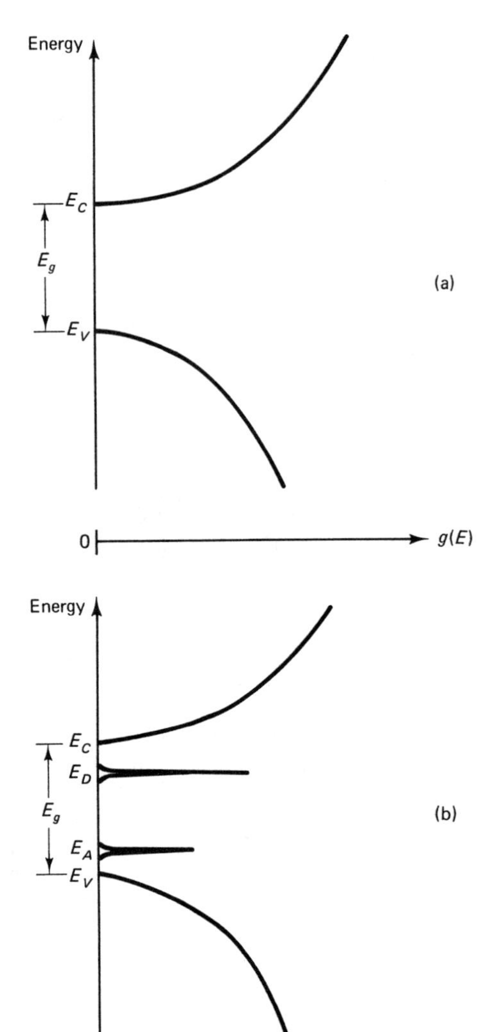

Figure 4.1 Variation of the density of states functions with energy: (a) for intrinsic material and (b) for extrinsic material; states due to donor and acceptor impurities are also shown.

$g(E)\,dE$ gives the ***density of available states*** and $f(E)$ gives the ***probability of their occupancy*** by electrons. It follows that the actual number of electrons in the conduction band, for example, can be computed by integrating the product of $g(E)\,dE$ and $f(E)$ over the energy range of the conduction band, that is,

$$n_0 = \int_{E_C}^{\text{top of band}} f(E)g(E)\,dE \tag{4.3}$$

Because the Fermi–Dirac function refers to the probability of state occupancy under conditions of thermal equilibrium, the electron concentration yielded by (4.3) is the thermal equilibrium concentration. The corresponding relationship for holes is

$$p_0 = \int_{\text{bottom of band}}^{E_V} [1 - f(E)]g(E)\, dE \tag{4.4}$$

where $[1 - f(E)]$ is the probability of a state at energy E being unoccupied by an electron which, in the valence band, can be interpreted as the probability of a state at energy E being occupied by a hole.

By assigning the values of $+\infty$ and $-\infty$ to the top and bottom limits of the integrals for n_0 and p_0, respectively, some progress in solving (4.3) and (4.4) can be made. This assignment can be justified by examining the form of $f(E)$ versus E, as given by (4.1) and plotted in Fig. 4.2. The curves are symmetrical about the

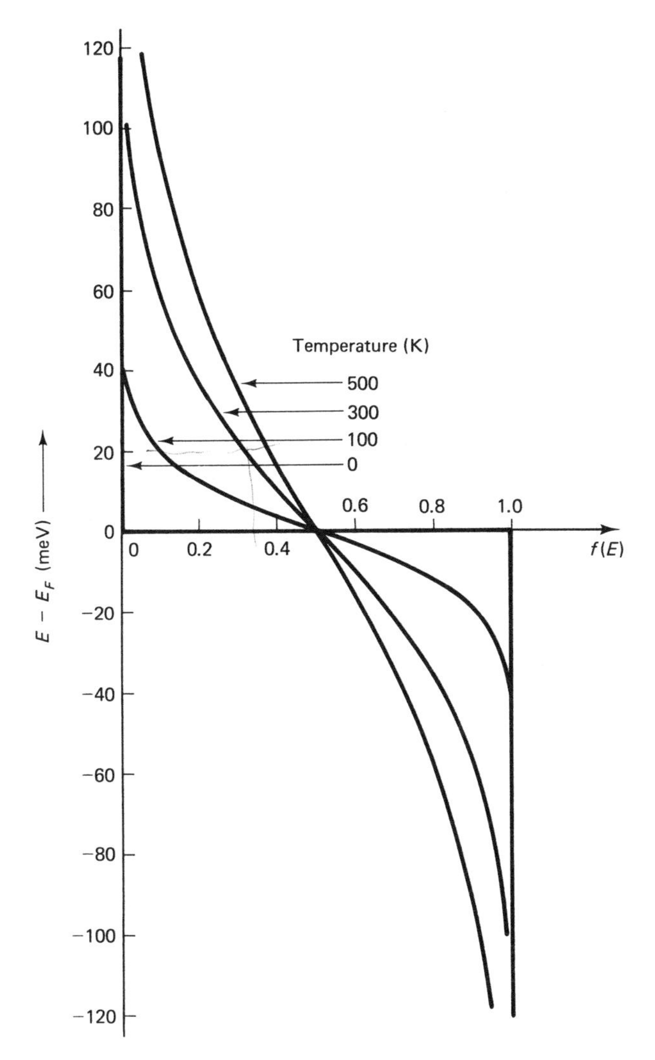

Figure 4.2 Fermi–Dirac distribution function for various temperatures.

reference energy level, the Fermi level E_F, and $f(E)$ changes rapidly on either side of E_F. Thus in the case of electrons, for example, provided that E_F is not located deep inside the conduction band, $f(E)$ will be essentially zero for all energies except those close to the bottom of the conduction band. Therefore, the integral for n_0 can be extended to infinity without incurring any appreciable error. Thus the assignment is justifiable on physical grounds and is most convenient from a mathematical point of view, as it allows (4.3) to be written in a concise form, namely

$$n_0 = 2N_C(\pi)^{-1/2}F_{1/2}(a_F) \tag{4.5}$$

where N_C, which is termed the ***effective density of states in the conduction band***, is given by

$$N_C = 2\left(\frac{2\pi m_e^* kT}{h^2}\right)^{3/2} \tag{4.6}$$

and $F_{1/2}(a_F)$ is a tabulated function [1], called the ***Fermi integral*** of order one-half, given by

$$F_{1/2}(a_F) = \int_0^\infty \frac{a^{1/2}\,da}{1 + \exp(a - a_F)} \tag{4.7}$$

where $a = (E - E_C)/kT$ and $a_F = (E_F - E_C)/kT$.

N_C is a temperature-dependent material constant. Values for silicon and gallium arsenide are given in Appendix 2. While values for the Fermi integral are available in tables or can be computed by numerical integration, the presence of the integral in (4.5) is not conducive to providing the ready insight into the factors determining the magnitude of n_0 that we are seeking. This insight can be gained, however, by making use of the approximate solution

$$F_{1/2}(a_F) = \frac{(\pi)^{1/2}}{2}\exp(a_F) \tag{4.8}$$

which allows (4.5) to be written in the concise form

$$n_0 = N_C \exp\left(\frac{E_F - E_C}{kT}\right) \tag{4.9}$$

This equation is extremely useful, as it relates the equilibrium electron carrier concentration to a single variable, namely the Fermi level. This allows n_0 to be computed easily. It also implies that the states in the conduction band can be viewed as being compressed into a single energy level E_C and represented by the effective density of states N_C. However, before (4.9) can be used with confidence, the question that needs to be addressed is: Under what circumstances is the approximation of (4.8) valid? The answer can be found from Fig. 4.3, where both the exact (4.5) and the approximate (4.9) equations are plotted. It appears that provided that E_F is more than about $2kT$ below E_C, the approximation is excellent.

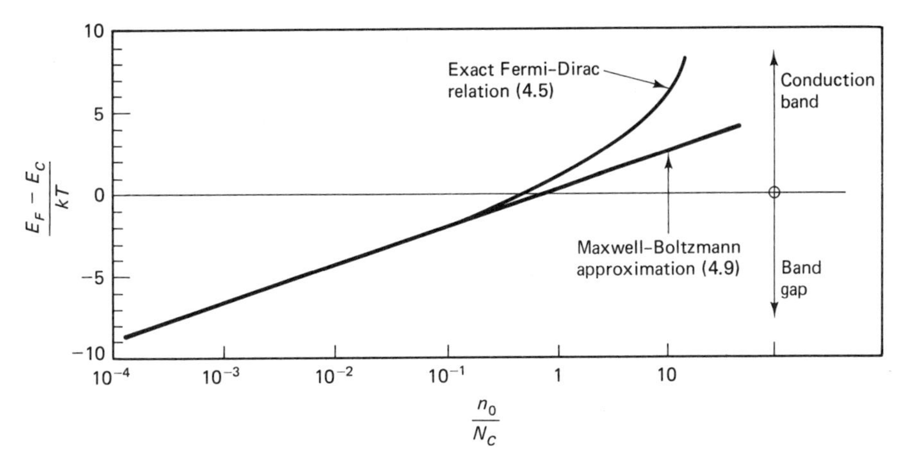

Figure 4.3 Comparison of the exact and approximate expressions for n_0. The approximation is excellent provided that $E_C - E_F > 2kT$.

The corresponding approximate equation for holes is

$$p_0 = N_V \exp\left(\frac{E_V - E_F}{kT}\right) \tag{4.10}$$

where N_V is termed the ***effective density of states in the valence band.*** Values of N_V for Si and GaAs are given in Appendix 2. Equation (4.10) is valid as long as E_F is more than about $2kT$ above E_V.

Although the approximations needed to derive (4.9) and (4.10) have been presented from a mathematical point of view, they also have a physical basis. These equations could have been arrived at by considering a ***Maxwell–Boltzmann***, rather than a Fermi–Dirac, distribution function. The essential difference between the two functions is that the former describes a situation in which states can be populated without regard to the restrictions imposed by the Pauli exclusion principle (i.e., any number of electrons can occupy a given state). However, if the number of particles is very small compared to the number of available states, the probability of more than one particle occupying the same state is very small. Under these circumstances, the Fermi–Dirac and Maxwell–Boltzmann distributions are essentially the same. Figure 4.3 indicates that for n-type material when $(E_C - E_F) > 2kT$, the number of electrons in the conduction band is sufficiently smaller than the number of available states, for it not to matter whether Fermi–Dirac or Maxwell–Boltzmann statistics are used. The corresponding situation regarding the number of holes in the valence band compared to the available number of filled states in the valence band arises in p-type material when $(E_F - E_V) > 2kT$.

In solid-state physics, the adjective ***degenerate*** is used to describe a situation in which more than one particle occupies a state of given energy level. When this is likely to occur in semiconductors, Fermi–Dirac statistics must be used as they, unlike Maxwell–Boltzmann statistics, take into account the correct rule (Pauli's exclusion principle) for population of the two states of each energy level. Thus a semiconductor in which E_F is within a $2kT$ of either band (or resides in the band) and for which, therefore, Fermi–Dirac statistics apply, is called degenerate. Correspondingly, when (4.9) and (4.10) are valid, the semiconductor is said to be ***nondegenerate***. As Fig. 4.3 shows, the crossover from nondegenerate to degenerate occurs for n-type material at room temperature when n_0 is about $0.4 \times N_C$ (i.e., about 10^{19} cm^{-3} for Si and about 10^{17} cm^{-3} for GaAs).

4.3 FERMI LEVEL IN ENERGY BAND DIAGRAMS

From the discussion above it should be apparent that the Fermi level can be viewed as a "bookkeeping index" for the carrier concentrations. Knowledge of the position of the Fermi level with respect to the band edges provides a ready means of calculating n_0 and p_0. Furthermore, the inclusion of E_F on an energy band diagram provides an indicator, or convenient visual aid, for ascertaining at a glance the approximate degree and type of doping in a semiconductor. The closer E_F is to the conduction band, the more n-type a material is. Similarly, the closer E_F is to the valence band, the more p-type a material is. It follows that in intrinsic material, E_F is close to the middle of the energy bandgap. These facts are illustrated in Fig. 4.4. Because the Fermi energy is the reference energy level for the distribution function of (4.1), as E_F moves along the energy axis of the band diagram it "pulls" $f(E)$ with it. Thus the curve shown in Fig. 4.2 slides up and down the band diagram. The degree of overlap of $f(E)$ with $g(E)$ is, in effect, a graphical evaluation of (4.3) and (4.4).

4.4 FINDING THE FERMI LEVEL IN INTRINSIC MATERIAL

Because the Fermi level is related to the carrier concentrations, the position of the Fermi level for a given situation must be consistent with the carrier concentrations calculated by any other means. This is a general truth and provides the starting point for any calculation of E_F. Specifically, a value of E_F is sought which, when used in (4.5) and its equivalent for holes, or in (4.9) or (4.10) if they are

---

Figure 4.4 Graphical representation of the computation of carrier concentrations. Using potential energy as a common axis, the Fermi–Dirac distribution function and the density of states functions are displayed for (a) intrinsic, (b) n-type, and (c) p-type material. The Fermi–Dirac function "slides" along the energy axis so that the Fermi level takes on the position required to give the correct magnitudes of n_0 and p_0 in each case. The shaded portions of the curves illustrate the area of overlap of the Fermi–Dirac function and the density of states functions. The magnitude of the carrier concentration is proportional to the extent of this overlap.

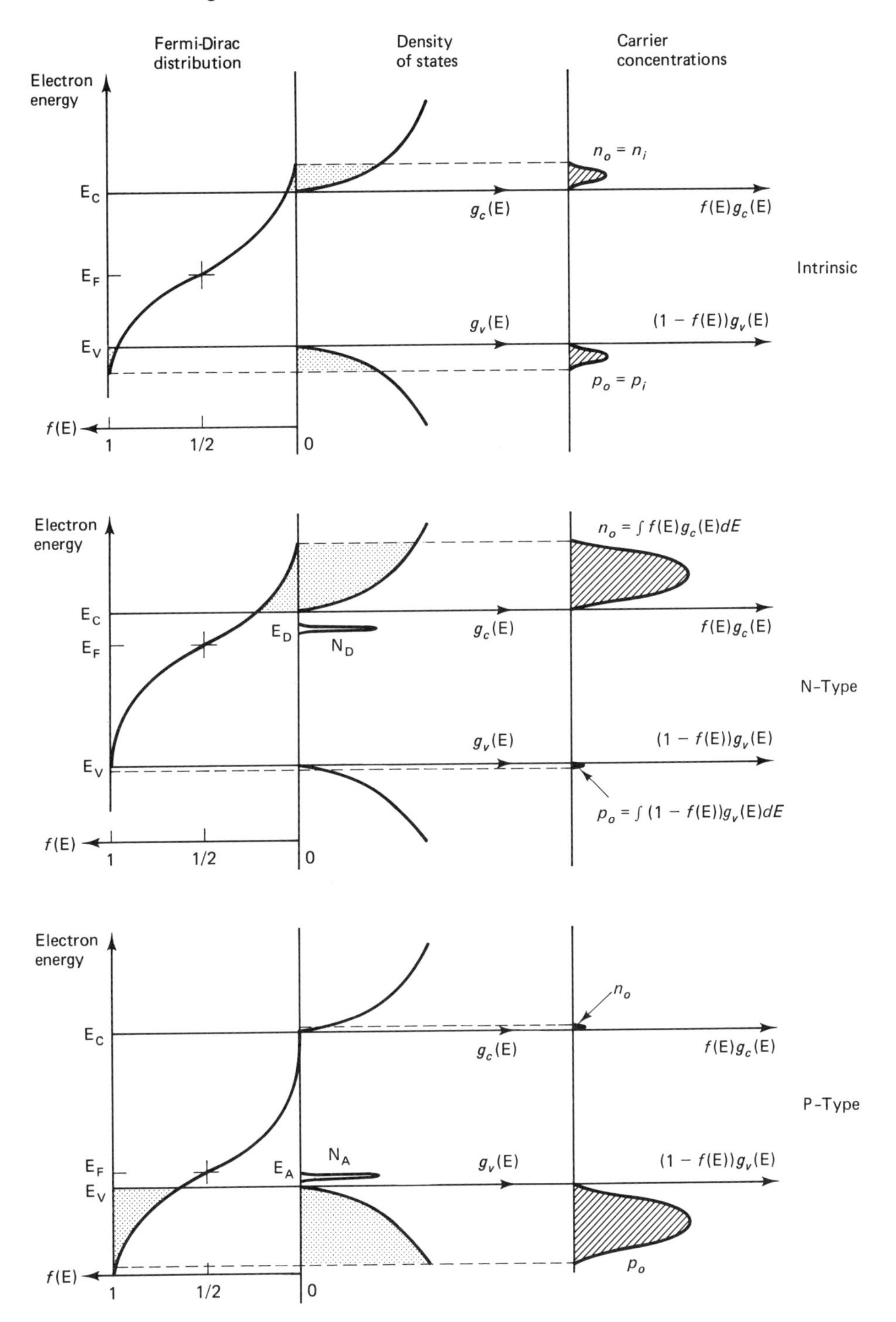
Fermi-Dirac distribution
Density of states
Carrier concentrations
Electron energy
E_C
E_F
E_V
f(E)
1
1/2
0
g_c(E)
g_v(E)
n_o = n_i
p_o = p_i
f(E)g_c(E)
(1 − f(E))g_v(E)
Intrinsic
E_D
N_D
n_o = ∫ f(E)g_c(E)dE
p_o = ∫ (1 − f(E))g_v(E)dE
N-Type
E_A
N_A
n_o
p_o
P-Type

appropriate, yields values for n_0 and p_0 that are consistent with Poisson's equation (3.29).

Taking the case of silicon as an example, the energy band gap (1.12 eV at 300 K) is equivalent to about $43kT$ at room temperature. As E_F is around midgap for intrinsic material (as in Fig. 4.4a) the approximations implicit in (4.9) and (4.10) are certainly valid in the intrinsic case. Assuming a homogeneous material with no voltages applied allows Poisson's equation to be reduced to the charge neutrality equation (2.8). There are no dopants present in intrinsic material, so $n_0 = p_0$. Thus equating (4.9) and (4.10), and rearranging, gives

$$E_F \text{ (intrinsic)} = \frac{E_g}{2} + \frac{3kT}{4} \ln \frac{m_h^*}{m_e^*} \tag{4.11}$$

There is still some uncertainty over the magnitudes of the effective masses for electrons and holes in semiconductors at room temperature. The difficulty arises because experimental measurements of effective mass are made at very low temperatures (4.2 K using the cyclotron resonance method) and the temperature dependence of the effective masses is not yet known exactly. The effective masses used in (4.11) are properly termed the ***density of states effective masses*** to distinguish them from the ***conductivity effective masses*** used in Chapter 3 to compute mobility [2]. Table 4.1 lists the proposed values that are used in this book. From these data the ratio of hole and electron effective masses for silicon at room temperature is 0.68 and so, at room temperature, the Fermi level is only 0.0073 eV below midgap. In intrinsic gallium arsenide the Fermi level is 0.040 eV above midgap. Thus, for most intents and purposes, and certainly in composing band diagrams, the Fermi level in intrinsic material can be considered to lie in the middle of the bandgap. It is sometimes convenient to use this energy level, rather than the band edges E_C or E_V, as a reference energy level, so it is given a specific label. In this book the symbol $\boldsymbol{E_{Fi}}$ is used.

TABLE 4.1 DENSITY OF STATES EFFECTIVE MASSES

	4.2 K measured	300 K Extrapolated[a]	300 K Proposed[b]
Si			
m_e^*/m_0	1.062	1.182	1.38
m_h^*/m_0	0.590	0.81	0.946
GaAs			
m_e^*/m_0	0.067	0.0655	
m_h^*/m_0	0.532	0.524	

[a] From Refs. 3 and 7.

[b] Proposed to make the result from (4.14) consistent with experimental data (see Section 4.4.1).

4.4.1 Computation of n_i, Proof of $n_0 p_0 = n_i^2$, and Statement of n_0 and p_0 in Terms of n_i

As an example of the use of E_{Fi} in calculations, we derive here an expression for n_i in the form that was merely quoted in (2.7). Using (4.9) and (4.10) for the intrinsic case gives

$$\begin{aligned} n_i &= N_C \exp\left(\frac{E_{Fi} - E_C}{kT}\right) \\ p_i &= N_V \exp\left(\frac{E_V - E_{Fi}}{kT}\right) \end{aligned} \tag{4.12}$$

Multiplying these two equations together gives

$$n_i p_i = N_C N_V \exp\left(\frac{E_V - E_C}{kT}\right) \tag{4.13}$$

As $n_i = p_i$, then, it follows that

$$n_i = (N_C N_V)^{1/2} \exp\left(\frac{-E_g}{2kT}\right) \tag{4.14}$$

which is the complete form of (2.7).

In passing, it is worth noting that for an arbitrary value of E_F, the product of n_0 and p_0, as evaluated by multiplying (4.9) and (4.10), yields the same expression as on the right-hand side of (4.13). This is a satisfying confirmation of (2.14), that is,

$$n_0 p_0 = n_i^2 \tag{2.14}$$

To emphasize the role of E_{Fi} as a reference energy level, divide (4.9) for n_0 by the expression for n_i given in (4.12). The result is

$$n_0 = n_i \exp\left(\frac{E_F - E_{Fi}}{kT}\right) \tag{4.15}$$

The corresponding result for holes is

$$p_0 = n_i \exp\left(\frac{E_{Fi} - E_F}{kT}\right) \tag{4.16}$$

It is a matter of personal preference whether these equations with the single reference energy level of E_{Fi} and the single coefficient of n_i are used instead of (4.9) and (4.10), which, between them, use two reference energy levels (E_C and E_V) and two coefficients (N_C and N_V). However, to make sure that both approaches give the same answer for the carrier concentration, n_i and N_C must be properly chosen. n_i is a measurable parameter, so, in principle, there should be little doubt about its value. The commonly used value of 1.45×10^{10} cm^{-3} [4]

at 300 K stems from the extrapolation of data from Hall-effect measurements made in the temperature range of 450 to 1100 K [5]. Later work using slightly purer silicon samples and a temperature range of 300 to 500 K, suggests a value for n_i at 300 K of 1.25×10^{10} cm^{-3} [6]. This is the value we have adopted in this book. To make the prediction of (4.14) consistent with this value at 300 K we need, if we accept that E_g at 300 K is 1.12 eV [7], $(N_C N_V)^{1/2} = 3.07 \times 10^{19}$ cm^{-3}. The proposed values for the effective densities of states given in Table 4.2 yield such a product. The actual individual values for N_C and N_V have been selected so that their ratio is consistent with the commonly accepted [3, 7] ratio of effective masses given in the center column of Table 4.1.

TABLE 4.2 EFFECTIVE DENSITIES OF STATES FOR Si AT 300 K

	Computed[a]	Proposed[b]
N_C (cm^{-3})	3.22×10^{19}	4.07×10^{19}
N_V (cm^{-3})	1.83×10^{19}	2.31×10^{19}

[a] From Refs. 3 and 7.

[b] Proposed to make the result from (4.14) consistent with experimental data (see Section 4.4.1).

4.5 FINDING THE FERMI LEVEL IN EXTRINSIC MATERIAL

As stated at the beginning of Section 4.4, the starting point for finding E_F is to seek a Fermi-level position that gives carrier concentrations from (4.9) and (4.10) (or their unapproximated equivalents) which are consistent with those appearing in Poisson's equation (3.29). Again, as in Section 4.4, if the semiconductor material is homogeneous and field free, Poisson's equation is the same as the charge neutrality equation, that is,

$$p_0 + N_D^+ = n_0 + N_A^- \tag{2.8}$$

Combining this equation with $n_0 p_0 = n_i^2$ yields, for n_0, for example,

$$n_0 = \frac{X + \sqrt{X^2 + 4n_i^2}}{2} \tag{4.17}$$

where $X = N_D^+ - N_A^-$.

Provided that the ionized doping densities are known, the value of n_0 from this equation can be used in (4.5), (4.9), or (4.15) to find the Fermi level. Usually, there is no problem in deciding which of these equations to use. If the problem is being solved numerically on a computer, (4.5) can be used. If an analytical solution by hand calculation is being sought, then either (4.9) or (4.15) should be used. If, in the latter instance, the resulting value of the Fermi level energy is such that E_F is within $2kT$ of the conduction band edge (for a calculation from the electron concentration), it must be appreciated that the value of E_F so calculated is inaccurate. The Fermi energy resulting from this approach can be written, for n-type material, as

$$E_F(n\text{-type}) = \begin{cases} E_C - kT \ln \dfrac{N_C}{N_D^+} \\ \quad \text{or} \\ E_{Fi} + kT \ln \dfrac{N_D^+}{n_i} \end{cases} \tag{4.18}$$

and for p-type material,

$$E_F(p\text{-type}) = \begin{cases} E_V + kT \ln \dfrac{N_V}{N_A^-} \\ \quad \text{or} \\ E_{Fi} - kT \ln \dfrac{N_A^-}{n_i} \end{cases} \tag{4.19}$$

4.5.1 Finding the Fermi Level When the Amount of Dopant Ionization Is Not Known

Before (4.18) and (4.19) can be used to compute the Fermi level, the concentrations of the ionized donors and acceptors must be known. Often, because intentional impurities have such low activation energies, E_D or E_A as in Figs. 2.16 and 2.18, for example, it is safe to assume 100% ionization (i.e., $N_D^+ = N_D$ and $N_A^- = N_A$). At low temperatures (see Fig. 3.8), or for "deeper" impurities (impurities with energy levels more than a few kT away from the band edges), it may not be accurate to make this assumption. In such cases the ionized dopant concentrations must first be calculated.

To do this, a distribution function describing the probability of occupation by electrons of states at the donor and acceptor energy levels must be employed. It might be thought that the required function is that of $f(E)$ given in (4.1). However this is not quite the case. The difference arises, in the case of group V donors in silicon, for example, because the localized energy level E_D associated with each donor only has room for one electron. Thus, if this state is filled by an electron (of either spin designation), there is no room for another electron, not even one of opposite spin. The Fermi–Dirac function of (4.1) was derived in accordance with Pauli's exclusion principle: namely, that each energy level has two states which can be populated by electrons of opposing spin. This is appropriate for the conduction and valence bands but not for the impurity energy levels. The correct form of the distribution function for the occupancy of energy levels due to donors with one electron to donate is [8]

$$f(E_D) = \frac{1}{1 + \frac{1}{2}\exp[(E_D - E_F)/kT]} \tag{4.20}$$

As the energy level for a particular impurity can usually be considered as single valued, an integration such as described by (4.3) is not required to find the electron concentration residing in the donor states. In the case of donors, ionization means the loss of an electron (to the conduction band), so N_D^+ is related to the probability of the energy level E_D being empty; thus

$$\frac{N_D^+}{N_D} = 1 - f(E_D) = \frac{1}{1 + 2\exp[(E_F - E_D)/kT]} \tag{4.21}$$

For holes the situation is different again because the electron accepted by an acceptor must be of such a spin so as to pair correctly with the electron, which is unpaired when the impurity is nonionized (see Fig. 2.17). The correct distribution for this case is [8]

$$f(E_A) = \frac{1}{1 + 2\exp[(E_A - E_F)/kT]} \tag{4.22}$$

However, for Si and GaAs, this equation is still not completely correct because of the details of the energy band structure at the top of the valence band (see Section 5.3). In the correct relationship, the 2 is replaced by a 4 [9], so

$$N_A^- = \frac{N_A}{1 + 4\exp[(E_A - E_F)/kT]} \tag{4.23}$$

For a real semiconductor system, the doping densities are determined by the processing conditions used in the fabrication of the device and can be taken as known. The energy levels for commonly used dopants are well known [10]. Therefore, (4.21) and (4.23) can be combined with (2.8), (2.14), and either (4.5) or (4.9) [or (4.15)] to constitute a set of five equations in the five unknowns E_F, N_D^+, N_A^-, n_0, and p_0. From the simultaneous solution of these equations, usually by employing iterative techniques, E_F can be determined, as we now demonstrate.

Example 4.1 Computation of the Fermi Level for the General Case

For this example, consider silicon at 300 K doped with 10^{17} cm^{-3} atoms of the group V donor arsenic, and 10^{16} cm^{-3} atoms of the group III acceptor aluminum. The activation energies for these dopants are shown in Fig. 4E.1. The problem is to find the position of the Fermi level without making the assumption that all the dopant atoms are ionized.

The five equations we need to solve are

$$p_o + N_D^+ = n_o + N_A^- \tag{2.8}$$

$$n_o p_o = n_i^2 \tag{2.14}$$

$$n_o = N_C \exp\left(\frac{E_F - E_C}{kT}\right) \tag{4.9}$$

or

$$n_o = n_i \exp\left(\frac{E_F - E_{\mathrm{Fi}}}{kT}\right) \tag{4.15}$$

$$N_D^+ = \frac{N_D}{1 + 2\exp\left(\dfrac{E_F - E_D}{kT}\right)} \tag{4.21}$$

$$N_A^- = \frac{N_A}{1 + 4\exp\left(\dfrac{E_A - E_F}{kT}\right)} \tag{4.23}$$

It does not matter whether (4.9) or (4.15) is chosen, provided that appropriate values for the effective masses, the effective densities of states, the energy band gap, and the intrinsic carrier concentration are used. This question of consistency in the values of these parameters was discussed at the end of Section 4.4.1. We use the values from that section, namely,

$$E_g = 1.12 \text{ eV}$$

$$n_i = 1.25 \times 10^{10} \text{ cm}^{-3}$$

$$N_C = 4.07 \times 10^{19} \text{ cm}^{-3}$$

$$N_V = 2.31 \times 10^{19} \text{ cm}^{-3}$$

$$\frac{m_h^*}{m_e^*} = 0.685$$

$$\frac{kT}{q} = 0.0259 \text{ V}$$

Note that in using (2.14) and (4.9) or (4.15), we are assuming that the material is nondegenerate (i.e., E_F is more than $2kT$ away from either band edge). If

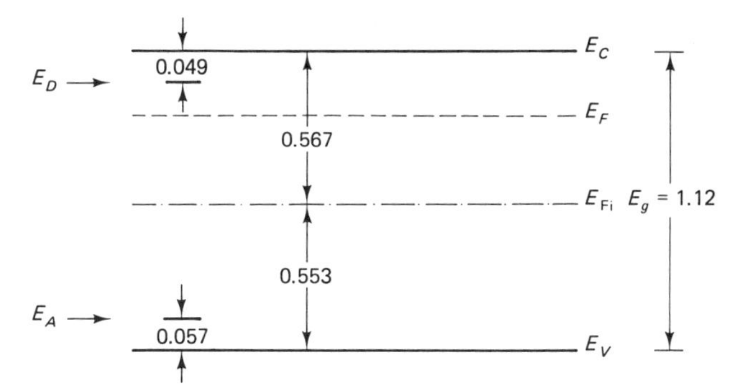

Figure 4E.1 Energy band diagram for Example 4.1. All energies are in eV.

our answer reveals that this is not the case, the calculation will have to be redone using (4.5) for n_0, and its equivalent for p_o instead of (2.14).

The last three equations of our set can be written concisely as

$$n_o = N_C A \tag{4E.1}$$

$$N_D^+ = \frac{N_D}{1 + BA} \tag{4E.2}$$

$$N_A^- = \frac{N_A}{1 + C/A} \tag{4E.3}$$

where

$$A = \exp\left(\frac{E_F - E_C}{kT}\right), \qquad B = 2\exp\left(\frac{E_C - E_D}{kT}\right),$$

$$C = 4\exp\left(\frac{E_A - E_V - E_g}{kT}\right)$$

As the donor and acceptor energy levels, and the energy bandgap, are known, substitution of A, B, and C, along with (2.14), into (2.8) yields a transcendental equation in A, the only unknown, that is,

$$\frac{n_i^2}{N_C A} + \frac{N_D}{1 + BA} = N_C A + \frac{N_A}{1 + C/A} \tag{4E.4}$$

To solve this equation we adopt an iterative approach and use as a first guess for A the value that would obtain if all the dopants were ionized. Thus, from (4.17), we find $n_0 = 0.9 \times 10^{16}$ cm^{-3} and, from (4.9), $A = 2.21 \times 10^{-3}$. Evaluation of (4E.4) using this number suggests that a slightly smaller value of A is needed for a correct solution. Judicious decrementing of A reveals 2.145×10^{-3} as being the appropriate value. This gives

$$(E_C - E_F) = 0.159 \text{ eV}$$

Note that this result is only 1 meV different from the value predicted by (4.17) for the case of all the impurities being ionized. This close agreement indicates that the assumption of complete ionization of shallow impurities at 300 K is a very good one. Recall that a shallow donor is one with an energy level close to the conduction band, and a shallow acceptor is one with an energy level close to the valence band. Thus the full, general method of solution for E_F outlined in this example is only necessary for use with shallow impurities at low temperatures, when full ionization cannot be guaranteed (see Fig. 3.8). On the other hand, the full approach should be taken for room-temperature calculations in the case of deep impurities with energy levels close to midgap.

Finally, note that $(E_C - E_F) \cong 6kT$ in this example, so the use of Maxwell–Boltzmann statistics, (4.9), is justified.

4.6 QUASI-FERMI LEVELS

The Fermi–Dirac and Maxwell–Boltzmann distribution functions apply only to thermal equilibrium conditions. Thus the usefulness of the Fermi level concept, in computing carrier concentrations and in interpreting energy band diagrams, can only be exploited in equilibrium situations. However, this usefulness is so appealing that it is desirable to render the Fermi-level concept applicable to nonequilibrium situations. Such situations arise very frequently in semiconductor devices, namely whenever excess carriers, external electric fields, or currents are present. Fortunately, it transpires that it is indeed possible to use entities analogous to the Fermi level in nonequilibrium conditions, but they must be distinguished from *the* Fermi level, which exists only in the case of thermal equilibrium. The Fermi levels used in nonequilibrium situations are called ***quasi-Fermi levels***.

A semiconductor system in thermal equilibrium is specified by a single Fermi level. Knowledge of the position of this level is sufficient to calculate both n_0 and p_0. However, in nonequilibrium situations, the carrier concentrations depend on external conditions, such as, for example, the intensity of ionizing radiation or the voltage applied to the junction between n-type and p-type material. The situation illustrated in Fig. 3.10, and reproduced in Fig. 4.5, is an example of injection of electrons into p-type material, such as occurs in a forward-biased pn junction (Section 6.3.1). Clearly, n is raised above n_0 in part of the device, so the equilibrium value of E_F, when used in (4.9) or (4.15), will underestimate the electron concentration close to the injecting junction. Thus a single quasi-Fermi level cannot be expected to describe adequately both the electron and hole carrier concentrations in nonequilibrium conditions. Accordingly, two quasi-Fermi levels are used: an ***electron quasi-Fermi level***, $\boldsymbol{E_{Fn}}$, which is related to the total electron concentration ($= n_0 + \hat{n}$), and a ***hole quasi-Fermi level***, $\boldsymbol{E_{Fp}}$, which is related to the total hole concentration. The relations are expressed, for nondegenerate material, as

$$
\begin{aligned}
E_{Fn} &= E_{\text{Fi}} + kT \ln \frac{n}{n_i} \\
E_{Fp} &= E_{\text{Fi}} - kT \ln \frac{p}{n_i}
\end{aligned}
\tag{4.24}
$$

These expressions are obviously based on the equations for the thermal equilibrium case [(4.15) and (4.16)]. Because of this, the two quasi-Fermi levels reduce to a single level, *the* Fermi level, when nonequilibrium conditions relax to thermal equilibrium. Although E_{Fn} and E_{Fp} are merely artifacts developed for convenience, they do have a physical basis, inasmuch as quasi-equilibrium usually holds in semiconductor devices. As E_{Fn} and E_{Fp} are defined by (4.24), it follows that prior knowledge of n and p, respectively, is needed before the quasi-Fermi levels can be specified.

The main employment of quasi-Fermi levels in this book is in energy band diagrams, where their positions relative to E_F give a ready indication of the extent of the nonequilibrium carrier concentrations. The example of electron injection into a p-type semiconductor depicted in Fig. 4.5 illustrates the point. The electron injection raises the electron concentration above its thermal equilibrium concentration. As the excess electrons diffuse away from the contact, they recombine with majority carrier holes and, after diffusing a distance equivalent to a few minority carrier diffusion lengths, their magnitude is effectively reduced to zero, and n becomes equal to n_0. Over the region in which $n > n_0$, E_{Fn} is raised above E_F in the direction of E_{Fi}, in accordance with (4.24). Note that E_{Fp} remains constant in this example. This is because low-level injection conditions, as discussed in Section 2.5.2, have been assumed and thus the hole concentration is essentially unaffected by the recombination events.

The diffusion of electrons away from the injecting contact constitutes a diffusion current. The presence of an electron current is conveniently indicated on the band diagram by the gradient of E_{Fn}. To corroborate this mathematically, we

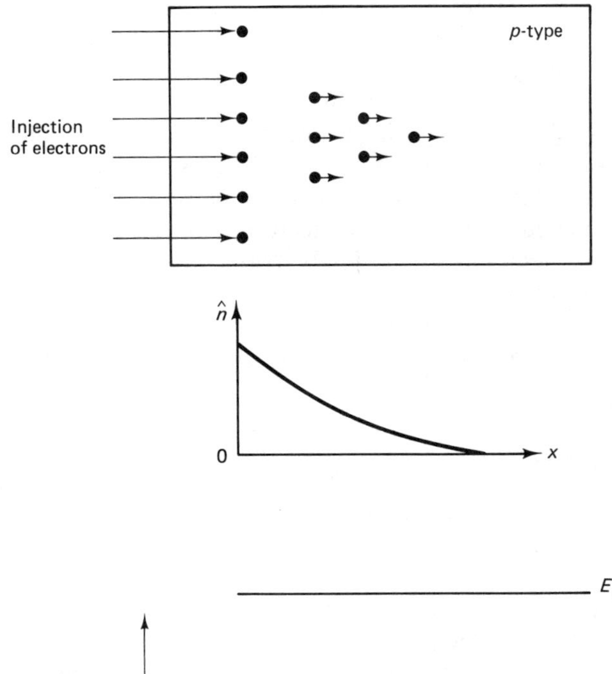

Figure 4.5 Example of quasi-Fermi levels on an energy band diagram. The elevation of the electron concentration above n_0 near the injecting contact is conveyed on the band diagram by the relative positions of E_{Fn} and E_F. As the excess electrons diffuse away from the contact, they recombine and their concentration decreases and, accordingly, E_{Fn} tends toward E_F.

obtain first an expression involving dE_{Fn}/dx by differentiating (4.24):

$$\frac{dn}{dx} = \frac{n}{kT}\left(\frac{dE_{Fn}}{dx} - \frac{dE_{\mathrm{Fi}}}{dx}\right) \tag{4.25}$$

The gradient of the intrinsic Fermi level, like that of the gradient of the other potential energy reference levels E_C and E_V (see Fig. 3.7), is related to the electric field by

$$\frac{1}{q}\frac{dE_{\mathrm{Fi}}}{dx} = \mathscr{E} \tag{4.26}$$

Recalling the Einstein relation and the electron transport equation, namely,

$$D = \frac{kT}{q} \tag{3.20}$$

and

$$J_e = q\mu_e n\mathscr{E} + qD_e\frac{dn}{dx} \tag{3.23}$$

and substituting (4.24), (4.25), and (3.20) into (3.23) gives the desired relationship:

$$J_e = n\mu_e\frac{dE_{Fn}}{dx} \tag{4.27}$$

The corresponding equation for hole current is

$$J_h = p\mu_h\frac{dE_{Fp}}{dx} \tag{4.28}$$

Equations (4.27) and (4.28) are true in general, as they account for charge movement due to both drift and diffusion. Thus, whenever a gradient of a quasi-Fermi level appears on an energy band diagram, it is evidence of the existence of a current. This very convenient way of indicating current is further justification for extending the Fermi level concept to the nonequilibrium situation. Conversely, whenever a system is in thermal equilibrium, the two quasi-Fermi levels coalesce into *the* single, flat Fermi level. This important feature is widely used as the starting point for constructing energy band diagrams for semiconductor devices.

4.7 CHAPTER SUMMARY

In this chapter the concept of the Fermi energy level has been introduced, along with some of its applications in computing carrier concentrations and in increasing the utility of energy band diagrams. The reader should now have an appreciation of the following topics:

1. ***Fermi–Dirac function.*** This ***distribution function*** describes the probability of ***quantum states*** in the conduction and valence bands being filled with electrons. The energy level at which the probability of occupancy is exactly one-half is called the ***Fermi energy level.***

2. ***Carrier concentrations and the Fermi level.*** In thermal equilibrium there is a precise relationship between the carrier concentrations and the Fermi level. If the equilibrium carrier concentrations are known, the position of the Fermi level can be computed, and vice versa. For ***nondegenerate*** material the Fermi level is in the bandgap at a position which is greater than $2kT$ away from either the valence band or the conduction band edges. Under these circumstances ***Fermi–Dirac statistics*** reduce to ***Maxwell–Boltzmann statistics***, and the carrier concentrations depend exponentially on the Fermi-level position, with a proportionality constant known as the ***effective density of states.***

3. ***Band diagrams and the Fermi level.*** The inclusion of the Fermi energy on an energy band diagram provides a convenient visual aid for ascertaining at a glance the approximate degree and type of doping in a semiconductor. For intrinsic material the Fermi level is very close to midgap (it would be located exactly in the middle of the bandgap if the ***effective masses*** of electrons and holes were the same). For increasingly n-type material the Fermi level shifts toward the conduction band; for increasingly p-type material the Fermi level shifts toward the valence band.

4. ***Population of dopant states.*** The distribution functions for the population of dopant states are slightly different from the Fermi–Dirac function used to compute the population of band states, but they include the Fermi level in their definition; therefore, if the Fermi level position is known, the amount of dopant ionization can be computed.

5. ***Quasi-Fermi levels.*** These allow the Fermi-level concept to be extended to ***nonequilibrium*** conditions. The position of the ***electron quasi-Fermi level*** enables the total (equilibrium + excess) concentration of electrons to be computed; the ***hole quasi-Fermi level*** performs a similar function for the total hole concentration. The ***gradient of the quasi-Fermi level*** is directly related to the particle current. We show in Section 6.2.2.1 that the ***separation of the quasi-Fermi levels*** is directly related to the applied voltage. Thus the quasi-Fermi levels perform the useful function of linking together external conditions (applied voltage) with internal conditions (carrier concentrations and particle currents).

4.8 REFERENCES

1. J. S. Blakemore, *Semiconductor Statistics*, pp. 346–353, Pergamon Press Ltd., Oxford, 1962.
2. J. P. McKelvey, *Solid-State and Semiconductor Physics*, Sec. 9.10, Harper & Row, Publishers, Inc., New York, 1966.

3. R. F. Pierret, *Advanced Semiconductor Fundamentals*, Chap. 4, Addison-Wesley Publishing Co., Inc., Reading, Mass., 1987.
4. S. M. Sze, *Physics of Semiconductor Devices*, 2nd ed., p. 21, John Wiley & Sons, Inc., New York, 1981.
5. F. J. Morin and J. P. Maita, "Electrical properties of silicon containing arsenic and boron," *Physics Review*, vol. 96, p. 28, 1954.
6. E. H. Putley and W. H. Mitchell, "The electrical conductivity and Hall-effect of silicon," *Proceedings of the Physical Society, London*, vol. 72, p. 193, 1958.
7. W. E. Beadle, J. C. C. Tsai, and J. D. Plummer, eds., *Quick Reference Manual for Silicon Integrated Circuit Technology*, Chap. 2, John Wiley & Sons, Inc., New York, 1985.
8. R. A. Smith, *Semiconductors*, 2nd ed., pp. 91–92, Cambridge University Press, Cambridge, 1978.
9. J. S. Blakemore, see Ref. 1, p. 120.
10. W. E. Beadle et al., see Ref. 7, Chap. 1.

PROBLEMS

4.1. Show that the probability of a state of energy $E = E_F + \Delta E$ being occupied is the same as the probability of a state of energy $E = E_F - \Delta E$ being empty.

4.2. Show that if $E_F - E \gg kT$, then $1 - f(E) \approx \exp[-(E_F - E)/kT]$.

4.3. If a state of energy E (where $E - E_F \gg kT$) has a probability f of being occupied, show that a state of energy $E + kT$ has a probability f/e of being occupied.

4.4. Calculate how far E_F is above E_V for silicon at 300 K doped with 5×10^{14} atoms/cm^3 of boron. Assume that all the acceptor states are occupied. Check the assumption of complete acceptor ionization by calculating the ratio N_A^-/N_A given that the energy of the boron states is 0.045 eV above the top of the valence band.

4.5. Silicon at 300 K is doped with 10^{16} cm^{-3} of donor atoms. Find the position of the Fermi level and the ionized donor concentration for the following two cases: **(a)** $E_C - E_D = 0.044$ eV; **(b)** $E_C - E_D = 0.44$ eV.

4.6. Calculate how far the intrinsic Fermi energy is separated from the midgap energy for Si and for GaAs at liquid nitrogen temperature (77 K).

4.7. A silicon sample is doped with 10^{16} atoms/cm^3 of phosphorus ($E_C - E_D = 0.044$ eV). Compute the concentration of free electrons in the sample at 77 K. Assume that N_C, N_V, and E_g are temperature independent.

4.8. A silicon wafer is doped with 10^{16} atoms/cm^3 of boron. Estimate the temperature at which the intrinsic carrier concentration becomes equal to the ionized boron concentration. The assumptions made in Problem 4.7 apply here also.

4.9. A large-diameter silicon wafer of thickness 200 μm contains an unknown concentration of donor atoms for which $E_C - E_D = 0.33$ eV. A four-point probe measurement yields a voltage of 550 mV for a current of 10 mA. Assuming that no other impurities are present and $T = 300$ K:

(a) Compute the position of the Fermi energy relative to the edge of the conduction band.

(b) Compute the concentration of donor atoms.

4.10. A silicon wafer is uniformly doped with both 10^{17} cm^{-3} of donor atoms and an acceptor concentration of N_A. The acceptor energy level is 0.5 eV above the valence band edge, and the donor energy level is 0.2 eV below the conduction band edge. What must N_A be for the semiconductor to be exactly compensated at 300 K?

4.11. A *p*-type silicon sample is doped with phosphorus to a concentration $N_D = 10^{15}$ cm^{-3} and with an unknown amount of boron. When the sample is cooled down to 77 K, Hall-effect measurements show that the hole concentration is 2×10^{16} cm^{-3}.

(a) What is the concentration of boron in the sample?

(b) What is the resistivity of the sample at room temperature (300 K)?

Note: It is not reasonable to assume that all the impurities are ionized at 77 K.

4.12. Consider a silicon sample doped with phosphorus to a concentration $N_D = 1 \times 10^{16}$ cm^{-3}. Estimate the highest and lowest temperatures at which semiconductor devices made from this material could function. To be more precise, specify the upper temperature limit as that at which $n_i = 0.1N_D$, and the lower temperature limit as that at which only half the dopant is ionized. (These specifications are arbitrary, but they do give a rough idea of the useful operating temperature range.)

Although N_C, N_V, and E_g all vary with temperature, it is reasonable to ignore their temperature dependence here.

5 Semiconductor Quantum Mechanics

5.1 INTRODUCTION

In Chapters 1 through 4, many of the important theoretical concepts needed to understand the behavior of semiconductors, such as the notion of allowed electron energy bands separated by forbidden gaps, and the existence of holes, were developed using intuitive arguments and analogies to chemistry. A complete understanding of semiconductors requires the use of quantum mechanics. This chapter provides a brief introduction to the application of quantum mechanics to semiconductors. The material covered is not straightforward, and might best be skipped by those with no previous background in semiconductor device physics who are reading this book for the first time. However, familiarity with the concepts developed in this chapter is essential for a full understanding of a number of important issues in the operation of semiconductor devices.

5.2 THE WAVELIKE ELECTRON

Every day in hundreds of laboratories around the world an experiment is carried out demonstrating that electrons can exhibit behavior more commonly associated with light waves than with particles. In this experiment, which is done using a transmission electron microscope (TEM), a collimated beam of electrons with energies of a few tens of keV is directed onto a thin solid sample, as suggested in Fig. 5.1. The electrons diffract as they pass through the sample in much the same way as x-rays would, yielding a ***diffraction pattern*** which might resemble

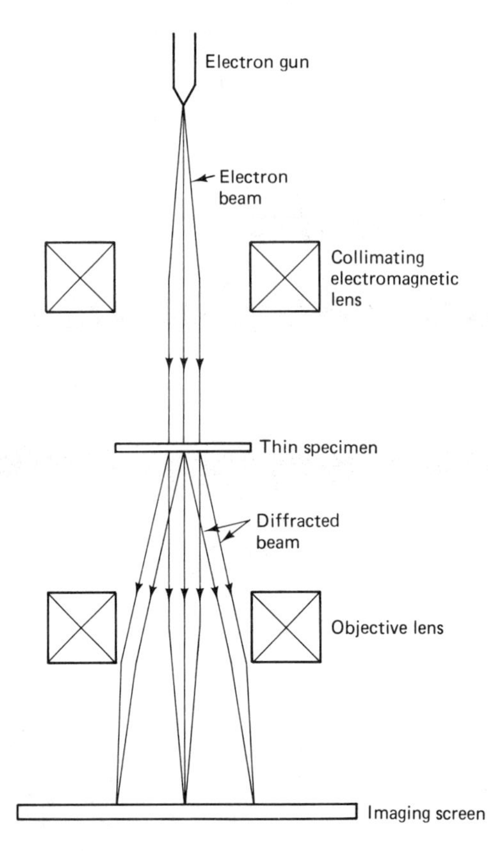

Figure 5.1 Schematic of transmission electron microscope configured to examine diffraction of electron beam by a thin solid sample.

that shown in Fig. 5.2. The spacing between the dots on the pattern varies with the energy of the incident beam, just as for x-rays. The diffraction pattern can be used to extract detailed information regarding the arrangement of atoms within the solid.

In quantum mechanics the wavelike nature of the electron is taken into account by assigning it a ***wavefunction*** $\Psi(\mathbf{r}, t)$, which in general is a complex-valued function. The physical meaning of the wavefunction is as follows: $|\Psi(\mathbf{r}, t)|^2 \triangleq \Psi(\mathbf{r}, t)\Psi^*(\mathbf{r}, t)$ is the probability of finding the electron in a unit volume centered around point $\mathbf{r}$ at time t. (Here * denotes complex conjugation.) The motion of the electron is described by determining the way in which its wavefunction changes with time. It is important to note that quantum mechanics can only predict the ***probability*** of finding an electron at a particular place at a particular time.

Naturally, the probability of finding the electron somewhere in the universe must be unity, so Ψ is subject to the following normalization condition:

$$\iiint_{\text{all space}} \Psi\Psi^* \, dx\, dy\, dz = 1 \tag{5.1}$$

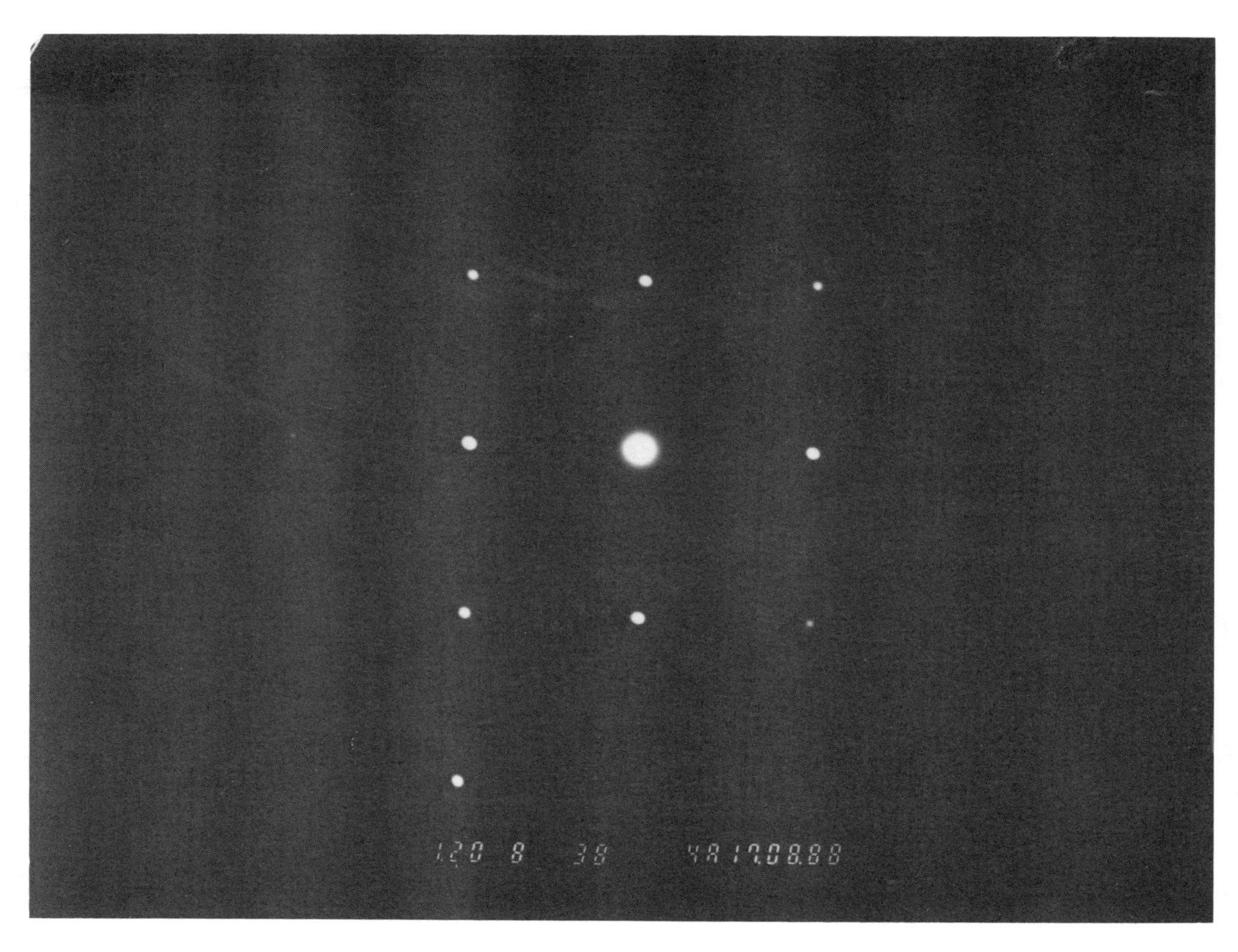

Figure 5.2 Diffraction pattern produced by electron beam passing through a thin (100) silicon wafer. Photograph courtesy of Dr. K. Kaluarachchi, University of British Columbia.

We are usually interested in electrons with precisely defined energies E. In this case, the wavefunction has the form

$$\Psi(\mathbf{r}, t) = \psi(\mathbf{r})e^{-j\omega t} \tag{5.2}$$

where

$$\omega = \frac{E}{\hbar} \tag{5.3}$$

$\psi(\mathbf{r})$ must satisfy the ***Schrödinger equation***

$$\frac{-\hbar^2}{2m_0}\nabla^2\psi + U(\mathbf{r})\psi = E\psi \tag{5.4}$$

Here $\nabla^2 = \partial^2/\partial x^2 + \partial^2/\partial y^2 + \partial^2/\partial z^2$ is the Laplacian operator, $U(\mathbf{r})$ is the potential energy the electron would have if it were located at position $\mathbf{r}$, m_0 is the electron mass and $\hbar = h/2\pi$, where h is ***Planck's constant***.

A crucial point must now be made: for a given potential energy function $U(\mathbf{r})$, only certain wavefunctions ψ and corresponding values of E will satisfy the equality in (5.4) at all points in space. The energies an electron is allowed to have are therefore restricted—whenever an attempt is made to measure the electron's energy, only values for which there are solutions to (5.4) will be found.

It should be stressed that the Schrödinger equation cannot be derived. Although a number of pieces of evidence led Erwin Schrödinger in 1925 to speculate that (5.4) really does describe the behavior of electrons, the Schrödinger equation was a guess nonetheless. That the guess was correct can be inferred from the fact that in the intervening 60 years no one has been able to devise an experiment whose results are at odds with the predictions of (5.4).

5.2.1 The Hydrogen Atom

Much of atomic and solid-state physics consists of finding solutions to (5.4) for specific potential energy distributions $U(\mathbf{r})$. The best known example of this procedure is provided by the hydrogen atom. The potential energy of an electron at a distance r from a stationary proton is given by

$$U(\mathbf{r}) = \frac{-q}{4\pi\epsilon_0 r} \tag{5.5}$$

where ϵ_0 is the permittivity of free space. Substituting this $U(\mathbf{r})$ into (5.4) yields an equation that, although difficult to solve, can be dealt with using analytic methods. It turns out that the valid solutions to (5.4) for this case can be written most easily in the form $\psi_{nlm}(r, \theta, \phi)$, where r, θ, and ϕ are spherical polar coordinates centered on the proton, and the parameters n, l, and m are ***quantum numbers***. n, l, and m must all be integers, with the additional conditions that $n \geq 1$, $0 \leq l \leq n - 1$, and $-l \leq m \leq l$. The allowed values of electron energy (measured relative to the energy of an electron infinitely far from the proton) are given by

$$E = \frac{-13.6 \text{ eV}}{n^2} \tag{5.6}$$

These allowed energies are represented on an ***energy-level diagram*** in Fig. 5.3. A few of the lowest-energy electron wavefunctions are

$$\psi_{100} = \frac{1}{\sqrt{\pi a^3}} e^{-r/a} \tag{5.7a}$$

$$\psi_{200} = \frac{1}{\sqrt{32\pi a^3}} \left(2 - \frac{r}{a}\right) e^{-r/a} \tag{5.7b}$$

and

$$\psi_{210} = \frac{1}{\sqrt{32\pi a^3}} r e^{-r/2a} \cos \theta \tag{5.7c}$$

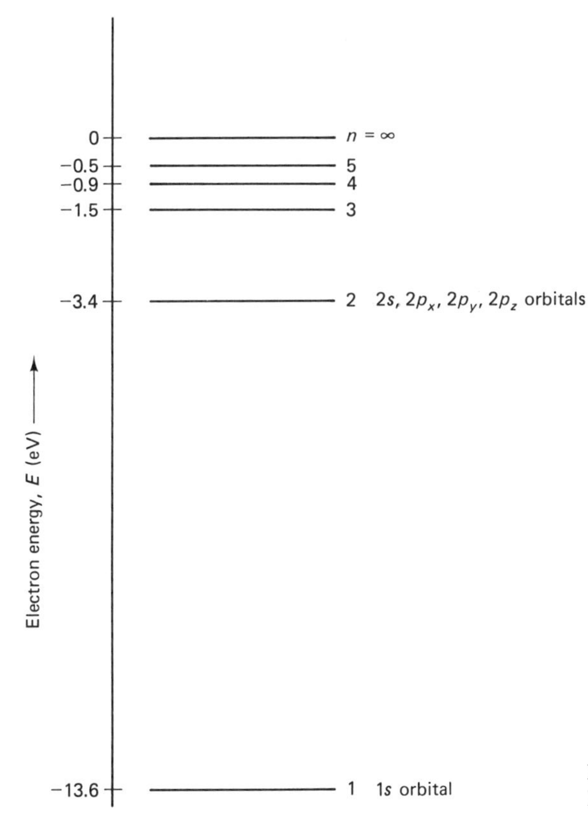

Figure 5.3 Allowed electron energy levels in the hydrogen atom.

where the ***Bohr radius***

$$a = \frac{\hbar^2}{m_0 q^2} = 0.053 \text{ nm} \tag{5.7d}$$

In chemistry the ψ_{100} wavefunction is called the 1*s* orbital, the ψ_{200} wavefunction the 2*s* orbital, and ψ_{210} the $2p_x$ orbital. These three wavefunctions are sketched in Fig. 5.4, with the shading indicating regions where the probability of finding an electron is high.

5.2.2 Free Electrons

For a free electron $U(\mathbf{r}) = 0$ everywhere by definition, so (5.4) becomes

$$\frac{-\hbar^2}{2m_0}\nabla^2\psi = E\psi \tag{5.8}$$

It can be verified by direct substitution that the solution to (5.8) is

$$\psi(\mathbf{r}) = Ce^{j\mathbf{k}\cdot\mathbf{r}} \tag{5.9}$$

where the normalization factor C and the ***wave vector*** $\mathbf{k} = (k_x, k_y, k_z)$ are constants. Inclusion of the time-dependent phase factor introduced in (5.2) gives for the

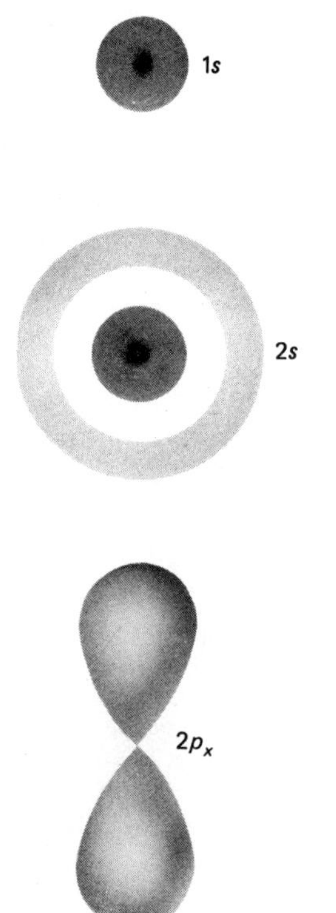

Figure 5.4 Some hydrogen atom wavefunctions. The shading is darkest where the probability of finding the electron is greatest.

complete wavefunction

$$\Psi(\mathbf{r}, t) = Ce^{j(\mathbf{k}\cdot\mathbf{r} - \omega t)} \tag{5.10}$$

where

$$E = \frac{\hbar^2 k^2}{2m_0} \tag{5.11}$$

Equation (5.10) can be recognized immediately as the equation of a traveling plane wave.

5.2.3 Wavepackets

In Section 5.2.2 it was found that the wavefunction describing a free electron with known energy E is a traveling plane wave. However, for the plane wave we have

$$\Psi\Psi^* = C^2 \tag{5.12}$$

irrespective of position $\mathbf{r}$—in other words, the probability of finding the electron is the same everywhere in the universe! This apparently ridiculous result is in fact an illustration of the ***Heisenberg uncertainty principle***, which states that it is never possible to determine the momentum and position of a particle simultaneously. (It is true that in the present problem we are considering energy and position, but for a free electron, momentum and energy are directly related.) If we know the electron's energy with infinite precision, we can say nothing regarding its position, and vice versa.

In real-world situations—for example, in describing the motion of an electron inside a semiconductor sample—we usually have at least a rough idea of where the electron is located, so this theoretical result is rather disconcerting. The gap between theory and reality can be bridged by introducing the concept of an electron ***wavepacket***. Applying Fourier transform techniques, we might try to represent a localized electron as a superposition of plane waves of the form given in (5.10), each with a slightly different wave vector and, therefore, energy. We would then have

$$\Psi(\mathbf{r}, t) = \iiint_{-\infty}^{\infty} C(k_x, k_y, k_z)e^{j(\mathbf{k}\cdot\mathbf{r} - \omega t)}\, dk_x\, dk_y\, dk_z \tag{5.13}$$

If we were to plot this wavefunction at some particular time, we would find that it resembles Fig. 5.5. Unlike a pure plane wave, Ψ now has appreciable values only in a limited region of space, and can be thought of as being enclosed by the envelope suggested in Fig. 5.5. The wider the range of energies used to construct the electron wavepacket in (5.13), the more tightly constrained in space the wavepacket will be.

The wavepacket concept applies to any wavelike motion, not just to the free electron wavefunctions we are considering here. For any wavepacket the center

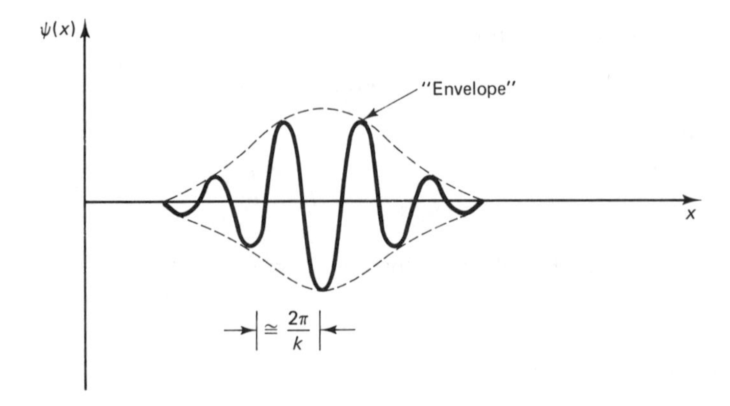

Figure 5.5 Snapshot at one instant in time of a "wavepacket" representing a localized electron.

of the packet—for our purposes, the place where it is most likely the electron will be found—moves with a ***group velocity*** **v** given by [1]

$$\mathbf{v} = \left(\frac{\partial \omega}{\partial k_x}, \frac{\partial \omega}{\partial k_y}, \frac{\partial \omega}{\partial k_z}\right) \triangleq \frac{\partial \omega}{\partial \mathbf{k}} \tag{5.14}$$

For an electron wave we therefore have

$$\mathbf{v} = \frac{1}{\hbar}\frac{\partial E}{\partial \mathbf{k}} \tag{5.15}$$

In many situations in solid-state physics, quantum mechanics is used to predict the motion of an electron assuming it has a precisely defined energy E. This approach often provides mathematical simplifications. The abstraction of a completely nonlocalized electron is then removed by viewing the electron as a wavepacket composed of plane waves with energies very close to E.

5.3 ELECTRONS IN A SOLID: ENERGY BAND STRUCTURE

In a crystalline solid, an electron moves through the potential energy wells associated with the positively charged ion cores that make up the lattice. Figure 5.6 gives a rough indication of the shape of the resulting potential energy function $U(\mathbf{r})$ along a line connecting a set of ion cores. For a perfect crystal, the ions are arranged in a regular pattern in space, so $U(\mathbf{r})$ must also repeat periodically in space. In particular, $U(\mathbf{r})$ must have the same form in every unit cell of the crystal. (The unit cells of the silicon and GaAs lattices are shown in Figs. 2.4 and 2.6.) Although the argument is too detailed to repeat here, it can be shown using Fourier analysis that when $U(\mathbf{r})$ repeats periodically in space in this way, the allowed

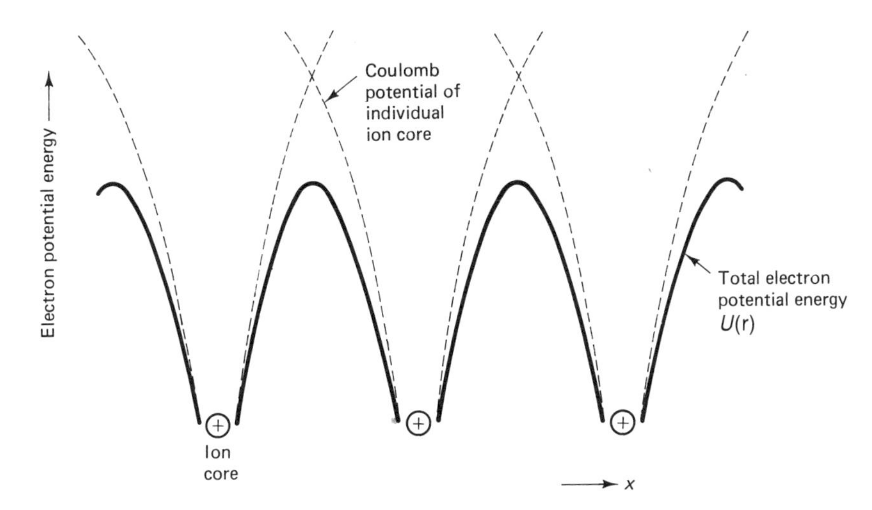

Figure 5.6 Electron potential energy $U(r)$ along a line connecting ion cores.

electron wavefunctions must have the form [2]

$$\Psi(\mathbf{r}, t) = u_{\mathbf{k}}(\mathbf{r})e^{j[\mathbf{k}\cdot\mathbf{r} - (E/\hbar)t]} \tag{5.16}$$

where $u_{\mathbf{k}}(\mathbf{r})$, like $U(\mathbf{r})$, has the same form in each unit cell. [The subscript **k** indicates that $u(\mathbf{r})$ has different functional forms for different values of **k**.] This statement is known as ***Bloch's theorem***, and the wavefunctions described by (5.16) are called ***Bloch waves***. The Bloch waves can be thought of as spatially modulated plane waves. The actual shape of the modulating envelope $u_{\mathbf{k}}(\mathbf{r})$ varies with electron energy and with the type of solid under study. For each Bloch wave solution to the Schrödinger equation, there is a corresponding allowed electron energy E.

The allowed solutions to the Schrödinger equation for a crystalline solid bear some resemblance to those for the hydrogen atom, in that only certain electron energies are allowed. On the other hand, the Bloch waves resemble the plane waves that represent free electrons in that they are not localized to any one region of space.

For electrons in the valence band of a semiconductor, we expect that the probability of finding an electron will be highest along the lines connecting the ion cores making up the crystal. This view ties in well with the picture of the valence electrons forming chemical bonds between atoms in the crystal. However, the wavefunctions of electrons in the conduction band are more like those of free electrons, and are not tightly localized to lines along which chemical bonds form.

A plot of the allowed electron energies E as a function of the wave vector **k** of the associated Bloch wave is known as the ***energy band structure*** of a solid. E versus **k** plots for the allowed states in the conduction and valence bands of silicon and GaAs are shown in Figs. 5.7 and 5.8. It is, of course, not possible to show the E versus **k** structure for any arbitrary direction of the wave vector **k** on

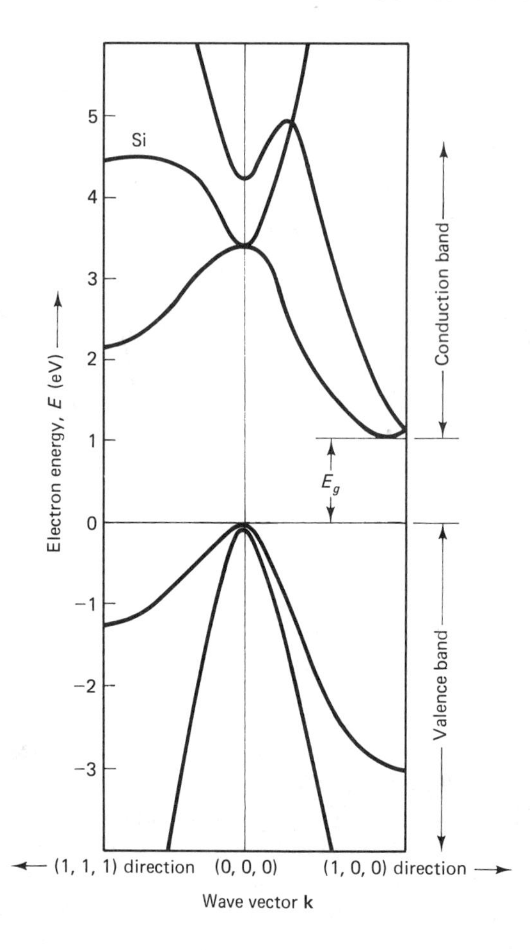

Figure 5.7 Energy band structure for silicon. (From Sze [3]. Reprinted with permission of the publisher, John Wiley & Sons, Inc.)

a single sheet of paper, so only **k** vectors lying in crystallographically important directions such as (1, 0, 0) and (1, 1, 1) are shown in Figs. 5.7 and 5.8.

A number of important features should be noted in Figs. 5.7 and 5.8. First, in agreement with the discussion of Section 2.3, we see that in both types of crystal there are bands of energies in which there are allowed Bloch wave solutions to the Schrödinger equation. These are, of course, the conduction and valence bands of the semiconductor. In between the allowed bands lie forbidden gaps—ranges of energy in which there are no solutions to the Schrödinger equation. For silicon we see that the forbidden gap is approximately 1.1 eV, while for GaAs it is 1.4 eV. It should also be noted that in both silicon and GaAs there are several valence bands, two of which are coincident at the highest energy. This twofold degeneracy gives rise to the extra factor of 2 in the expression for the ionized acceptor concentration in (4.23).

A surprisingly large amount of additional information concerning the properties of silicon and GaAs—or of any other solid—can be predicted from its band

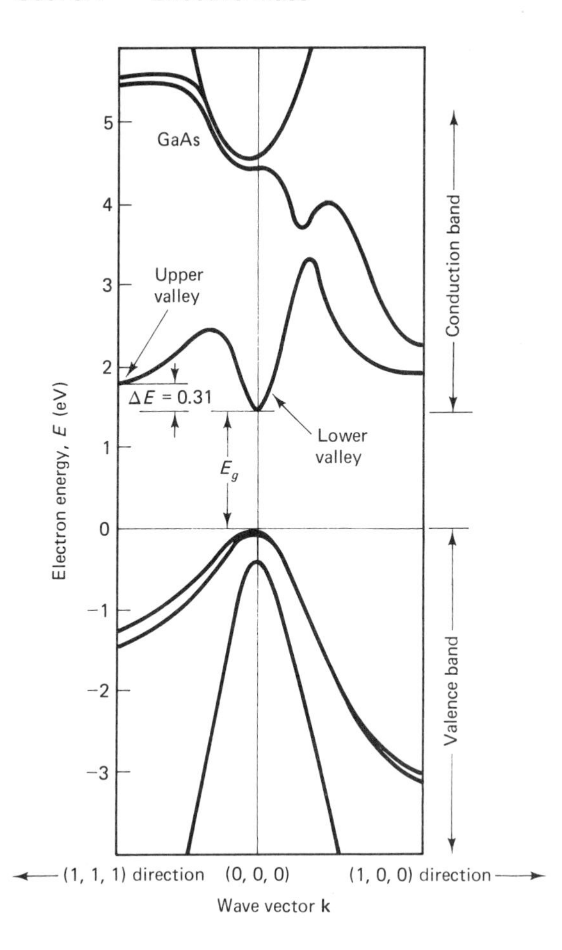

Figure 5.8 Energy band structure for GaAs. (From Sze [3]. Reprinted with permission of the publisher, John Wiley & Sons, Inc.)

structure. How some of this information can be extracted forms the subject matter of Sections 5.4 through 5.6.

5.4 EFFECTIVE MASS

So far in applying quantum mechanics to analyze the motion of electrons in a solid, we have assumed that the only forces acting on the electrons are those associated with their attraction to the positive ion cores in the crystal lattice. In this case we have found that the allowed solutions to the Schrödinger equation are Bloch waves, which we might try and visualize as modulated plane waves moving freely through the crystal.

From a device application viewpoint, we also need to know how the electrons will move if we subject the semiconductor to an externally applied electric field. (This might be done, for example, by forming metal electrodes on two sides of a crystal and applying a bias V between them.) The external field will produce a force $\mathbf{F}_{ext}$ that tends to accelerate an electron, causing it to make transitions to higher energy levels in the band structure. (We will assume that the band we are dealing with is not completely filled, so that there are some empty states into which the electron can transfer.) The rate at which the electron gains energy from the field is given by

$$\frac{dE}{dt} = \mathbf{F}_{ext} \cdot \frac{d\mathbf{r}}{dt} = \mathbf{F}_{ext} \cdot \mathbf{v} \tag{5.17}$$

Substituting for the group velocity of the electron wavepacket from (5.15), we obtain

$$\frac{dE}{dt} = \mathbf{F}_{ext} \cdot \frac{1}{\hbar} \frac{\partial E}{\partial \mathbf{k}} \tag{5.18}$$

However,

$$\frac{dE}{dt} = \frac{\partial E}{\partial \mathbf{k}} \cdot \frac{d\mathbf{k}}{dt} \tag{5.19}$$

so

$$\frac{dE}{dt} = \frac{1}{\hbar} \frac{\partial E}{\partial \mathbf{k}} \cdot \hbar \frac{d\mathbf{k}}{dt} \tag{5.20}$$

which leads us to conclude that

$$\mathbf{F}_{ext} = \hbar \frac{d\mathbf{k}}{dt} \tag{5.21}$$

This is a most interesting result. For a free electron represented by a traveling plane wave, it can be shown that the momentum $\mathbf{p}$ is given by

$$\mathbf{p} = \hbar \mathbf{k} \tag{5.22}$$

In this case we must also have

$$\mathbf{F} = \hbar \frac{d\mathbf{k}}{dt} \tag{5.23}$$

where $\mathbf{F}$ is now the ***total*** force acting on the electron.

In summary, for an electron in a crystal the quantity $\hbar\mathbf{k}$ changes in response to an externally applied force $\mathbf{F}_{ext}$ in the same way the momentum of a free electron would change in response to total force. $\hbar\mathbf{k}$ is therefore referred to as the ***crystal momentum*** of the electron. It is important to note that crystal momentum and total momentum are *not* the same quantity; the total momentum of an electron in a solid might change in response to an externally applied force in a rather complicated way that would include effects from its attraction to the ion cores in the lattice. The interaction between the electron and the ion cores is hidden in the crystal momentum, resulting in a simple equation of motion.

For simplicity, before proceeding further with our analysis of an electron's response to an externally applied force, we will specialize to the case of motion in one dimension. The wavevector **k** then becomes a scalar parameter k. This approach provides considerable mathematical simplification, and the results obtained can easily be generalized to the three-dimensional case.

From (5.15) we have

$$a = \frac{dv}{dt} = \frac{1}{\hbar}\frac{\partial^2 E}{\partial k^2}\frac{dk}{dt} = \frac{1}{\hbar^2}\frac{\partial^2 E}{\partial k^2}\frac{d\hbar k}{dt} \tag{5.24}$$

Therefore,

$$a = \left(\frac{1}{\hbar^2}\frac{\partial^2 E}{\partial k^2}\right) F_{\text{ext}} \tag{5.25}$$

Equation (5.25) has the form $F = m^*a$ if we identify

$$m^* = \frac{1}{(1/\hbar^2)(\partial^2 E/\partial k^2)} \tag{5.26}$$

m^* is known as the ***effective mass*** of the electron and is generally not equal to the free electron mass m_0. The value of m^* is determined by the band structure, and therefore varies from material to material. In general, m^* is least where the curvature $\partial^2 E/\partial k^2$ of the E–k surface is greatest.

Although (5.25) is simple in appearance, it expresses a truly remarkable result. It states that an electron moving in a solid under the combined influence of an externally applied force and the forces associated with the lattice ion cores responds to the external force just as if it were a free particle with mass m^*. In other words, the effects of electron attraction to the ion cores, which might be expected to produce a very complicated trajectory, can be accounted for simply by modifying the apparent mass of the electron.

From Figs. 5.7 and 5.8 it can be seen that the curvature of the E versus **k** surface near the bottom of the conduction band in GaAs is considerably greater than that in silicon. In consequence, electrons in GaAs have a lower effective mass than those in silicon, undergo greater acceleration in response to a given applied electric field, and therefore have a higher mobility. This means that, other factors being equal, a GaAs transistor should be faster than a silicon transistor.

For an electron near the top of the valence band in silicon or GaAs, we see from Fig. 5.7 or 5.8 that $\partial^2 E/\partial k^2$ is negative. This implies that the electron has a ***negative*** effective mass—in other words, if we apply an external force to the electron, it will move in the opposite direction to the force. This result seems bizarre, but actually leads to the model of positively charged holes as the current carriers in a nearly full valence band. First, we know that a full band carries no current, so to compute the net current carried by a nearly full band, we need only determine the contribution to the current an empty state would have made, and then subtract that contribution from zero. A negative effective mass electron will accelerate in the direction of an applied electric field $\mathscr{E}$, just as a positively charged particle would. The electron would normally make a contribution $-q\mathbf{v}$ to the current density carried by the band, so a missing electron gives a current density contribution of $q\mathbf{v}$. The negative mass missing electron therefore adds to the current carried by the band just as would a positively charged, positive-mass particle—a hole.

5.5 DIRECT AND INDIRECT BANDGAPS

As Figs. 5.7 and 5.8 show, there are several conduction and valence bands in silicon and GaAs. However, each material has a unique energy gap E_g given by the separation of the lowest conduction band minimum and highest valence band maximum. When these extrema occur at the same value of wave vector $\mathbf{k}$, as in GaAs, the bandgap is said to be ***direct***. Otherwise, as in silicon, the bandgap is described as ***indirect***. The difference is significant as the nature of the bandgap has important consequences regarding the recombination of carriers and the optoelectronic properties of semiconductors.

To appreciate this, recall from (5.21) that the crystal momentum $\hbar\mathbf{k}$ of an electron in a solid serves much the same role as the true momentum of a free electron. In particular, when an electron in a solid makes a transition from one allowed energy level to another, it must do so in such a way that the crystal momentum is conserved. Ordinarily, an electron changes its level in the band structure by exchanging energy and crystal momentum either with a photon or with the lattice itself. A quantum mechanical analysis of the vibrational motion of the lattice reveals that only a discrete set of motions is allowed, each with a distinct energy and crystal momentum [4]. These possible modes of vibration are termed ***phonons***, and when an electron exchanges energy and momentum with the lattice, it is said to have collided with a phonon.

Whenever an electron–hole pair is generated or recombines, energy and crystal momentum must be conserved. This is usually accomplished by having the electron and hole exchange momentum and energy with one or more photons and/or phonons. In general, the momentum of a photon with suitable energy to engage in a recombination or generation process is negligible compared to that of an electron or phonon. For this reason, ***direct radiative*** recombination–generation processes involving only the emission or absorption of a photon are common only in direct bandgap materials, where an electron can make a transition between the bottom of the conduction band and the top of the valence band with little or no change in its wave vector $\mathbf{k}$.

A direct radiative recombination process is illustrated in Fig. 5.9. The light emitted is nearly monochromatic, with a characteristic wavelength determined by the bandgap energy of the semiconductor. The radiative recombination process

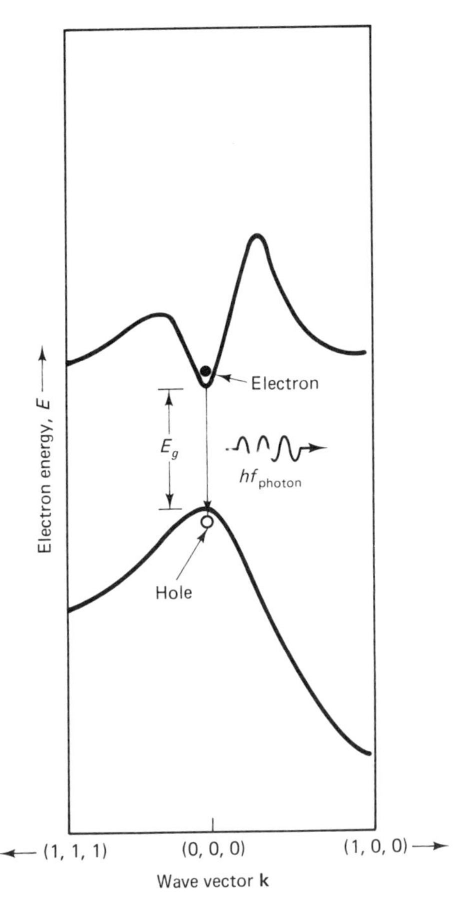

Figure 5.9 Direct radiative recombination process in GaAs.

forms the basis for the operation of light-emitting diodes and semiconductor lasers. When direct radiative generation is possible, it usually leads to very efficient absorption of light with photon energies greater than the forbidden gap. For this reason, direct bandgap materials such as GaAs absorb light much more strongly than do indirect gap materials.

In indirect bandgap materials such as silicon, the direct transfer of an electron from the bottom of the conduction band to a state in the valence band would either require the release of a photon with an energy far greater than the forbidden gap, or the simultaneous transfer of some momentum to a phonon, as suggested in Fig. 5.10. A three-body interaction of this kind is highly unlikely. Instead, recombination and generation are likely to occur via processes involving midgap energy levels, as suggested in Fig. 5.11. These midgap recombination–generation centers are frequently associated with transition metal impurities such as copper or iron, and with crystal defects. They are localized energy levels, meaning that an electron occupying onc is confined to a relatively small area centered on the impurity or defect, and is not free to move through the crystal. In a recombination

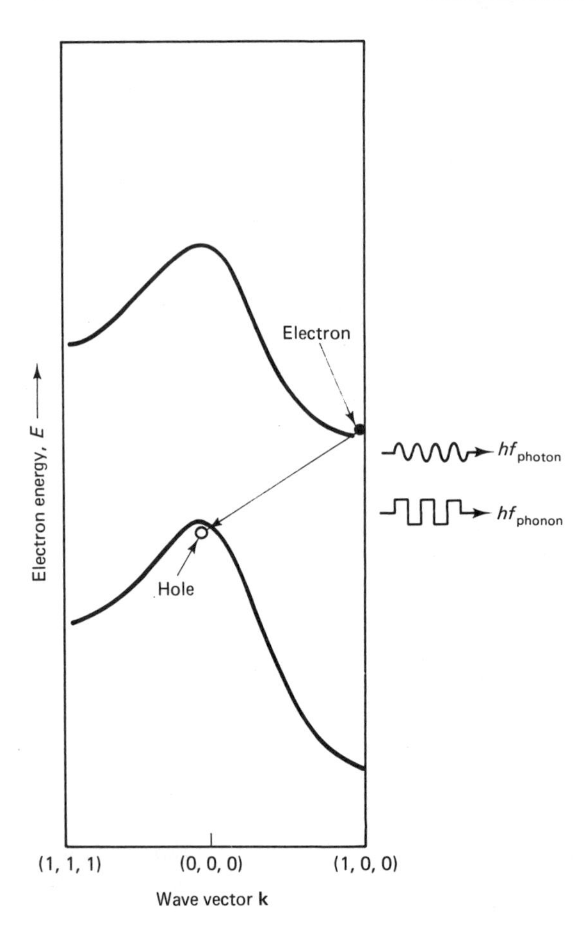

Figure 5.10 Radiative recombination process in silicon involving the simultaneous emission of a photon and a phonon. Processes of this kind are very rare.

process, a conduction band electron is captured by a recombination–generation center, with energy and momentum being released in the form or one or more phonons. Since recombination–generation processes involving midgap energy levels do not normally result in the release of photons, indirect gap materials are not suitable as substrates for light-emitting diodes or lasers.

Because direct, radiative recombination processes are highly unlikely in indirect bandgap materials, the minority carrier lifetime tends to be far higher in indirect than in direct bandgap semiconductors. For example, the minority carrier lifetime in GaAs is rarely more than a microsecond, while lifetimes of as much as a millisecond are not uncommon in carefully prepared silicon samples.

Generation processes in indirect gap materials basically follow the recombination sequence in reverse, with a valence band electron obtaining sufficient energy from a phonon to be promoted to a midgap energy level. Another phonon can then provide the energy needed to release the trapped electron to the conduction band.

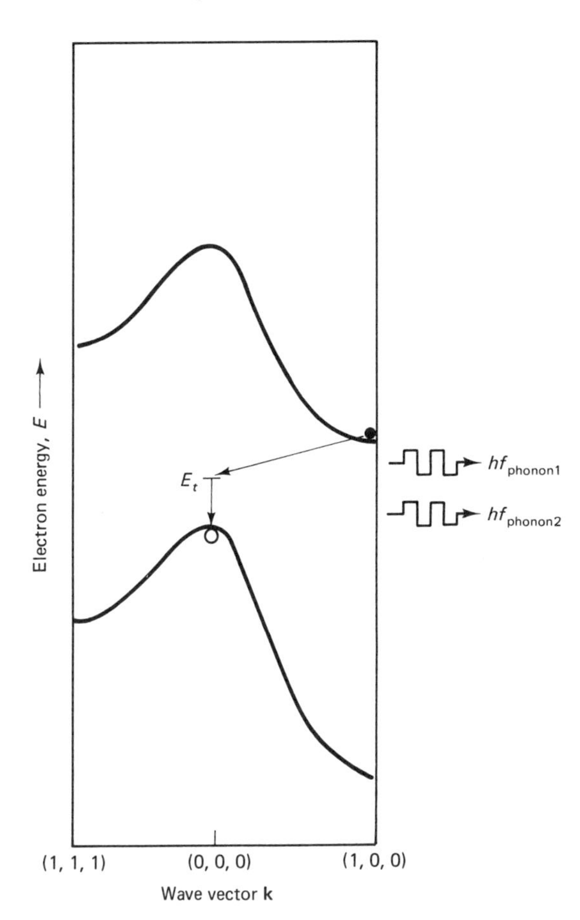

Figure 5.11 Recombination via a midgap energy level E_t in silicon. Energy and crystal momentum are conserved by the release of phonons.

5.6 NEGATIVE DIFFERENTIAL CONDUCTIVITY IN GaAs

As Fig. 5.8 shows, the conduction band in GaAs has two minima or "valleys" which are separated by only a small energy difference (0.31 eV). The value of $\partial^2 E/\partial k^2$ at the lower minimum is considerably greater than that at the higher minimum, indicating that electrons in the upper valley have considerably higher effective mass than those in the lower valley. The related mobilities are, for lightly doped material, about 500 and 8500 $cm^2\ V^{-1}\ s^{-1}$, respectively.

In equilibrium in a GaAs sample, nearly all the conduction band electrons will lie near the bottom of the lower valley. If, however, a sufficiently large electric field is applied to the crystal, some of the electrons may gain sufficient energy between collisions to scatter into the upper valley. Since crystal momentum must also be conserved in a process of this kind, the electron must gain a large amount of momentum through collision with a phonon in order to make the transition.

Once a substantial number of electrons have been transferred to the upper

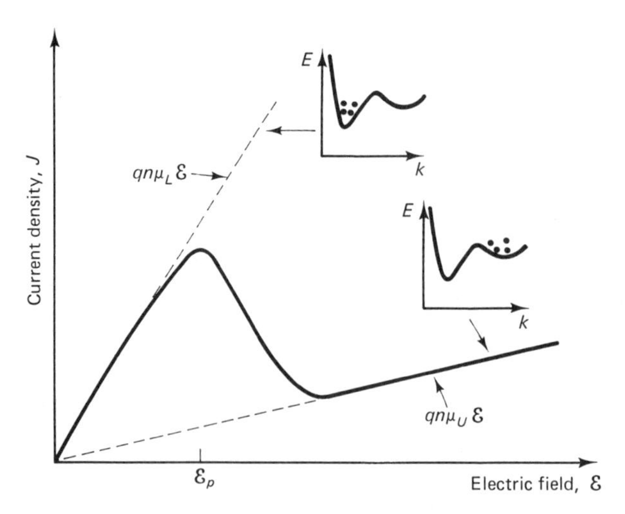

Figure 5.12 Conduction characteristics for a transferred-electron device. At low fields the upper valley in the conduction band (see top inset) is empty, so the mobility (μ_L) is determined by the shape of the E–k plot around the floor of the lower valley. After field-induced, electron transfer to the upper valley, where the mobility (μ_U) is less, the conductivity decreases. The peak $\mathcal{E}_p$ is 3.2 kV cm^{-1} in GaAs.

valley, the conductivity of the crystal will drop. The consequence of this, as illustrated in Fig. 5.12, is a region at intermediate values of $\mathcal{E}$ where the differential conductivity $dJ/d\mathcal{E}$ is negative. The existence of this region of negative differential conductivity is exploited in GaAs oscillators and microwave amplifiers, and is also of importance in the operation of MESFETs.

5.7 DENSITY OF STATES

For a crystal of infinite size, any wave vector $\mathbf{k}$ would have associated with it a valid Bloch wave solution to the Schrödinger equation. Although real crystals are usually extremely large compared to atomic dimensions—the smallest silicon crystal visible with an optical microscope would contain about 10,000 silicon atoms on a side—they are still finite in size, and in consequence only a discrete set of Bloch wave vectors $\mathbf{k}$ is allowed.

To see why this is so, we will consider an idealized cubic crystal of side length L in which the potential energy function $U(\mathbf{r})$ associated with electron attraction to the positive ion cores is very weak—so weak, in fact, that $U(\mathbf{r})$ can effectively be set to zero. Artificial as this condition may seem, for many purposes it actually provides a useful description of the behavior of electrons in real metals [5].

Weak as the crystal potential $U(\mathbf{r})$ may be, a potential energy barrier must exist at the surfaces of the crystal to hold the electrons in. For simplicity, we will assume that the surface potential energy barrier is so large that the electron wavefunction Ψ is zero outside the crystal. The allowed solutions to the Schrödinger equation inside the crystal subject to the boundary condition that $\Psi = 0$ at the surfaces are standing waves of the form

$$\Psi = \left(\frac{2}{L}\right)^{3/2} \sin k_x x \sin k_y y \sin k_z z \, e^{-j\omega t} \tag{5.27}$$

where

$$k_x = \frac{\pi n}{L}, \quad k_y = \frac{\pi m}{L}, \quad \text{and} \quad k_z = \frac{\pi l}{L} \tag{5.28}$$

n, m, and l being positive integers. The electron energy is still given by (5.11).

The standing waves represented by (5.27) can be constructed by superposing the traveling-wave solutions to the Schrödinger equation appropriate for free electrons. For example, the sum of a wave $\Psi = e^{j(k_x x - \omega t)}$ traveling in the positive x-direction and a wave $\Psi = e^{j(-k_x x - \omega t)}$ traveling in the negative x-direction is a standing wave of the form $\Psi = \sin(k_x x)e^{-j\omega t}$.

The most important point to note about (5.28) is that the allowed electron wave vectors are now restricted. It is true that as the crystal becomes larger and L increases the allowed **k** values become more closely spaced, but as long as L is finite only a discrete set of **k** values is allowed.

In a semiconductor crystal the electron potential energy function $U(\mathbf{r})$ resulting from electron attraction to the ion cores is generally *not* negligible. However, it can be shown that the allowed values of **k** in the Bloch wave solutions to the Schrödinger equation are still restricted to the values given in (5.28) [6].

We can now compute the density of states in energy. First, from (5.28) the spacing between the k_x components of allowed states is π/L. Naturally, the same spacing is found between the k_y and k_z components. The volume in (k_x, k_y, k_z) space occupied by each state is therefore π^3/L^3. Next consider those states with wave vector magnitudes between k and $k + \Delta k$ wave vector magnitudes between k and Δk. The number of states in this group must be $2 \cdot L^3/\pi^3 \cdot 4\pi k^2 \Delta k/8$. (The factor of 2 at the front of this expression accounts for the fact that it is possible for a single energy level to accommodate two electrons of opposite spin.)

Using (5.11), the energy range covered between wave vector magnitudes k and $k + \Delta k$ is given by

$$\Delta E = \frac{\hbar E^{1/2} 2^{1/2}}{m_0^{1/2}} \Delta k \tag{5.29}$$

The density of states per unit volume of crystal per unit increment in energy is then given by

$$g(E) = \frac{m_0 (2m_0)^{1/2} E^{1/2}}{\pi^2 \hbar^3} \tag{5.30}$$

For a semiconductor crystal, it is not reasonable to approximate $U(\mathbf{r})$ as being zero in determining the allowed electron wavefunctions. However, in many cases for states near the bottom of the conduction band it is possible to approximate the E versus $\mathbf{k}$ surface as being parabolic; that is,

$$E = \frac{\hbar^2 k^2}{2m^*} \tag{5.31}$$

where m^* is the effective mass [6]. In this case (5.30) still applies, but the free electron mass m_0 must be replaced by m^*.

5.8 TUNNELING

The tunneling of electrons through a potential energy barrier is a purely quantum mechanical phenomenon that has no classical analog. Tunneling can perhaps best be understood by reference to an example. Figure 5.13 shows a situation in which an electron is approaching two metal electrodes held at a high negative potential. Figure 5.14 shows the electron's potential energy as a function of position along the path it is attempting to follow between the electrodes. Since we have chosen the electron's total energy to be less than the height of the barrier it faces, classical physics predicts that it will be reflected and has no chance of passing through the region between the electrodes.

Quantum mechanics views the situation in a very different light, represented in Fig. 5.15. For simplicity, we will treat the problem as one-dimensional and approximate the barrier as being square. The electron approaching the barrier is thought of as a traveling wave whose amplitude at a particular location relates to the probability of finding the electron there. Within the barrier, the Schrödinger equation (5.4) becomes

$$\frac{-\hbar^2}{2m_0}\frac{d^2\psi}{dx^2} = (E - U_0)\psi \tag{5.32}$$

Since we are assuming that $E - U_0$ is negative, the general solution to (5.32) is an exponential,

$$\psi(x) = Ce^{kx} \tag{5.33}$$

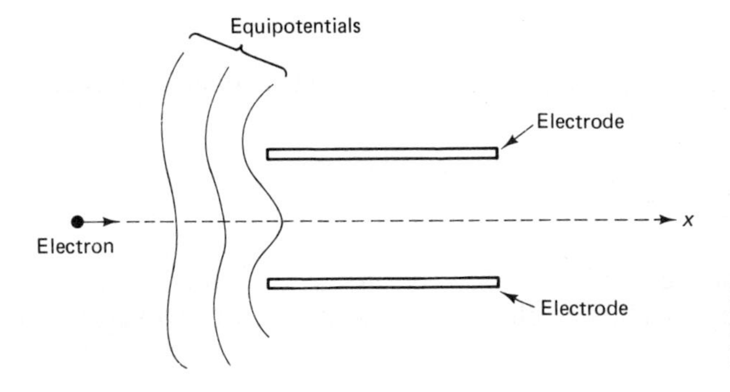

Figure 5.13 Electron attempting to pass between two electrodes held at a large negative potential.

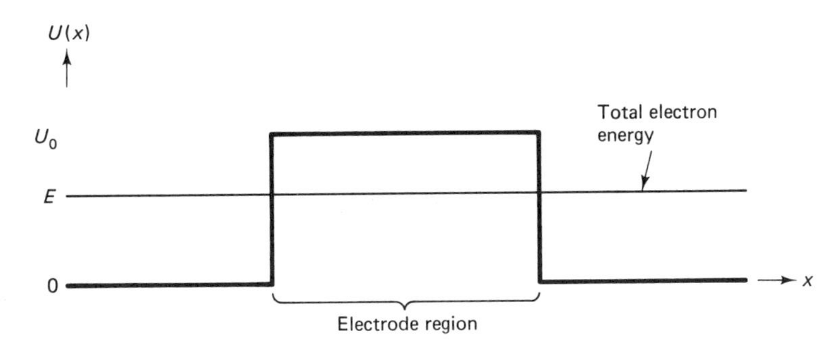

Figure 5.14 Plot of electron potential energy $U(x)$ as a function of position between the electrodes.

where C, as usual, is a normalization factor. Only a negative value of k is physically meaningful, since it seems clear that the probability of finding the electron must drop upon moving into the barrier. On the right-hand side of the barrier, the electron is once again viewed as a traveling wave. The amplitude of this transmitted wave is less than that of the incident wave, indicating that the probability of the electron passing through the barrier is not large but is still nonzero. The main point is that quantum mechanics allows for a nonzero probability of an electron penetrating a potential energy barrier that classical physics predicts it can never cross. This is the essence of the phenomenon of tunneling. In practice, the exponential decay of the electron wavefunction within the barrier usually implies that the probability of tunneling will be negligible unless the thickness of the barrier is comparable to atomic dimensions.

Quantum mechanical tunneling is encountered in a number of situations involving semiconductor devices. One of the simplest of these is illustrated in Fig. 5.16, which represents two semiconductor crystals separated by a very thin insulator. We have said that within an insulator or semiconductor there are no allowed states in the forbidden gap. This is not strictly true; there are Bloch wave

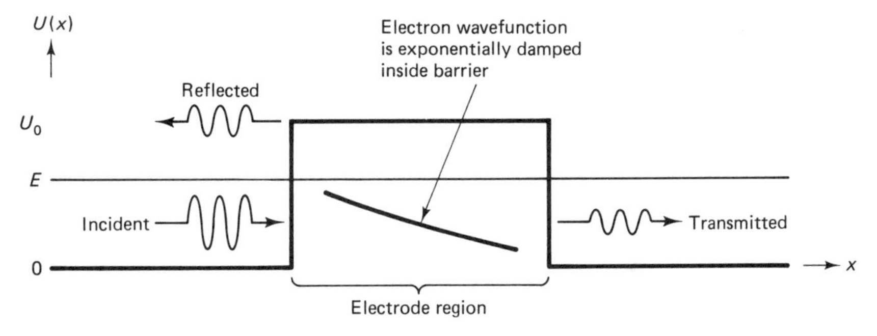

Figure 5.15 How quantum mechanics views the tunneling of an electron through the potential energy barrier between the electrodes in Fig. 5.14. The incident electron is represented by a wavepacket traveling to the right.

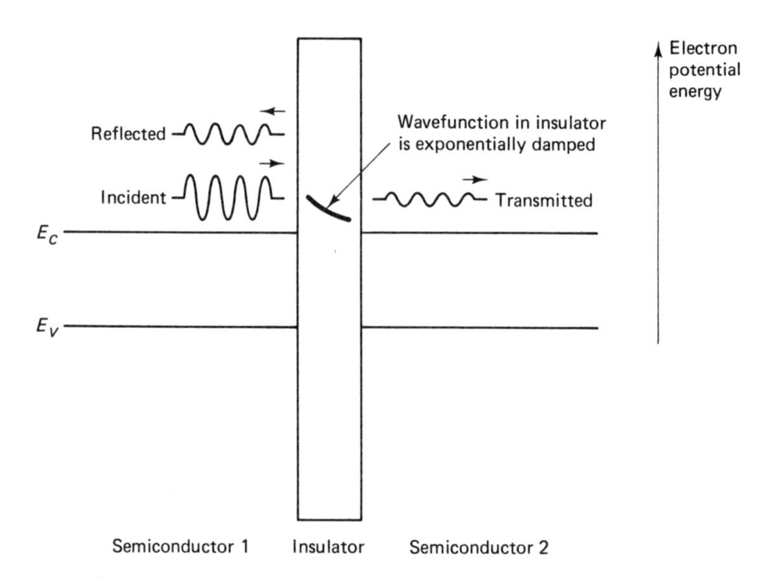

Figure 5.16 Quantum mechanical view of a conduction band electron tunneling through a thin interfacial insulator separating two semiconductors.

solutions to the Schrödinger equation corresponding to energies in the gap, of the form given in (5.16), but these have imaginary wave vectors **k**. In consequence, the actual wavefunctions have an exponential dependence on position. Ordinarily, solutions to the Schrödinger equation of this kind would not be physically acceptable, since they would require ψ to grow without bound in one direction. However, in a confined region of space such as the interfacial insulator, an exponential dependence of the wavefunction on position can be accommodated. Figure 5.16 illustrates how quantum mechanics views an attempt by a conduction band electron in the left-hand semiconductor to penetrate the insulator and enter the right-hand electrode. The wavefunction of the electron approaching from the left is a traveling Bloch wave. Part of this wavefunction is reflected and returns to the left, while part enters the insulator and continues to propagate to the right. Although the wavefunction in the insulator is exponentially damped, it still gives rise to a traveling wave propagating into the right-hand electrode. Since the probability of finding an electron is proportional to the square of the amplitude of its wavefunction, the probability of the electron tunneling through the insulator is given by the square of the ratio of the transmitted wave's amplitude to the incident wave's amplitude.

5.9 CHAPTER SUMMARY

In this chapter some of the properties of semiconductors which cannot be satisfactorily explained using classical physics have been examined from the viewpoint of the quantum theory of solids. The chapter began with a brief outline of the view of the universe taken by quantum mechanics, with emphasis on the wavelike properties of the electron and on the use of the ***Schrödinger equation*** to describe an electron's motion. Solutions to the Schrödinger equation for the hydrogen atom and for free electrons were examined as a preliminary to consideration of the case of electrons in a crystalline solid. The allowed electron energy states in a solid were found to take the form of traveling ***Bloch waves***, and it was further found that only certain bands of electron energies give rise to acceptable solutions to the Schrödinger equation. The bands of allowed energies are separated by ***forbidden gaps***.

It was found that many of the properties of a semiconductor can be predicted from knowledge of its ***band structure***—specifically, the relationship between the electron energy E and the Bloch wave vector $\mathbf{k}$. For example, the curvature $\partial^2 E/\partial k^2$ of the E versus $\mathbf{k}$ surface was found to determine the ***effective mass*** of electrons in the semiconductor, and thereby the electron mobility. It was noted that electrons in GaAs have higher mobility than those in silicon, since the curvature of the E versus $\mathbf{k}$ surface at the bottom of the GaAs conduction band is significantly greater than that in silicon.

Electrons near the top of a valence band were found to have ***negative effective mass***. This strange property was seen to tie in directly with the ability to describe a missing electron in a state near the top of the valence band as a positively charged particle of positive mass known as a ***hole***.

The band structure of a semiconductor was found to be closely linked to its optoelectronic properties and to its minority carrier lifetime. Only in ***direct bandgap*** materials is there a significant probability of direct, radiative recombination of electrons and holes, so only these materials are suitable for the formation of solid-state light-emitting diodes and lasers. Conversely, direct bandgap materials absorb light with photon energies above the bandgap much more strongly than do ***indirect gap*** materials. The possibility of direct radiative recombination generally makes the minority carrier lifetime in a direct bandgap material far shorter than in an indirect gap material.

The ***density of states*** in energy function $g(E)$ was derived by examining the allowed wavefunctions for electrons constrained to move in a finite volume of solid.

Finally, the quantum mechanical phenomenon of ***tunneling***, in which an electron passes through a thin energy barrier that classical physics predicts it should not be able to penetrate, was considered.

5.10 REFERENCES

1. J. D. Jackson, *Classical Electrodynamics*, pp. 208–211, John Wiley & Sons, Inc., New York, 1962.
2. N. W. Ashcroft and N. D. Mermin, *Solid State Physics*, Chap. 8, Holt, Rinehart and Winston, New York, 1976.

3. S. M. Sze, *Physics of Semiconductor Devices*, 2nd ed., p. 13, John Wiley & Sons, Inc., New York, 1981.
4. N. W. Ashcroft and N. D. Mermin, see Ref. 2, Chap. 23.
5. N. W. Ashcroft and N. D. Mermin, see Ref. 2, Chap. 2.
6. N. W. Ashcroft and N. D. Mermin, see Ref. 2, Chap. 28.

PROBLEMS

5.1. The goal of this problem is to determine the probability of an electron tunneling through a square potential energy barrier of the kind shown in Fig. 5P.1. An electron approaching the barrier from the left is represented by a traveling plane wave of the form

$$\psi_{\text{inc}} = Ae^{j(k_1x - \omega t)} \tag{5P.1}$$

where

$$k_1 = \frac{\sqrt{2m_0E}}{\hbar} \tag{5P.2}$$

The incident wave gives rise to a reflected wave of the form

$$\psi_{\text{ref}} = Be^{-j(k_1x + \omega t)} \tag{5P.3}$$

returning to the left. If the electron energy E is less than the barrier height V_0, then within the barrier

$$\psi = Ce^{-k_2x}e^{-j\omega t} + De^{k_2x}e^{-j\omega t} \tag{5P.4}$$

where

$$k_2 = \frac{\sqrt{2m_0(V_0 - E)}}{\hbar} \tag{5P.5}$$

Finally, a transmitted wave

$$\psi_{\text{tran}} = Fe^{j(k_1x - \omega t)} \tag{5P.6}$$

emerges on the right-hand side of the barrier.

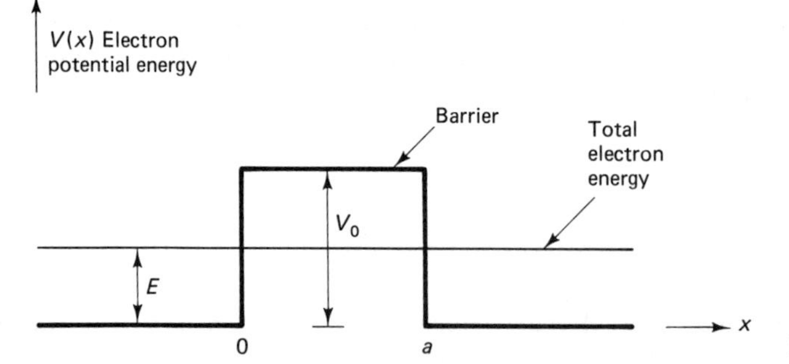

Figure 5P.1

Quantum mechanics requires that the wavefunction ψ and its derivative $d\psi/dx$ be continuous everywhere. Applying these conditions at the interfaces $x = 0$ and $x = a$ generates a set of four equations that can be used to determine the amplitudes of the transmitted and reflected wavefunction components exactly. However, for most problems of practical interest the electron wavefunction decays very rapidly on moving into the barrier, so that the coefficients D and F must be negligible compared to A, B, and C. In this case at interface $x = 0$ we have

$$A + B = C \quad \text{and} \quad jk_1A - jk_1B = -k_2C \tag{5P.7}$$

Similarly, at $x = a$,

$$Ce^{-k_2a} + De^{k_2a} = Fe^{jk_1a} \quad \text{and} \quad k_2Ce^{-k_2a} - k_2De^{k_2a} = -jk_1Fe^{jk_1a} \tag{5P.8}$$

(a) Using (5P.7) and (5P.8), show that

$$\frac{F}{A} = \frac{4}{2 + j(k_2/k_1 - k_1/k_2)} e^{-k_2a}e^{-jk_1a} \tag{5P.9}$$

(b) The probability density of finding an electron is proportional to $|F/A|^2$. Using this result, show that the probability P of an electron tunneling through the barrier is given by

$$P = \frac{16}{4 - (k_2/k_1 - k_1/k_2)^2} e^{-2k_2a} \tag{5P.10}$$

6
PN-Junction Diodes

6.1 STRUCTURE

To form a *pn*-junction diode in an *n*-type silicon wafer, for example, all that is required is to create a *p*-type region and make contact to the front and back of the wafer. The doping can be accomplished by ion implantation or diffusion, as described in Sections 12.2.2 and 12.2.4. Using photoresist–silicon dioxide masking procedures, as described in Section 12.2.9, many diodes can be defined on a single wafer. Dicing of the wafer then yields a multitude of discrete devices (see Fig. 6.1).

A silicon *pn*-junction diode for use in integrated circuits is fabricated a little differently, principally because of the need to make contact to both regions of the diode from the same side of the wafer, and to isolate the diode from other devices on the chip. A possible fabrication sequence is shown in Fig. 6.2. The "island" of *n*-type epitaxial material, within which the diode is to be fabricated, is grown as described in Section 12.2.12 and then defined by *p*-type isolation diffusions, as shown in Fig.6.2b. The *p* region for the actual diode is then diffused or implanted (see Fig. 6.2c). To facilitate making ohmic contact to the *n* region, a heavily doped layer, designated as n^+, is formed, resulting, after metallization with aluminum, in the structure of Fig. 6.2d.

Compared to silicon, gallium arsenide bipolar technology is immature and the starting wafers are expensive. For these reasons GaAs diodes are not found in the same wide variety of applications as their silicon counterparts. The main applications of GaAs *pn*-junction diodes are in devices that exploit material properties not possessed by silicon, most notably that of photon emission (see Fig.

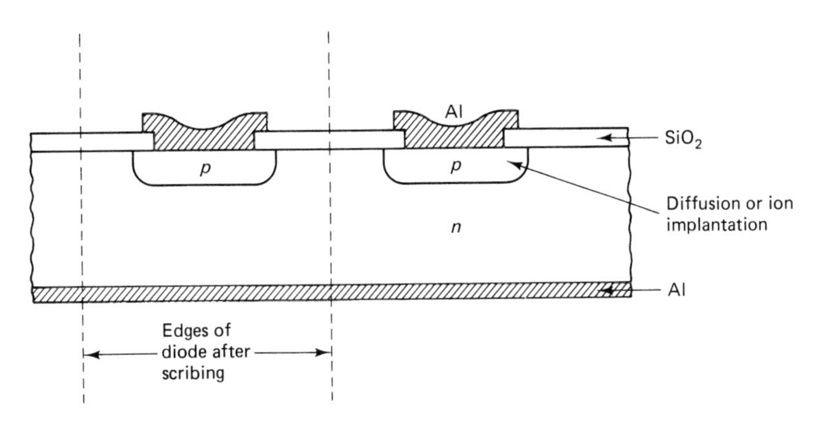

Figure 6.1 Structure of discrete *pn* diodes.

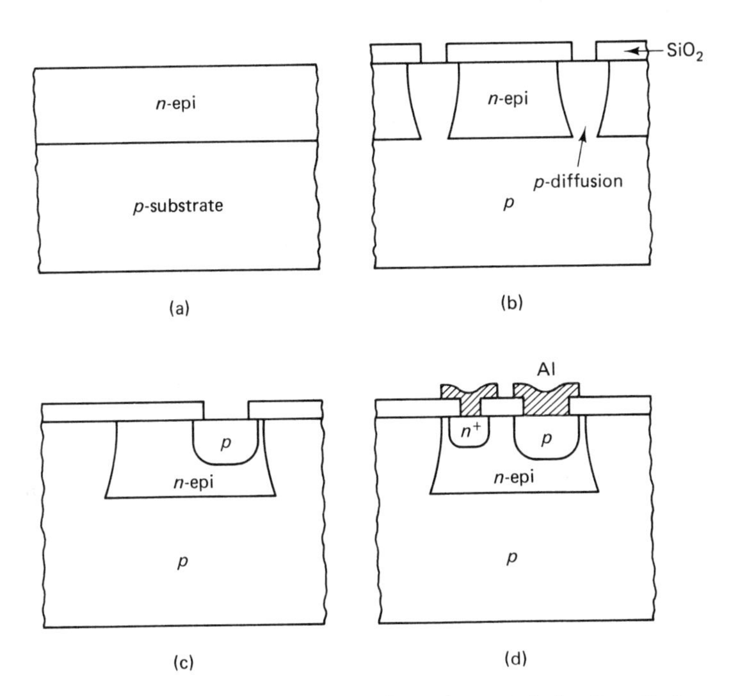

Figure 6.2 Fabrication sequence for an integrated circuit *pn* diode.

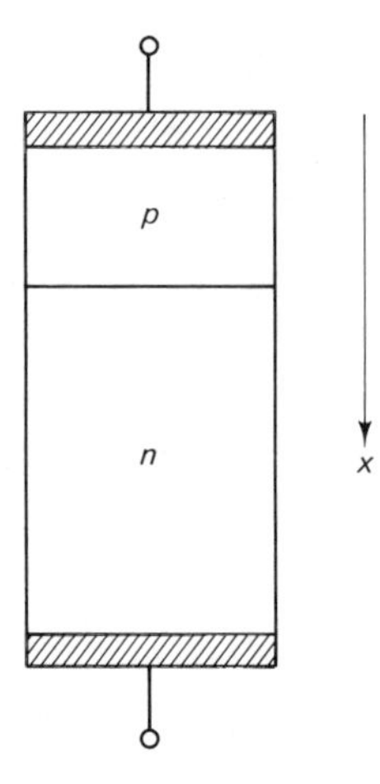

Figure 6.3 Idealized one-dimensional representation of a *pn*-junction diode.

2.23 and Section 5.5). Light-emitting diodes (LEDs) and junction lasers are examples of such devices.

6.2 PRINCIPLES OF OPERATION

The electric fields and charge flows in a *pn* junction are essentially one-dimensional in nature. It is thus permissible, for descriptive and analytical purposes, to represent the "real" diodes of Figs. 6.1 and 6.2 by the simple structure of Fig. 6.3. The circuit symbol for a diode is shown in Fig. 6.4. As a further simplification we consider the *pn* junction to be ***abrupt***; that is, the transition between *p*-type doping and *n*-type doping takes place over a distance that is negligible compared to any of the other distances, such as the depletion region width or the minority carrier diffusion length, which are important in characterizing the diode. To appreciate the scale involved, we note that the smallest values for these parameters which are encountered in the examples in this chapter are 0.017 μm and 0.6 μm, respectively. Because the *pn* diode is the first device to be described in this book, and because it forms an integral part of the major transistors described in later chapters, we treat the physics of its operation in considerable detail.

6.2.1 Thermal Equilibrium Conditions

Consider first the *p*-type region close to the metallurgical boundary with the *n*-type material. Here the concentration of holes (majority carriers) is much greater

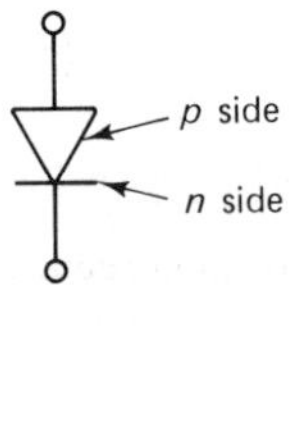

p side
n side

Figure 6.4 Circuit symbol for a diode.

than the concentration of holes (minority carriers) in the n region just to the right of the junction in Fig. 6.5. This concentration gradient provides the driving force for the diffusion of holes to the right. This constitutes a hole current. However, if the situation is one of thermal equilibrium, we know that there can be no net current of holes. It follows, therefore, that in thermal equilibrium, there is also a flow of holes to the left which produces a current which exactly cancels that due to the diffusion of holes to the right. Clearly, the direction of the concentration gradient of holes is such that this current cannot be due to diffusion. It is, in fact, a drift current. The electric field to drive this current arises from the acceptor ions, which give the region bereft of holes to the left of the junction a negative

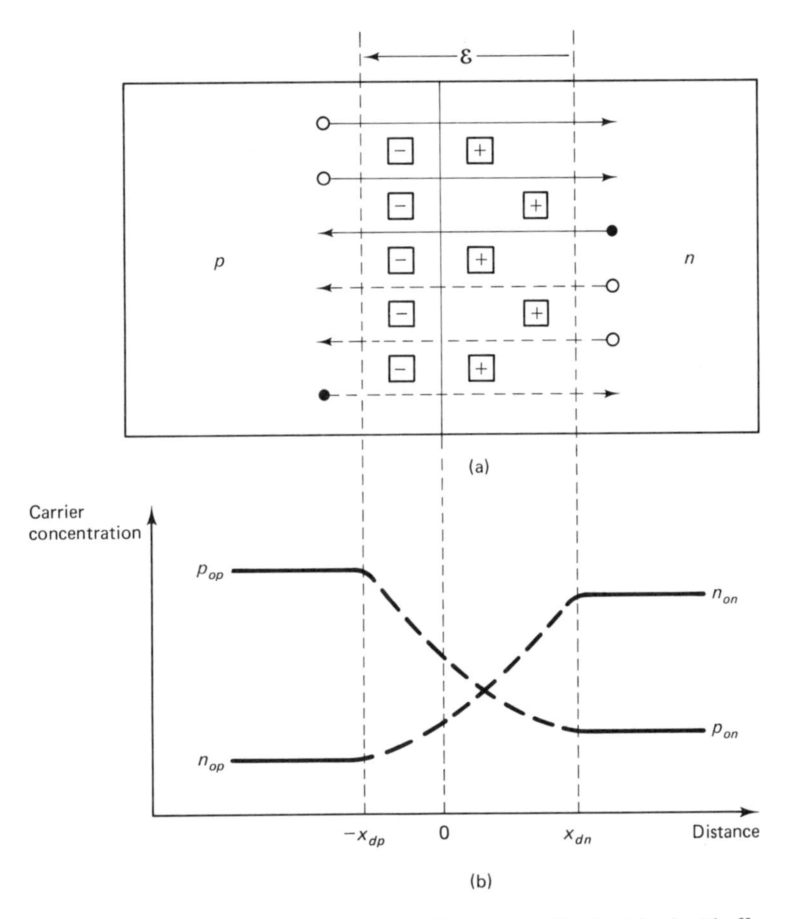

Figure 6.5 The pn junction at thermal equilibrium. (a) The field in the "buffer zone" (the depletion region) causes a flow of minority carriers (dashed lines) which constitutes a drift current. This cancels exactly the current due to the diffusive flow of majority carriers (solid lines). (b) Carrier concentration profiles in thermal equilibrium. At all points $p_0 n_0 = n_i^2$, but in the depletion region p is less than p_{0p} (the equilibrium hole concentration in the p material) and n is less than n_{0n}.

space charge, and from the donor ions which create a corresponding positive space charge in the region to the right of the junction that is depleted of electrons. These regions of ***ionic space charge*** are known collectively as the ***depletion region***.

In summary, the loss of mobile carriers from the depletion region is a result of the initial diffusion of electrons and holes which occurs on formation of the *pn* junction. As the electric field builds up, due to the presence of uncompensated donor and acceptor ions, only the more energetic majority carriers can diffuse against the action of the field. An exact balance between the diffusion current and the drift current for each type of carrier is reached at thermal equilibrium. In other words, the depletion region is a "buffer zone" across which, at thermal equilibrium, there is no net flow of either electrons or holes. The situation is depicted in Fig. 6.5. The larger the gradient in carrier concentrations, the larger the electric field must be at equilibrium. This can be seen from evaluation of the transport equations (3.22) and (3.23) under conditions of no net hole or electron current, that is,

$$\mathscr{E} = \frac{kT}{qp}\frac{dp}{dx}$$
$$\mathscr{E} = -\frac{kT}{qn}\frac{dn}{dx} \tag{6.1}$$

where use has been made of Einstein's relation (3.20).

6.2.1.1 Depletion region. The depletion region is such an important part of a *pn* junction that we now devote considerable space to formulating equations to compute its important parameters, that is, its width, the electric field within it, and the potential difference across it. We also show how to represent the depletion region on an energy band diagram.

The electric field due to the space charge in the depletion region can be computed by integrating Poisson's equation (3.29), that is,

$$\frac{-dV}{dx} = \mathscr{E} = \frac{q}{\epsilon}\int [p(x) + N_D^+(x) - n(x) - N_A^-(x)]\, dx \tag{6.2}$$

Equation (6.2) is exact. Some progress can be made toward solving it analytically by making some approximations based on the relative magnitudes of the carrier concentrations and the ionized impurity concentrations in the space charge region. Considering first the region to the left of the metallurgical boundary, it can be inferred from Fig. 6.5b that $n(x)$ and $p(x)$ are everywhere much less than N_A^-. In other words, in this region the mobile charge carriers are depleted below their thermal equilibrium values. Carrying this notion to the limit that $n(x)$ and $p(x)$ can be ignored in comparison to N_A^- is known as invoking the ***depletion approximation***. As there are no donors in this depletion region, (6.2) simplifies to

$$\mathscr{E}(x) = \frac{q}{\epsilon}\int -N_A^-\, dx \qquad -x_{dp} < x < 0 \tag{6.3}$$

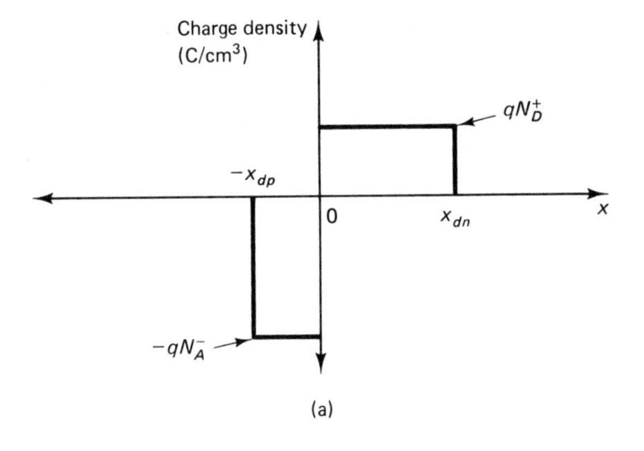

(a)

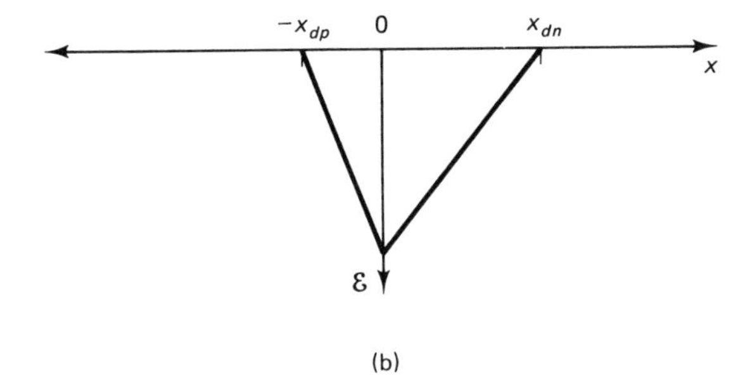

(b)

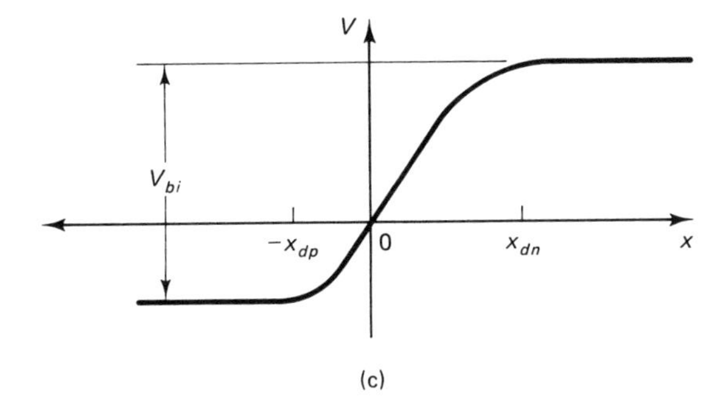

(c)

Figure 6.6 Variation of the space charge density, the electric field, and the potential in the depletion region. The junction is assumed to be abrupt, as are the edges of the depletion regions. Also, there is assumed to be no contribution to the depletion region space charge from mobile carriers.

Performing the integration, and making use of the fact that there is no electric field in the neutral region outside the depletion region, yields

$$\mathscr{E}(x) = \frac{-q}{\epsilon} N_A^-(x + x_{dp}) \qquad -x_{dp} < x < 0 \tag{6.4}$$

A similar analysis of the depletion region to the right of the metallurgical boundary gives

$$\mathscr{E}(x) = \frac{q}{\epsilon} N_D^+(x - x_{dn}) \qquad 0 < x < x_{dn} \tag{6.5}$$

These equations reveal that the electric field in the depletion region has a triangular distribution (see Fig. 6.6b). However, further information is needed before $\mathscr{E}$ can be evaluated because the magnitudes of x_{dp} and x_{dn}, the extremities of the depletion region, are not known. By comparing (6.4) and (6.5) when evaluated at the common point of $x = 0$, the relative magnitudes of x_{dp} and x_{dn} can be found, that is,

$$x_{dp}N_A^- = x_{dn}N_D^+ \tag{6.6}$$

This result could have been anticipated from the requirement that the magnitude of the space charge must be the same on either side of the junction; that is, the field lines originating at the sources of positive charge (ionized donors in the depletion approximation) must terminate on an equal number of negative charges (ionized acceptors). Equation (6.6) does not give the absolute magnitude of either x_{dp} or x_{dn}, so it does not provide the information we are lacking in order to be able to estimate the electric field intensity. What is needed to circumvent this apparent dead end to our calculation is a description of fields and carrier concentrations which is arrived at by an approach that is different from, yet complementary to, the foregoing approach. The required approach is via the Fermi-level concept which was discussed at length in Chapter 4.

From (4.16), for example, expressions for the hole concentrations at both edges of the depletion region can be written down immediately, namely,

$$\begin{aligned} p_0(-x_{dp}) &= n_i \exp\left(\frac{E_{\mathrm{F}ip} - E_F}{kT}\right) \\ p_0(x_{dn}) &= n_i \exp\left(\frac{E_{\mathrm{F}in} - E_F}{kT}\right) \end{aligned} \tag{6.7}$$

where the subscripts p and n in $E_{\mathrm{F}ip}$ and $E_{\mathrm{F}in}$ refer to the p side and the n side, respectively, of the *pn* junction. As the situation is one of thermal equilibrium and no net currents are present, the Fermi level E_F for the system must be constant (see Section 4.6). The constancy of the Fermi level is the starting point for constructing the energy band diagram of the junction (see Fig. 6.7). Ignoring for the moment the form of the energy bands in the depletion region, it is still possible to infer from the figure the ratio of hole concentrations at the edges of the depletion region, that is, via (6.7):

$$\frac{p_0(-x_{dp})}{p_0(x_{dn})} = \exp\left(\frac{E_{\mathrm{F}ip} - E_{\mathrm{F}in}}{kT}\right) \tag{6.8}$$

The intrinsic Fermi level is just a reference energy level, so the difference in E_{Fi} across the junction represents a difference in potential energy of the mobile carriers. As potential energy is the product of charge and potential, this difference is also a difference in potential. Explicitly, building on (4.26) we have

$$\mathscr{E} = \frac{1}{q}\frac{dE_{\mathrm{Fi}}}{dx} = \frac{-dV}{dx} \tag{6.9}$$

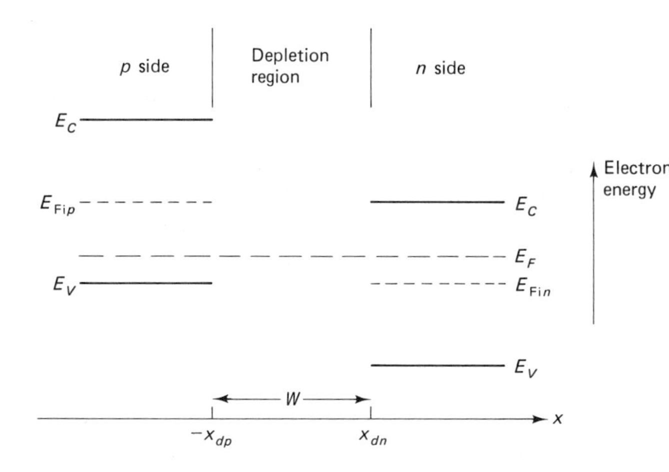

Figure 6.7 Incomplete energy band diagram for the *pn* junction at thermal equilibrium, showing only the energy levels in the neutral regions. The important feature is the constancy of the Fermi level, which is a consequence of the absence of a net current in thermal equilibrium (see Section 4.6).

Using this relation, (6.8) can be rewritten in terms of the potential difference across the junction, that is,

$$\frac{p_0(-x_{dp})}{p_0(x_{dn})} = \exp\left(\frac{qV_{bi}}{kT}\right) \tag{6.10}$$

The entity V_{bi} is the potential difference that is built in across the junction (see Fig. 6.6c) and arises because of the differences in doping density. Its magnitude is such that the associated electric field produces drift currents which are of exactly the same value as the diffusion currents. An expression for the ***built-in potential*** can be derived by taking logarithms in (6.10), that is,

$$V_{bi} = \frac{kT}{q} \ln \frac{p_0(-x_{dp})}{p_0(x_{dn})} \tag{6.11}$$

Further, by using (2.14) and ignoring the contribution of the minority carriers to the charge density, a very convenient expression for the built-in potential emerges, namely,

$$V_{bi} = \frac{kT}{q} \ln \frac{N_A^- N_D^+}{n_i^2} \tag{6.12}$$

Thus the Fermi-level approach yields a relationship between potential and either the carrier concentrations (6.10) or the ionized doping densities. It follows that if our earlier expressions for electric field, (6.4) and (6.5), can be developed to describe potentials, the link we are seeking between carrier concentrations, field, and depletion region geometry may be forged. The required development is simply one of integrating (6.4) and (6.5). The result is shown graphically in Fig. 6.6c. Mathematically, taking the *p*-side case as an example, we have

$$\int -\,dV = \frac{-qN_A^-}{\epsilon} \int (x + x_{dp})\,dx \tag{6.13}$$

Thus

$$-V(p\text{-side}) = \frac{-qN_A^-}{\epsilon}\left(\frac{x^2}{2} + x_{dp}x\right) + B \qquad -x_{dp} < x < 0 \tag{6.14}$$

where B is a constant of integration. Because potential can be measured with respect to any arbitrary reference, we are at liberty to choose the point at which $V = 0$. It is convenient to set $V = 0$ at $x = 0$, as this eliminates B. A similar operation on the integration of (6.5) yields the potential in the depletion region on the n-side of the metallurgical boundary:

$$-V(n\text{-side}) = \frac{qN_D^+}{\epsilon}\left(\frac{x^2}{2} - x_{dn}x\right) \qquad 0 < x < x_{dn} \tag{6.15}$$

The potential distributions described in (6.14) and (6.15) are shown in Fig. 6.6c. On multiplying these potentials by the charge on an electron $(-q)$, the variation of the potential energy of an electron in the depletion region is specified. As the energy level E_C represents the potential energy of an electron in the conduction band, the missing part of the energy band diagram in Fig. 6.7 can now be filled in (see Fig. 6.8). E_V must follow E_C in a parallel fashion as the energy

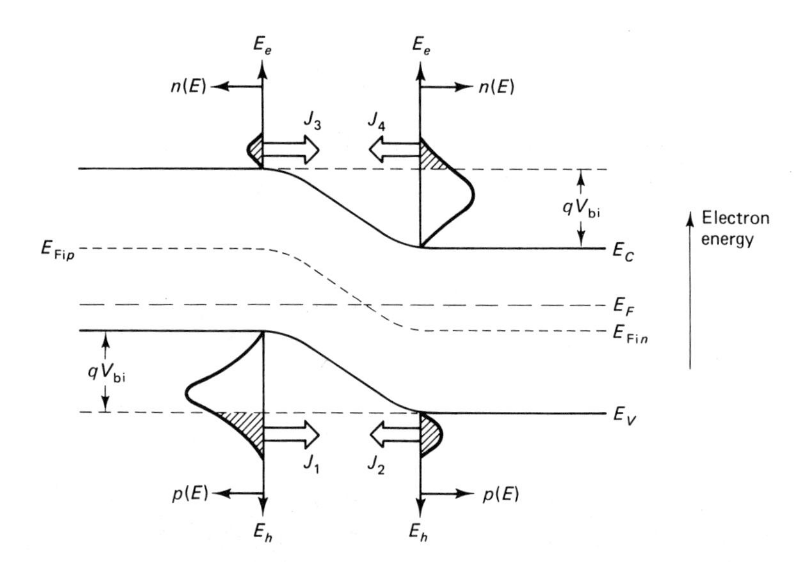

Figure 6.8 Complete energy band diagram for the pn junction at thermal equilibrium. The energy bands in the depletion region follow from the potential profile shown in Fig. 6.6c. Also shown are the distributions in energy of the mobile carriers (see Fig. 4.4), and the particle flows resulting from drift and diffusion. J_1 is the current due to diffusion of majority carrier holes. Only those holes with sufficient energy to surmount the junction barrier can participate in this process. J_2 is the current due to the drift of minority carrier holes. J_3 and J_4 are the corresponding particle flows for electrons.

band gap E_g, which is a characteristic of the semiconductor, obviously cannot change if both sides of the junction are made from the same semiconductor material.

Setting $x = -x_{dp}$ in (6.14) and $x = x_{dn}$ in (6.15), gives, on subtraction of the two equations, an expression for the total potential difference across the junction, that is, the built-in potential, namely,

$$V_{\text{bi}} = \frac{q}{2\epsilon}(N_D^+ x_{dn}^2 + N_A^- x_{dp}^2) \tag{6.16}$$

V_{bi}, as specified by (6.12) and (6.16), is truly "built-into" the junction. It cannot be measured by a voltmeter connected to the ends of the diode as a voltmeter responds to differences in Fermi energy.

Noting that the total depletion region width is

$$W = x_{dp} + x_{dn} \tag{6.17}$$

allows (6.16) to be written as

$$V_{\text{bi}} = \frac{q}{2\epsilon}\frac{N_A^- N_D^+}{N_A^- + N_D^+} W^2 \tag{6.18}$$

Comparison of our two equations for V_{bi} [(6.18) and (6.12)] gives W, from which, via (6.6) and (6.17), x_{dp} and x_{dn} can be ascertained. This completes the computation of all the unknown variables in the expression for the electric field as given by (6.4) or (6.5). From this knowledge of $\mathscr{E}(x)$, the drift current can be directly calculated. Also, via (6.1), the carrier concentrations in the depletion region can be calculated. From these the magnitude of the diffusion current follows. A good check on any such calculation would be to confirm that the drift and diffusion currents are equal and opposite. We now perform such a check.

6.2.1.2 Calculation of the drift and diffusion hole flows. As an example, we compute here the hole drift and diffusion flows at equilibrium in an abrupt junction *pn* diode with a *p*-side doping density of 10^{17} acceptors/cm^3 and an *n*-side doping density of 10^{15} donors/cm^3. We assume that at the temperature of interest (300 K), all the impurities are ionized (i.e., $N_A^- = N_A$ and $N_D^+ = N_D$). If these specifications referred to a junction made by the diffusion or ion implantation of acceptors into an *n*-type substrate, it should be appreciated that the *p*-side region would contain both donors and acceptors. However, in this instance where the acceptor doping is 100 times that of the substrate background doping, it is reasonable to take the net doping density in the *p* region as N_A. This approximation would be even better for a junction made by epitaxy, where the donor concentration in the *p*-type epilayer would comprise only the relatively few donors which outdiffused from the substrate during film growth.

As the carrier concentrations vary with position across the depletion region, we need to specify the point at which we wish to compute the currents. We select

$x = 0$, for no other reason than the fact that the electric field is largest at the metallurgical boundary (see Fig. 6.6).

We start by evaluating the hole drift current density using the appropriate form of (3.11), namely,

$$J_{h\mathrm{dr}}(0) = qp(0)\mu_h \mathscr{E}(0) \tag{6.19}$$

To compute this current density, we need to know $\mathscr{E}(0)$ and $p(0)$. The latter can be found from (4.16) evaluated at $x = 0$, that is,

$$p(0) = n_i \exp\left[\frac{E_{\mathrm{Fi}}(0) - E_F}{kT}\right]$$

The position of the Fermi level could be calculated using the procedures given in Section 4.5, but we will avoid doing this by dividing the expression for $p(0)$ by that given in (6.7) for $p(-x_{dp})$. The result is

$$p(0) = p_{0p} \exp\left[\frac{qV(-x_{dp})}{kT}\right] \tag{6.20}$$

where $p(-x_{dp})$ has been written as p_{0p}, the hole equilibrium concentration on the p side, and $(1/q)[E_{\mathrm{Fi}}(0) - E_{\mathrm{Fi}p})$ has been written as $V(-x_{dp})$. Because we have chosen to take $V = 0$ at $x = 0$, $V(-x_{dp})$ is actually the potential difference across the p side of the depletion region. It is also given by (6.14) evaluated at $x = -x_{dp}$. To compute this potential difference, we need to find the extent of the depletion region on the p side of the junction. From (6.6) and (6.17) we have

$$x_{dp} = \frac{N_D W}{N_A + N_D} \tag{6.21}$$

We can put numbers into this expression after W has been determined. This can be achieved by using (6.18), after substituting for V_{bi} from (6.12). Thus

$$V_{\mathrm{bi}} = \frac{kT}{q} \ln \frac{N_A N_D}{n_i^2} \tag{6.12}$$

$$= 0.026 \ln \frac{10^{17} \times 10^{15}}{(1.25 \times 10^{10})^2}$$

$$= 0.7 \text{ V}$$

Note that kT/q has the dimensions of volts. It is often referred to as the thermal voltage. To get its correct magnitude, Boltzmann's constant must be expressed in joules per kelvin (i.e., $k = 1.38 \times 10^{-23}$ J/K).

Rearranging (6.18) gives

$$W = \left[\frac{2V_{bi}\epsilon}{q}\left(\frac{1}{N_A} + \frac{1}{N_D}\right)\right]^{1/2} \tag{6.22}$$

$$= \left[\frac{2 \times 0.7 \times 11.9 \times 8.85 \times 10^{-14}}{1.6 \times 10^{-19}}(10^{-17} + 10^{-15})\right]^{1/2}$$

$$= 0.96 \times 10^{-4} \text{ cm}$$

We can now find x_{dp} directly from (6.21):

$$x_{dp} = 0.96 \times 10^{-4}\left(\frac{10^{15}}{10^{17} + 10^{15}}\right)$$

$$= 0.95 \times 10^{-6} \text{ cm}$$

Knowing x_{dp} allows $V(-x_{dp})$ to be computed from (6.14):

$$V(-x_{dp}) = \frac{1.6 \times 10^{-19} \times 10^{17}}{11.9 \times 8.85 \times 10^{-14}} \times \frac{(-0.95 \times 10^{-6})^2}{2}$$

$$= -6.86 \text{ mV}$$

After these preliminaries, we can return to (6.20) and achieve our first objective, the computation of $p(0)$:

$$p(0) = 10^{17} \exp\left(\frac{-6.86 \times 10^{-3}}{0.026}\right)$$

$$= 7.68 \times 10^{16} \text{ cm}^{-3}$$

Because x_{dp} is known we can also accomplish our second objective, computation of $\mathscr{E}(0)$. This follows directly from (6.4):

$$\mathscr{E}(0) = \left(\frac{-1.6 \times 10^{-19} \times 10^{17}}{11.9 \times 8.85 \times 10^{-14}}\right) 0.95 \times 10^{-6}$$

$$= -1.47 \times 10^{4} \text{ V/cm}$$

All that remains to be done to compute the hole drift current density is to find the appropriate value for the hole mobility. This can be read off Fig. 3.5 or computed from (3.9). Using the former for an impurity concentration of 10^{17} atoms/cm^3, the value is found to be about 200 cm^2/(V·s). We now have values for all the parameters in (6.19), thus

$$J_{hdr} = 1.6 \times 10^{-19} \times 7.68 \times 10^{16} \times 200 \times (-1.47 \times 10^{4})$$

$$= -3.61 \times 10^{4} \text{ A/cm}^2$$

This is an enormous current density. If the n-region of the diode were 100 μm long and the diode area were 10^{-4} cm^2, the heat dissipation in this part of the diode would be $\simeq$ 6 kW! However, there is no overheating problem in the junction under consideration because, as we have repeatedly said, in equilibrium the net current is zero. To verify this, we now compute the hole diffusion current density at $x = 0$ using the appropriate form of (3.19), namely,

$$J_{h\text{dif}} = -qD_h \frac{dp}{dx} \tag{6.23}$$

dp/dx can be found by differentiating (4.16) and can be related to the electric field via (6.9). The resulting expression is a restatement of (6.1):

$$\frac{dp}{dx} = \frac{q}{kT} p(x) \mathscr{E}(x) \tag{6.24}$$

Substituting the values for the hole concentration and the electric field at $x = 0$, as computed above, gives

$$\left.\frac{dp}{dx}\right|_{x=0} = \frac{1}{0.026} \times 7.68 \times 10^{16} \times (-1.47 \times 10^4)$$

$$= -4.34 \times 10^{22} \text{ cm}^{-4}$$

To complete the calculation we estimate D_h at the doping density of 10^{17} cm^{-3} from the Einstein relation (3.20), and substitute the acquired value of 5.2 cm^2/s into (6.23), giving

$$J_{h\text{dif}} = -1.6 \times 10^{-19} \times 5.2 \times -4.34 \times 10^{22}$$

$$= 3.61 \times 10^4 \text{ A/cm}^2$$

A similar calculation can be performed to verify that the electron current densities due to drift and diffusion are equal at thermal equilibrium.

6.2.2 Nonequilibrium Conditions

The application of voltages to the contacts of a diode disturbs the system from thermal equilibrium. The first question that has to be asked in analyzing the resulting nonequilibrium situation is: Where is the applied potential difference dropped in the device? In reality there will be voltage drops across the metal-semiconductor contact regions, the *p*-type and *n*-type regions, and the depletion region. A complete analysis would take account of all these. However, contact potential drops are usually small and can be ignored. Therefore, the situation can be likened to that of a resistive potential divider made out of semiconductor material. Two elements of the divider, the *p*- and *n*-type regions, contain copious numbers of mobile charge carriers and therefore have a very low resistance compared to the space charge region, which is depleted of mobile carriers. Thus, for a first approximation, we can assume that all the applied voltage is dropped across the depletion region.

The applied voltage drop either aids or opposes the built-in potential difference, causing the electric field in the depletion region to change. As the electric field strength is related to the dipole created by the charge of the ionized donors and acceptors in the depletion region, it follows that the amount of space charge in the depletion region must also change. As the donors and acceptors are immobile, the only way that the space charge can change is by alteration of the width of the depletion region. An increase in the potential difference across the depletion region leads to a larger field and the need for more space charge to support this field (i.e., to a wider depletion region). Contrarily, a decrease in the potential difference across the junction results in a lower field than in the equilibrium case and, consequently, to a narrower depletion region width. The effects on the depletion region width are described mathematically by amending (6.22) to take account of the applied voltage V_A, that is,

$$W = \left[\frac{2\epsilon}{q}(V_{\text{bi}} - V_A)\left(\frac{1}{N_A} + \frac{1}{N_D}\right)\right]^{1/2} \tag{6.25}$$

where, because we have chosen to write $(V_{\text{bi}} - V_A)$ rather than the equally correct but less conventional $(V_{\text{bi}} + V_A)$, a positive value must be attributed to V_A when the polarity of the applied voltage is such as to shrink the depletion region width. Under these conditions, the diode is said to be ***forward biased***. Conversely, ***reverse bias*** refers to the case when the polarity of the applied voltage is negative (i.e., the depletion region is expanded). These changes in W lead to changes in x_{dp} and x_{dn}, which allow the associated changes in electric field to be computed via (6.4) and (6.5).

The effects noted above are summarized in Fig. 6.9. Here the ground reference of zero potential has been taken to be the contact to the n side of the device. Relative to this, a positive voltage to the p side gives forward bias conditions. The alliteration in "p" provides a useful mnemonic. Figure 6.9 also conveys some information on the changes in Fermi level that must accompany the application of voltage to the diode. The Fermi level, which represents the system potential energy, must reflect the differences in potential energy between the two sides of the device, brought about by the application of voltage. In forward bias, for example, the positive voltage on the p side reduces the electron potential energy of that side of the device relative to the n side by an amount $(-qV_A)$, that is, E_F is lowered by this amount relative to its position on the n side. The position of the Fermi level relative to the band edges E_C and E_V determines the carrier concentrations. If, for the moment, we make the reasonable assumption that these concentrations will be affected only by the applied voltage in or close to the depletion region, where this voltage is dropped, it follows that elsewhere the carrier concentrations will maintain their equilibrium values. Thus, in forward bias, the depression of E_F on the p side is accompanied by a like change in E_C and E_V. In reverse bias, the band edges on the p side are elevated on an electron energy band diagram by the amount $(-q)(-V_A)$.

In recognition of the fact that the overall system equilibrium is disturbed by the application of bias, the Fermi levels on Fig. 6.9 have been depicted as quasi-Fermi levels (recall Section 4.6). To complete the band diagrams in this figure it is necessary to determine the variation of E_{Fn} and E_{Fp} in the depletion region. This necessitates finding out how the carrier concentrations in this region are affected by the applied voltage. We will deal with the forward bias case first.

Figure 6.10 is still incomplete but it does indicate how the magnitudes of the drift and diffusion currents are changed by the forward bias. The reduction

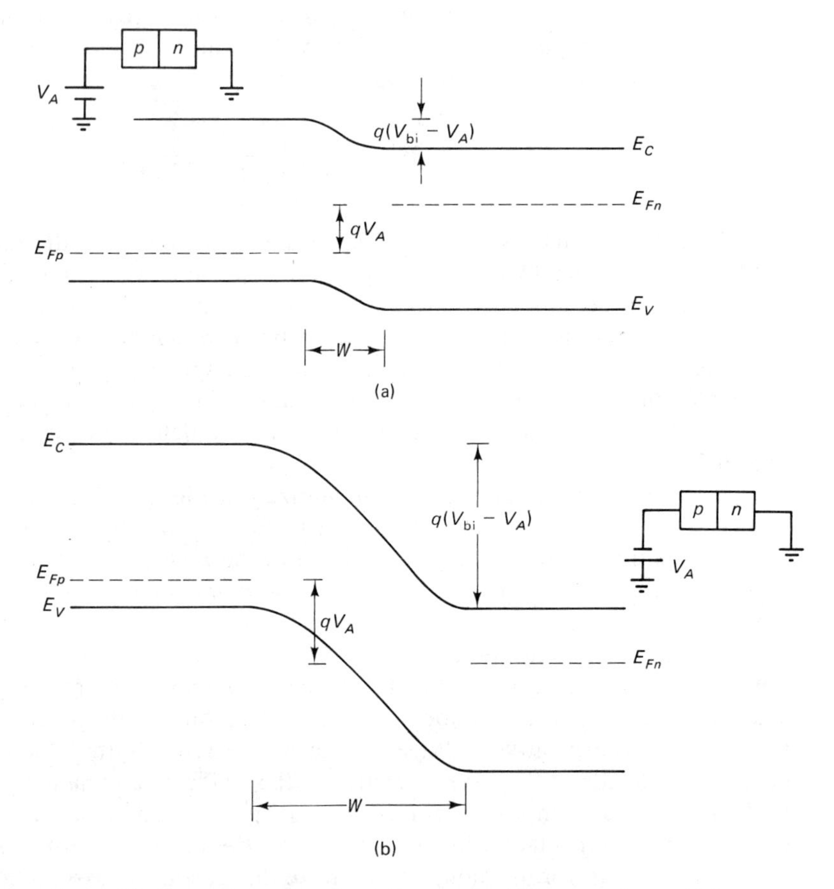

Figure 6.9 Incomplete energy band diagrams for (a) forward bias, and (b) reverse bias. The applied voltage is dropped across the depletion region causing W to either shrink (forward bias when V_A opposes V_{bi}) or expand (reverse bias when V_A aids V_{bi}). Note: V_A is taken to be negative in reverse bias.

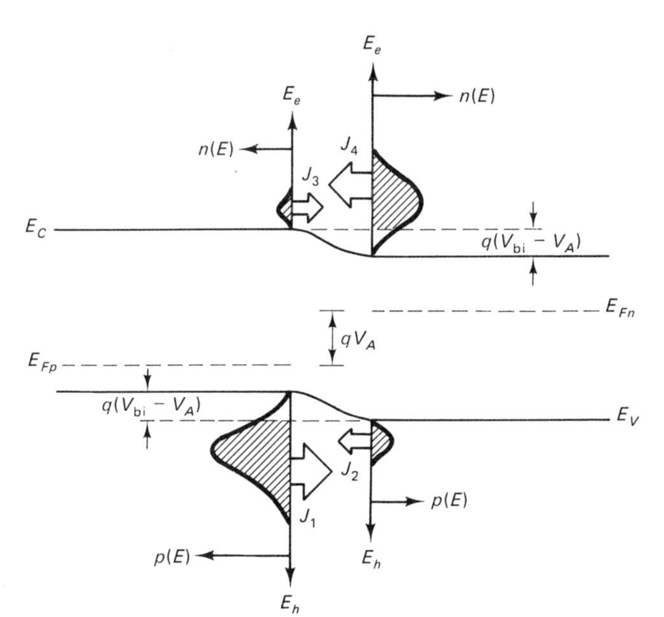

Figure 6.10 Band diagram for forward bias illustrating the increase in diffusion currents (J_4 and J_1) relative to the equilibrium case shown in Fig. 6.8.

of the height of the potential barrier at the junction allows many more majority carriers from the neutral regions to diffuse across the junction. The holes diffuse to the right and the electrons diffuse to the left, producing, on summation, a large diffusion current. On the other hand, the drift currents due to the flow of minority carriers from the neutral regions are reduced somewhat by the diminution in electric field in the junction. These phenomena result in a net current. Furthermore, as the relationship between carrier concentrations and potential energy is an exponential one [see (4.47)], the number of majority carriers able to surmount the junction potential barrier increases rapidly with forward bias, leading to a current that grows exponentially with applied voltage (see Fig. 1.1a).

The reverse-bias case is illustrated in Fig. 6.11. Here the polarity of the applied voltage is such that the junction potential barrier is enhanced. This leads to fewer majority carriers being of sufficient energy to pass over the barrier and thus diffuse into the adjoining material. Because of the exponential relationship between carrier concentrations and potential energy, it does not take much reverse bias before the diffusion current due to majority carriers is effectively reduced to zero. On the other hand, the increase in the field in the depletion region leads to an enhancement of the drift current. Because the depletion region width increases as the square root of the reverse-bias voltage for $|V_A| \gg V_{bi}$ [see (6.25)], the electric field, which is related to V/W, also increases with voltage to about this same degree. Thus the reverse-bias drift current does not increase nearly so rapidly as the forward-bias diffusion current. This explains the asymmetry of the diode current–voltage characteristic shown in Fig. 1.1a.

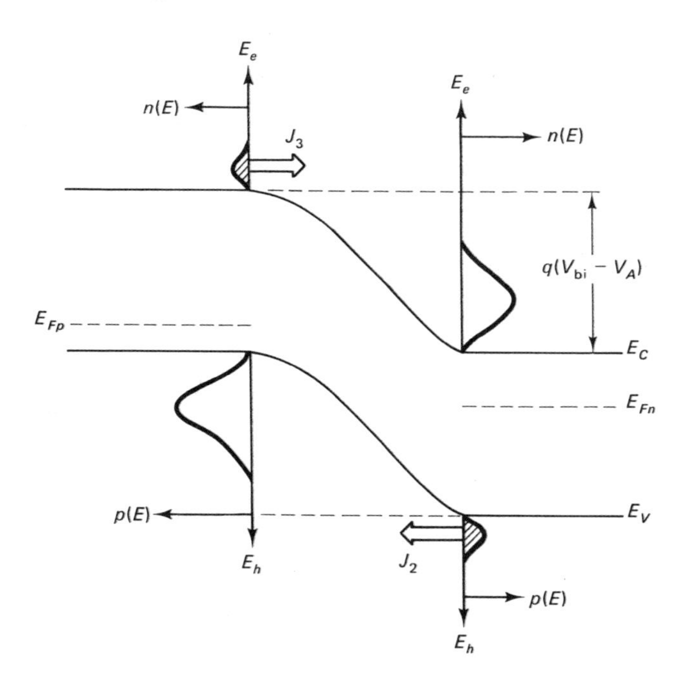

Figure 6.11 Band diagram for reverse bias. There are no diffusion currents because the energy barrier $[q(V_{bi} - V_A)]$ at the junction is too high for any majority carriers to surmount. Note that V_A is taken to be negative in reverse bias.

6.2.2.1 Quasi-equilibrium. Whereas the preceding two paragraphs give a qualitative explanation of the salient features of the diode I–V characteristic, they do not directly provide the information on carrier concentrations in the depletion region which is needed to draw in the missing quasi-Fermi levels on the energy band diagrams of Figs. 6.9 to 6.11. This information can be obtained exactly only by a full numerical solution of the equations of state, namely, Poisson's equation (3.29), the transport equations (3.22) and (3.23), and the continuity equations (3.27) and (3.28). However, an approximation which is excellent for many practical applications, and which allows an analytical solution to be effected, does exist. It is called the ***quasi-equilibrium approximation***. To appreciate the foundation for the approximation, consider the magnitude of the particle flows in thermal equilibrium as computed in Section 6.2.1.2. Compared to these, the net currents in real devices are very small. For example, a diode designed specifically for handling large currents, such as one from Fairchild's FRP1600 series, has a maximum current density rating of only 0.6% of that of the equilibrium particle flows computed in Section 6.2.1.2. The implication of this is that the system thermal equilibrium, at least as regards currents, is disturbed barely at all by the net currents that arise in response to applied voltages. The situation of a diode in forward or reverse bias can thus be described as one of quasi-equilibrium. One of the tenets of true equilibrium is that the product of the equilibrium carrier concentrations is a constant [see (2.14)]. The postulation of quasi-equilibrium is that this same rule, with a different constant, applies in nonequilibrium situations. It is a hypothesis that greatly simplifies semiconductor device calculations.

To develop this postulate, consider the product of $p(x)$ and $n(x)$ derived from (4.24), namely,

$$pn = n_i \exp\left(\frac{E_{\text{Fi}} - E_{Fp}}{kT}\right) \times n_i \exp\left(\frac{E_{Fn} - E_{\text{Fi}}}{kT}\right) \\ = n_i^2 \exp\left(\frac{E_{Fn} - E_{Fp}}{kT}\right) \tag{6.26}$$

Constancy of the pn product demands that the separation of the quasi-Fermi levels remains constant. One way of accomplishing this is to allow the quasi-Fermi levels to remain flat at their neutral region values right through the depletion region. This is intuitively satisfying, as it mimics the constancy of the true Fermi level in true equilibrium conditions. The separation of the quasi-Fermi levels in the space charge region is, under these circumstances, equal to the applied voltage multiplied by the electronic charge (see Figs. 6.9 to 6.11). Thus

$$pn = n_i^2 \exp\left(\frac{qV_A}{kT}\right) \qquad -x_{dp} < x < x_{dn} \tag{6.27}$$

The implication of this under forward-bias conditions, where V_A is positive, is that $pn > n_i^2$ in the space charge region. As the recombination rate of electrons and holes depends directly on the pn product, at least for direct band-to-band recombination [see (2.27)], it follows that recombination is a significant process in the depletion region of a forward-biased diode. As such, it can affect the forward-bias current in diodes (see Section 6.3.1). For reverse-bias conditions, where V_A is negative, $pn < n_i^2$ in the space charge region. This means that the recombination rate is low and, therefore, that the generation rate of electrons and holes in the depletion region can significantly contribute to the carrier concentrations, and thus the current, in this region of a reverse-biased diode.

6.2.2.2 Diffusion in the neutral regions. The postulation of quasi-equilibrium helps to fill in significant portions of the quasi-Fermi levels which are missing from Figs. 6.9 through 6.11. To complete the band diagram, it is necessary to determine the manner in which E_{Fn} tends to E_{Fp} in the p region, and the manner in which E_{Fp} tends to E_{Fn} in the n region. We will consider the situation in the n-type material just to the right of the depletion region. From (4.24) we can write

$$p(x) = n_i \exp\left(\frac{E_{\text{Fi}} - E_{Fp}}{kT}\right) \tag{6.28}$$

Differentiating this, we obtain

$$\frac{dp}{dx} = \frac{p}{kT}\left(\frac{dE_{\text{Fi}}}{dx} - \frac{dE_{Fp}}{dx}\right) \tag{6.29}$$

As the derivative of the intrinsic Fermi level is related directly to the electric field [see (6.9)], and because we have assumed that $\mathscr{E} = 0$ outside the depletion region, the gradient of the quasi-Fermi level is

$$\frac{dE_{Fp}}{dx} = \frac{-(dp/dx)kT}{p} \qquad x > x_{dn} \tag{6.30}$$

As we know E_{Fp} at $x = x_{dn}$, (6.30) can be used to find the quasi-Fermi level information we are seeking. But first, it is necessary to compute p and dp/dx in the region of interest. This calls for a solution of the continuity equation under steady-state conditions in a region where there is no field, no optical generation of carriers, and the doping density is uniform. These conditions are precisely those considered in instance 5 at the end of Section 3.5.1. The solution for the excess carrier concentration is [see (3.39)]

$$\hat{p}(x) = A \exp\left(\frac{-x}{L_h}\right) + B \exp\left(\frac{x}{L_h}\right) \tag{6.31}$$

where A and B are constants and L_h is the hole minority carrier diffusion length. To evaluate A and B it is convenient to shift the origin of the x-axis to the depletion layer boundary on the n-side (i.e., to the point defined previously as x_{dn}). As x increases, $\hat{p}$ must decrease due to recombination of the excess holes with some of the large number of majority carrier electrons in the n-type region. From these physical considerations applied at $x = \infty$, it follows that $B = 0$ in (6.31). Obviously, diodes are not infinitely long in practice. However, from the minority carrier point of view, they can be considered to be so if the neutral regions are longer than several minority carrier diffusion lengths. The other boundary condition, namely $\hat{p}(x) = \hat{p}(0)$ at $x = 0$, gives A. Thus the expression for $\hat{p}(x)$ becomes

$$\hat{p}(x) = \hat{p}(0) \exp\left(\frac{-x}{L_h}\right) \qquad x > 0 \text{ (old } x_{dn}) \tag{6.32}$$

Differentiating, we obtain the intermediate result we are seeking, namely,

$$\frac{d\hat{p}}{dx} = \frac{-\hat{p}}{L_h} \qquad x > 0 \tag{6.33}$$

Because the thermal equilibrium hole concentration is constant in the n-region, it follows that

$$\frac{d\hat{p}}{dx} = \frac{dp}{dx}$$

Using this identity in (6.33) and comparing with (6.30) gives the relationship we need:

$$\frac{dE_{Fp}}{dx} = \frac{kT}{L_h}\left(1 - \frac{p_{n0}}{p(x)}\right) \qquad x > 0 \tag{6.34}$$

As $p(x \cong 0) \gg p_{n0}$ E_{Fp} tends toward E_{Fn} initially in a linear fashion. This fact, and the corresponding relationship for E_{Fn} on the p side of the depletion region, allows the energy band diagrams of Figs. 6.10 and 6.11 to be completed (see Figs. 6.12 and 6.13).

6.2.2.3 Law of the junction. Also shown on Figs. 6.12 and 6.13 are the minority carrier concentrations at the edges of the space charge region, namely, $p(x_{dn})$ and $n(-x_{dp})$. These are important quantities, as they strongly influence the operation of diodes. Considering the n side as an example, elevation of $p(x_{dn})$ above p_0 in forward bias emphasizes the injection of minority carrier holes into the n region. Conversely, the depression of $p(x_{dn})$ below p_0 in reverse bias emphasizes the extraction of minority carriers by the large electric field in the depletion region. An expression for $p(x_{dn})$ can be derived starting from (6.8), that is,

$$\frac{p_0(-x_{dp})}{p_0(x_{dn})} = \exp\left(\frac{qV_{\text{bi}}}{kT}\right) \tag{6.35}$$

Extending this equation to nonequilibrium conditions, we have

$$\frac{p(-x_{dp})}{p(x_{dn})} = \exp\left(\frac{q}{kT}(\mathrm{V}_{\text{bi}} - \mathrm{V}_A)\right) \tag{6.36}$$

Dividing these two equations and recognizing that there are essentially no excess

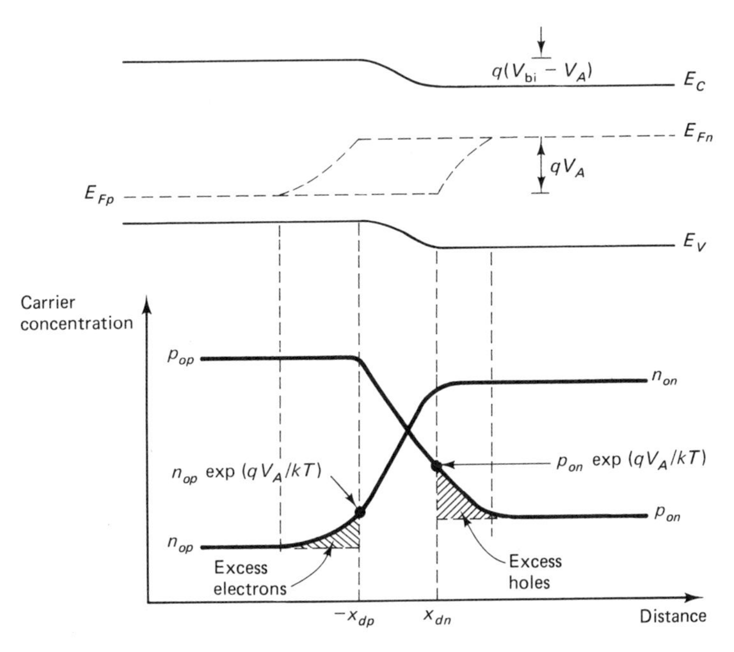

Figure 6.12 Complete energy band diagram for forward bias (V_A positive). The lower figure shows the carrier concentrations and emphasizes the injection of minority carriers into the neutral regions.

holes at $x = -x_{dp}$ gives the useful relationship

$$p(x_{dn}) = p_{0n} \exp\left(\frac{qV_A}{kT}\right) \tag{6.37}$$

This equation neatly ties together the internal physical phenomena of minority carrier enhancement or diminution at the edge of the depletion region, with their external electrical cause, the polarity of the applied voltage. Equation (6.37) is known as the ***law of the junction***.

6.2.2.4 Neutrality outside the depletion region. The carrier concentration plots in Figs. 6.12 and 6.13 suggest that the zones just outside the depletion region are not completely neutral. Considering the n side of the diode close to the junction in Fig. 6.12, for example, the injected excess holes would appear to

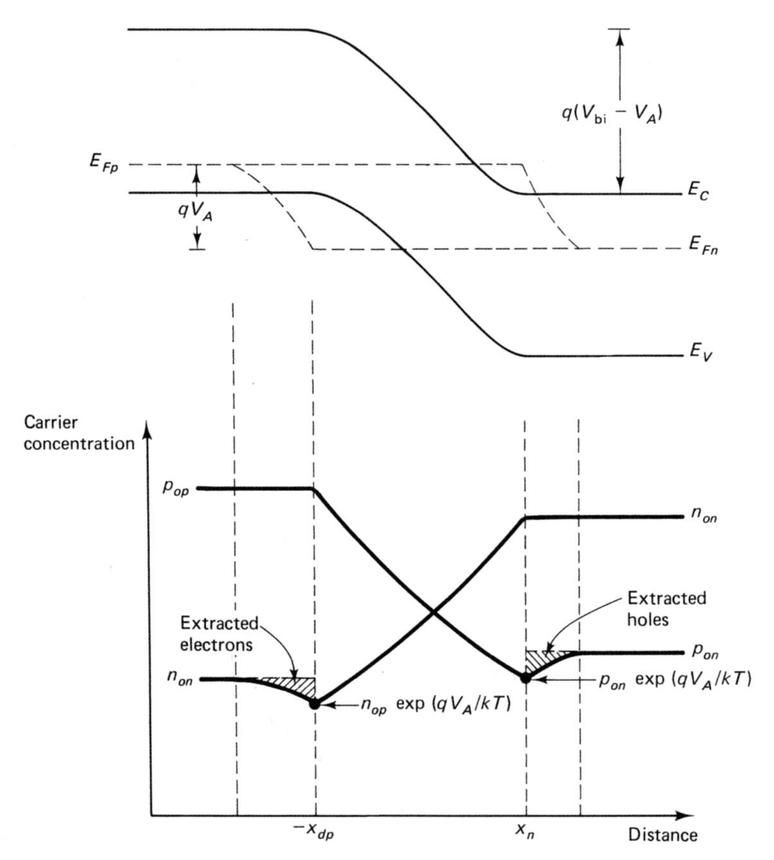

Figure 6.13 Complete energy band diagram for reverse bias (V_A negative). The lower figure shows the carrier concentrations and emphasizes the extraction of minority carriers to supply the drift current.

give this region a net positive charge. However, if such a space charge were to exist, the electric field associated with it would rapidly draw in electrons from elsewhere in the n-type material. Thus the space charge would disappear. Hence, even in nonequilibrium situations, it is correct to refer to the material outside the depletion region as being neutral. The concomitant increase in electron concentration in the n material close to the depletion region is not depicted on Fig. 6.12 because it is so small in magnitude compared to the majority carrier concentration. In other words, we have assumed that low-level injection conditions apply.

A numerical example readily demonstrates the orders of magnitude involved. Taking n_{0n} to be 10^{16} cm^{-3}, it follows that p_{0n} is 1.56×10^4 cm^{-3}, and for a forward bias of 0.5 V, $p(x_{dn})$ is 3.8×10^{12} cm^{-3}. The fractional change in minority carrier density is 2.4×10^8 times, whereas the corresponding change in majority carrier density, due to the accumulation of an equal number of electrons, is only 3.8×10^{-4} times.

Note that in examples of minority carrier injection appearing earlier in this book (see Figs. 3.10 and 4.5), the emphasis was on diffusion and recombination of the minority carriers. For this reason, the fact that a concentration gradient of minority carriers automatically involves a concentration gradient of majority carriers was not introduced. It can be appreciated now that such an association of concentration gradients must exist in order to maintain charge neutrality.

6.3 DC ANALYSIS

In this section we apply the continuity equation to derive expressions for the J–V characteristic of pn-junction diodes. Before commencing this task, some comments on the band diagrams of Figs. 6.12 and 6.13 are in order.

These diagrams show a gradient of quasi-Fermi level only in the regions abutting the depletion region. As current is directly proportional to this gradient, the band diagrams imply that a current exists only in two parts of the device. Clearly, in a single-loop series circuit such as is the case here, this cannot be true. This inconsistency can be resolved, at least as far as the regions of constant quasi-Fermi level outside the depletion region are concerned, by noting that in these regions the majority carrier concentrations are high. As the current is actually proportional to the product of the carrier concentration and the quasi-Fermi level gradient [see (4.27) and (4.28)], a reasonably sized current can be supported with need only for a very small gradient in quasi-Fermi level. This gradient is usually so small compared to that in the zones close to the depletion region that it is not shown on energy band diagrams.

Even though the carrier concentrations in the depletion region are not as high as the majority carrier concentrations in the neutral regions, their magnitudes are sufficient for the assumption of constant quasi-Fermi levels in the depletion region to be a very good approximation to the truth. The beauty of the approximation is that it leads to a tractable analytic solution to the equations of state.

Such a solution is worth seeking because the result of the exercise is an expression for the current which relates explicitly to material and device properties. A complete solution to the equations of state, using iterative methods of solution in a numerical analysis, would not need to make any assumptions, but the dependencies of the current on the properties of interest to the device designer would be hidden.

6.3.1 Forward Bias

Comparing the carrier concentration profiles in forward bias (Fig. 6.12) with those in equilibrium (Fig. 6.5), and noting the remarks on neutrality in Section 6.2.2.4, it can be appreciated that in the depletion region and in the regions immediately adjacent to it, the equilibrium concentration is exceeded in forward bias by both types of carrier. These excess carriers eventually recombine and so must be replaced by carriers injected from the contacts to the diode. The situation is illustrated in Fig. 6.14. By computing the rate of recombination in the various regions of the device, the total current can be found by superposition.

To elaborate: Consider the left-hand contact in Fig. 6.14. The positive potential at the contact metal due to its connection to the external power supply encourages the extraction of electrons from the contact. This extraction of electrons from the metal-semiconductor interface provides the charge carriers that flow in the wire connecting the external circuit. This removal of electrons from the *p*-type semiconductor is manifest at the end of the semiconductor as the creation of holes. These holes move into the device and replenish the holes lost by recombination in the various regions of the device. The flow of holes into the

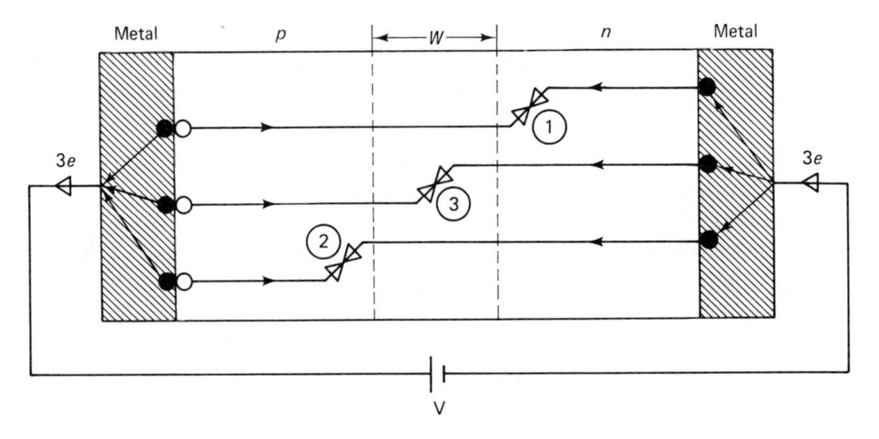

Figure 6.14 Charge flow in forward bias. Recombination of excess carriers occurs in the neutral regions (1 and 2) and in the depletion region (3). The excess carriers consumed in recombination are replenished by charges from the contacts that constitute the forward-bias current.

semiconductor, and therefore the flow of extracted electrons in the external circuit (i.e., the current), are governed by the rates at which recombination takes place. The three regions of recombination are identified in Fig. 6.14. In region 1, the excess holes, injected across the junction, diffuse into the n-type region and recombine with majority carrier electrons. In region 2 it is electrons which are the injected carriers, so the flow of holes to this area is just that which is needed to replace the majority carrier holes used up in the recombination process. In region 3, both electrons and holes are injected and recombination takes place throughout the depletion region. The holes to fuel all these recombination processes are supplied from the left-hand contact. This rate of supply of holes to the semiconductor is the same as the rate of supply of electrons to the external circuit, and thus its computation, via a summation of the recombination rates in the semiconductor, gives the total current. Note that the recombination events could also be viewed from the perspective of the electrons rather than the holes. The electrons to fuel the recombination processes are injected from the right-hand contact, where they have an elevated potential energy by virtue of the contact to the negative electrode of the power supply. Note further that these injected electrons are, in essence, the same electrons that were extracted from the left-hand contact. Thus to compute the total current we can calculate either the rate of injection or the rate of extraction. However, we must not add the two together, as that would overestimate the current by a factor of 2. In the calculation that follows we consider the recombination events from the hole point of view.

In each of the three regions marked on Fig. 6.14, the hole current is to be computed from the hole continuity equation (3.31), which, for steady-state conditions and no optical generation, is

$$dJ_h = -q\frac{\hat{p}}{\tau_h}\,dx \tag{6.38}$$

In performing the calculation for region 1, we set $x = 0$ at the boundary of the depletion region and the n region. The excess hole concentration is given by (6.32), so the appropriate integration, assuming a diode of long length, is

$$\int_{J_{h1(0)}}^{J_{h1}(\infty)} dJ_{h1} = \frac{-q}{\tau_h}\int_{x=0}^{x=\infty} \hat{p}(0)\exp\left(\frac{-x}{L_h}\right) dx \tag{6.39}$$

The minority carrier lifetime, under the low-level injection conditions we are considering, is inversely proportional to either the majority carrier concentration (for direct recombination) or the trap density (for indirect recombination) (see Section 2.5.2.2). We assume that N_t is constant and, in region 1, n_{n0} is certainly constant. Thus, taking τ_h outside the integral, as we have done in (6.39), is justifiable. At $x = \infty$ the excess hole concentration will have been reduced to zero by recombination; thus $J_{h1}(\infty) = 0$. Using this fact in performing the integration yields

$$J_{h1}(0) = \frac{qL_h}{\tau_h} \hat{p}(0) \tag{6.40}$$

Recalling that $\hat{p} = p - p_0$ [see (2.21)], and taking $\hat{p}(0)$ from (6.37), gives the final expression for the current of holes which is supplied from the left contact in order to fuel the recombination in region 1, that is,

$$J_{h1} = \frac{qL_h}{\tau_h} p_{0n} \left[\exp\left(\frac{qV}{kT}\right) - 1\right] \tag{6.41}$$

where V is the applied voltage.

A similar expression to this pertains to the rate of supply of holes which are needed to replenish those lost to recombination in region 2. This current must be equal to that current due to the recombination of electrons injected into region 2. The latter is just the electron equivalent of J_{h1}; thus

$$J_{h2} = \frac{qL_e}{\tau_e} n_{0p} \left[\exp\left(\frac{qV}{kT}\right) - 1\right] \tag{6.42}$$

To estimate the holes lost to recombination in the depletion region, we must first examine the carrier concentrations in region 3. The information on the carrier profiles shown in Fig. 6.12 is repeated in more detail in Fig. 6.15. We consider first the zone to the right of the point at which $n = p$ (i.e., zone Z), the region in which holes are the minority carriers. Redefining $x = 0$ as the point where $n = p$, we have, from (6.27),

$$p(0) = n_i \exp\left(\frac{qV}{2kT}\right) \tag{6.43}$$

At the right-hand edge of the depletion region (i.e., at $x = x_Z$ in Fig. 6.15), we have, from (6.37),

$$p(x_Z) = p_{0n} \exp\left(\frac{qV}{kT}\right) \tag{6.44}$$

This hole concentration is the source of the concentration gradient that drives the diffusion current in region 1. Therefore, the holes participating in the recombination events in zone Z of region 3 are given by the integral from $x = 0$ to $x = x_Z$ of the difference between $p(x)$ and $p(x_Z)$ (see the shaded area on Fig. 6.15). For values of V smaller than about $(V_{bi} - 10kT/q)$, $p(x_Z) \ll p(0)$, so it is reasonable to set $p(x_Z) = 0$. To find the excess hole carrier concentration, we need an estimate of the equilibrium carrier concentration in the depletion region. When $p = n$ in thermal equilibrium, then $p = n_i$. The point at which this occurs will be very close to the point we have labeled $x = 0$ in Fig. 6.15. The excess hole concentration $\hat{p}(0)$ is, therefore, approximately equal to $n_i[\exp(qV/2kT) - 1]$.

We now proceed to accommodate all these assumptions and conditions in

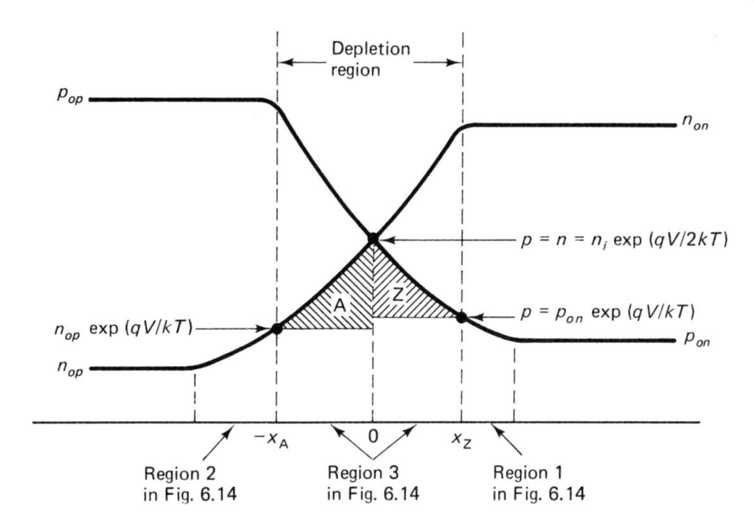

Figure 6.15 Carrier concentrations in forward bias. The two shaded areas A and Z, wherein occurs recombination of minority carrier electrons and holes, respectively, are both considered to be triangular in shape.

an expression for the hole current due to recombination in region 3Z. We start, as before, with the relevant form of the continuity equation:

$$\int_{J_{h3Z}(0)}^{J_{h3Z}(x_Z)} dJ_{h3Z} = \frac{-q}{\tau_h} \int_{x=0}^{x=x_Z} \hat{p}(x)\, dx \tag{6.45}$$

Placing τ_h outside the integral may seem improper because the equilibrium majority carrier concentration of electrons is position dependent in the depletion region (see Fig. 6.12). However, the concentration of majority carriers is, everywhere in the depletion region, smaller than the majority carrier concentrations elsewhere in the device, so the direct recombination of electrons and holes will be an unlikely event. Instead, indirect recombination is likely to be the dominant recombination mechanism. The minority carrier lifetime in this case depends on the concentration of traps [see (2.31)], for which the postulation of positional independency is quite reasonable. Thus (6.45) is justifiable. A ready solution to this equation can be achieved by making yet another assumption, namely that the shaded profile in Fig. 6.15 can be represented by a triangular distribution. Thus

$$J_{h3Z} = \frac{qn_i x_Z}{2\tau_h} \left[\exp\left(\frac{qV}{2kT}\right) - 1 \right] \tag{6.46}$$

An equation similar to this pertains to the supply of holes needed to replace those lost to recombination in region 3A (Fig. 6.15), where the minority carriers are electrons. Recombination in this region can be computed from the point of view of electrons in a manner which is exactly analogous to that used to derive

(6.46). The required hole current is, therefore, just the electron equivalent of (6.46), namely,

$$J_{h3A} = \frac{qn_i x_A}{2\tau_e}\left[\exp\left(\frac{qV}{2kT}\right) - 1\right] \tag{6.47}$$

Addition of J_{h3A} and J_{h3Z} gives the hole current needed to fuel the recombination in the depletion region. Thus the summation of (6.46), (6.47), (6.41), and (6.42) gives the total current in forward bias:

$$\begin{aligned} J_h = &\left[\frac{qn_i}{2}\left(\frac{x_Z}{\tau_h} + \frac{x_A}{\tau_e}\right)\right]\left[\exp\left(\frac{qV}{2kT}\right) - 1\right] \\ &+ q\left(\frac{L_e}{\tau_e} n_{0p} + \frac{L_h}{\tau_h} p_{0n}\right)\left[\exp\left(\frac{qV}{kT}\right) - 1\right] \end{aligned} \tag{6.48}$$

This J–V relationship is displayed on the log-linear plot of Fig. 6.16. The characteristic has two distinct parts, corresponding to the ranges of dominance of the two components of the forward bias current. The preexponential coefficient in the term for the recombination current in the depletion region is greater than the corresponding coefficient for the other recombination current; thus at low values of V, recombination in the depletion region dominates. The factor of $\frac{1}{2}$ in the exponent for this current dictates that the slope of the ln J–V curve will be one-half of that which is appropriate to higher voltages, when the concentrations of minority carriers injected from the depletion region are so great that recombination in the neutral regions of the device is dominant. The voltage at which the transition between the two slopes occurs is strongly temperature dependent, mainly because of the different dependencies on n_i of the two current components.

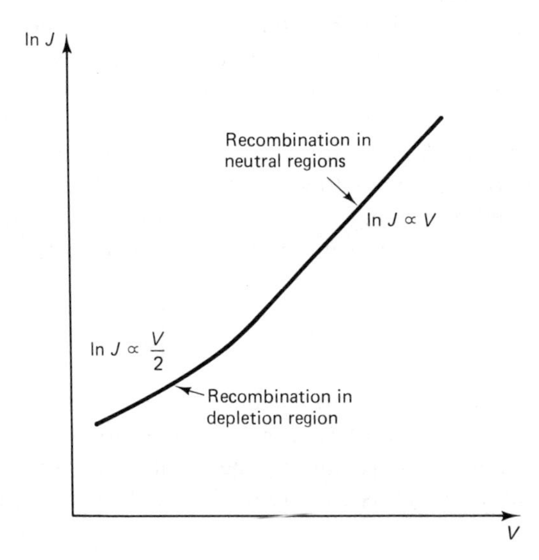

Figure 6.16 Forward-bias J–V characteristic, on a log-linear plot, from (6.48), showing the different slopes corresponding to recombination in the different regions of the device.

Example 6.1 provides an opportunity to compute I–V characteristics for silicon diodes under forward-bias conditions.

Example 6.1 Forward-Bias Characteristics of *PN* Diodes: The Effect of Doping Density

To carry out this exercise, we need first to cull from the previous material all the equations that relate the forward-bias current to the doping densities in the p- and n-type regions of the diode. The forward-bias current comprises two components: one due to recombination in the neutral regions of the device, and one due to recombination in the depletion region. Thus we can write

$$I = I_{\mathrm{NR}} + I_{\mathrm{DR}} \tag{6E.1}$$

Taking A as the cross-sectional area of the diode and using (6.41) and (6.42) together with the relationships

$$p_0 n_0 = n_i^2 \tag{2.14}$$

and

$$L = \sqrt{D\tau} \tag{3.40}$$

we have

$$I_{\mathrm{NR}} = Aqn_i^2 \left(\frac{D_e}{L_e} \frac{1}{N_A} + \frac{D_h}{L_h} \frac{1}{N_D} \right) \left[\exp\left(\frac{qV}{kT} \right) - 1 \right] \tag{6E.2}$$

which can be abbreviated as

$$I_{\mathrm{NR}} = I_{\mathrm{SNR}} \left[\exp\left(\frac{qV}{kT} \right) - 1 \right] \tag{6E.3}$$

For the depletion region current, we assume that the minority carrier lifetime of the holes is equal to that of the electrons. This enables (6.46) and (6.47) to be combined to give

$$I_{\mathrm{DR}} = \frac{Aqn_i W}{2\tau_e} \left[\exp\left(\frac{qV}{2kT} \right) - 1 \right] \tag{6E.4}$$

The depletion width W is voltage dependent, and the width of the depletion region in equilibrium, which we now label as W_0, depends on the doping densities, that is,

$$W = \left[\frac{2\epsilon}{q} (V_{\mathrm{bi}} - V) \left(\frac{1}{N_A} + \frac{1}{N_D} \right) \right]^{1/2} \tag{6.25}$$

and

$$W_0 = \left[\frac{2\epsilon}{q} V_{\mathrm{bi}} \left(\frac{1}{N_A} + \frac{1}{N_D} \right) \right]^{1/2} \tag{6.18}$$

Combining these equations and substituting in (6E.4) gives

$$I_{\mathrm{DR}} = I_{\mathrm{SDR}} \left(1 - \frac{V}{V_{\mathrm{bi}}}\right)^{1/2} \left[\exp\left(\frac{qV}{2kT}\right) - 1\right] \tag{6E.5}$$

where

$$I_{\mathrm{SDR}} = \frac{Aqn_i}{2\tau_e} \left[\frac{2\epsilon}{q} V_{\mathrm{bi}} \left(\frac{1}{N_A} + \frac{1}{N_D}\right)\right]^{1/2} \tag{6E.6}$$

The doping densities appear explicitly in (6E.2) and (6E.5), and implicitly via the parameters D, τ, and V_{bi}. The hidden doping density dependencies are revealed by the following:

$$\mu_e = 88 + \frac{1252}{1 + 6.984 \times 10^{-18} N_A} \tag{3.8}$$

$$\mu_h = 54.3 + \frac{407}{1 + 3.745 \times 10^{-18} N_D} \tag{3.9}$$

$$D = \frac{kT}{q} \mu \tag{3.20}$$

$$\tau_e = \frac{5 \times 10^{-7}}{(1 + 2 \times 10^{-17} N_A)} \tag{2.33}$$

$$\tau_h = \frac{5 \times 10^{-7}}{(1 + 2 \times 10^{-17} N_D)} \tag{2.33}$$

$$V_{\mathrm{bi}} = \frac{kT}{q} \ln \frac{N_A N_D}{n_i^2} \tag{6.12}$$

The equations presented in this example give all the relationships required to compute the diode forward-bias current as a function of doping density. For this example we choose an n^+p diode with a donor doping

TABLE 6E.1 NUMERICAL VALUES OF PARAMETERS USED TO COMPUTE DIODE CURRENTS

	n^+	p			Unit
	$N_D = 10^{19}$	$N_A = 10^{15}$	10^{17}	10^{19}	cm^{-3}
D_h	1.68				$\mathrm{cm}^2\ \mathrm{s}^{-1}$
τ_h	2.5×10^{-9}				s
D_e		34.5	21.3	2.7	$\mathrm{cm}^2\ \mathrm{s}^{-1}$
τ_e		4.9×10^{-7}	1.7×10^{-7}	2.5×10^{-9}	s
V_{bi}		0.82	0.94	1.06	V
W_0		1.04×10^{-4}	1.12×10^{-5}	1.67×10^{-6}	cm
I_{SNR}		2.1×10^{-12}	2.8×10^{-14}	1.5×10^{-15}	A
I_{SDR}		2.1×10^{-9}	6.5×10^{-10}	6.7×10^{-9}	A

density of 10^{19} cm^{-3} and with various acceptor doping densities. Table 6E.1 summarizes the numerical data. In compiling this table, a temperature of 300 K has been assumed, which leads to $kT/q = 0.0259$ V and $n_i = 1.25 \times 10^{10}$ cm^{-3}. The diode area has been taken as 0.01 cm^2 (1 mm × 1 mm). The relative permittivity of silicon is 11.9, giving a permittivity of 1.05×10^{-12} F/cm.

The resulting forward-bias current components for two n^+p diodes are shown in Fig. 6E.1. Note the following:

(a) The graphs are of the log-linear variety in order to represent adequately the exponential dependence of the current on forward-bias voltage.

(b) The depletion region current dominates at low bias because I_{SDR} is greater than I_{SNR}.

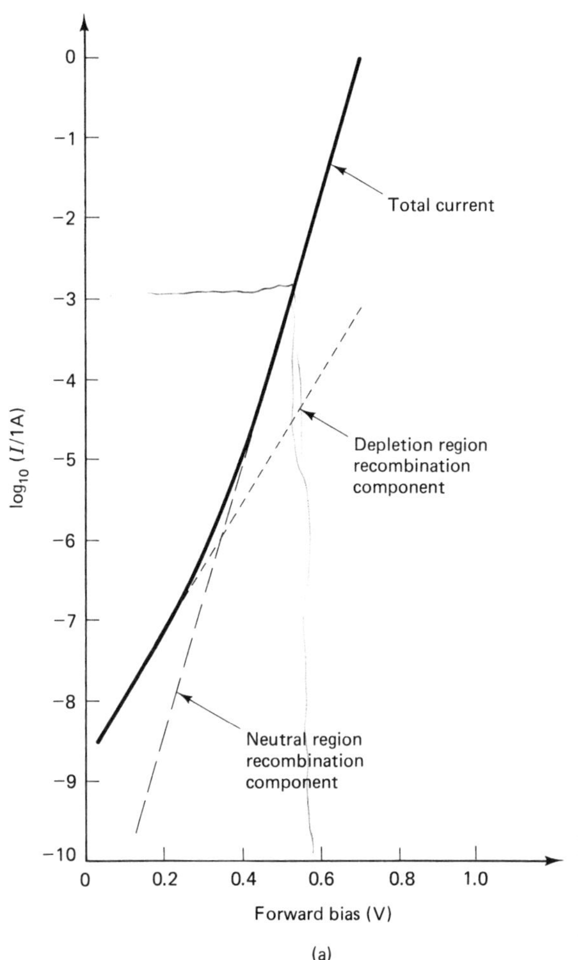

Figure 6E.1 Computed forward-bias $\log_{10}$–V characteristics for n^+p diodes. The n-side doping density is 10^{19} cm^{-3} in both cases. For (a) $N_A = 10^{15}$ cm^{-3}; for (b) $N_A = 10^{19}$ cm^{-3}.

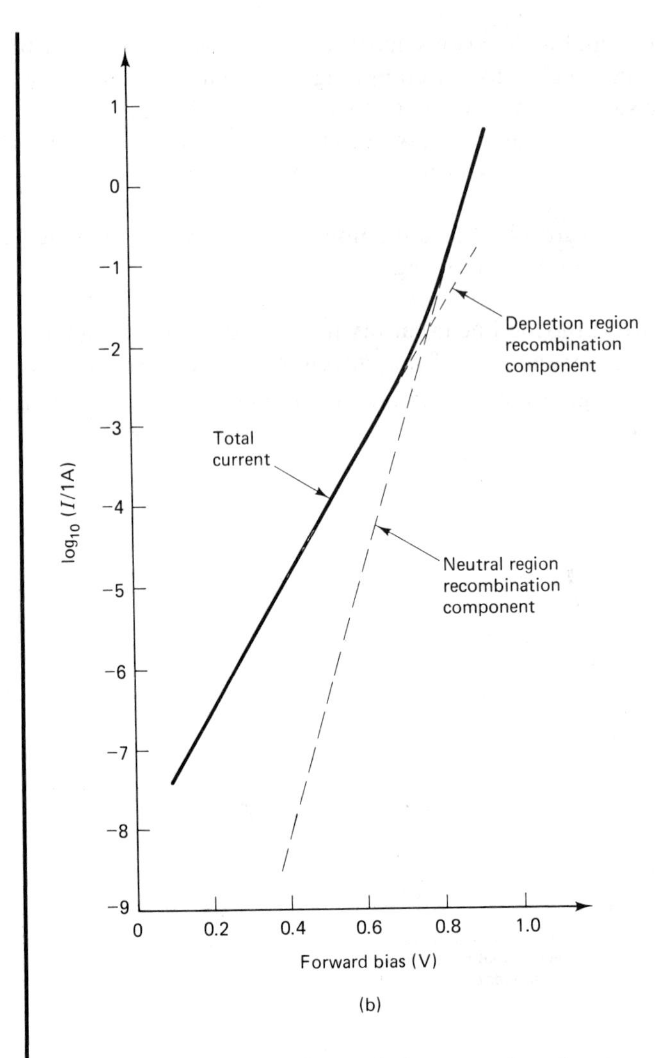

Figure 6E.1 *(continued)*

(c) The steeper slope of the neutral region current causes it to eventually dominate at higher bias voltages.

(d) The "crossover point" depends on the doping density. Surprisingly, perhaps, the depletion region current dominates for longer in the high-doping-density case, even though the depletion region is smaller in this instance. The reason for this is the choice of τ_e as the appropriate lifetime to use in computing I_{SDR}. As τ_e decreases with the acceptor doping density, I_{SDR} increases. τ_e appears in the denominator of I_{SNR} also, but only as a square-root term. The doping density dependence of this current is therefore determined mainly by the explicit reciprocal doping density terms.

6.3.2 Reverse Bias

Referring to the carrier concentrations in reverse bias (Fig. 6.13), and noting the remarks on neutrality in Section 6.2.2.4, it follows that in the depletion region and in the regions immediately adjacent to it, the concentrations of both of the carriers are reduced below their equilibrium values. This diminution of the carrier concentrations reduces the possibility of recombination and leads to generation being the dominant mechanism in determining the current. Carriers generated in the depletion region of the device are separated by the electric field and drift to the neutral regions of the device (see Fig. 6.17). Here, the new carriers are of the majority type and they drift under the influence of the small electric field in the neutral region to the contacts of the diode. The electrons pass into the external circuit. The holes move to the other contact, where they combine with electrons from the metal. These electrons are, in essence, the ones supplied to the external circuit from the other electrode. Thus the generation of one electron–hole pair in the depletion region leads to one electronic charge flowing in the external circuit. Similarly, current in the external circuit results from generation in all other regions of the device where there is not an associated equal rate of recombination. The situation is depicted in Fig. 6.17. The total current can be computed from a summation of the currents arising from generation in the three regions indicated. The procedure for estimating the current is, therefore, the same as that employed in the forward-bias case, the only difference being the consideration of generation rather than recombination. Again, we choose to consider hole flow in order to compute the current–voltage relation.

Commencing with generation in region 1 (Fig. 6.17), the minority carriers diffuse to the depletion region, where they are swept across by the junction field to become majority carriers in the p-type region. The situation of minority carrier

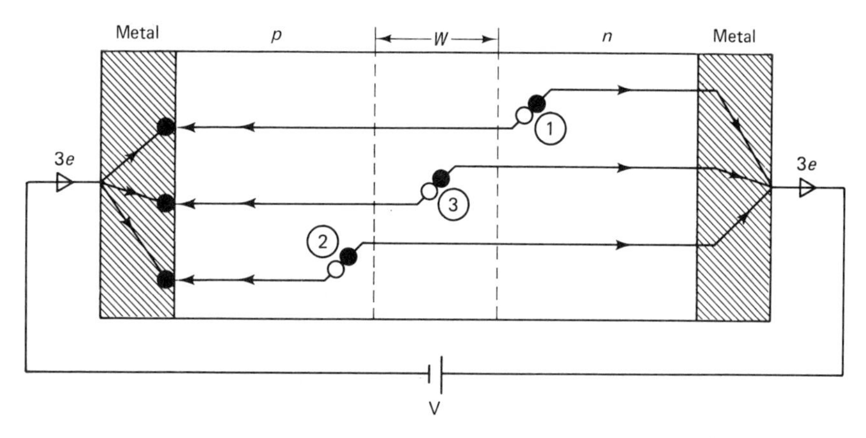

Figure 6.17 Charge flow in reverse bias. Generation of carriers exceeds recombination in the neutral regions (1 and 2) and in the depletion region (3). The generated carriers flow to the contacts and constitute the reverse-bias current.

diffusion in a field-free region, under steady-state conditions with no optical generation, gives a minority carrier distribution as computed in instance 5 of Section 3.5.1, namely,

$$\hat{p}(x) = A \exp\left(\frac{-x}{L_h}\right) + B \exp\left(\frac{x}{L_h}\right) \tag{3.39}$$

Designating the interface between region 1 and the depletion region as $x = 0$ gives the following description of the boundary conditions:

$$p(0) = p_{0n} \exp\left(\frac{qV}{kT}\right) \quad \text{and} \quad p(\infty) = p_{0n} \tag{6.49}$$

or, in terms of excess carrier concentrations,

$$\hat{p}(0) = p_{0n}\left[\exp\left(\frac{qV}{kT}\right) - 1\right] \quad \text{and} \quad \hat{p}(\infty) = 0 \tag{6.50}$$

Applying these to (3.39) gives an expression for the excess hole concentration in region 1, namely,

$$\hat{p}(x) = p_{0n} \exp\left(\frac{-x}{L_h}\right)\left[\exp\left(\frac{qV}{kT}\right) - 1\right] \tag{6.51}$$

The reverse-bias voltage V is negative, so for $|V| > 3kT/q$, $p(0)$ is essentially equal to $-p_{0n}$. The negative sign renders positive the term $-\hat{p}/\tau_h$ in the continuity equation, as is consistent with the notion of a net rate of generation. The appropriate form of the continuity equation is

$$\int_{J_{h1}(0)}^{J_{h1}(\infty)} dJ_{h1} = \frac{-q}{\tau_h}\int_{x=0}^{x=\infty} p_{0n}\left[\exp\left(\frac{qV}{kT}\right) - 1\right]\exp\left(\frac{-x}{L_h}\right) dx \tag{6.52}$$

which gives the hole current resulting from generation in region 1:

$$J_{h1} = \frac{(qp_{0n}L_h)[\exp(qV/kT) - 1]}{\tau_h} \tag{6.53}$$

A similar expression to this pertains to the current resulting from the generation of electron–hole pairs in region 2 of Fig. 6.17. Electrons are the minority carriers in this region, so the equation for the current is just the electron equivalent of (6.53). This current, even though it is expressed in terms of electronic properties, can be assigned to J_{h2} because the generation of one hole accompanies the generation of one electron and only the flow of one of these has to be followed to compute the external current correctly (see Fig. 6.17). Thus

$$J_{h2} = \frac{(qn_{0p}L_e)[\exp(qV/kT) - 1]}{\tau_e} \tag{6.54}$$

To estimate the current due to electron–hole pairs generated in the depletion region, we follow the procedure used in the forward-bias case and partition the depletion region into two zones with a common interface at the point at which $n = p$. Figure 6.18 gives an enlarged version of the depletion region conditions originally presented in Fig. 6.13.

Considering zone Z first, for computational convenience we locate the x-origin at the point at which $n = p$ (i.e., using the quasi-equilibrium approximation, the point at which $p = n_i \exp[qV/(2kT)]$). Assuming, as in the forward-bias case, that the equilibrium hole concentration at the boundary is n_i, the excess hole concentration at $x = 0$ is $n_i[\exp(qV/2kT) - 1]$. At the other boundary, $p(x_Z) = p_{0n}[\exp(qV/kT)]$, which, because of the reverse-bias voltage, is effectively equal to zero. The excess hole concentration at $x = x_Z$ is thus $-p_{0n}$. The shaded area shown under the hole profile in Fig. 6.18 represents the number of holes generated in region Z. Because $p_{0n} \ll n_i$, it is legitimate to take the excess hole concentration as varying from $n_i[\exp(qV/2kT) - 1]$ to zero over this region. Approximating the resulting excess hole concentration profile as triangular allows formulation of the continuity equation for the hole current originating from generation in this zone as

$$\int_{J_{h3Z}(0)}^{J_{h3Z}(x_Z)} dJ_{h3Z} = \frac{-q}{\tau_h} \int_{x=0}^{x=x_Z} n_i \left[\exp\left(\frac{qV}{2kT}\right) - 1 \right] \left(1 - \frac{x}{x_Z}\right) dx \tag{6.55}$$

where, as in the forward-bias case, τ_h has been taken outside the integral because

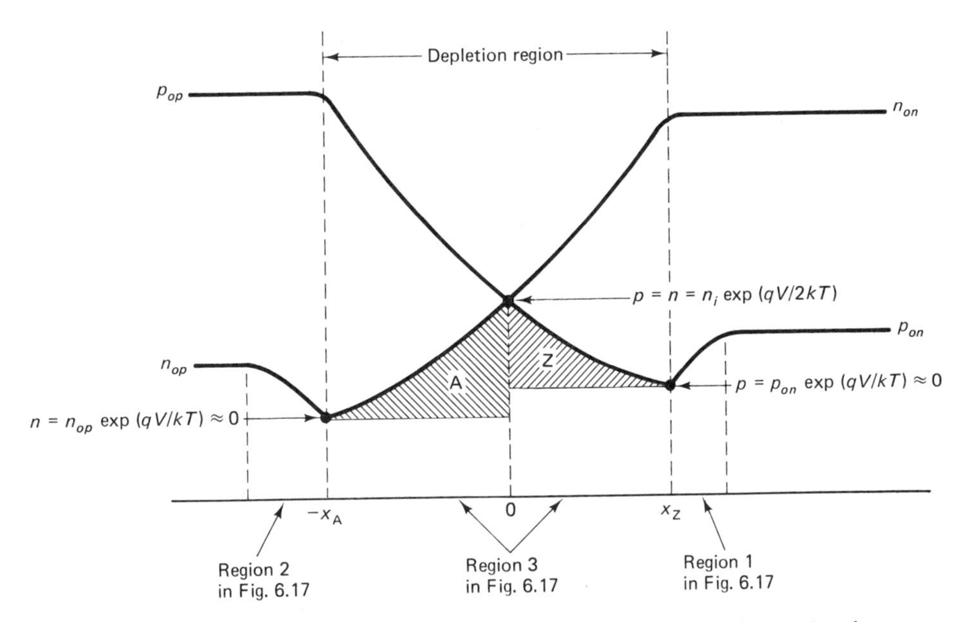

Figure 6.18 Carrier concentrations in reverse bias. The two shaded areas A and Z, wherein generation occurs of minority carrier electrons and holes, respectively, are considered to be triangular in shape.

of its assumed positional independency. Performing the integration, we obtain:

$$J_{h3Z} = \frac{qn_i}{2\tau_h} x_Z \left[\exp\left(\frac{qV}{2kT}\right) - 1\right] \tag{6.56}$$

In zone A of the depletion region the minority carriers are electrons. Therefore, the expression for the current resulting from the generation of electron–hole pairs in this zone is just the electron equivalent of (6.56), namely,

$$J_{h3A} = \frac{qn_i}{2\tau_e} x_A \left[\exp\left(\frac{qV}{2kT}\right) - 1\right] \tag{6.57}$$

Addition of J_{h3A} and J_{h3Z} gives the hole current due to generation of electron–hole pairs in the depletion region. Thus the summation of (6.56), (6.57), (6.53), and (6.54) gives the total current in reverse bias, namely,

$$\begin{aligned} J_h = {} & \left[\frac{qn_i}{2}\left(\frac{x_A}{\tau_e} + \frac{x_Z}{\tau_h}\right)\right]\left[\exp\left(\frac{qV}{2kT}\right) - 1\right] \\ & + q\left(\frac{p_{0n}L_h}{\tau_h} + \frac{n_{0p}L_e}{\tau_e}\right)\left[\exp\left(\frac{qV}{kT}\right) - 1\right] \end{aligned} \tag{6.58}$$

Not surprisingly, this equation is exactly the same as the one derived from considerations of recombination [see (6.48)]. This has to be because no limitation has been placed on the polarity of V. With V positive, the terms [exp $(qV/2kT)$ $-$ 1] and [exp (qV/kT) $-$ 1] in (6.48) are positive, giving a positive forward-bias current. With V negative, these terms in (6.58) are negative, giving the desired negative reverse-bias current.

The J–V relationship for reverse bias is displayed in Fig. 6.19. Note that the current changes so slowly with voltage that the data can be presented in a linear-linear plot, unlike the case for forward bias (Fig. 6.16). The exponential voltage terms affect the current only over the voltage range from 0 to about

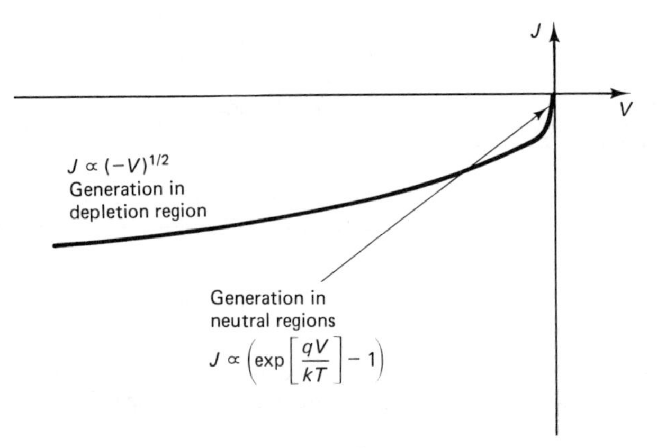

Figure 6.19 Reverse-bias J–V characteristic, from (6.58).

$4kT/q$ volts (i.e., for reverse biases up to about 100 mV). Thus the major voltage dependence of the reverse-bias current stems from the variation with bias of x_A and x_Z in the term for generation in the depletion region. This is a square-root dependence [see (6.25)], which accounts for the relative constancy of the reverse-bias current when compared to the forward-bias current, which has an exponential dependence on bias. This feature of ***asymmetry in currents***, or ***rectification*** as it is so-called, is the principal characteristic of diodes. It is illustrated in Fig. 1.1a.

Calculations of reverse-bias characteristics for silicon diodes can be performed by following Example 6.2.

Example 6.2 Reverse-Bias Characteristics of *PN* Diodes: The Effect of Doping Density

The various terms in the diode I–V equation which are affected by the doping densities in the diode have been compiled in Example 6.1. We use the data of that exercise, as summarized in Table 6E.1, to compute the reverse-bias currents for two n^+p diodes, each with a donor doping density of 10^{19} cm^{-3}, but with different acceptor doping densities (10^{19} cm^{-3} and 10^{15} cm^{-3}). Again, we take $T = 300$ K and the diode area as 0.01 cm^2. The results are plotted in Fig. 6E.2. Note the following:

(a) The results can be displayed on a linear-linear plot because, for all

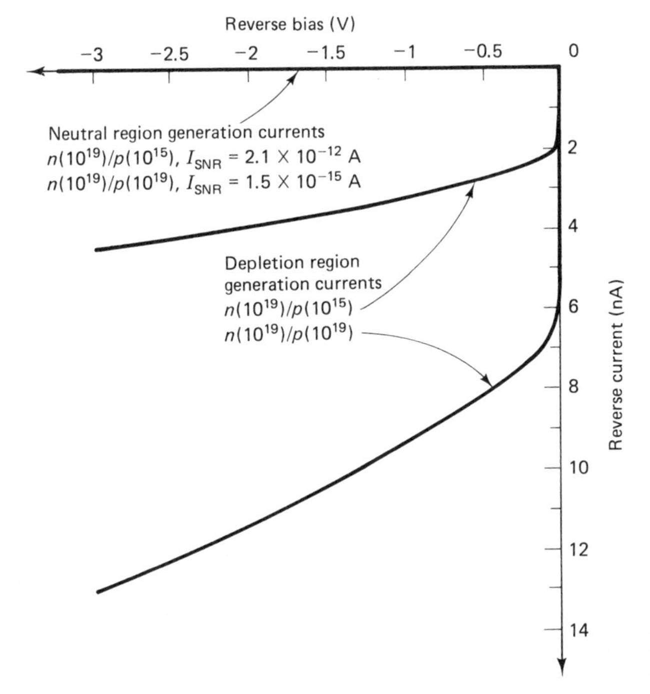

Figure 6E.2 Computed reverse bias I–V characteristics for n^+p diodes.

practical purposes, the exponential voltage terms in (6.58) are essentially zero.

(b) The neutral region generation currents are negligible compared to the depletion region generation currents. This is because of the relative magnitudes of I_{SNR} and I_{SDR} (see Table 6E.1).

(c) The depletion region generation current increases with the p-type doping density. This is a consequence of the doping density dependence of the lifetime that appears in the denominator of I_{DR}.

(d) The depletion region current increases with reverse bias. This is due to the widening of the depletion region, which follows an approximate $(-V)^{1/2}$ relationship [see (6.25)].

6.3.3 Ideal Diode Equation

In some practical diodes, either the doping densities of the device or the magnitude of the applied voltage are such that the components of current due to recombination and generation in the depletion region are negligible. A diode with such properties is termed an ***ideal diode***. The ideal diode equation is much used in diode analysis because of its simplicity when compared to the more complete equations we have developed in the previous two sections. By ignoring the depletion region recombination–generation current components in (6.48) and (6.58) the ideal equation materializes:

$$J = q\left(\frac{L_e n_{0p}}{\tau_e} + \frac{L_h p_{0n}}{\tau_h}\right)\left[\exp\left(\frac{qV}{kT}\right) - 1\right] \tag{6.59}$$

This equation is usually abbreviated to

$$J = J_0\left[\exp\left(\frac{qV}{kT}\right) - 1\right] \tag{6.60}$$

where J_0 is called the ***saturation current density***, in recognition of the fact that for reverse biases in excess of a few kT/q, J is effectively constant in an ideal diode at a value $= -J_0$.

6.4 DC CIRCUIT MODELS

The equations presented in Section 6.3 serve the purpose of linking the external observables of current and voltage to the internal material parameters of the semiconductor device. As such, the equations are of use to device designers who try to meet a particular I–V specification for a diode by determining the required doping densities. On the other hand, the equations would not normally be used by solid-state circuit designers, as they are rarely concerned with how the diode has come to have a particular I–V characteristic. These engineers are more interested in how a diode can interact with other components to produce a circuit that performs a specific function. For them, a circuit model of the diode, which represents the I–V characteristic of the real diode, is often a valuable design aid.

Three dc models for the diode are shown in Fig. 6.20. The simplest (Fig. 6.20a) draws on the highly asymmetrical character of the I–V curve and models the diode as a ***perfect switch*** (i.e., with $I_D = 0$ for $V_D < 0$ and $V_D = 0$ for $I_D > 0$). A slight improvement on this is shown in Fig. 6.20b, where the forward conduction of the diode is delayed by a reverse bias until a voltage threshold called V_{TO}, the ***turn-on voltage***, has been exceeded. The resulting I–V curve bears a little more resemblance to the real curve than does the one for the perfect switch. V_{TO} is often taken to be about 0.5 V for silicon diodes and about 0.75 V for gallium arsenide *pn* junctions. Note that these values refer to the junction voltage and are somewhat arbitrary as the value of V_{TO} really depends on the scale of observation of the current. In practice, some voltage is dropped across the contacts to the diode and the neutral regions of the device. These effects are accounted for by raising $\boldsymbol{V_{TO}}$ to about ***0.7 V for Si*** and ***0.9 V for GaAs***, and displaying the

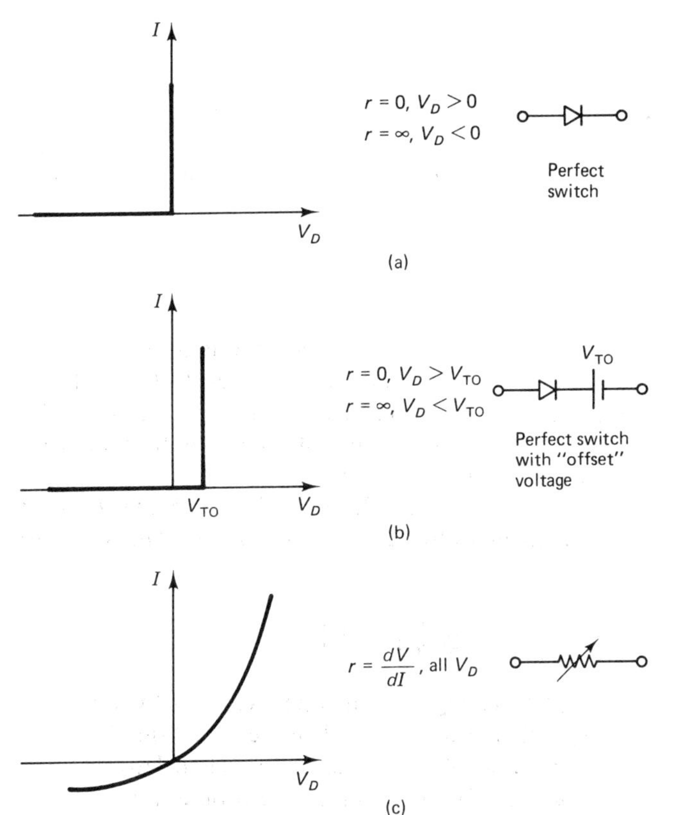

Figure 6.20 Dc equivalent circuit models for the *pn*-junction diode.

applied voltage, V_A, on the abscissa of the characteristic, rather than the diode junction voltage, V_D. Note that this is the first time in this book that it has been necessary to distinguish between these two voltages. In all the preceding text V_A and V_D have been taken as equal and labeled as V_A or simply as V.

In the final dc model shown in Fig. 6.20c, the diode is modeled by a voltage-dependent resistor of resistance $r(V)$. The value of this resistance at any voltage is computed from the inverse of the slope of the I–V curve at that voltage, that is,

$$r(V) = \frac{dV}{dI} \tag{6.61}$$

In reverse bias when the current is due primarily to generation in the depletion region, it can be deduced from (6.58) that $r(V_R)$ is large and increases only slowly with voltage (approximately in proportion to the square root of the applied voltage). On the other hand, for forward bias, $r(V_F)$ rapidly becomes very small as the voltage increases. For example, in a silicon diode when the junction voltage is in excess of about 0.4 V, such that the current is dominated by recombination in the neutral regions, it follows from (6.48) that

$$r(V_F) = \frac{kT}{qI} \qquad V_F > 0.4 \text{ V} \tag{6.62}$$

Thus, in this regime of operation, r decreases exponentially with V. This wide range of voltage-controlled resistance is often exploited in circuits that use diodes for signal attenuators (see Example 6.5). Diodes used in this fashion as variable resistors are called ***varistors***.

6.4.1 SPICE Model

The most widely used circuit analysis program in the semiconductor field was developed at the University of California, Berkeley and is called SPICE. This is an acronym for ***s***imulation ***p***rogram with ***i***ntegrated ***c***ircuit ***e***mphasis. A brief description of the program appears in Appendix 1. Many of the examples and problems presented in this book make use of this program.

The dc model for the diode employed by SPICE is shown in Fig. 6.21. It comprises a resistor RS and a voltage-controlled current source. The value of diode current I_D is determined by

$$I_D = \mathsf{IS}\left[\exp\left(\frac{qV_D}{\mathsf{N}kT}\right) - 1\right] \tag{6.63}$$

where V_D is that portion of the applied voltage which appears across the junction. This equation bears a close resemblance to the ideal diode equation (6.60). In fact, by setting IS $= I_0$ and N $= 1$, an ideal diode can be modeled. However, IS and N are ***model parameters*** and can be arbitrarily set by the program user to any values. For example, IS could be set according to (6.46) and (6.47) and N could

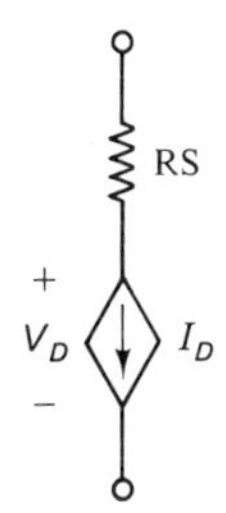

Figure 6.21 Dc equivalent circuit for a diode, as used in SPICE. The dependent current source is described by (6.63).

be set equal to 2. These conditions would allow modeling of a diode in which recombination–generation in the depletion region was the dominant component of current, although it should be noted that IS is a constant and so does not properly model the voltage dependence of the depletion region current, which is due to the varying width of the depletion region.

The third model parameter RS is used to take account of real resistances associated with the neutral regions and contacts, and any equivalent resistance associated with high-level injection (see Section 6.5.1).

Example 6.3 demonstrates how to compute I–V characteristics for pn diodes with series resistance using SPICE.

Example 6.3 Forward-Bias Characteristics of *PN* Diodes: The Effect of Series Resistance

For this example we consider the effect of series resistance on the forward-bias characteristic of a diode. Series resistance can be attributed to the neutral regions of the device and to the metal-semiconductor contacts at which connections to external circuitry are made. At high currents the voltage drop across these resistive regions becomes nonnegligible, causing the voltage appearing at the actual junction of the diode to be significantly less than the applied voltage.

Specifically, we consider the effect of 1 and 10 Ω of series resistance on the forward-bias characteristic of a n^+p diode which has donor and acceptor doping densities of 10^{19} and 10^{17} cm^{-3}, respectively. As in previous examples, we take $T = 300$ K and the area as 0.01 cm^2. Because series resistance effects are manifest at high currents, at which the neutral region is likely to be the dominant source of recombination current (see Fig. 6E.1), we neglect the depletion region current. The relevant diode current is given, therefore, by (6E.3), that is,

$$I = I_{\text{SNR}} \left[\exp\left(\frac{qV}{kT}\right) - 1 \right]$$

This equation is easily matched to the SPICE diode equation (6.63) by noting that

$$\text{IS} = I_{\text{SNR}} = 2.8 \times 10^{-14}\ \text{A} \qquad \text{(from Table 6E.1)}$$

$$\text{N} = 1$$

The additional model parameter RS takes on the values of 0, 1, or 10 Ω, as desired.

Three diodes, corresponding to the three instances of series resistance, can be simulated simultaneously using the circuit of Fig. 6E.3. The SPICE program input listing is given below.

```
*EFFECT OF RS ON DIODE FORWARD IV CHARACTERISTICS
VDD 4 0
VI1 4 1
VI2 4 2
VI3 4 3
D1 1 0 D170
D2 2 0 D171
D3 3 0 D1710
.MODEL D170 D IS=2.8E-14 N=1
.MODEL D171 D IS=2.8E-14 N=1 RS=1
.MODEL D1710 D IS=2.8E-14 N=1 RS=10
.DC VDD 0.1 0.9 0.05
.WIDTH OUT=80
.PRINT DC I(VI1) I(VI2) I(VI3)
.END
```

The zero-value independent voltage sources VI1, VI2, and VI3 act as ammeters, each measuring the current in one of the diodes. The notation for the diodes uses 17 to refer to the *p*-type doping density of 10^{17} cm^{-3}, and a following number to indicate the series resistance. Note in the listing that the dc voltage source VDD is varied over the range 0.1 to 0.9 V in 0.05-V increments, and an instruction is issued on the PRINT line to print the current in each of the diodes. The SPICE program output listing automatically lists these currents alongside each *VDD* value. The results are shown graphically in Fig. 6E.4. Note the following:

(a) A series resistance of 1 Ω in a 0.01-cm^2 device only begins to affect the diode *I–V* characteristic at a forward-bias current of about 10^{-2} A (i.e., at a current density of 1 A cm^{-2}).

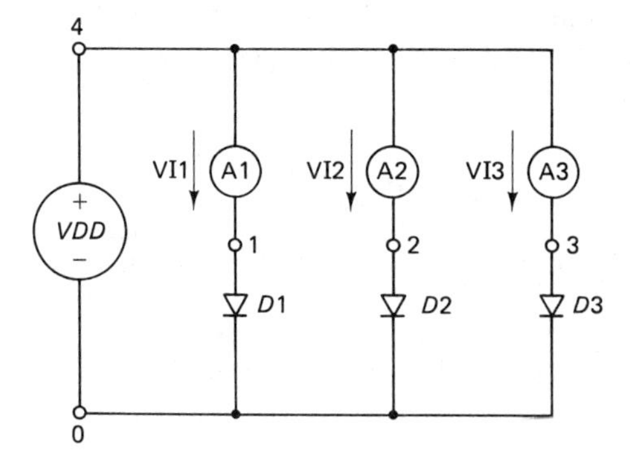

Figure 6E.3 Circuit for the SPICE simulation of diode forward-bias characteristics. *Note:* A voltage source with no assigned value and denoted by *VI* is used as an ammeter in the SPICE circuits in this book.

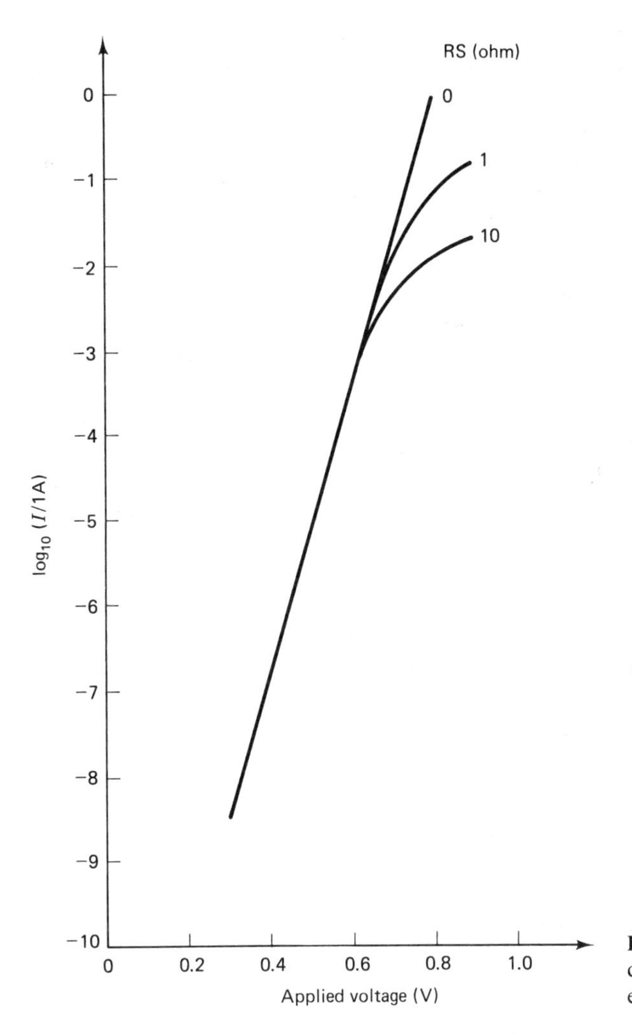

Figure 6E.4 Computed forward-bias diode characteristics, illustrating the effect of series resistance.

(b) The horizontal displacement of the curve for diode 171, with respect to the curve for diode 170, is 0.1 V at 0.1 A. This confirms that the difference between the two curves is due to the voltage drop across the series resistance.

6.5 SECONDARY EFFECTS IN REAL DIODES

6.5.1 High Current Effects

A departure of the forward bias I–V curve at high current densities from its ideal relationship, in a manner similar to that depicted in Fig. 6E.4, can also be brought about by ***high-level injection***. This phenomenon occurs when the minority carrier concentrations injected into the neutral regions are so great that low-level injection conditions no longer apply (i.e., the excess carrier concentrations are comparable to the equilibrium majority carrier concentrations). The excess majority carrier concentrations, which materialize to neutralize the charge of the excess minority carriers (see Section 6.2.2.4), cannot be ignored in high-level injection. They substantially increase the conductivity of the diode, leading to a condition called ***conductivity modulation***.

The presence in the same part of the device of large concentrations of both excess minority carriers and excess majority carriers leads to increased recombination. The situation bears some resemblance to the forward-bias, low-level injection case in the depletion region of a diode. We saw in Section 6.3.1 that this situation leads to a depletion region current that varies as exp $(qV/2kT)$. In high-level injection the neutral region current has a similar voltage dependency. As the neutral region current dominates the diode $I-V$ characteristic at high bias, the onset of high-level injection leads to a reduction in the slope of the ln $I-V$ curve.

6.5.2 Avalanche Breakdown

At high reverse voltages, the $I-V$ curve for real diodes departs markedly from the theoretically predicted relationship (see Fig. 6.22). This departure is due to carrier multiplication effects which occur within the depletion region when a critical value of electric field, called the breakdown field, is exceeded. To understand the origin of this effect, consider the energy band diagram of Fig. 6.23, which has been drawn for the case of a large reverse bias, and follow the path of an electron generated close to the left-hand edge of the depletion region.

The electron is accelerated by the electric field in the depletion region and gains kinetic energy. Recall from Fig. 2.12 that the energy level E_C represents potential energy, so the drawing of an electron in the conduction band with

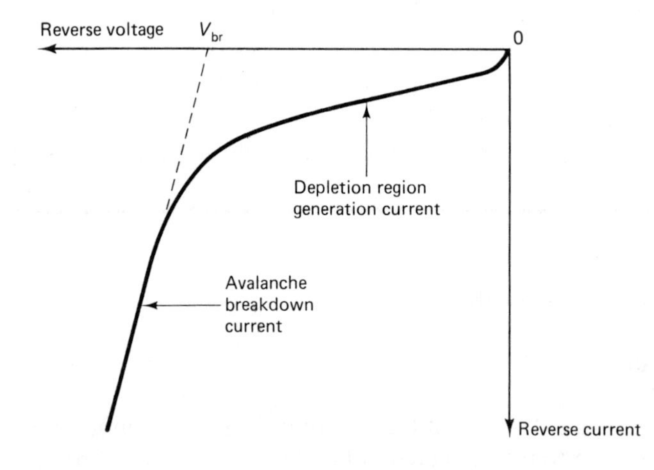

Figure 6.22 Diode $I-V$ characteristic for reverse bias, showing the effect of avalanche breakdown.

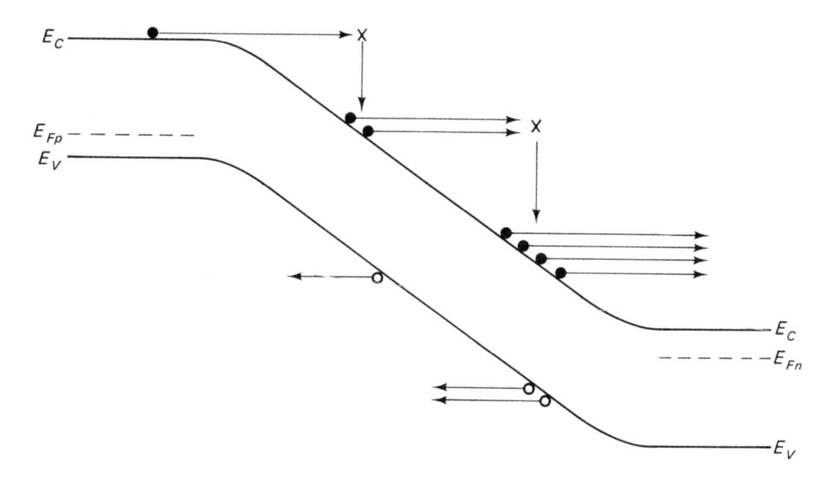

Figure 6.23 Energy band diagram for large reverse bias, showing avalanche multiplication. Extra electron–hole pairs can result from an electron–atom collision, provided that the electron has gained kinetic energy in excess of E_g since its previous collision.

$E > E_C$ implies that the electron has kinetic energy. If this electron can gain a kinetic energy that is at least equal to the energy bandgap before it collides with a lattice atom, then, when it finally does collide in an inelastic collision, the kinetic energy it loses can be sufficient to break a bond completely. The result is that the original electron is momentarily stationary, with only potential energy (E_C), but is joined by a second electron, namely, the one released in the collision. These two electrons can now be accelerated by the field and, on gaining K.E. $> E_g$, can generate two more electron–hole pairs on collisions with lattice atoms. And so the process can continue, causing the current to grow in the fashion of an avalanche accreting snow as it slides down a mountainside.

Avalanching in a diode is usually referred to as breakdown; however, this does not mean that the process is necessarily destructive. The large currents generate heat through Joulean heating (I^2R losses), and it is the dissipation of this heat which is critical to the survival of the device. If the construction and packaging of the diode is such that a large temperature rise cannot be avoided, then destructive breakdown, usually in the form of the vaporization of the metallic contacts to the diode, can occur. Operation in the avalanche breakdown mode for short periods, or for longer times with adequate heat sinking, may not be harmful to the device. IMPATT diode oscillators and avalanche photodiodes are examples of devices that normally operate in the avalanche breakdown mode [1].

For avalanching to occur, a reverse-bias voltage in excess of E_g/q has to be applied and the depletion region width must be greater than several electron mean-free-path lengths. These conditions are amalgamated in the specification of a ***breakdown field strength*** which is characteristic of the semiconductor material.

The values of $\mathscr{E}_{br}$ for silicon and gallium arsenide are about 3×10^5 and 4×10^5 V/cm, respectively.

An appreciation of the dependence of the breakdown voltage on device properties can be acquired by substituting $\mathscr{E}_{br}$ for the maximum field in the depletion region. This field occurs at the metallurgical boundary of the junction (Fig. 6.6) and has a magnitude that can be deduced from (6.4):

$$|\mathscr{E}_{max}| = qN_A x_{dp} \tag{6.64}$$

Substituting for x_{dp} from (6.21) and (6.25), and assuming that the applied voltage is much greater than the built-in potential, as it assuredly will be at breakdown, gives the following expression for the ***breakdown voltage***:

$$|V_{br}| = \mathscr{E}_{br}^2 \frac{\epsilon}{2q}\left(\frac{1}{N_D} + \frac{1}{N_A}\right) \tag{6.65}$$

From this equation it can be appreciated that the only way a device designer can increase significantly the breakdown voltage is by reducing the lesser of the two doping densities in the diode.

6.5.3 Zener Breakdown

Another type of electrical breakdown can occur on the reverse biasing of a special type of *pn*-junction diode called a Zener diode. The I–V characteristic of a Zener diode is shown in Fig. 6.24. Its distinctive feature is the abrupt rise in current at a reverse-bias voltage V_z. This voltage, which is usually in the neighborhood of 3 to 6 V, is generally much lower than that at which current increase due to avalanche breakdown occurs. The rapid increase in current is due to the onset of tunneling. The phenomenon is often called ***Zener breakdown***, after Clarence Zener, one of the early workers in the field of electronic excitations in semiconductors.

Tunneling, as Section 5.8 relates, can take place when charge carriers at a particular energy are separated from empty allowed states at the same energy level by a short physical distance (i.e., <5 nm). These conditions can arise in a reverse-biased *pn* diode which has been fabricated such that the doping density on both sides of the junction is high. The high doping densities ensure that the depletion region width is small [see (6.25)]. Although this width increases with reverse bias, the actual separation between the conduction and valence bands decreases (see Fig. 6.25). When this separation reaches about 5 nm the energy levels populated by electrons in the valence band of the p material will be aligned with the empty levels in the conduction band of the n material. Under these conditions the electrons can tunnel from the p side to the n side of the junction. The voltage at which this occurs is the ***Zener voltage***, V_z. The onset of tunneling is abrupt because for voltages of less than V_z, the energy levels full of electrons in the valence band of the p-type material are either opposite the forbidden energy levels of the bandgap of the n-type material, or are separated from the empty

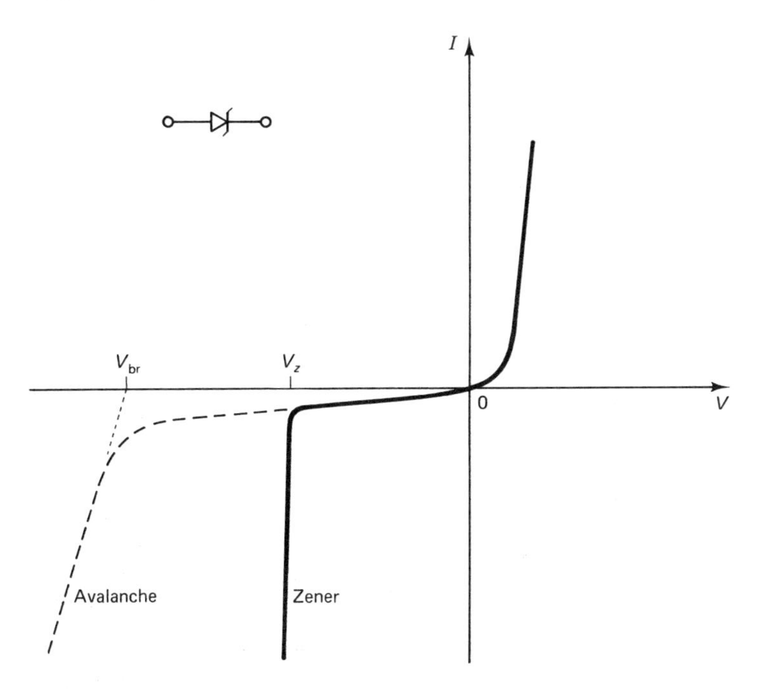

Figure 6.24 *I–V* characteristic of a Zener diode. The onset of breakdown is usually much "sharper" than in the avalanche case, and the conductance is generally higher. The circuit symbol for a Zener diode is shown in the top left.

allowed levels of the *n*-type conduction band by a distance in excess of about 5 nm. In each of these instances, the electrons cannot transfer by tunneling, so the reverse-bias current is limited to the small magnitude that is representative of generation in the depletion region.

It follows from the description above that Zener breakdown will occur in any *pn* junction which can be reverse biased such that regions of full and empty allowed states at the same energy are separated by the required short distance. Note, however, that if the doping densities are such as to render the zero-bias depletion region width large, it may not be possible to increase the reverse bias to V_z and so obtain the necessary physical separation of the energy levels, because avalanche breakdown will occur first.

Relative to avalanche breakdown, the conductance dI/dV in Zener breakdown is high and enables Zener diodes to be used effectively in voltage regulator circuits.

6.5.4 Quasi-Neutral Regions

In most diodes the more heavily doped region, often called the ***emitter***, is formed by diffusion or ion implantation into a uniformly doped ***base*** region, as discussed in Sections 12.2.1 through 12.2.4. The resulting doping profile is sketched in Fig.

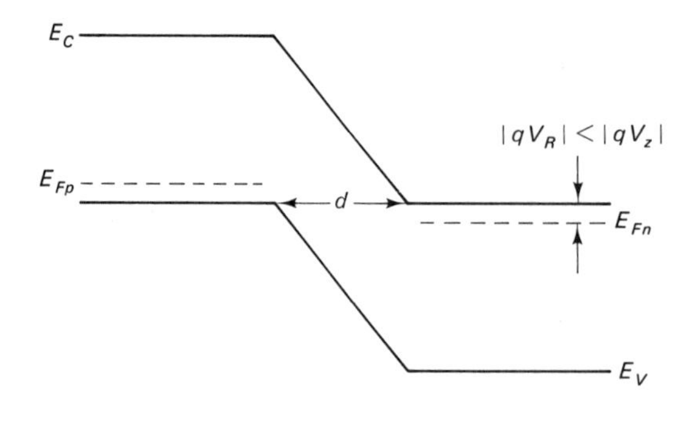

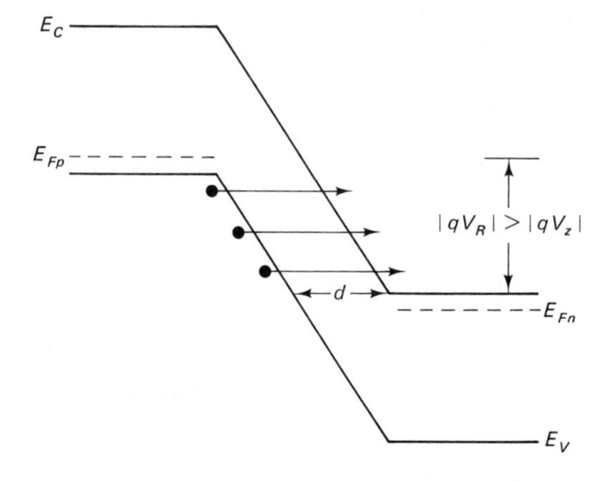

Figure 6.25 Electron tunneling in Zener breakdown. As the reverse bias increases, the depletion region width increases approximately as $(-V_R)^{1/2}$, but $E_{Fp} - E_{Fn}$ increases as $|V_R|$. Thus the physical distance d (horizontal separation) between the bands in the depletion region decreases as the reverse bias gets larger. When d reaches about 5 nm, tunneling can occur.

6.26; the important point to note is that the emitter is not uniformly doped. In Section 6.2.1 it was shown that the difference in doping type between the emitter and base leads to the formation of a depletion region in which there are effectively no free carriers. The depletion region contains a net charge density associated with uncompensated donor and acceptor ions, and consequently has a large built-in electric field. This depletion region centers on the base–emitter metallurgical junction at $x = x_j$.

It is fairly easy to show that within the emitter itself there must also be a ***built-in electric field*** and a net charge density to support this field. From Fig. 6.26 it can be seen that the donor concentration is highest near the surface of the emitter and falls off toward the depletion region boundary. As a result, the electron concentration drops on moving through the emitter. In response to this concentration gradient, electrons diffuse away from the surface of the emitter, creating a region in which there is a surplus of donor ions over compensating electrons, giving a net positive charge. The electrons that have diffused away from the surface congregate closer to the edge of the depletion layer, creating another region in which there is a slight surplus of electrons, leaving a net negative charge. The presence of these two charged regions sets up a built-in electric field, as shown in Fig. 6.26. As expected, this field is in the direction opposing the diffusion

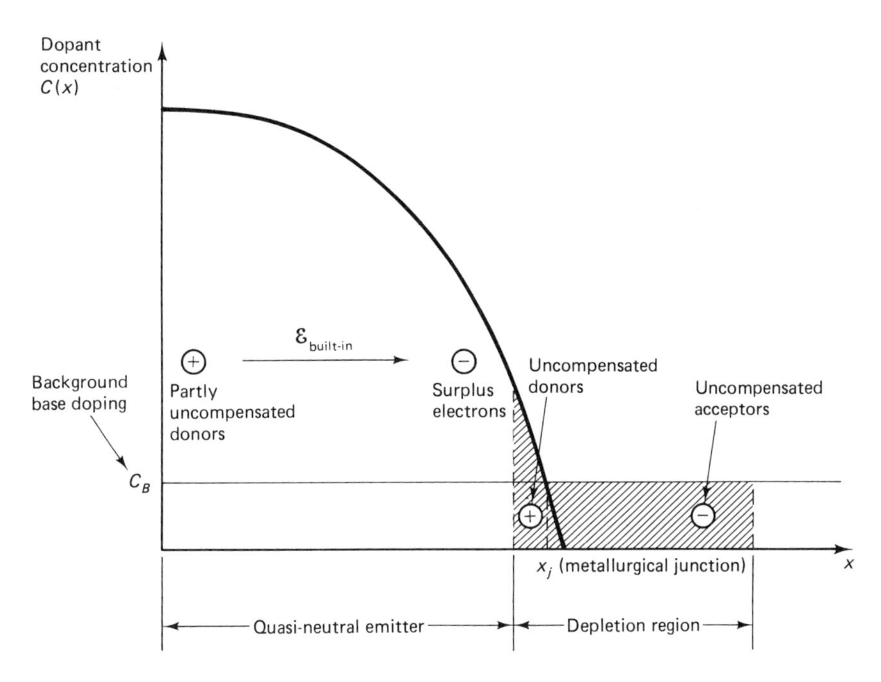

Figure 6.26 The nonuniform doping profile in a diffused or implanted emitter produces a built-in electric field.

of electrons from the surface; in equilibrium, the electron drift current resulting from the field exactly balances the electron diffusion current. The emitter is ***quasi-neutral*** rather than being precisely electrically neutral; the nonuniform doping profile leads to a slight imbalance between the concentration of dopant ions and free charge carriers, producing a built-in electric field. In contrast, in equilibrium the uniformly doped base is actually neutral, having no net charge density.

It is of interest to estimate the extent of the charge imbalance for a typical emitter doping profile. This is done in Problem 6.16, where it is found that the fractional difference between the dopant charge and free carrier charge is usually very small.

6.6 TRANSIENT ANALYSIS

One of the important uses of diodes in electronic circuits is in switching. The highly conductive state of forward bias and the poorly conductive state of reverse bias have obvious associations with the two states, on and off, of a switch. The time taken for the diode to change state has considerable practical importance in the realization of high-speed switching circuits. The response of the diode to an abrupt change in applied voltage or current is determined by performing a transient analysis of the circuit. As far as the diode is concerned, its response to a transient signal is determined by the time taken to establish new charge distributions within the device. Changes in charge are associated with changes in the width of the depletion region, and with changes in the number of minority carriers stored in the neutral regions close to the depletion region. For example, a change in applied voltage that takes the diode from forward bias to reverse bias will cause a widening of the depletion region and a reduction in the number of minority carriers stored near to the junction. This can be seen by comparing Figs. 6.12 and 6.13.

Overall the device will remain neutral while the charge redistribution is taking place. In the example cited, as electrons are drawn out of one terminal to increase the depletion width on the n side of the junction, an equal number of electrons must flow in at the other terminal to ionize the acceptors and extend the depletion region on the p side of the junction.

In general, the current in a device will not attain its new steady-state value until the charge redistribution is complete. In principle, the time taken for the charge redistribution, and the current during this time, should be computed by solving the full, time-dependent equations of state: the transport equations (3.22) and (3.23), the continuity equations (3.27) and (3.28), and Poisson's equation (3.29). Unfortunately, solving these coupled partial differential equations in a time-dependent mode is feasible in only a few very simple situations. For this reason, a much simpler approach to transient analysis is often adopted. This so-called ***quasi-static approach*** is based on the assumption that at any instant the charge distribution within the device is a function only of the present terminal voltages and not of any past history. The quasi-static approach is valid as long as circuit elements external to the device (particularly load capacitors) constrain the terminal voltages to change slowly compared to the time required for the internal charge redistribution.

Quasi-static models are used exclusively in the SPICE circuit simulator, and provide acceptable accuracy in describing the transient response of most modern bipolar and field-effect integrated circuits. In this book we avail ourselves of the results obtained from SPICE to investigate the transient response of diodes and transistors. SPICE implements the quasi-static method in the manner described in Appendix 1. The analysis employs numerical, iterative methods to solve for the nodal voltages and currents in an ***equivalent circuit*** representation of the device. Voltage-dependent current sources represent the dc I–V characteristic of the device, and capacitors in the equivalent circuit are associated with the space charge in the depletion region and with the excess minority carrier charge in the neutral regions in forward bias. We now proceed to calculate the relevant capacitances in a pn junction and to accommodate them in an equivalent circuit for the diode.

6.6.1 Junction Capacitance

The ***junction capacitance***, often also called the depletion or transition capacitance, is associated with the charge dipole formed by the ionized donors and acceptors in the junction space charge region. Specifically, the junction capacitance relates the changes in charge at the edges of the depletion region to changes in the junction voltage. The situation is illustrated in Fig. 6.27 and the capacitance is described by

$$C_J = \frac{dQ}{dV_D} \tag{6.66}$$
$$= \frac{d}{dV_D}(qN_D x_{dn}) = \frac{d}{dV_D}(-qN_A x_{dp})$$

where C_J and Q refer to unit area values of capacitance and charge, respectively, and all the impurities in the depletion region are assumed to be ionized. The boundaries of the depletion region are related to the depletion width W, and W is voltage dependent. Making the appropriate substitutions from (6.21) and (6.25), and differentiating gives

$$C_J = \frac{\epsilon}{[(2\epsilon/q)(1/N_A + 1/N_D)(V_{bi} - V_D)]^{1/2}} \tag{6.67}$$

The voltage dependence of the junction capacitance, as described by this equation, is plotted in Fig. 6.28. Under reverse-bias conditions, the combination

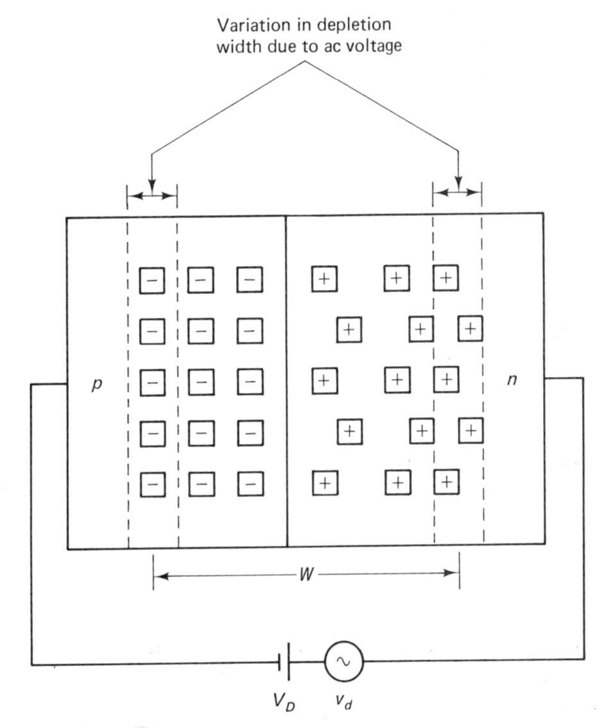

Figure 6.27 Junction capacitance. The variation in charge at the edges of the depletion region in response to the ac signal v_d gives the depletion region a resemblance to a parallel-plate capacitor of thickness W. If $v_d \ll V_D$, W is determined by the dc bias voltage.

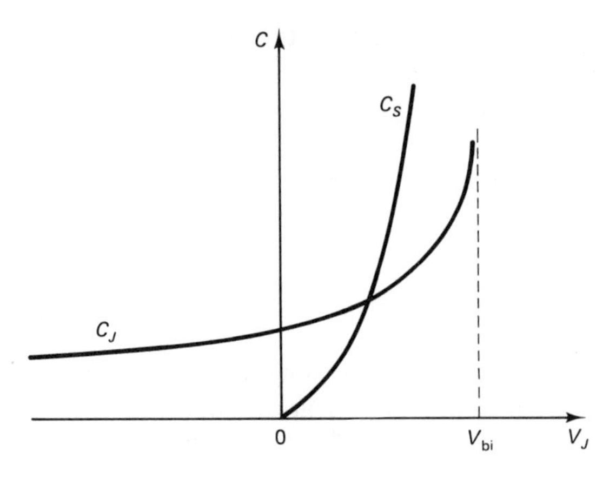

Figure 6.28 Variation of the storage capacitance and the junction capacitance with voltage.

of voltage-dependent capacitance and high resistance gives the diode the useful property of a low-current-drain reactance. Diodes that exploit this property are called ***varactors*** and are widely used in the electronic tuning stages of televisions and radios.

The junction capacitance described by (6.67) indicates a dependence on the square root of the junction voltage. This is true for an abrupt junction diode. For a linearly graded diode, one in which the transition from p-type to n-type material occurs in a linear fashion, the dependence is as the cube root of the voltage [2]. Practical diodes usually exhibit a voltage dependence which is well described by an index lying between these two limits.

By comparing (6.67) and (6.25), it will be appreciated that the junction capacitance can also be written as

$$C_J = \frac{\epsilon}{W} \tag{6.68}$$

This is the familiar expression for a parallel-plate capacitor. Its applicability here might seem surprising as the charge in the depletion region is distributed between the "plates" rather than located at them. In fact, there is no inconsistency because the charge of relevance to the capacitance is the new charge in the depletion region which arises due to the change in voltage. This charge does appear at the edges of the depletion region (see Fig. 6.27), so the parallel-plate analogy is understandable. Because the relevant charge is this incremental charge, C_J is given by (6.68) for any charge distribution in the depletion region, not just the uniform doping case used above.

6.6.2 Storage Capacitance

The ***storage capacitance***, often also called the diffusion capacitance, is associated with the excess minority carrier charge injected into the neutral regions under forward-bias conditions (see Fig. 6.12). Considering, for example, the excess hole charge stored per unit area in the neutral n region during forward bias, we have, from (6.32),

$$\begin{aligned} Q_h &= \int_0^{\infty} q\hat{p}(0) \exp\left(\frac{-x}{L_h}\right) dx \\ &= qL_h\hat{p}(0) \end{aligned} \tag{6.69}$$

The excess hole concentration at $x = 0$ (i.e., at the edge of the depletion region) is voltage dependent as described by (6.37), the law of the junction. Thus the capacitance per unit area associated with this charge density is given by

$$\begin{aligned} C_S &= \frac{dQ_h}{dV_F} \\ &= \frac{d}{dV_F}\{qL_h p_{0n}[\exp(qV_F/kT) - 1]\} \\ &= \frac{q^2}{kT} L_h p_{0n} \exp\left(\frac{qV_F}{kT}\right) \end{aligned} \tag{6.70}$$

where V_F, the forward-bias voltage, is explicitly used because it is only under this polarity of applied voltage that minority carrier charge is injected across the junction.

Note that even though we have referred to charge being stored, there is no actual space charge outside the depletion region. The injected minority carrier charge is always neutralized by excess majority carrier charge drawn from the contacts. The flow of excess minority carriers across the depletion region and the flow of excess majority carriers across the neutral regions constitute the ***transient current***. The net result of this current is a new steady-state charge distribution of carriers in the neutral regions close to the depletion region. In low-level injection, only the excess minority carrier charge is significant. The establishment of this charge can be viewed as the charging of a capacitor, and it is in this context that the term ***storage capacitance*** has meaning.

C_S is plotted in Fig. 6.28. Its exponential dependence on voltage causes it to be the dominant capacitance over most of the forward-bias range. As such, it imposes a limit on the speed with which the pn diode can be switched from forward bias to reverse bias (i.e., from "on" to "off"). Discharging this capacitance involves removal of the excess holes, either by diffusion to the depletion region or by recombination. The time taken for these processes to alter the excess hole profile from its form given in Fig. 6.12 to that shown in Fig. 6.13 is often called the ***storage time***. It is related to the minority carrier lifetime. Thus to fabricate high-switching-speed diodes, it is necessary to use semiconductor material with a short minority carrier lifetime. Gallium arsenide, for which minority carrier lifetimes are generally several orders of magnitude shorter than in equivalently doped silicon, would appear to be a promising material in this regard. However, the well-established processing technology for silicon devices results in most high-switching-speed *pn* diodes being made from this semiconductor material. The silicon for these diodes is intentionally doped with "lifetime killer" impurities such as gold. Gold increases the density of traps with energy levels close to the middle of the silicon energy bandgap, and thus increases the opportunity for recombination (see Section 2.5.2.2). As in most instances in engineering design, the adjustment of one parameter to improve some feature of performance leads to the deterioration of some other aspect of performance. In the case of reducing the lifetime, the debit side is that the reverse-bias current increases [see (6.58)], rendering the "off" state more conductive.

The correspondence between the storage capacitance and the minority carrier lifetime can be brought out by accommodating J_{h1}, from (6.41), in (6.70) by substituting for $p_{0n} \exp[qV/kT]$:

$$C_S = \frac{q}{kT} J_{h1}\tau_h \tag{6.71}$$

where it has been assumed that $qV_F \gg 5kT$, so the "-1" term in (6.41) can be neglected. In a *p-n* diode operating at a forward-bias voltage in excess of about $V_{bi}/2$ (see Fig. 6E.1), J_{h1} will be the dominant component of current. Under these circumstances the storage capacitance is directly proportional to the total current.

6.6.3 SPICE Model

The junction capacitance and the storage capacitance add together to form the reactive branch of the complete equivalent circuit for the *pn* junction diode (see Fig. 6.29). In SPICE, the total capacitance C_D for an abrupt junction diode is given by

$$C_D = \frac{q}{\mathrm{N}kT} \mathrm{TT\ IS} \exp\left(\frac{qV_D}{\mathrm{N}kT}\right) + \mathrm{CJO}\left(1 - \frac{V_D}{\mathrm{VJ}}\right)^{-1/2} \tag{6.72}$$

The first term represents the storage capacitance and the second term is the junction capacitance. The latter is merely a different way of writing (6.67), as is easily proved when it is realized that CJO is just the junction capacitance at zero

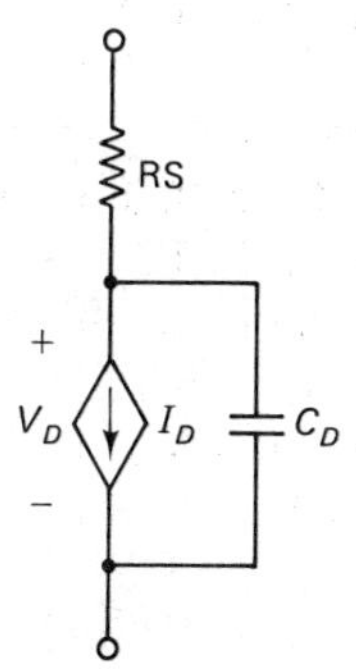

Figure 6.29 Large-signal ac equivalent circuit for a diode, as used in SPICE for transient analyses.

applied voltage. The expression for the storage capacitance is very similar to that appearing in (6.71). The differences arise because of the slight differences between the ideal diode equation (6.60), on which our expression for C_S is based, and the empirical diode equation (6.63) used in SPICE. Because of the empirical nature of the SPICE equation, the parameter TT appearing in (6.72) is not strictly identical to the minority carrier lifetime that appears in (6.71). The new model parameters needed to describe the capacitance C_D are, therefore, TT, CJO, and VJ, the built-in potential (equivalent to V_{bi} used elsewhere in this book).

For our purposes the diode model in SPICE comprises six parameters: IS, N, RS, TT, CJO, and VJ. For given doping densities in the p and n regions of a diode, estimates for IS and CJO can be obtained by evaluating (6.59) and (6.67), respectively. TT can be approximated by the minority carrier lifetime. N falls within the range 1 to 2 and is related to the dominant mechanism of charge flow in forward bias (see Sections 6.3.1 and 6.3.2). RS can be estimated from the deviation of the actual diode voltage from the ideal exponential characteristic at a specific current (see Fig. 6E.4).

The model as used in transient analyses is termed a large-signal model because there are no restrictions placed on the magnitude of the applied voltage. The general procedure followed by SPICE in carrying out a transient analysis is outlined in Appendix 1.

Example 6.4 illustrates how SPICE can be used to evaluate the response of a pn diode to an abrupt change in applied voltage.

Example 6.4 Switching Characteristics of *PN* Diodes: The Turn-Off Transient

For this example we investigate the response of a pn diode to a step function of applied voltage that seeks to turn the diode from its "on" state (forward bias) to its "off" state (reverse bias).

We consider an n^+p diode with donor and acceptor doping densities of 10^{19} and 10^{17} cm^{-3}, respectively. The diode area is 0.01 cm^2 and the temperature is 300 K. The dominant component of current is taken to be recombination/generation in the neutral regions; series resistance is ignored. The five appropriate model parameters can be obtained as follows (all from Table 6E.1):

$$\text{IS} = I_{SNR} = 2.8 \times 10^{-14}\ \text{A}$$

$$\text{N} = 1$$

$$\text{TT} = \tau_e = 1.7 \times 10^{-7}\ \text{s}$$

$$\text{VJ} = V_{bi} = 0.94\ \text{V}$$

$$\text{CJO} = \frac{\epsilon A}{W_0} = 9.49 \times 10^{-10}\ \text{F}$$

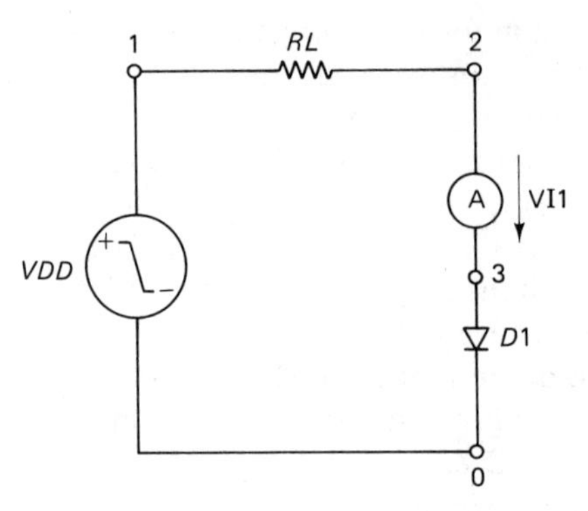

Figure 6E.5 Circuit for the SPICE simulation of diode transient performance.

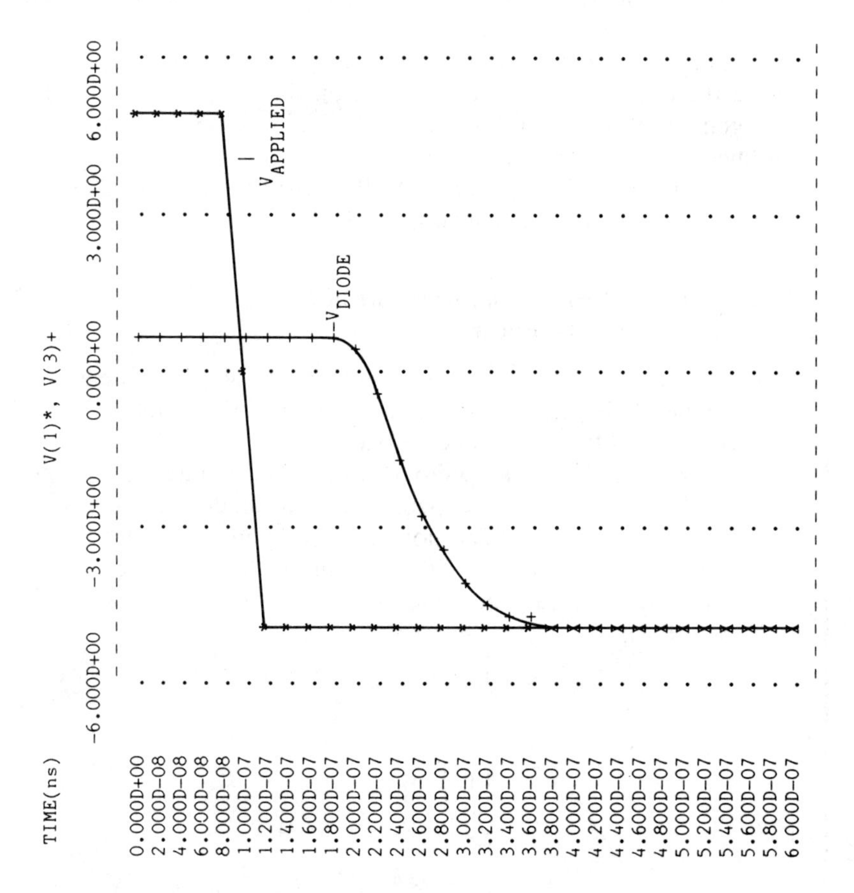

Figure 6E.6 Computed turn-off response for the diode in the circuit of Fig. 6E.5 with TT = 170 ns.

These values are entered in the model line of the SPICE input listing (see below).

```
*DIODE TURN-OFF TRANSIENT
VDD 1 0 PWL (0NS 5 90NS 5 110NS -5 600NS -5)
VI1 2 3
RL 1 2 100
D1 3 0 D17
.MODEL D17 D IS=2.8E-14 N=1 TT=1.7E-7 VJ=0.94 CJO=9.49E-10
.WIDTH OUT=80
.TRAN 20NS 600NS
.PLOT TRAN I(VI1) V(3)
.END
```

The circuit for this simulation is shown in Fig. 6E.5. The applied voltage *VDD* is specified in a piecewise-linear fashion and switches from +5 V to −5 V in 20 ns after the forward bias has been applied for 90 ns. The control

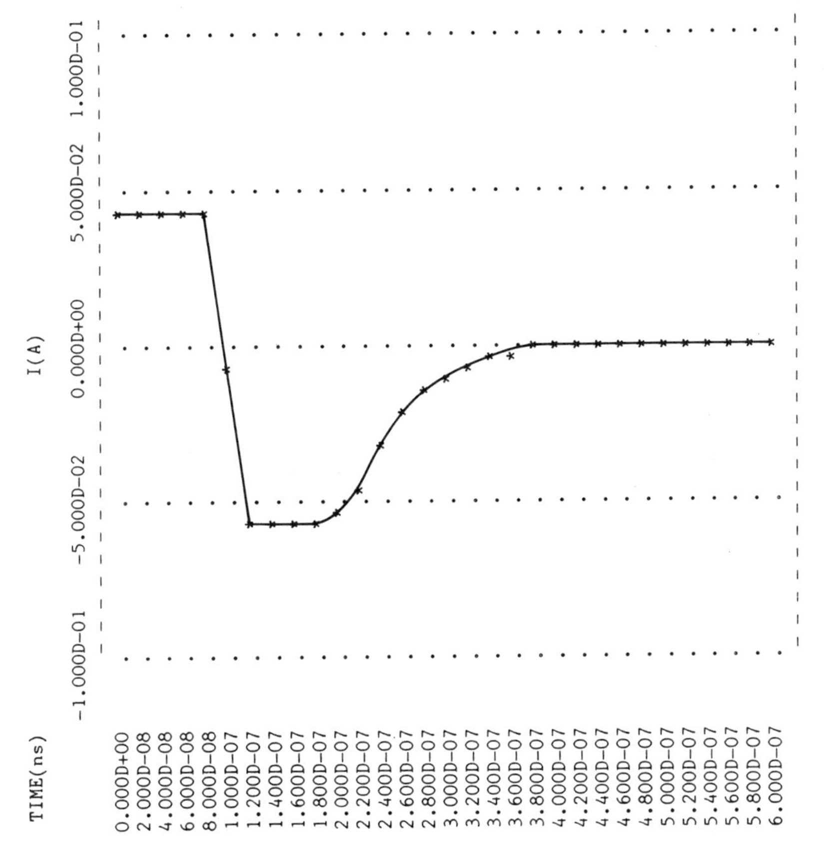

Figure 6E.6 *(continued)*

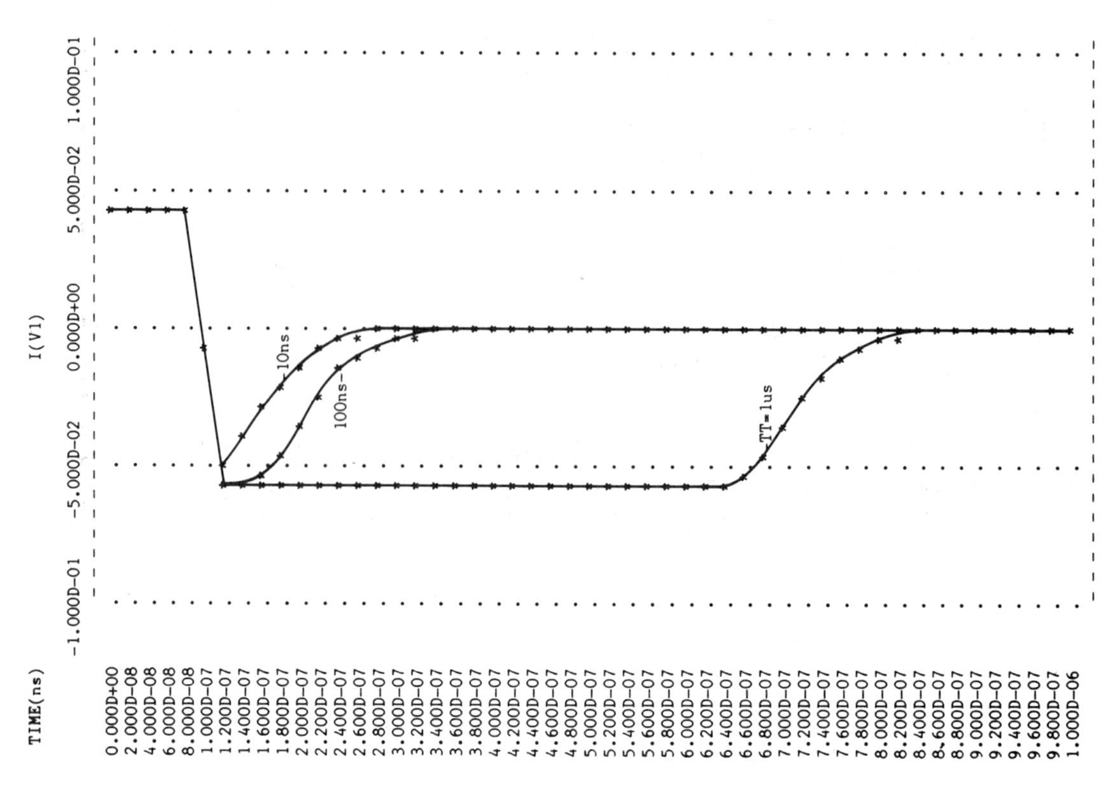

Figure 6E.7 Computed results for the effect of minority carrier lifetime on the turn-off time of the diode in the circuit of Fig. 6E.5.

line in the listing instructs SPICE to perform a transient analysis and obtain a solution every 20 ns over a 600-ns period. The plot line calls for a display of the current through the diode and the voltage across the diode (i.e., between nodes 3 and 0). The results are shown in Fig. 6E.6. Note the following:

(a) In response to the voltage change, the current switches initially to a value given by $(V_{\text{applied}} - V_{\text{diode}})/R_L$ (i.e., about -50 mA).

(b) The voltage across the diode does not respond immediately to the change in applied voltage.

(c) When the diode voltage does start to respond, the diode current commences its change to the low value expected for reverse-bias operation.

This interesting behavior is a result of the time needed for the charge distributions in the diode to alter from their forward-bias forms to their reverse-bias forms. Specifically, the most important change is that of the profile of the minority carriers stored in the neutral regions. As a comparison of Figs. 6.12 and 6.13 reveals, the change is significant. It can be accomplished only by allowing the minority carriers a chance either to diffuse to the depletion region and be extracted by the field therein, or to recombine with majority carriers in the neutral regions. Both of these processes depend on the magnitude of the minority carrier lifetime. While these events are taking place, the diode remains in forward bias and so the voltage drop across the junction is small. Thus, most of the applied voltage appears across the resistor in the circuit, so determining the magnitude of the current. It is only after the excess carrier concentrations have been reduced to zero that the diode can enter reverse bias. This is a high-resistance state, so the applied voltage starts to be dropped predominantly across the diode and the circuit current is determined by the diode's reverse-bias conduction characteristic.

The dependence of the turn-off response time of the diode on the minority carrier lifetime can be illustrated by running SPICE with different values of TT in the diode model line. Figure 6E.7 gives some results and confirms the importance of using diodes with short minority carrier lifetimes to obtain good high-speed-switching performance. Several ways of obtaining such diodes in practice were mentioned in Section 6.6.2.

6.7 AC ANALYSIS

6.7.1 Small-Signal, Linearized Model

A common analog application of *pn* diodes is in voltage-tunable resonator circuits (see Fig. 6.30). In this circuit the diode is reverse biased by the dc source V_{DD} at some operating point (V_D in this example). This voltage determines the junction

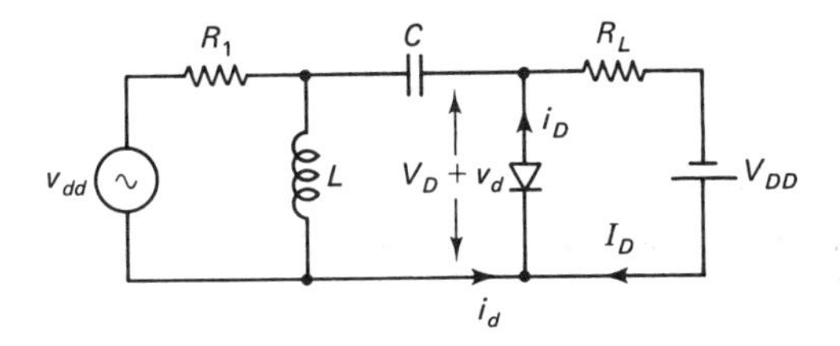

Figure 6.30 Example of diode circuit for which a small-signal analysis is appropriate.

capacitance of the diode (see Fig. 6.28), and hence controls the resonant frequency of the LC circuit. Circuits such as this, where both dc and ac voltages are present, can often be analyzed by the small-signal approach which we now discuss.

We consider a sinusoidal signal v_d and assume that its frequency is sufficiently low so that the charge carriers in the diode can reach steady-state in a time which is short compared to the period of the ac signal. This limits the validity of our analysis to frequencies less than about 100 MHz. The attraction of the assumption is that it allows the instantaneous total current i_D at some combined bias $V_{dc} + v_{ac}$ to be represented satisfactorily by the dc diode equation evaluated at $V = V_{dc} + v_{ac}$. For example, using the ideal diode relationship (6.60), with $V_{dc} = V_D$ and $v_{ac} = v_d$:

$$i_D = I(V_D + v_d) = I_0 \left\{ \exp \left[\frac{q}{kT} (V_D + v_d) \right] - 1 \right\} \tag{6.73}$$

The ac signal current i_d is given by

$$i_d = i_D - I(V_D) = I(V_D + v_d) - I(V_D) \tag{6.74}$$

The first term on the right-hand side can be expanded in a Taylor series. If we make the assumption that the ac signal is small compared to the dc bias voltage (i.e., $v_d \ll V_D$), we need retain only the first two terms of the Taylor series, thus:

$$\begin{aligned} i_d &= I(V_D) + \frac{dI}{dV_D} v_d - I(V_D) \\ &= \frac{dI}{dV_D} v_d \end{aligned} \tag{6.75}$$

dI/dV_D is evaluated at the dc operating point and has the dimensions of conductance. This suggests the nature of the circuit element which may be used in an equivalent circuit to represent the linear relationship between the signal current and voltage, as expressed by (6.75). The appropriate equivalent circuit is known as the ***linearized, small-signal model*** and is shown in Fig. 6.31.

The reciprocal of dI/dV_D is the resistance used in the dc circuit model of the diode. For moderate forward bias it is given by kT/qI [see (6.62)]. Using this in (6.75), we have

$$i_d = v_d \frac{q}{kT} I_D \tag{6.76}$$

To preserve the validity of our small-signal analysis, we must have $i_d \ll I_D$. This implies that $v_d \ll kT/q$ (i.e., $v_d \ll 26$ mV at 300 K).

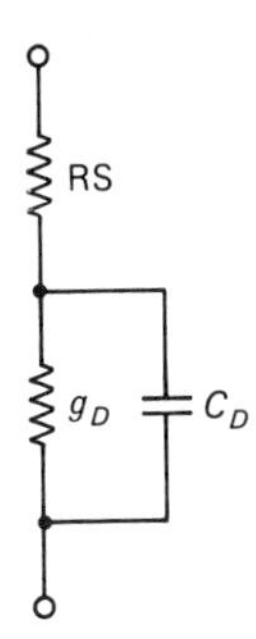

Figure 6.31 Small-signal ac equivalent circuit for a diode, as used in SPICE.

6.7.2 SPICE Model

Figure 6.31 is the model of the *pn* diode used in SPICE for performing ac analyses. The conductance g_D is given by the derivative of (6.63):

$$g_D = \frac{\partial I_D}{\partial V_D} = \frac{q}{\mathrm{N}kT}\,\mathrm{IS}\,\exp\left(\frac{qV_D}{\mathrm{N}kT}\right) \tag{6.77}$$

The small-signal capacitance C_D is given by (6.72). Both g_D and C_D are evaluated at the relevant dc operating point. The operating point is computed in SPICE, before the ac analysis is initiated, by performing a dc analysis with all capacitors open-circuited and all inductors short-circuited.

It can be seen that no new model parameters are needed to perform an ac analysis using SPICE. Example 6.5 illustrates how SPICE can be used to perform a small-signal analysis of diode circuits.

Example 6.5 Small-Signal Analysis of *PN* Diodes: A Varistor Circuit

When operating in forward bias, the diode has the property that its effective resistance is inversely proportional to the dc current [see (6.62)]. This variable resistance (varistor) property can be exploited, in circuits such as that shown in Fig. 6E.8, to provide a dc voltage-controlled attenuation of an ac signal. The dc bias point is set via *VDD* and *R*2; the capacitors *C*1 and *C*2 prevent the dc signal from appearing at the input and output terminals; *R*1 and the resistance of the diode form the potential divider which attenuates the ac signal.

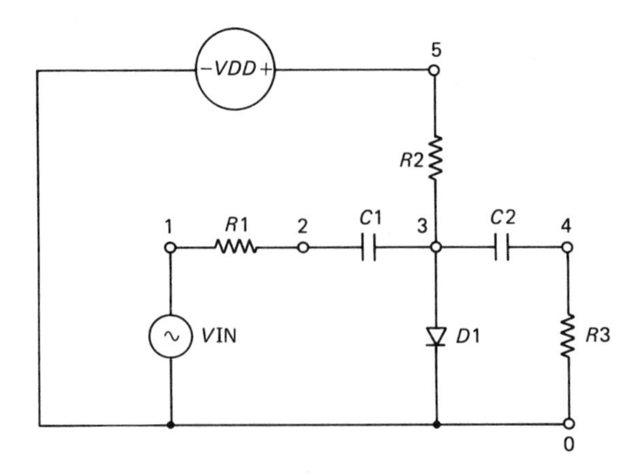

Figure 6E.8 Circuit for the SPICE simulation of diode small-signal ac performance.

The SPICE input listing for this circuit is given below.

```
*DIODE VARISTOR (SMALL-SIGNAL)
VDD 5 0 DC 10
VIN 1 0 AC 0.01
R1 1 2 100
R2 5 3 10K
R3 4 0 100MEG
C1 2 3 1UF
C2 3 4 1UF
D1 3 0 D17
.MODEL D17 D IS=2.8E-14 N=1 TT=1.7E-7 VJ=0.94 CJO=9.49E-10
.AC DEC 1 100 1MEG
.WIDTH OUT=80
.PRINT AC VM(4)
.END
```

The model parameters for the diode are the same as used in Example 6.4. The control card instructs the program to perform a small-signal analysis over the frequency range 10^2 to 10^6 Hz at intervals of one decade. The print instruction calls for the display of the magnitude of the ac voltage at the output (node 4).

The analysis commences by performing a dc analysis of the circuit to find the operating point of the diode. This point and the associated equivalent resistance and capacitance of the diode automatically appear in the SPICE output listing. The small-signal analysis is then carried out. Results are presented in Fig. 6E.9. Note the following:

(a) As *VDD* increases, the diode current increases, so reducing the diode resistance and further attenuating the ac signal.

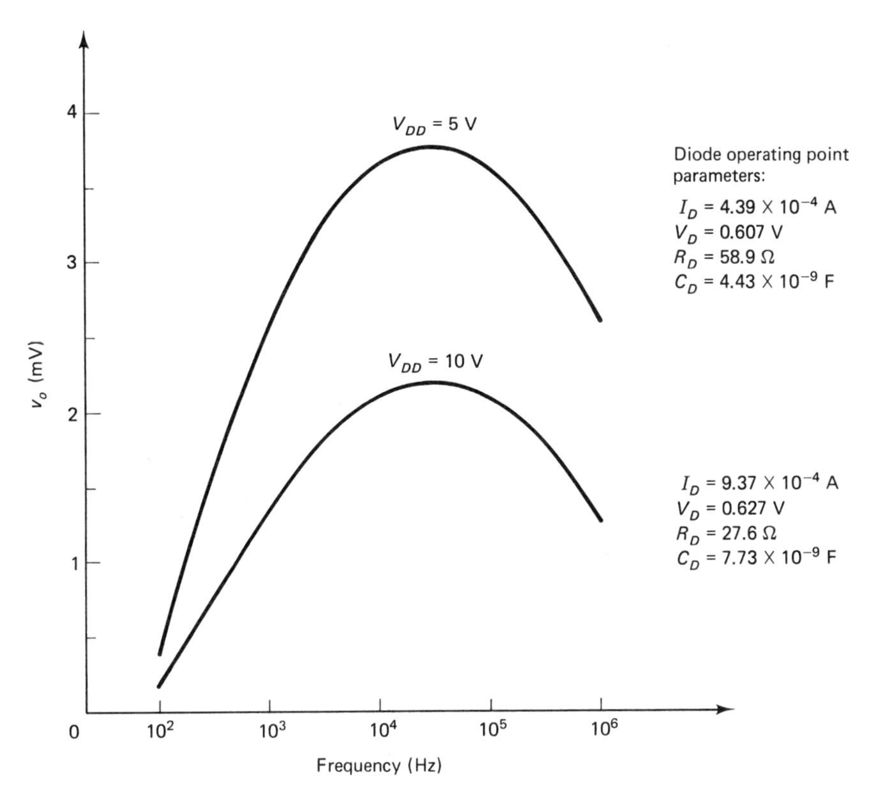

Figure 6E.9 Computed output voltage magnitude (at node 4 of the circuit of Fig. 6E.8) for a sine-wave input signal of amplitude 10 mV.

(b) The diode resistance agrees with the value predicted by (6.62) (i.e., $R = kT/qI$).

(c) The output is frequency dependent, as is to be expected from the presence of the blocking capacitors $C1$ and $C2$ and the diode capacitance C_D.

(d) The diode capacitance is predominantly storage capacitance, for example, 65% for $VDD = 5$ V and 79% for $VDD = 10$ V, as computed from (6.71), that is, $C = qI\text{TT}/kT$.

6.8 CHAPTER SUMMARY

In this chapter detailed descriptions of the principles of operation of the *pn* diode and its electrical performance under both dc and ac conditions have been presented. The reader should now have an appreciation of the following topics:

1. ***Depletion region.*** This is the region of ***ionic space charge*** at the metallurgical junction between the *n*- and *p*-type portions of the diode. An electrostatic potential difference, the ***built-in potential*** exists across the junction at equilibrium. The magnitude of this potential difference is such that the drift of minority carriers across the junction exactly balances the counterdirected diffusion of majority carriers in thermal equilibrium.

2. ***Quasi-equilibrium.*** Application of an external voltage (***bias***) disturbs the equilibrium, but over the usual operating range of diodes, the resulting current is small compared to either of the equilibrium carrier flows. Under these circumstances the diode is in a state of ***quasi-equilibrium*** which can be represented by a constancy across the depletion region of the quasi-Fermi levels. This characterization greatly simplifies the derivation of the ***dc I–V characteristic*** of the *pn* diode.

3. ***I–V characteristic.*** The *I–V* characteristic is asymmetrical with respect to the polarity of the applied voltage. In ***forward bias*** the diode current increases exponentially with applied voltage, with a magnitude that depends on whether the recombination of ***injected*** excess carriers occurs mainly within the depletion region or predominantly just outside the depletion region. In ***reverse bias*** the current is either essentially constant, or increases directly with the square root of the voltage, depending on whether generation of carriers occurs predominantly outside or inside the depletion region, respectively. Deviations from the relationships above occur when quasi-equilibrium no longer holds. At high forward-bias currents this is because the low-level injection condition is violated; at high reverse-bias voltages it is because of ***avalanche multiplication of extracted carriers***.

4. ***Diode modeling.*** In analyzing circuits containing diodes it is useful to be able to model the diode in terms of its ***equivalent circuit*** components. For dc operation the diode is modeled as a voltage-dependent current source (representing the junction) in series with a resistor (representing the parasitic resistances of the neutral *p* and *n* regions). For ***large-signal ac conditions***, the model must include capacitors associated with the ionic space charge of the depletion region, and, under forward bias, with the ***storage*** just outside the depletion region of injected minority carriers. The charging and discharging of these capacitors determines the ***switching speed*** of the diode, as we illustrate via a SPICE simulation of the diode ***turn-off transient***. These capacitors also appear in the ***small-signal ac*** representation of the diode and play an important role in determining the ***frequency response*** of the diode.

6.9 REFERENCES

1. S. M. Sze, *Physics of Semiconductor Devices,* Chap. 10, Sec. 13.4, John Wiley & Sons, Inc., New York, 1981.
2. A. S. Grove, *Physics and Technology of Semiconductor Devices,* p. 190, John Wiley & Sons, Inc., New York, 1967.

PROBLEMS

6.1. An abrupt junction silicon *pn* diode has a *p*-layer acceptor doping density of 10^{18} cm^{-3} and an *n*-side donor doping density of 10^{15} cm^{-3}. Assume that all the dopants are fully ionized. For this junction in equilibrium at 300 K:

(a) Compute the position of the Fermi level (with respect to the conduction band edge) on both sides of the junction.

(b) Sketch the band diagram (with the energy axis drawn to scale) and from it estimate the built-in potential.

(c) Compute V_{bi} directly from the doping densities and the intrinsic concentration, and compare the result with that from part (b).

(d) Calculate the widths of the depletion regions on either side of the junction.

(e) Calculate the maximum electric field in the depletion region.

6.2. Consider the diode of Problem 6.1 to have an area of 0.005 cm^2 and to be rated at a forward current of 1 A. Neglecting recombination/generation in the depletion region, compute the forward-bias voltage required to obtain the rated current.

6.3. (a) Compute the breakdown voltages [using (6.65)] for the three diodes cited in Example 6.1.

(b) To investigate the effect of avalanche breakdown on the reverse-bias characteristic, first follow the procedure used in Example 6.2 and plot the depletion region generation current in reverse bias for the $10^{19}/10^{17}$ diode to a voltage equal to the avalanche breakdown voltage. Then account for avalanche breakdown by assuming that the diode exhibits a resistance of 5 Ω when in the breakdown mode.

6.4. An abrupt *pn* junction is made in *n*-Si (donor density = 10^{16} cm^{-3}) by diffusing-in acceptor atoms which have an energy level 0.200 eV above the valence band edge. It is desired that the diode have a breakdown voltage of 200 V. Assume that the acceptor concentration is uniform throughout the *p*-side of the diode. The donors are all ionized at 300 K, but do not assume that this is necessarily the case for the acceptors. Calculate the concentration of acceptor atoms.

6.5. The *n* region of a p^+n diode is much longer than the hole minority carrier diffusion length of 1 μm. When operating in forward bias the ratio of hole current to electron current in the steady state at $x = 0$ (the edge of the depletion region on the *n*-type side) is 100. Calculate the ratio of the hole current to the electron current in the *n* region at $x = 1$ μm under steady-state conditions.

6.6. Some of the terms in the equations presented in Example 6.1 are temperature dependent, notably n_i, D, μ, L, and V_{bi}. These dependencies can be deduced from (4.14), (3.20), (3.6), (3.7), and (6.12). Use these equations, together with those needed from Example 6.1, to compute the forward-bias log I–V curve for the $n^+(10^{19})$ – $p(10^{15})$ diode cited in Example 6.1 at temperatures within the normal operating range of diodes (e.g., −50, 27, and 150°C). Comment on the trends revealed.

6.7. Consider the possibility of using a n^+p diode with a long *p* region as the temperature sensor in an electronic thermometer. The current I is to be measured at a constant forward bias of 500 mV.

(a) Derive an expression for the fractional change in current $\Delta I/I$ resulting from a temperature change ΔT. Evaluate this expression at $T = 300$ K. Assume that the diode current is dominated by recombination of electrons in the neutral p region, and for simplicity, consider the temperature dependence to be due solely to the factors that have an exponential dependence on temperature.

(b) Suppose that part of the diode current results from recombination in the depletion region. How will this affect the sensitivity of the thermometer?

6.8. A silicon pn junction is formed by diffusing phosphorus into a uniformly boron-doped substrate of resistivity 2 Ω·cm. The J–V characteristic for the completed diode is shown in Fig. 6P.1.

(a) Using the graph, estimate the electron diffusion length and the electron lifetime in the p region. Assume that the n region is much more heavily doped than the p region and that the latter is much longer than L_e.

(b) Estimate the bias voltage at which high-level injection sets in in the base (p-region). Is the shape of the J–V curve consistent with your calculation?

6.9. The more heavily doped region of a diode is often referred to as the ***emitter***, and the lighter doped region as the ***base*** (see Fig. 6P.2). In computing the current due to minority carrier recombination in the base region of a long base diode in forward bias, we made use of the principle that every carrier injected from the emitter must recombine in the base. Equation (6.41), rewritten for the case of a n^+p diode, gives

$$J_e = \frac{qL_e}{\tau_e} n_{op} \left[\exp\left(\frac{qV}{kT}\right) - 1 \right] \tag{6P.1}$$

This component of current can be computed in a different way; namely, assume that the current is due to diffusion and then use the continuity equation to find the steady-state electron distribution in the p region. The current then follows from (3.18), evaluated at the depletion region edge on the p side of the junction. Confirm that the result obtained by this approach is the same as that given in (6P.1).

6.10. Figure 6P.2 shows the structure of a short-base diode, that is, a diode in which the length L_B of the base region is less than the electron diffusion length L_e in the base.

(a) Write down the differential equation governing the steady-state electron distribution in the base. Assume that low-level injection conditions apply.

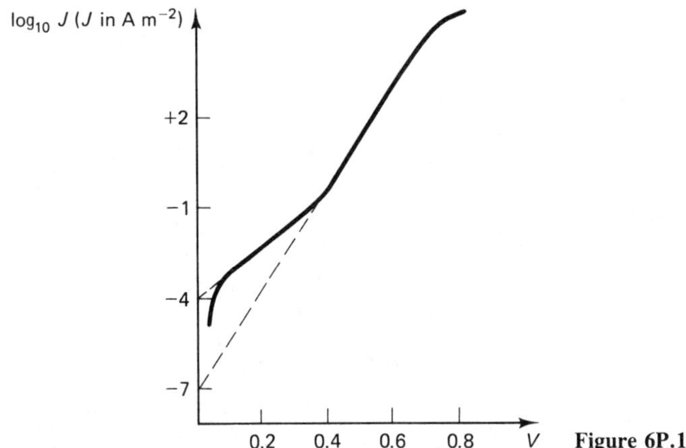

Figure 6P.1

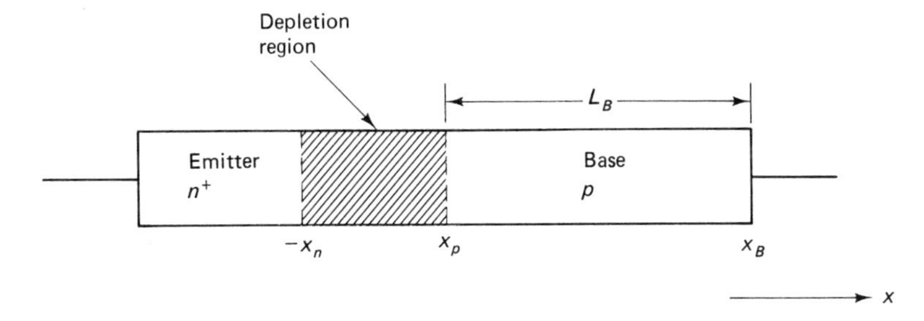

Figure 6P.2

(b) Show that

$$n_p(x) - n_{0p} = A \cosh \frac{x}{L_e} + B \sinh \frac{x}{L_e} \tag{6P.2}$$

where A and B are constants to be determined, is a valid general solution to the equation found in (a). Shifting the coordinate origin so that $x = 0$ at the point x_p, for convenience, evaluate A and B. Assume that the back contact is ohmic; that is, the excess carrier concentrations are zero at x_B.

(c) Show that the electron current density $J_e(x_p)$ is given by

$$J_e(x_p) = \frac{qD_e n_i^2}{L_e N_A} \coth \frac{L_B}{L_e} \left[\exp\left(\frac{qV}{kT}\right) - 1 \right] \tag{6P.3}$$

6.11. A *pn* diode with a J–V characteristic that obeys the ideal equation when operating in the dark, is irradiated such that the optical generation rate G_{op} is uniform throughout the volume of the diode. Show that the current density in the diode is given by

$$J = J_0 \left(\exp \frac{qV}{kT} \right) - J_0 - J_L$$

where

$$J_L = qG_{op}(W + L_e + L_h),$$

J_0 is the saturation current density [see (6.60)] and W is the depletion layer width.

6.12. The junction capacitance of a 10^{-3}cm^2 abrupt junction n^+p diode in reverse bias has the following experimental values:

C (pF)	V (V)
3.849	−0.5
3.288	−1
2.626	−2
2.253	−3
2.008	−4
1.826	−5

Plot these data in such a fashion that a linear relationship results, from which the p-type doping density and the built-in potential can be ascertained.

6.13. A pn junction has a heavily doped emitter and a lightly doped base. The base doping profile is nonuniform. The capacitance–voltage characteristic for the diode in reverse bias is sketched in Fig. 6P.3. Using this characteristic, determine the dopant concentration as a function of position in the base.

6.14. A p^+n diode is fabricated in n-type silicon of resistivity 10 $\Omega\cdot$cm. For reverse-bias voltages greater than a few hundred millivolts, the current density in the diode is constant at 10^{-10} A cm^{-2}.

(a) Calculate the storage capacitance of this diode when the forward current is 1 mA.

(b) If this diode is to be used as a varactor, what must be the diode area if the capacitance is to be 100 pF at a reverse bias of 50 V?

6.15. Figure 6P.4 shows the structure of a silicon power diode that has a two-part base doping profile. The part of the base adjacent to the metallurgical junction has doping $N_A = 10^{15}$ cm^{-3}, while the rest of the base is more heavily doped ($N_A = 10^{16}$ cm^{-3}). This arrangement gives a high breakdown voltage and a low series resistance.

(a) When a reverse bias of 40 V is applied to the diode, the junction capacitance is 3.48×10^{-5} F m^{-2}. Show that the lightly doped part of the base is completely depleted at this bias point.

(b) Where in the depletion region does the electric field reach its maximum value?

(c) What is the maximum electric field for the bias condition in part (a)?

6.16. A real pn junction diode formed using the diffusion techniques discussed in Section 12.2.2 might have an emitter doping profile given by

$$N_D(x) = N_{D0}e^{-x^2/2\sigma^2}$$

where $N_{D0} = 10^{20}$ cm^{-3} and $\sigma = 0.5$ μm. According to the arguments of Section 6.5.4, an emitter region of this kind should be quasi-neutral. This problem checks the accuracy of the quasi-neutrality assumption.

(a) Assuming quasi-neutrality holds, the electron concentration in the emitter is

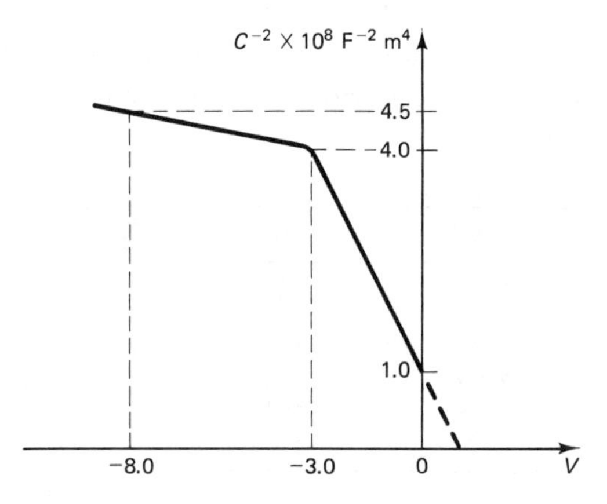

Figure 6P.3

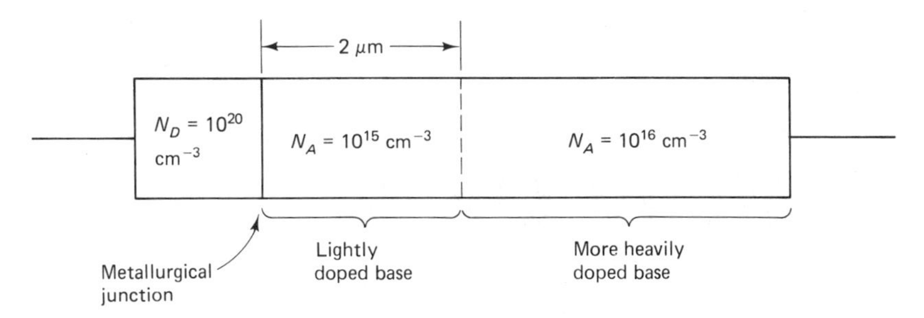

Figure 6P.4

given by $n(x) = N_D(x)$. This being the case, show that a built-in electric field

$$\mathscr{E}(x) = \frac{kT}{q}\frac{d\ln(n)}{dx}$$

must exist in the emitter.

(b) From Gauss' law, the electric field in the emitter is related to the local charge density $\rho(x)$ by

$$\rho(x) = \epsilon\frac{d\mathscr{E}}{dx}.$$

Using this result, show that the magnitude of the charge is given by

$$|\rho(x)| = \epsilon\frac{kT}{q}\frac{x}{\sigma^2}.$$

(c) Make a table showing $\rho(x)/q$ and the electron concentration $n(x)$ as a function of depth into the emitter for $x = 0.5, 1.0, \ldots, 2.0$ μm. By what percentage does the electron concentration actually differ from the donor concentration at each depth?

6.17. By simply reversing the polarity of the applied voltage in the circuit of Fig. 6E.5, the response of a diode to a transient which seeks to change the diode from an off-state (reverse bias) to an on-state (forward bias) can be investigated. Perform such an investigation by making the appropriate change to the input file given in Example 6.4 and running SPICE. Comment on why the diode response does not appear to be determined by the minority carrier lifetime, unlike the situation for the turn-off transient.

6.18. Use SPICE to analyze the tunable resonator circuit of Fig. 6.30. Specifically, examine the effect of the dc voltage-dependent capacitance of the diode on the ac voltage across the diode. Take $C = 1$ μF, $L = 1$ μH, $R_1 = 100$ Ω, and $R_L = 10$ kΩ. Using the same diode model parameters as in Example 6.5, perform a SPICE small-signal analysis over a range of frequencies (e.g., 1 to 100 MHz) and dc source voltages (e.g., -1 to -10 V). Confirm the results using the equations of Section 6.6.1 for the diode junction capacitance.

7 Metal-Oxide-Semiconductor Field-Effect Transistors

7.1 STRUCTURE

The basic structure of an n-channel ***m***etal-***o***xide-***s***emiconductor ***f***ield-***e***ffect ***t***ransistor (MOSFET) is shown in Fig. 7.1. The device consists of two n^+ regions spaced close together in a lightly doped p-type silicon substrate. In the gap between these two regions a thin layer of silicon dioxide has been grown on the silicon surface. This oxide layer is overlaid with a conducting electrode called the ***gate***. The basic principle of operation of the device is simple. The diodes formed between the n^+ regions and the substrate are always held at zero or reverse bias, so that ordinarily only the very small leakage current of a reverse-biased pn junction exists in these regions. If, however, a sufficiently large positive voltage is applied to the gate, electrons are attracted to the surface under the gate. These electrons form a shallow conducting layer or ***channel*** connecting the n^+ regions, so that charge can flow if a voltage difference is established between them. Electrons flow from the more negatively biased n^+ region, called the ***source***, to the more positively biased region, or ***drain***. This electron flow is equivalent to a current we will term the ***drain current*** I_D. The MOSFET is thus in essence a voltage-controlled solid-state switch. Ordinarily, MOSFETs are made so that the source and drain are identical in structure and can be interchanged electrically.

Although, as the discussion above implies, transistor action in the MOSFET is due principally to the control of current at the drain and source terminals by the voltage on the gate, the voltage applied to the substrate or ***body*** of the device also has some influence on the characteristics. The MOSFET is therefore inherently a four-terminal device.

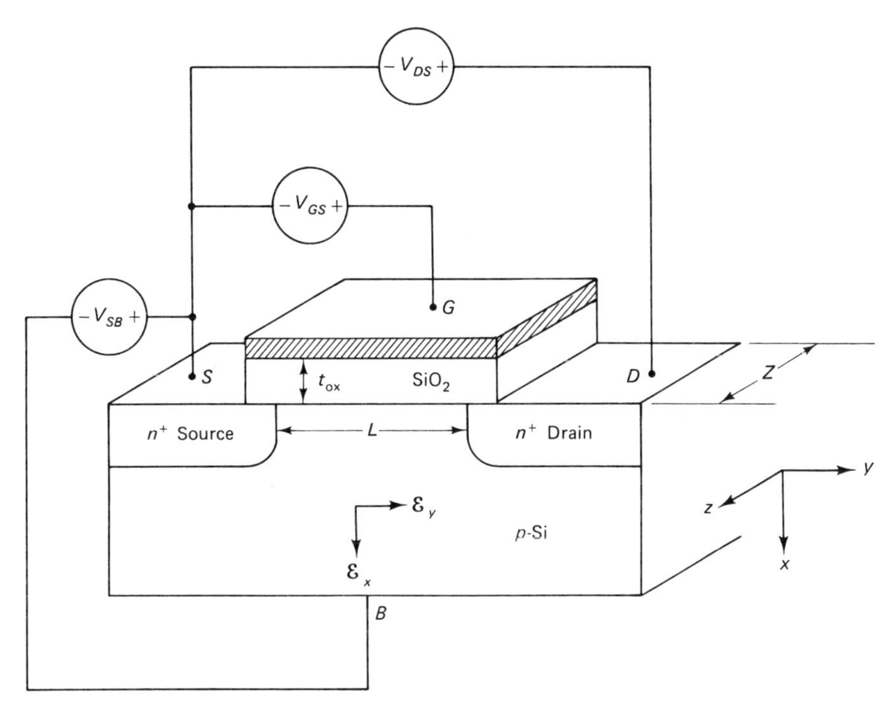

Figure 7.1 MOSFET structure and biasing.

The length of the MOSFET channel—that is, the distance between the source and drain—is denoted by L in Fig. 7.1. The width of the channel—that is, the dimension of the channel in the direction perpendicular to that of the charge flow— is denoted by Z. In integrated circuit design the symbol W is frequently used for the channel width, but Z will be used here to avoid confusion with the width of the depletion region. It is worth noting that in many MOSFETs the channel width is greater than the length, but the convention that the length is the dimension parallel to the current is adhered to nonetheless. The thickness of the oxide layer separating the gate electrode from the substrate will be denoted as t_{ox}.

All MOSFETs made commercially today are formed on silicon substrates. The main reason for this is that the electrical properties of the oxidized silicon surface are well understood and easily controlled. For many other semiconductors the surface properties all but prohibit the formation of a usable MOSFET; one example of such a material is GaAs. It is, however, possible to fabricate a useful MOSFET-like device on an indium phosphide (InP) substrate using a deposited dielectric layer (typically, SiO_2) rather than a thermally grown oxide to form the gate insulator. Devices of this kind are usually referred to not as MOSFETs but rather as MISFETs, for ***m***etal-***i***nsulator-***s***emiconductor ***f***ield-***e***ffect ***t***ransistor. Generically, MOSFETs and MISFETs are sometimes termed *i*nsulated *g*ate *f*ield-*e*ffect *t*ransistors, or IGFETs.

Early MOSFETs had gates made of aluminum. Today the vast majority of MOSFETs produced commercially have gates of heavily doped polycrystalline silicon (also known as ***polysilicon*** or simply "poly"). Despite this use of a semiconductor as a gate material, the transistors are still referred to as MOSFETs.

The operation of diodes and bipolar transistors is governed primarily by events taking place within a semiconducting crystal, with the crystal surface having only a secondary effect on device performance. Exactly the opposite situation holds in a MOSFET, where the properties of the surface are central to the operation of the device. For this reason, a brief discussion of the properties of solid surfaces is in order before a theory of the current–voltage relationship in the MOSFET can be developed.

7.2 SEMICONDUCTOR SURFACES

7.2.1 Electron Affinity and Work Function

At the surface of any solid there is an energy barrier holding the electrons in. This barrier exists simply because it is energetically more favorable for electrons to be close to the positively charged ion cores in the solid's interior. The energy required to take an electron from inside the solid and release it to the exterior is an important property of any material. Naturally, since electrons in the solid can have a wide range of energies, a more precise definition of this energy barrier is required. For a semiconductor, it is useful to work with the ***electron affinity*** $q\chi_s$, which is defined as the energy needed to take an electron from the bottom of the conduction band and deposit it at rest in vacuum just outside the solid. (Although the electron affinity is defined in terms of an energy difference, it is normal practice to write this quantity as $q\chi_s$, where χ_s must therefore have dimensions of volts. As usual in this text, q is the electron charge—1.6×10^{-19} C.) When dealing with a metal, it is more convenient to work with the energy required to remove an electron at the Fermi energy and place it at rest outside. This energy difference is known as the ***work function*** $q\phi_m$ of the metal. Since there are a large number of electrons in the metal with energies close to E_F, $q\phi_m$ is a good measure of the minimum energy needed to free an electron from the metal. It is also possible to define the work function $q\phi_s$ of a semiconductor in an analogous fashion, but this quantity varies with the position of the Fermi energy relative to the band edges and hence with the doping level, and is therefore less useful than $q\chi_s$.

Evidence for the existence of a surface energy barrier can be obtained using the experimental setup shown schematically in Fig. 7.2. Here a voltage difference is established between two metal electrodes in vacuum. In the absence of light there is effectively no current between the plates. However, if the negative electrode is illuminated with photons of energy greater than $q\phi_m$, electrons near the

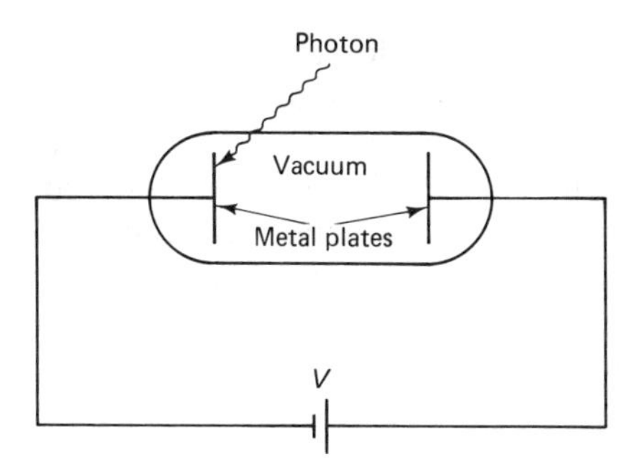

Figure 7.2 Experimental setup used to demonstrate the existence of a work function energy barrier at a solid's surface.

Fermi energy will be emitted from that plate into the vacuum, and charge will flow between the plates.

Energy band diagrams for the surface regions of both a metal and a semiconductor are shown in Fig. 7.3a and b. In the figure the "vacuum level" represents the energy of an electron at rest just outside the solid. To draw a band diagram for the MOSFET, it is also necessary to consider the available energy levels in the silicon dioxide. These are shown in Fig. 7.3c. The band structure of SiO_2 resembles that of a semiconductor, except that the band gap is very wide and the conduction band lies relatively close to the vacuum level.

7.2.2 Accumulation, Depletion, and Inversion

Before attempting to analyze the current–voltage characteristics of the MOSFET, it is useful first to consider a simpler device: the MOS capacitor. Although MOS

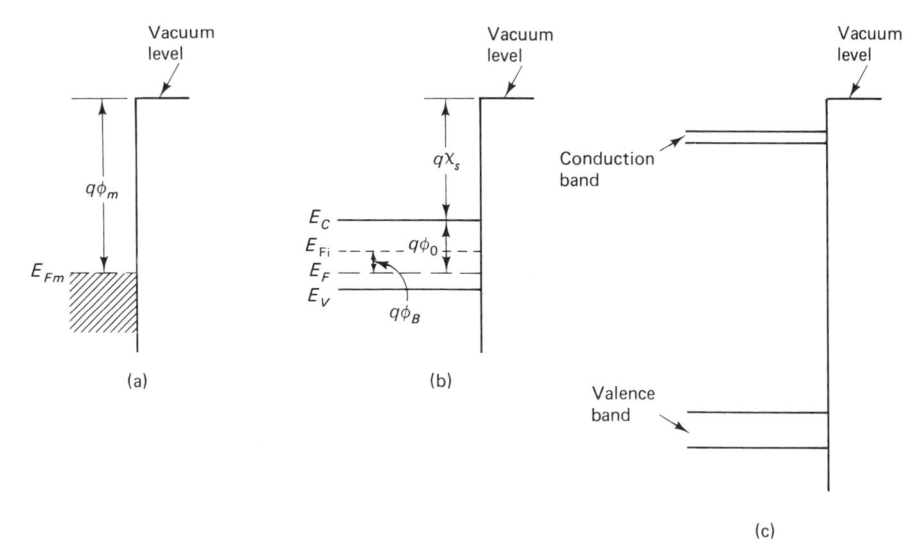

Figure 7.3 Energy band diagrams for the surface regions of (a) a metal, (b) silicon, and (c) silicon dioxide.

capacitors find some application in analog MOS ICs, they are not widely used in circuits. However, because they are very simple and very easily made, these devices are used extensively as monitors of oxide quality for process control in integrated circuit fabrication. This application is considered in Section 7.9.

The basic structure of a MOS capacitor is shown in Fig. 7.4. The device consists of a metal or polysilicon gate overlying an oxide layer; it can be viewed as a MOSFET lacking source and drain regions. A MOS capacitor can be made by simply evaporating a metal film on an oxidized silicon wafer, as discussed in Section 12.2.8.

A complete energy band diagram for a MOS capacitor formed on a p-type silicon substrate is shown in Fig. 7.5a. p-type substrates are used to fabricate n-channel MOSFETs, which are more widely used than their p-channel counterparts, so we will consider only devices formed on p-type material throughout the following unless specified otherwise. For most purposes it is possible to represent the capacitor with the simplified band diagram of Fig. 7.5b, which shows only the conduction and valence band edges in the silicon dioxide. When using the simplified band diagram, the metal work function $q\phi_m$ is replaced with a modified work function $q\phi_m'$ defined by

$$\phi_m' = \phi_m - \chi_{SiO_2} \tag{7.1}$$

Similarly, the semiconductor electron affinity $q\chi_s$ is modified to

$$\chi_s' = \chi_s - \chi_{SiO_2} \tag{7.2}$$

Before proceeding further with the analysis of the MOS capacitor, it is useful to introduce two parameters related to the doping level in the substrate. The first of these is the energy difference $q\phi_0$ between the conduction band edge and the Fermi level measured deep in the ''bulk'' substrate, well away from the influence

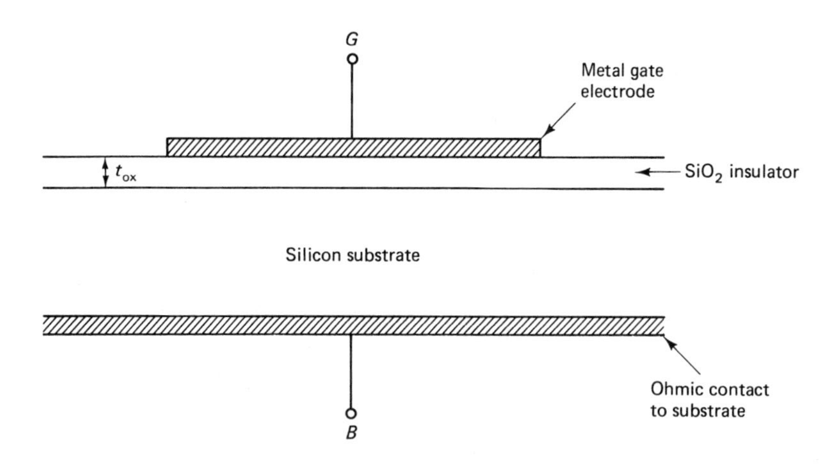

Figure 7.4 Structure of the MOS capacitor.

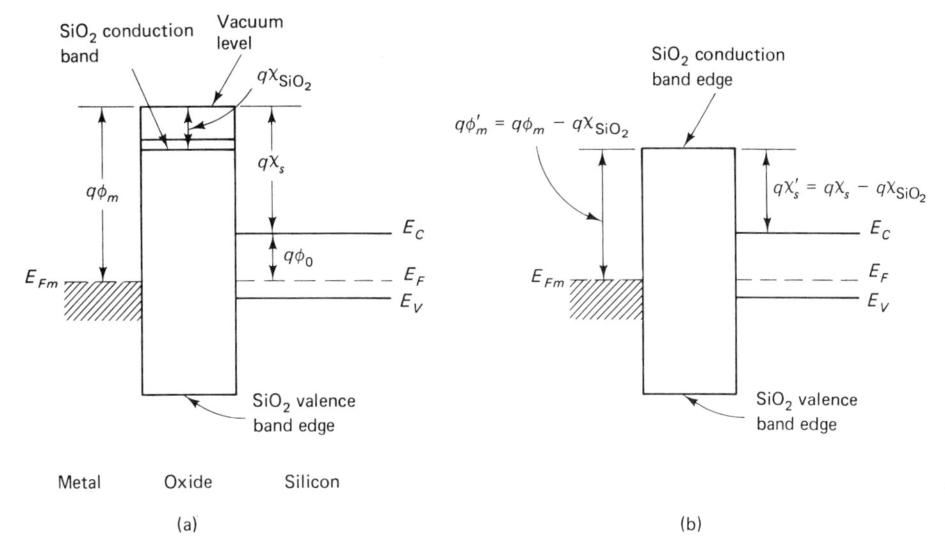

Figure 7.5 (a) Complete band diagram for MOS capacitor, showing energy band structure of SiO_2. (b) Simplified band diagram.

of the gate electrode. From (4.9),

$$\phi_0 = \frac{kT}{q} \ln \frac{N_c}{n_0} \tag{7.3}$$

where n_0 is the equilibrium electron concentration in the substrate. ϕ_0 is illustrated in Fig. 7.3b. Technically, the work function of a semiconductor is defined by

$$\phi_s = \chi_s + \phi_0 \tag{7.4}$$

which emphasizes the dependence of ϕ_s on doping.

A second useful parameter is the energy difference $q\phi_B$ between the actual Fermi level E_F and the intrinsic Fermi level E_{Fi}. For a p-type substrate with acceptor doping N_A, ϕ_B is given by

$$\phi_B = \frac{kT}{q} \ln \frac{N_A}{n_i} \tag{7.5}$$

Similarly, for an n-type substrate with donor doping N_D,

$$\phi_B = \frac{kT}{q} \ln \frac{n_i}{N_D} \tag{7.6}$$

It should be noted that in this case ϕ_B has been defined as a negative quantity; this will allow the same equations to be used to describe both n- and p-channel devices in later analysis. ϕ_B is also illustrated in Fig. 7.3b.

The band diagram of Fig. 7.5 applies to the situation in which $V_{GB} = 0$, so that the Fermi energies in the gate and substrate coincide. In drawing this figure it has been assumed that the work functions of the metal and semiconductor are equal, and that there is no charge trapped in the silicon dioxide layer. In this special case there is no electric field in the silicon and hence no band bending across the silicon at zero applied bias. Although this could happen in a real capacitor, it is rather unlikely in practice. Modifications to the theory of MOSFET operation which must be made when this condition does not hold are considered in Section 7.4.

When a positive bias is applied to the gate, holes are repelled from the semiconductor surface, so the bands bend as shown in Fig. 7.6. As the holes leave the surface, ionized acceptor ions are left behind, forming a surface depletion region. The applied bias V_{GB} can be viewed as splitting up into two parts: a potential drop ψ_s across the semiconductor and a potential drop ψ_{ox} across the insulator. These quantities are defined to be positive when the direction of band bending is that shown in Fig. 7.6. Adding up energy differences on moving from left to right in Fig. 7.6, we find that

$$\phi_m' + \psi_{ox} = V_{GB} + \phi_0 - \psi_s + \chi_s' = V_{GB} + \phi_s' - \psi_s \tag{7.7}$$

so, since we are assuming that $\phi_m' = \phi_s'$,

$$V_{GB} = \psi_s + \psi_{ox} \tag{7.8}$$

as expected.

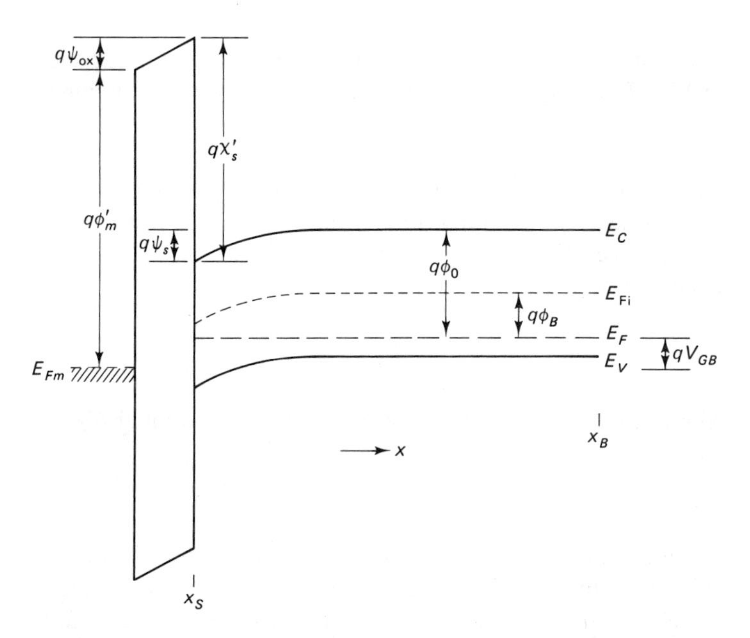

Figure 7.6 MOS capacitor band diagram for positive applied bias V_{GB}.

It is important to note that the Fermi energy E_F has been drawn flat across the semiconductor in Fig. 7.6. This comes about since there can be no current through the insulator, and, in consequence, no steady-state current in the substrate. From (4.27), a lack of current implies that there is no gradient in Fermi energy.

The potential drop ψ_{ox} across the oxide can be related to the ***total charge per unit area Q_s stored in the semiconductor*** through Gauss's law. Figure 7.7 shows an imaginary Gaussian surface enclosing part of the MOS capacitor. Deep in the semiconductor substrate the electric field should go to zero, so there is no contribution to flux through the Gaussian volume from the bottom surface. Also, there is no flux contribution from the sides of the Gaussian volume, since the electric field lines run parallel to the sides. Letting $\mathscr{E}_{ox}$ be the electric field in the oxide, we have

$$\epsilon_{ox}\mathscr{E}_{ox} = -Q_s \tag{7.9}$$

where ϵ_{ox} is the permittivity of the oxide. (Here $\mathscr{E}_{ox}$ has been defined to be positive when pointing in the positive x-direction in Fig. 7.6). Since $\psi_{ox} = \mathscr{E}_{ox}t_{ox}$, we find

$$\psi_{ox} = -\frac{t_{ox}}{\epsilon_{ox}}Q_s = -\frac{Q_s}{C_{ox}} \tag{7.10}$$

where

$$C_{ox} = \frac{\epsilon_{ox}}{t_{ox}} \tag{7.11}$$

C_{ox} is usually termed the ***oxide capacitance***; it is the capacitance per unit area one would measure in the MOS capacitor if the semiconductor were replaced by a

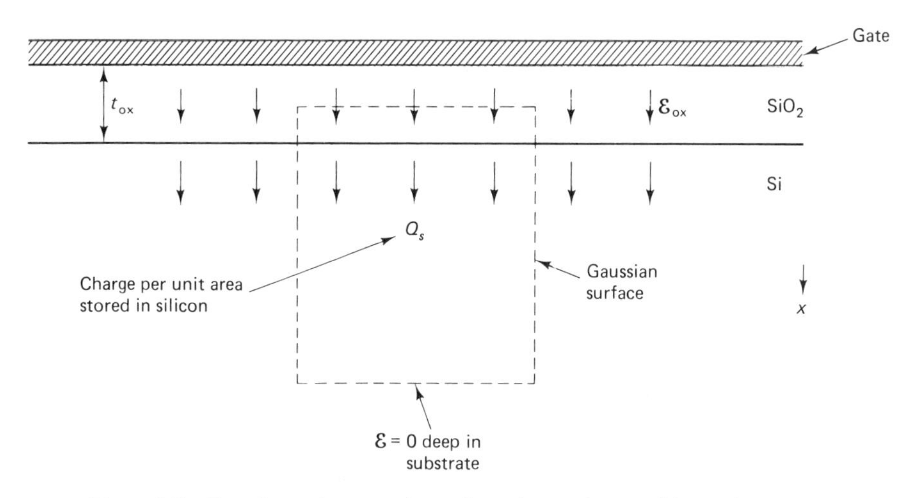

Figure 7.7 Gaussian volume used to relate charge Q_s stored in semiconductor to field $\mathscr{E}_{ox}$ in oxide.

metal electrode. (It is worth pointing out that when dealing with MOS capacitors, it is normal practice to write all equations in terms of capacitance per unit area rather than total capacitance.)

If V_{GB} is increased further, the bands in the semiconductor will continue to bend downward, as suggested in Fig. 7.8a. Eventually, E_F at the surface will lie midway between the conduction and valence band edges, as shown in Fig. 7.8b. At this point the surface electron concentration n_s is equal to the surface hole concentration p_s. Further increases in V_{GB} will make n_s greater than p_s, so this bias point is said to mark the onset of ***inversion***. At an inverted semiconductor surface the minority carrier concentration is greater than the majority carrier concentration.

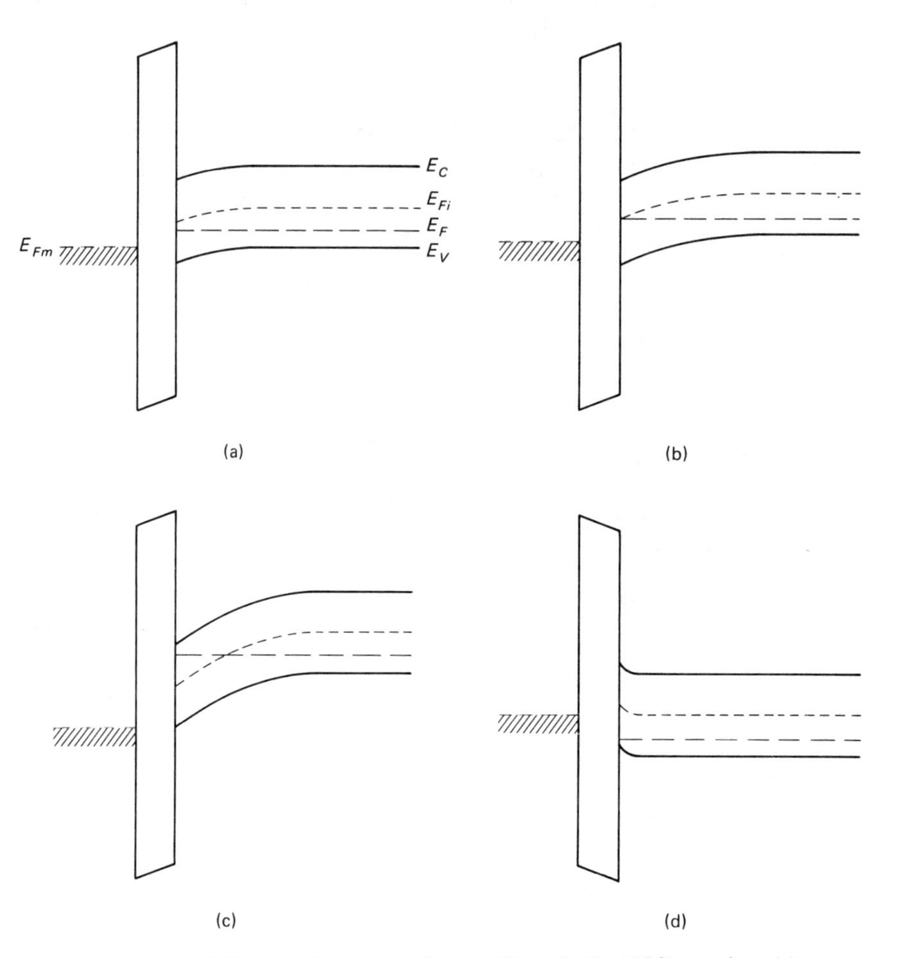

Figure 7.8 Band diagrams for various bias conditions in the MOS capacitor: (a) depletion, $0 < \psi_s < \phi_B$; (b) onset of inversion, $\psi_s = \phi_B$; (c) strong inversion, $\psi_s = 2\phi_B$; (d) accumulation, $\psi_s < 0$.

The surface electron concentration is given by

$$n_s = N_C e^{-[E_c(x_s) - E_F]/kT} \tag{7.12}$$

where $E_C(x_S)$ is the energy of the conduction band edge at the semiconductor surface x_S. But from Fig. 7.6, $E_C(x_S) = E_C(x_B) - q\psi_s$, where $E_C(x_B)$ is the conduction band energy at the substrate contact x_B, so

$$n_s = N_C e^{-[E_c(x_B) - q\psi_s - E_F]/kT} = N_C e^{-[E_C(x_B) - E_F]/kT} e^{q\psi_s/kT} = n_{0p} e^{q\psi_s/kT} \tag{7.13}$$

where n_{0p} is the equilibrium electron concentration in the p-type substrate. Since inversion requires that $n_s \geqslant n_i$, at the onset of inversion,

$$n_i = n_{0p} e^{q\psi_s/kT} = \frac{n_i^2}{N_A} e^{q\psi_s/kT} \tag{7.14}$$

therefore

$$\psi_s = \frac{kT}{q} \ln \frac{N_A}{n_i} = \phi_B \tag{7.15}$$

If V_{GB} is increased still further, the bands in the semiconductor will bend still more, the conduction band edge at the surface will move closer to E_F, and the surface electron concentration will increase. The total charge per unit area Q_s stored in the semiconductor must therefore split into a part $\boldsymbol{Q_B}$ associated with ionized acceptor ions in the depletion region, and a part $\boldsymbol{Q_n}$ associated with electrons near the surface; that is,

$$Q_s = Q_n + Q_B \tag{7.16}$$

The ionized acceptor ions are, of course, fixed in the lattice, but the electrons are mobile and can move in response to an electric field applied parallel to the silicon surface. This fact is central to the operation of the MOSFET.

As V_{GB} is increased, the electron concentration near the surface will at first be insignificant compared to the acceptor ion concentration. In this regime, the semiconductor is said to be in ***weak inversion***. In depletion or weak inversion, it is possible to compute the width W of the depletion region using the depletion approximation applied in the analysis of the pn-junction diode in Section 6.2.1.1. The only change that needs to be made in the analysis is the replacement of the built-in potential drop V_{bi} across the pn junction with ψ_s; thus we have

$$W = \sqrt{\frac{2\epsilon_s \psi_s}{qN_A}} \tag{7.17}$$

where ϵ_s is the permittivity of the semiconductor. However, if V_{GB} is increased sufficiently, a point will eventually be reached at which the surface electron concentration exceeds the acceptor ion concentration, as suggested in Fig. 7.8c. This bias point is said to mark the onset of ***strong inversion***, and the region at the semiconductor surface with this high concentration of electrons is called an ***inversion layer***.

Once strong inversion sets in, further increases in V_{GB} are absorbed almost entirely as changes in the potential drop ψ_{ox} across the oxide while the band bending ψ_s across the semiconductor remains relatively constant. The depletion region width therefore saturates at a maximum value W_{max}. The reason for this is as follows. Equation (7.13) shows that a relatively small change in ψ_s produces a large change in n_s. In fact, it is easy to show that an increase in ψ_s of only 60 mV will increase n_s by an order of magnitude at room temperature. In strong inversion, a large increase in n_s implies a large increase in the charge Q_s stored in the semiconductor. However, since ψ_{ox} is related to Q_s by (7.10), a relatively small increase in ψ_s must therefore produce a large increase in ψ_{ox}. Since V_{GB} must split between ψ_s and ψ_{ox}, once strong inversion is reached further increases in V_{GB} are absorbed almost entirely as changes in ψ_{ox} while ψ_s increases very little. Physically, the electrons in the inversion layer can be thought of as screening the semiconductor interior from changes in potential at the gate electrode, just as electrons at the surface of a metal container will redistribute to screen out electric fields in the interior.

The value of V_{GB} that must be applied to just create a condition of strong inversion at the silicon surface is known as the ***threshold voltage*** V_T, and is one of the most important parameters describing any MOS capacitor. To obtain strong inversion, we need $n_s = N_A$, or

$$N_A = \frac{n_i^2}{N_A} e^{q\psi_s/kT} \tag{7.18}$$

so

$$\psi_s = \frac{2kT}{q} \ln \frac{N_A}{n_i} = 2\phi_B \tag{7.19}$$

The depletion region width at threshold is therefore given by

$$W_{max} = \sqrt{\frac{2\epsilon_s(2\phi_B)}{qN_A}} \tag{7.20}$$

so the ionized acceptor charge per unit area Q_B stored in the depletion region is given by

$$Q_B = -qN_A W_{max} = -\sqrt{2\epsilon_s qN_A(2\phi_B)} \tag{7.21}$$

Although at threshold the electron concentration at the silicon surface is equal to the acceptor ion concentration, throughout most of the depletion region n is negligible compared to N_A. At threshold it is therefore reasonable to assume

$Q_s \simeq Q_B$, so

$$V_T = \psi_s + \psi_{ox} = 2\phi_B + \frac{\sqrt{2\epsilon_s q N_A (2\phi_B)}}{C_{ox}} \tag{7.22}$$

If V_{GB} is made negative, holes are attracted to the silicon surface, and the band diagram appears as shown in Fig. 7.8d. In this case the surface is said to be ***accumulated***, with the excess holes forming an ***accumulation layer***. The surface hole concentration is given by

$$p_s = p_0 e^{-q\psi_s/kT} \tag{7.23}$$

where $p_0 \simeq N_A$ is the equilibrium hole concentration in the bulk. Just as in the case of strong inversion, once an accumulation layer forms, further increases in the magnitude of V_{GB} lead to very little change in ψ_s. The high density of charge in the accumulation layer screens the interior of the semiconductor, so that increases in V_{GB} are absorbed entirely by changes in ψ_{ox} rather than ψ_s.

At this point it is worth summarizing the surface conditions that may be encountered in the MOS capacitor. For $\psi_s < 0$, the surface is in accumulation. If $0 < \psi_s < \phi_B$, the surface is depleted. The surface becomes inverted once $\psi_s > \phi_B$. For $\phi_B < \psi_s < 2\phi_B$, the surface is only in weak inversion, with free electrons making a negligible contribution to Q_s. For $\psi_s > 2\phi_B$ the surface becomes strongly inverted.

7.3 DC ANALYSIS

In this section we extend the concepts developed in analyzing the MOS capacitor to derive an expression for the drain current I_D as a function of the biases applied to the four terminals of the MOSFET. Obtaining a simple, yet reasonably accurate analytic solution for the current–voltage characteristics of the MOSFET is difficult because of the device's inherent two-dimensional nature. Here this problem will be avoided by assuming that the electric field is much larger in the ***transverse direction*** (the direction perpendicular to the silicon surface, labeled as the x-direction in Fig. 7.1) than in the ***longitudinal direction*** (the direction parallel to the silicon surface along a line connecting the source and drain, labeled as the y-direction in Fig. 7.1). This is known as the ***gradual channel approximation***. If this condition holds, it is possible to obtain an accurate estimate of the amount of mobile electron charge stored in the channel by imagining that the MOSFET consists of a large number of very thin MOS capacitors strung together to fill the region between the source and drain, as suggested in Fig. 7.9. If the longitudinal field is negligible compared to the transverse field, the charges Q_n and Q_B stored in each capacitor can be computed by applying the one-dimensional electrostatic

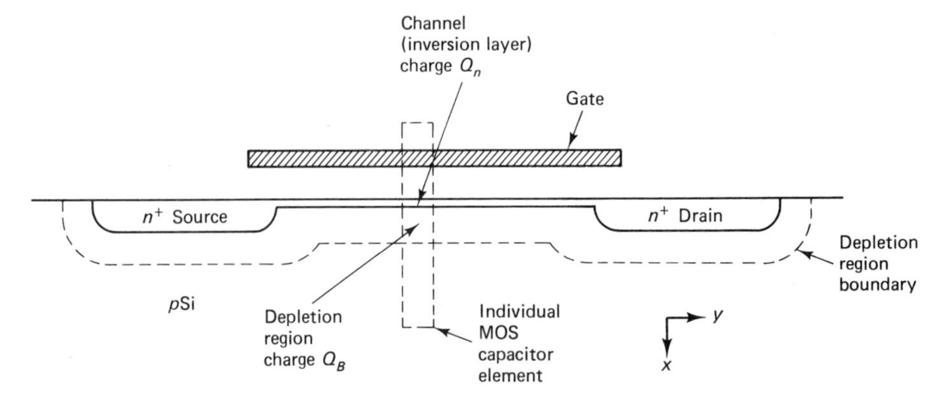

Figure 7.9 How the gradual channel approximation views the MOSFET as a set of MOS capacitors connecting source and drain.

analysis developed in the preceding section. The stored charge will in general vary from capacitor to capacitor, since the application of biases between the source, drain, and substrate means that the carrier concentrations can no longer be described by a single, position-independent Fermi energy. How the electron charge Q_n moves in response to the small longitudinal field can then be treated as a separate, independent problem. Viewed in a slightly different way, the gradual channel approximation amounts to ignoring the longitudinal electric field when Gauss's law is used to compute the stored charge in the channel. This should be permissible as long as the transverse field is much greater than the longitudinal field.

Throughout this section, we will make frequent reference to ***long-channel*** MOSFETs. For the present, a long-channel device can be thought of as one to which the gradual change approximation applies, at least over a reasonable range of operating bias. A more detailed discussion of the differences between long- and short-channel transistors is given in Section 7.6.

It should be noted that in normal operation electrons and holes are never found together in high concentrations in the MOSFET. As a result, recombination does not play an important role in the operation of the device. The only significant current in an n-channel MOSFET results from electron flow from the source to the drain, while the only current of importance in a p-channel transistor results from hole flow from source to drain. For this reason, MOSFETs are sometimes classed along with JFETs, MESFETs, and Schottky diodes as ***unipolar*** devices—that is, devices in which only one carrier type contributes significantly to the current. In contrast, both electrons and holes are required for the operation of the bipolar junction transistor.

7.3.1 Channel Voltage

Application of biases to the source and drain relative to the substrate creates conditions in which it is no longer reasonable to assume that the electron and hole quasi-Fermi energies coincide or are constant across the semiconductor. However, some simplifying assumptions regarding the behavior of E_{Fn} and E_{Fp} can still be made. In normal operation of the MOSFET the only major current results from the motion of electrons from the source to the drain. Since the hole current is negligible everywhere, the hole quasi-Fermi energy E_{Fp} should be constant, at least in regions in which the hole concentration is appreciable. Similarly, E_{Fn} should be constant in the direction perpendicular to the surface, at least in the surface region where n is large.

Application of a positive bias V_{SB} to the source relative to the substrate lowers E_{Fn} in the source relative to E_F at the back contact by an amount qV_{SB}. This effect is illustrated in the band diagram of Fig. 7.10a, which plots the energy levels in the MOSFET as a function of distance from the surface in a region near the source. (Frequently, MOSFETs are operated with the source and substrate shorted, but we will allow for nonzero V_{SB} here.) Similarly, the drain–source bias V_{DS} lowers E_{Fn} in the drain relative to the source, as shown in the band diagram of Fig. 7.10b. It is this difference in E_{Fn} between the source and drain that drives the electrons down the channel.

In determining the magnitude of the drain current, it will prove convenient to define a quantity $V_{CB}(y)$ which we will term the ***channel voltage***. V_{CB} will be defined as the difference between the electron quasi-Fermi energy E_{Fn} at the semiconductor surface and the Fermi energy at the substrate (back) contact, as shown in Fig. 7.11. At the source $V_{CB} = V_{SB}$, while at the drain $V_{CB} = V_{SB} + V_{DS}$. The increase in V_{CB} from source to drain can also be viewed as the driving force for electron motion down the channel.

7.3.2 Threshold Voltage

In order to have appreciable electron flow between the source and drain in the MOSFET, it seems clear that an inversion layer must be present at the semiconductor surface. The threshold voltage V_T of the MOSFET is defined to be that value of V_{GS} (*not* V_{GB}) which must be applied just to produce a surface inversion layer in the absence of V_{DS}. In the surface region of the MOSFET, E_{Fn} is lower by an amount qV_{CB} compared to the case of the MOS capacitor. The electron concentration is therefore reduced by a factor $e^{-qV_{CB}/kT}$, so

$$n_s = n_{0p} e^{q\psi_s/kT} e^{-qV_{CB}/kT} = \frac{n_i^2}{N_A} e^{q\psi_s/kT} e^{-qV_{CB}/kT} \tag{7.24}$$

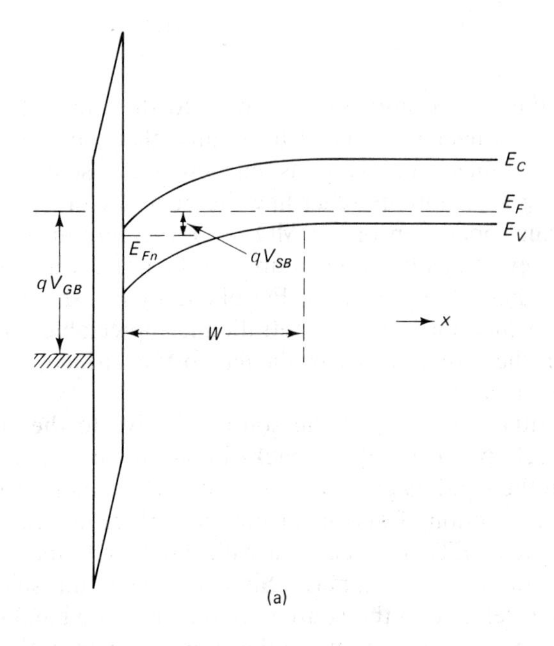

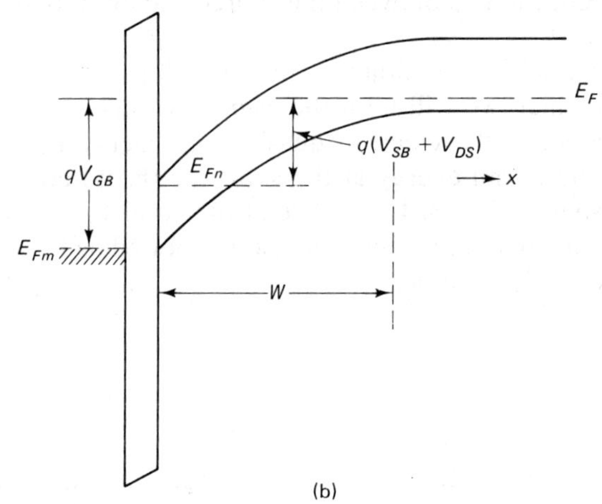

Figure 7.10 Energy band diagrams for the MOSFET at (a) the source end and (b) the drain end of the channel.

The condition for strong inversion then becomes

$$N_A = \frac{n_i^2}{N_A} e^{q\psi_s/kT} e^{-qV_{CB}/kT} \tag{7.25}$$

so

$$\psi_s = \frac{2kT}{q} \ln \frac{N_A}{n_i} + V_{CB} = 2\phi_B + V_{CB} \tag{7.26}$$

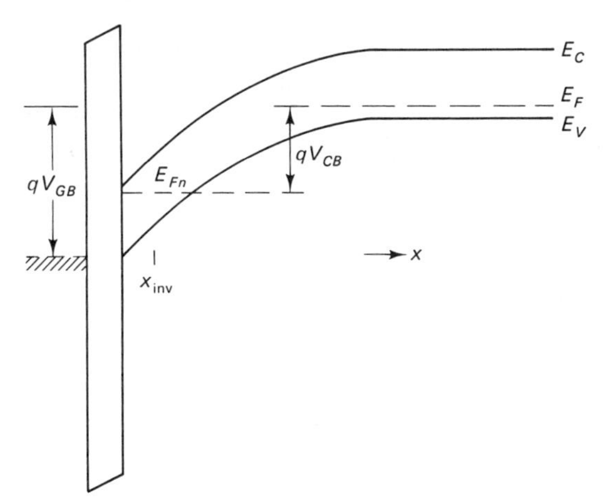

Figure 7.11 Band diagram illustrating the definition of the channel voltage V_{CB}. *Note:* $E_{Fn} + qV_{CB} = E_F$.

If $V_{DS} = 0$, then $V_{CB} = V_{SB}$ and the formation of an inversion layer requires

$$\psi_s = 2\phi_B + V_{SB} \tag{7.27}$$

At threshold, the depletion region width is therefore given by

$$W = \sqrt{\frac{2\epsilon_s(2\phi_B + V_{SB})}{qN_A}} \tag{7.28}$$

so

$$Q_B = -\sqrt{2\epsilon_s qN_A(2\phi_B + V_{SB})} \tag{7.29}$$

Assuming again that Q_n is negligible at threshold, so $Q_s \simeq Q_B$, we have

$$V_{GB} = \psi_s + \psi_{\text{ox}} = 2\phi_B + V_{SB} + \frac{\sqrt{2\epsilon_s qN_A(2\phi_B + V_{SB})}}{C_{\text{ox}}} \tag{7.30}$$

Noting that $V_{GS} = V_{GB} - V_{SB}$, we finally obtain

$$V_T = 2\phi_B + \frac{\sqrt{2\epsilon_s qN_A(2\phi_B + V_{SB})}}{C_{\text{ox}}} \tag{7.31}$$

It should be stressed once again that V_T refers to the value of V_{GS}, not V_{GB}, needed to produce strong inversion.

7.3.3 Equation for the Drain Current

To compute the magnitude of the drain current when $V_{GS} > V_T$, it is necessary to determine the amount of mobile electron charge in the channel. In the analysis of the MOS capacitor, it was argued that once strong inversion set in, the potential drop across the silicon should remain effectively fixed at $\psi_s = 2\phi_B$ regardless of the value of V_{qB}. A similar argument holds for the MOSFET, but now at each point along the channel ψ_s is pinned at a value of $2\phi_B + V_{CB}$ irrespective of V_{GS}. Applying the same argument used to obtain (7.29), the depletion region charge is given by

$$Q_B(y) = -\sqrt{2\epsilon_s q N_A(2\phi_B + V_{CB}(y))} \tag{7.32}$$

From (7.10), the total charge stored in the silicon is given by

$$Q_s(y) = -C_{ox}\psi_{ox} = -C_{ox}(V_{GB} - \psi_s) = -C_{ox}(V_{GB} - 2\phi_B - V_{CB}(y)) \tag{7.33}$$

Therefore,

$$Q_n(y) = -C_{ox}(V_{GB} - 2\phi_B - V_{CB}(y)) + \sqrt{2\epsilon_s q N_A(2\phi_B + V_{CB}(y))} \tag{7.34}$$

It is important to note that, since V_{CB} is a function of position along the channel, so is Q_n. This can be seen directly from (7.34), but it is instructive to examine in detail why this should be so. On moving from the source to the drain, V_{CB} increases, and therefore so does the potential drop ψ_s across the semiconductor required to maintain strong inversion; as Fig. 7.12 suggests, the depletion region is wider near the drain than near the source. Since V_{GB} is fixed and $V_{GB} = \psi_s + \psi_{ox}$, this in turn implies that ψ_{ox} is smaller near the drain than near

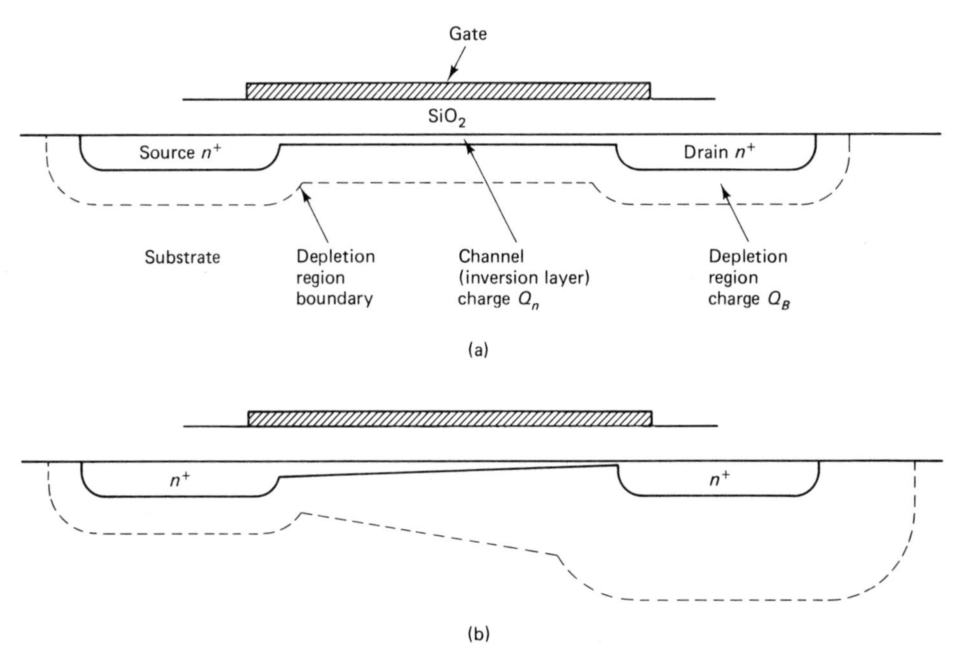

Figure 7.12 Cross section through MOSFET showing how thicknesses of channel and depletion region vary on moving from source to drain: (a) $V_{DS} = 0$; (b) $V_{DS} > 0$.

the source, and therefore, from (7.33), so is the magnitude of the total charge stored in the semiconductor. If the total stored charge is smaller in magnitude, and the depletion region charge is larger, the mobile electron charge must be smaller.

The driving force for electron flow from source to drain is, as always, the gradient of the electron quasi-Fermi energy level E_{Fn}:

$$J_n = \mu_e n \frac{dE_{Fn}}{dy} = -q\mu_e n \frac{dV_{CB}}{dy} \tag{4.27}$$

Integrating (4.27) from the surface at $x = 0$ to the bottom edge of the inversion layer at $x = x_{\text{inv}}$ and multiplying by the channel width Z gives the total current I_n carried by electrons flowing from source to drain:

$$I_n = -Z\mu_e \frac{dV_{CB}}{dy} \int_0^{x_{\text{inv}}} qn(x)\, dx = Z\mu_e Q_n \frac{dV_{CB}}{dy} \tag{7.35}$$

(The location of x_{inv} is shown in Fig. 7.11.) Here we have used the fact that the integral of the electron concentration over the depth of the channel must equal the total number of elecrons per unit area in the channel.

To complete the evaluation of the drain current, we note that since there is normally no appreciable hole flow in the MOSFET, under steady-state conditions I_n must be constant all the way along the channel from source to drain. (If this were not the case, there would be a rapid buildup of elecrons in one part of the channel, which could not be sustained.) Therefore, $I_D = -I_n$, and is taken by convention to the positive. Also,

$$-\int_0^L I_n(y)\, dy = I_D L \tag{7.36}$$

so

$$\begin{aligned} I_D &= \frac{Z\mu_e C_{\text{ox}}}{L} \int_0^L \left[V_{GB} - 2\phi_B - V_{CB} - \frac{\sqrt{2\epsilon_s q N_A (2\phi_B + V_{CB})}}{C_{\text{ox}}} \right] \frac{dV_{CB}}{dy}\, dy \\ &= \frac{Z\mu_e C_{\text{ox}}}{L} \int_{V_{SB}}^{V_{SB}+V_{DS}} \left[V_{GB} - 2\phi_B - V_{CB} - \frac{\sqrt{2\epsilon_s q N_A (2\phi_B + V_{CB})}}{C_{\text{ox}}} \right] dV_{CB} \end{aligned} \tag{7.37}$$

Evaluation of the integral in (7.37) gives finally

$$\begin{aligned} I_D = \frac{Z\mu_e C_{\text{ox}}}{L} \Bigg\{ &\left(V_{GS} - 2\phi_B - \frac{V_{DS}}{2} \right) V_{DS} \\ &- \frac{2\sqrt{2\epsilon_s q N_A}}{3C_{\text{ox}}} [(V_{SB} + 2\phi_B + V_{DS})^{3/2} - (V_{SB} + 2\phi_B)^{3/2}] \Bigg\} \end{aligned} \tag{7.38}$$

Equation (7.38) is sometimes referred to as the ***Ihantola and Moll model***, after its developers. It is also known as the ***bulk charge model***, since it computes the mobile electron charge as the difference between the total charge stored in the semiconductor and the "bulk charge"—in other words, the charge stored in the depletion region.

7.3.4 Saturation

In deriving (7.38) it was tacitly assumed that an inversion layer was present all along the silicon surface from source to drain. If $V_{GS} > V_T$, this will be the case for sufficiently small V_{DS}. However, as V_{DS} is increased, E_{Fn} near the drain moves lower, the potential drop across the silicon required to maintain strong inversion increases, and the charge stored in the depletion region grows. Furthermore, as argued above, if ψ_s increases with V_{GB} held constant, the magnitude of the total charge stored in the silicon drops. In consequence, increasing V_{DS} lowers the electron charge in the channel near the drain. This effect is illustrated in Fig. 7.13a. Eventually, if V_{DS} is made large enough, the gradual channel approximation model predicts Q_n drops to zero. Under these conditions, the channel is said to be ***pinched off*** near the drain, as suggested in Fig. 7.13b. This might lead one to conclude that the drain current would abruptly drop to zero. This seems absurd, and in fact the current does continue to exist. Once the electron charge in the channel becomes very small, the gradual channel approximation itself breaks down; it is no longer possible to compute Q_n using one-dimensional electrostatics, and (7.38) no longer applies.

When the channel pinches off in a long-channel MOSFET, the drain current saturates—that is, becomes effectively constant irrespective of further increases in V_{DS}. We will denote the value of V_{DS} required to just reach this condition of pinch-off as $V_{DS,\text{sat}}$. Qualitatively, the saturation of I_D can be explained by reference to (4.27) or (7.35). From these equations it follows that to maintain a given electron current in a region of low electron concentration, it is necessary to have a very large gradient of electron quasi-Fermi energy level. If V_{DS} is increased above $V_{DS,\text{sat}}$, the channel will pinch off at a point close to the drain where $V_{CS} = V_{DS,\text{sat}}$. It is important to remember that the pinch-off point marks the place where the gradual channel approximation first begins to fail, not the place where the electron concentration falls to zero. Between the pinch-off point and the drain the electron concentration is very small, and a drop in channel voltage equal to $V_{DS} - V_{DS,\text{sat}}$ is required to force the current through this region of very high effective resistance to electron flow. This is illustrated in Fig. 7.13c and in the band diagram of Fig. 7.14. If V_{DS} is increased still further, the pinch-off point is displaced slightly toward the source, but the general situation remains unchanged.

Since we have defined $V_{DS,\text{sat}}$ to be that value of V_{CS} at which the gradual channel approximation gives $Q_n = 0$ for some fixed value of V_{GS}, we have from (7.34)

$$-C_{\text{ox}}(V_{GS} - 2\phi_B - V_{DS,\text{sat}}) + \sqrt{2\epsilon_s q N_A(2\phi_B + V_{SB} + V_{DS,\text{sat}})} = 0 \tag{7.39}$$

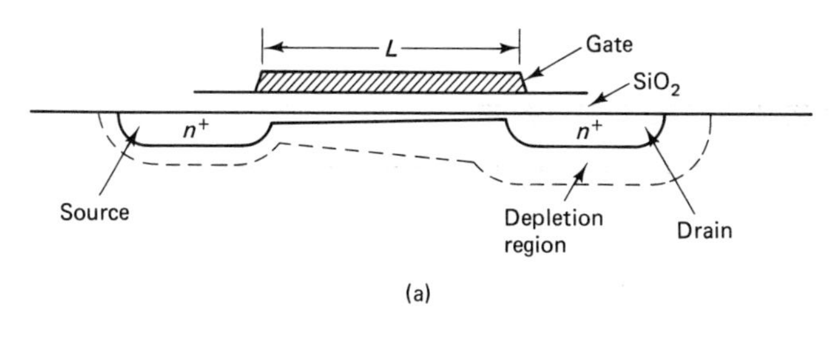

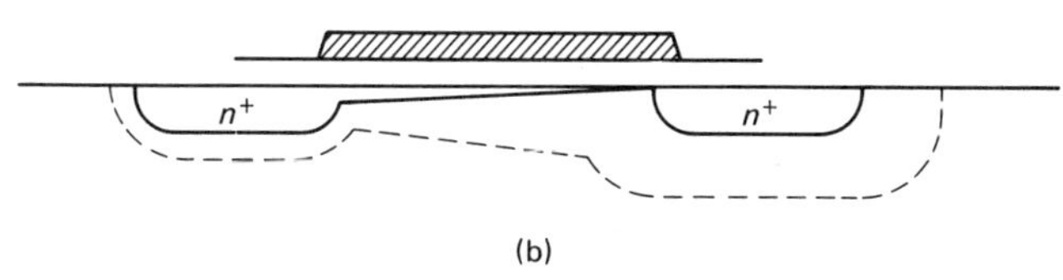

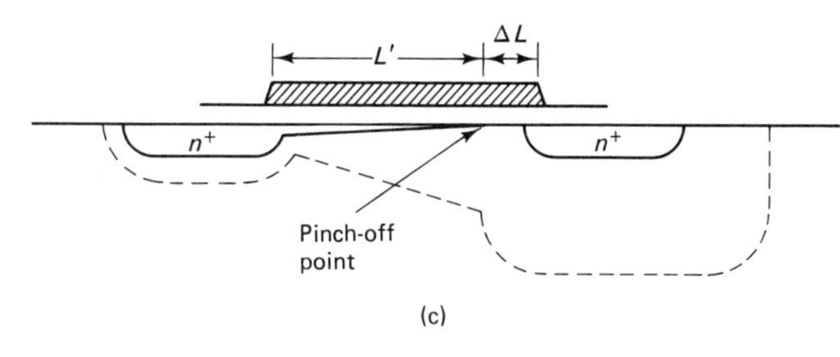

Figure 7.13 How the channel pinches off as V_{DS} is increased. (a) $V_{DS} < V_{DS,\text{sat}}$; an inversion layer connects the source and drain. (b) $V_{DS} = V_{DS,\text{sat}}$; at the onset of saturation the channel pinches off at the drain. (c) $V_{DS} > V_{DS,\text{sat}}$; as V_{DS} is increased above $V_{DS,\text{sat}}$, the pinch-off point is displaced slightly toward the source.

Solving for $V_{DS,\text{sat}}$ we find that

$$V_{DS,\text{sat}} = V_{GS} - 2\phi_B + \frac{\epsilon_s q N_A}{C_{\text{ox}}^2}\left(1 - \sqrt{1 + \frac{2C_{\text{ox}}^2 V_{GB}}{\epsilon_s q N_A}}\right) \tag{7.40}$$

In obtaining (7.40) from (7.39), the negative root was taken in the quadratic formula. This can be justified as follows. We know that

$$V_{GB} = \psi_s + \psi_{\text{ox}} \tag{7.41}$$

so

$$V_{GB} \geqslant \psi_s = V_{DS} + V_{SB} + 2\phi_B \tag{7.42}$$

and therefore

$$V_{DS} \leqslant V_{GB} - V_{SB} - 2\phi_B = V_{GS} - 2\phi_B \tag{7.43}$$

7.3.5 Operating Regimes for the MOSFET

Figure 7.15 plots I_D as a function of V_{DS} for the MOSFET for different values of V_{GS}. For $V_{DS} < V_{DS,\text{sat}}$, the MOSFET is said to be operating in the ***triode*** regime, since the characteristics resemble those of a triode vacuum tube. For $V_{DS} > V_{DS,\text{sat}}$, the device is in ***saturation***.

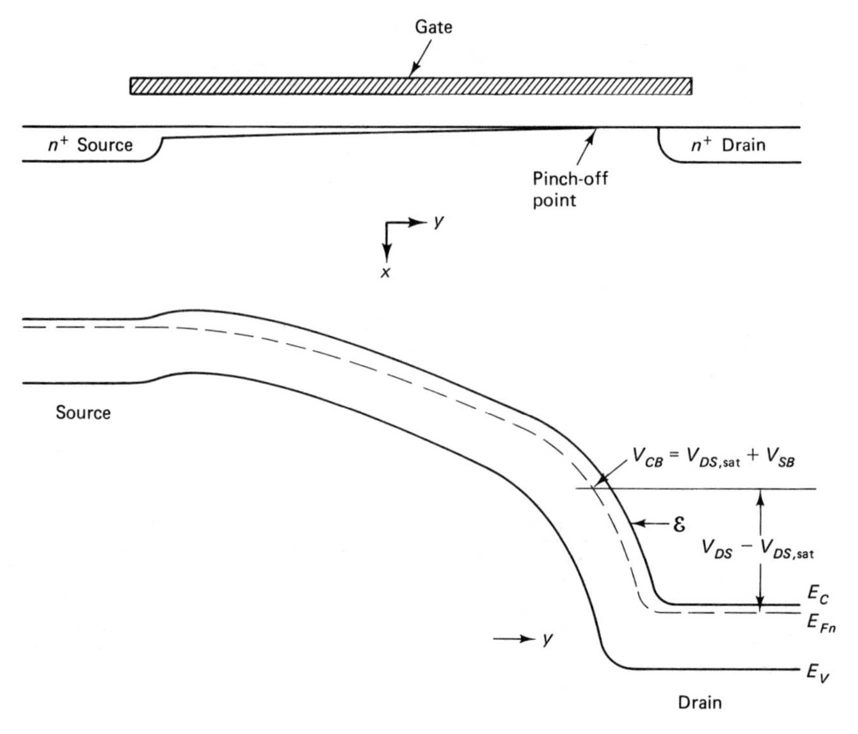

Figure 7.14 Energy band diagram along the channel for a MOSFET in saturation. Note the high electric field and large drop in channel voltage V_{CB} across the pinch-off region.

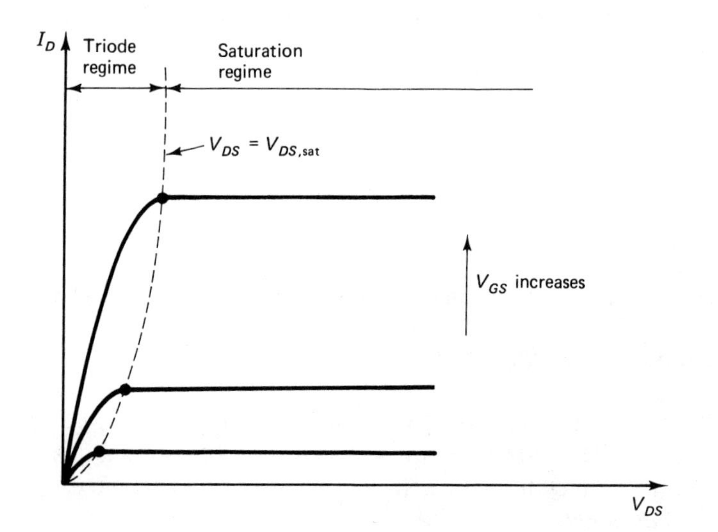

Figure 7.15 Drain characteristics of MOSFET.

The theory we have developed so far assumes that $I_D = 0$ when $V_{GS} < V_T$, so it would seem sensible to define this bias range as the cutoff regime for the MOSFET. In fact, a small but nonzero drain current exists even when $V_{GS} < V_T$, so it is more appropriate to speak of a ***subthreshold*** regime. The subthreshold current depends roughly exponentially on V_{GS}, and is almost independent of V_{DS} for $V_{DS} > 2kT/q$. Further details on the computation of I_D in the subthreshold regime are given in Ref. 1.

The logic that must be followed to determine the operating regime and therefore the correct equation to use when computing the drain current in the MOSFET is summarized in Fig. 7.16.

7.3.6 Square-Law Model

Equation (7.38) provides a relatively accurate description of current in the long-channel MOSFET. However, this equation is rather cumbersome to use in hand calculations and slow to evaluate when incorporated in circuit simulation programs such as SPICE. A simplified version of (7.38) that is considerably easier to work with can be obtained if $2\phi_B + V_{SB} \gg V_{DS}$. Using the binomial expansion,

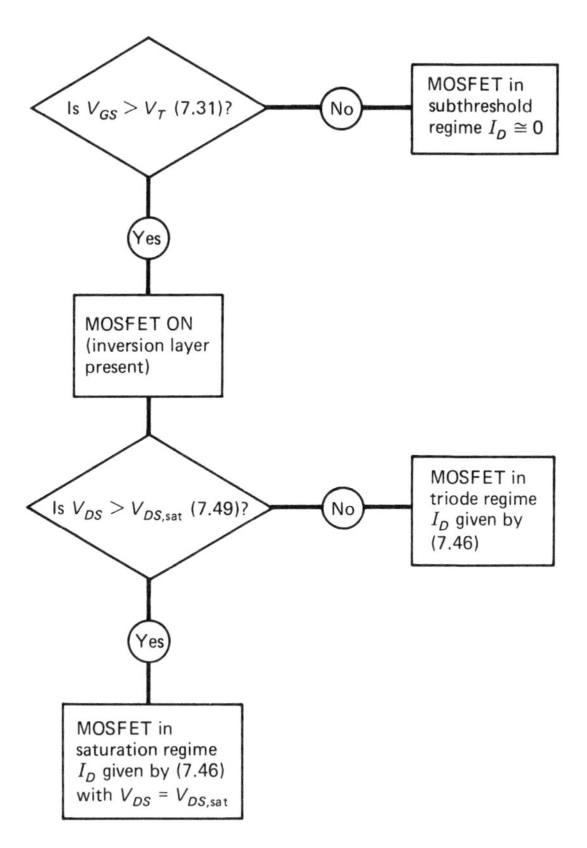

Figure 7.16 Determining the operating regime for a MOSFET.

we then have

$$(2\phi_B + V_{SB} + V_{DS})^{3/2} = (2\phi_B + V_{SB})^{3/2}\left(1 + \frac{V_{DS}}{2\phi_B + V_{SB}}\right)^{3/2}$$

$$\simeq (2\phi_B + V_{SB})^{3/2} + \frac{3}{2}(2\phi_B + V_{SB})^{1/2}V_{DS} \tag{7.44}$$

so

$$I_D = \frac{Z\mu_e C_{ox}}{L}\left[V_{GS} - 2\phi_B - \frac{\sqrt{2\epsilon_s qN_A(2\phi_B + V_{SB})}}{C_{ox}} - \frac{V_{DS}}{2}\right]V_{DS} \tag{7.45}$$

which can be written more succinctly as

$$I_D = \frac{Z\mu_e C_{ox}}{L}\left(V_{GS} - V_T - \frac{V_{DS}}{2}\right)V_{DS} \tag{7.46}$$

For very small V_{DS}, (7.46) can be further approximated as

$$I_D = \frac{Z\mu_e C_{ox}}{L}(V_{GS} - V_T)V_{DS} \tag{7.47}$$

A plot of I_D versus V_{GS} with V_{DS} held at a small, constant value should therefore be a straight line with a voltage axis intercept equal to V_T. In fact, as shown in Fig. 7.17, the $I_D - V_{GS}$ characteristic departs from a straight line in the subthreshold regime and at high values of V_{GS}. Plotting I_D versus V_{GS} is, however, still one of the most accurate methods for determining threshold voltage.

If $2\phi_B + V_{SB} \gg V_{DS}$, the condition for saturation becomes

$$-C_{ox}(V_{GS} - 2\phi_B - V_{DS,sat}) + \sqrt{2\epsilon_s qN_A(2\phi_B + V_{SB})} = 0 \tag{7.48}$$

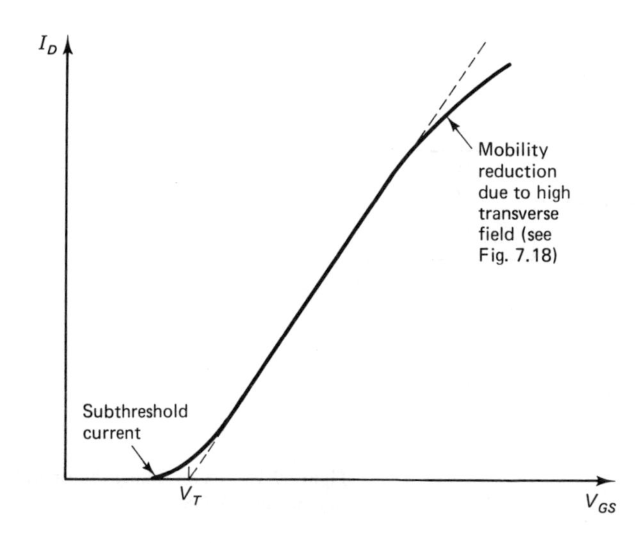

Figure 7.17 The I_D versus V_{GS} characteristic can be used to determine V_T.

so

$$V_{DS,\text{sat}} = V_{GS} - V_T \tag{7.49}$$

In saturation, the drain current is given by (7.46) evaluated at $V_{DS} = V_{DS,\text{sat}}$; that is,

$$I_D = \frac{Z\mu_e C_{\text{ox}}}{L} \frac{(V_{GS} - V_T)^2}{2} \tag{7.50}$$

Since (7.50) predicts a current that is proportional to $(V_{GS} - V_T)^2$, (7.46) is sometimes referred to as the ***square-law model*** for the MOSFET.

Physically, the square-law model can be arrived at by assuming that, all along the channel, the depletion region has the same width as it does near the source. Near the source,

$$W = \sqrt{\frac{2\epsilon_s(2\phi_B + V_{SB})}{qN_A}} \tag{7.51}$$

so

$$Q_B = -\sqrt{2\epsilon_s qN_A(2\phi_B + V_{SB})} \tag{7.52}$$

$$\begin{aligned} Q_n &= -C_{\text{ox}}\left[V_{GB} - 2\phi_B - V_{CB} - \frac{\sqrt{2\epsilon_s qN_A(2\phi_B + V_{SB})}}{C_{\text{ox}}}\right] \\ &= -C_{\text{ox}}(V_{GB} - V_T - V_{CB}) \end{aligned} \tag{7.53}$$

and therefore

$$I_D = \frac{Z\mu_e C_{\text{ox}}}{L} \int_{V_{\text{SB}}}^{V_{\text{SB}}+V_{\text{DS}}} (V_{GB} - V_T - V_{CB})\, dV_{CB} \tag{7.54}$$

Integrating (7.54) from source to drain gives (7.46).

It is easy to show that the square-law model always overestimates the drain current for given bias conditions. We have argued that the width of the depletion region must increase on moving from the source to the drain, so by assuming that the width of this region is everywhere the same as at the source, the square law model underestimates the magnitude of the depletion region charge Q_B. Since the mobile electron charge Q_n is the difference between the total charge Q_s stored in the silicon and Q_B, the square law model therefore consistently overestimates the magnitude of Q_n. Overestimating the electron concentration in the channel naturally leads to an overestimate of I_D. The error involved can be quite large; a 20 to 30% overestimate of I_D is not unusual for typical devices and operating conditions.

7.3.7 Transconductance

Perhaps the most important figure of merit for any field-effect transistor is its ***transconductance*** g_m, which is defined as

$$g_m = \left.\frac{\partial I_D}{\partial V_{GS}}\right|_{V_{DS}, V_{SB} \text{ const}} \tag{7.55}$$

The transconductance is therefore a measure of how well the gate–source bias can control the drain current. Since I_D scales in proportion to the channel width, the transconductance per unit channel width is the most appropriate parameter to use when comparing the performance of two FETs.

A particularly simple expression for g_m in the MOSFET can be obtained from the square-law model. For a device operating in saturation, I_D is given by (7.50), so

$$g_{m,\text{sat}} = \frac{Z\mu_e C_{\text{ox}}}{L}(V_{GS} - V_T) \tag{7.56}$$

7.3.8 MOSFET Models in SPICE

A variety of models for the MOSFET is used in current versions of the SPICE circuit simulation program. The two most common are known as the Level 1 and Level 2 models. The Level 1 model is essentially the square-law model embodied in (7.31), (7.46), and (7.49), while the Level 2 model is the bulk charge model of (7.38) and (7.40). The Level 2 model also incorporates an approximate treatment of some of the short-channel effects discussed in Section 7.6, and is capable of computing the subthreshold current.

The basic structural parameters required to describe a MOSFET when using either the Level 1 or Level 2 model are the substrate doping, the gate oxide thickness, and the channel length and width. SPICE computes the threshold voltage from N_A and t_{ox} using (7.31). Alternatively, the threshold voltage **for the case of $V_{SB} = 0$** can be input directly by supplying a value to the model parameter VTO, in which case the internally computed threshold voltage is overridden.

Example 7.1 MOSFET Characteristics

This example illustrates the I_D–V_{DS} characteristics for an n-channel MOSFET with a gate oxide thickness t_{ox} of 50 nm and a substrate doping N_A (NSUB in SPICE) of 10^{16} cm^{-3}. These values of oxide thickness and substrate doping are fairly representative of those used in present-day commercial MOS ICs. The threshold voltage VTO is set equal to 1.4 V, in accord with (7.31). The channel length and width are arbitrarily set to 10 μm, and the device is assumed not to be subject to short-channel effects. The circuit used in the simulation is shown in Fig. 7E.1. The SPICE input file listed below generates the I_D–V_{DS} characteristics using the Level 1 (square-law) model.

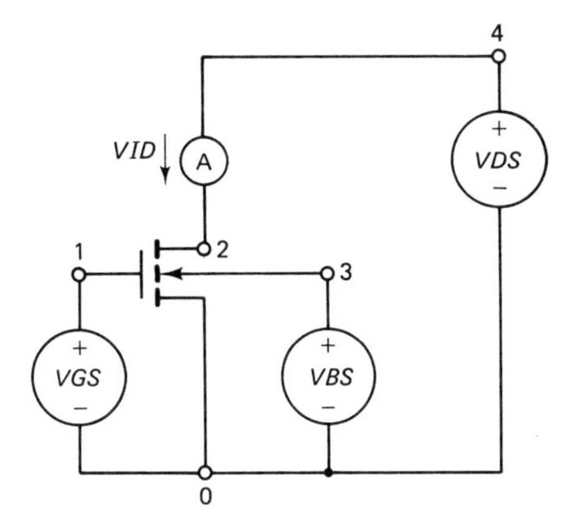

Figure 7E.1 Circuit for Example 7.1.

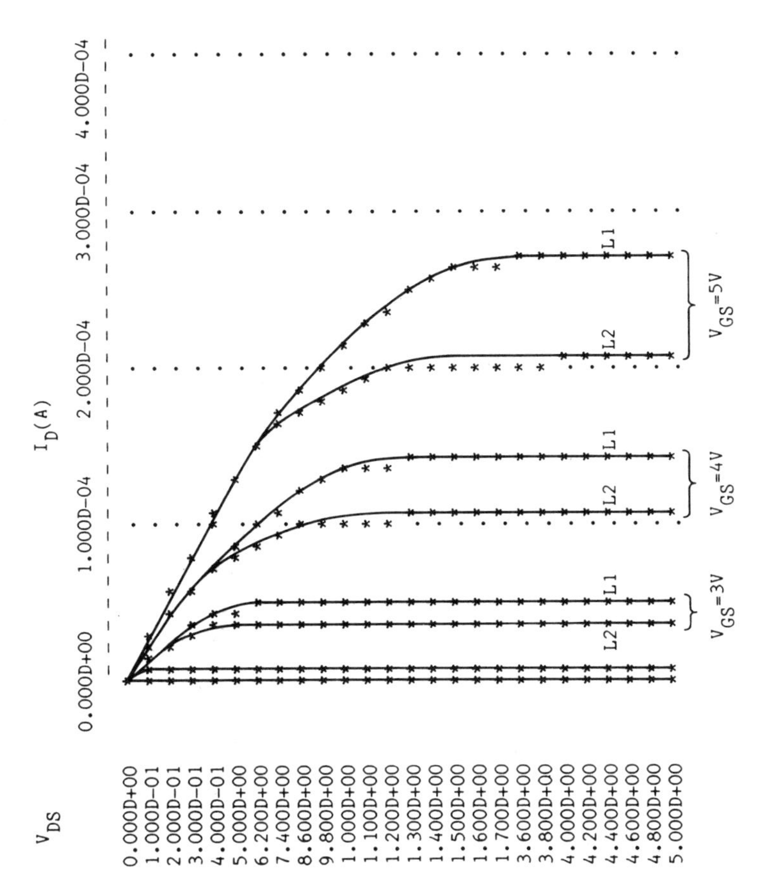

Figure 7E.2 I_D–V_{DS} characteristics predicted by Level 1 (solid line) and Level 2 (dashed line) models.

```
*MOSFET ID-VDS CHARACTERISTICS: LEVEL 1 MODEL
VDS 4 0
VGS 1 0
VBS 3 0 DC 0
VID 4 2 DC 0
MOS1 2 1 0 3 NMOS1 L=10U W=10U
.MODEL NMOS1 NMOS LEVEL=1 NSUB=1E16 TOX=50E-9 VTO=1.4
.WIDTH OUT=80
.DC VDS 0 5 0.2 VGS 0 5 1
.PLOT DC I(VID)
.END
```

The characteristics predicted by the Level 2 (bulk charge) model can be obtained simply by changing the .MODEL line to read

```
.MODEL NMOS1 NMOS LEVEL=2 NSUB=1E16 TOX=50E-9 VTO=1.4
```

The characteristics generated by the two models are compared in Fig. 7E.2. As expected, the predictions of the two models coincide for small V_{DS}, but the more accurate Level 2 model predicts a substantially smaller current at large V_{DS}.

7.3.9 Electron Mobility in the MOSFET Channel

According to (7.47), by measuring the slope of the $I_D - V_{GS}$ characteristic at fixed V_{DS}, it is possible to determine the mobility of electrons in the MOSFET channel. When this is done, it is found that the value of μ_e which should be used in (7.38) is only one-half to two-thirds of the electron mobility in bulk silicon. Further, μ_e varies with the field normal to the silicon surface, as shown in Fig. 7.18. (This accounts for the bending over of the I_D characteristic in Fig. 7.17 at high V_{GS}.)

The fact that the electron mobility in a MOSFET channel is considerably lower than that in bulk silicon is usually explained by noting that electrons in the channel are confined to a region very close to the silicon surface, and make frequent collisions with the surface as suggested in Fig. 7.19. Following the arguments of Section 3.2.1, an enhanced collision frequency gives a lower mobility.

In the SPICE Level 1 model, the effective electron mobility is assumed fixed. The Level 2 model allows for a fall-off in μ_e at high V_{GS} through an empirical formula. This feature of the Level 2 model is, however, invoked only if the parameters UCRIT and UEXP are specified. Further details are given in the SPICE reference manual and associated documentation [2, 4].

7.3.10 Body Effect

In the analysis of the MOSFET presented above, allowance was made for the presence of a source–substrate bias V_{SB}. V_{SB} is often referred to as the ***backgate***

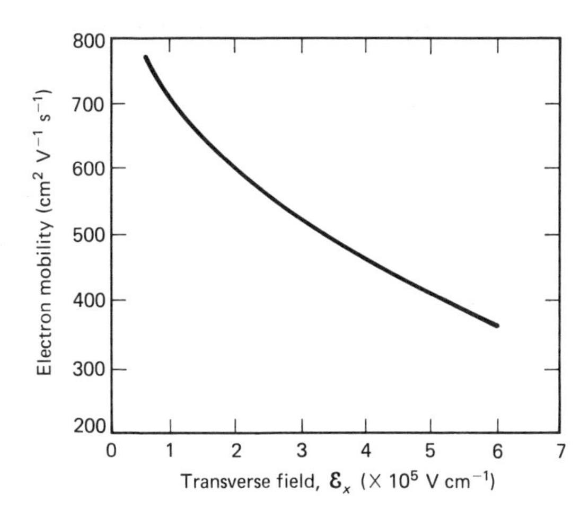

Figure 7.18 Electron mobility in the MOSFET channel as a function of transverse field. (From Sabnis and Clemens [3]. Reprinted with permission of the publisher, IEEE, Inc.)

bias, the idea being that the substrate terminal exerts some control over current in the channel, and therefore behaves to some extent like a second or backside gate. From (7.31) it can be seen that a positive V_{SB} raises the threshold voltage of an n-channel device. This is often referred to as the ***body effect***. Equations (7.38) and (7.40) show that application of backgate bias also reduces the drain current for given V_{DS} and V_{GS}, and lowers $V_{DS,\text{sat}}$.

Physically, the body effect arises because application of a reverse bias to the source–substrate junction lowers the electron quasi-Fermi energy level in the source. This in turn means that more band bending is required to create a condition of strong inversion at the silicon surface, so more charge is stored in the depletion region. A larger value of ψ_s also implies a smaller value of ψ_{ox} and hence, from (7.10), less total charge stored in the silicon. With $|Q_s|$ reduced and $|Q_B|$ increased, the magnitude of the mobile electron charge in the channel must drop. In consequence, applying a backgate bias removes electrons from the channel and reduces the current in the MOSFET.

The effect can also be thought of as an increase in the reverse bias of the n-channel/p-substrate junction diode at the source end of the channel. The associated increase in depletion region width narrows the n-channel. Thus a higher V_{GS} is required to maintain the channel current.

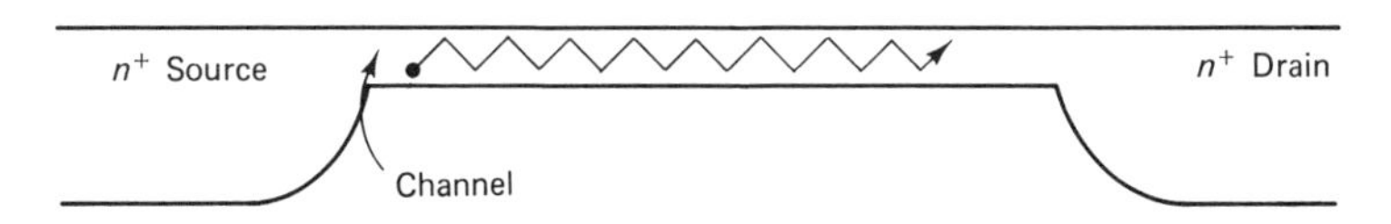

Figure 7.19 How an electron moving from the source to the drain undergoes frequent collisions with the surface. This surface scattering results in a low effective electron mobility in the channel.

Example 7.2 The Body Effect

This example illustrates how application of a source–substrate bias alters the characteristics of an n-channel MOSFET. The circuit used to generate the characteristics is the same as that used in Example 7.1, and is shown in Fig. 7E.1. The SPICE input file listed below generates $I_D - V_{GS}$ characteristics, from which V_T can be extracted.

```
*MOSFET ID-VGS CHARACTERISTICS: THE EFFECT OF VSB
VDS 4 0 DC 0.1
VGS 1 0
VBS 3 0
VID 4 2 DC 0
MOS1 2 1 0 3 NMOS1 L=10U W=10U
.MODEL NMOS1 NMOS LEVEL=2 NSUB=1E16 TOX=50E-9 VTO=1.4
.WIDTH OUT=80
.DC VGS 0 5 0.2 VBS 0 -5 -1
.PLOT DC I(VID)
.END
```

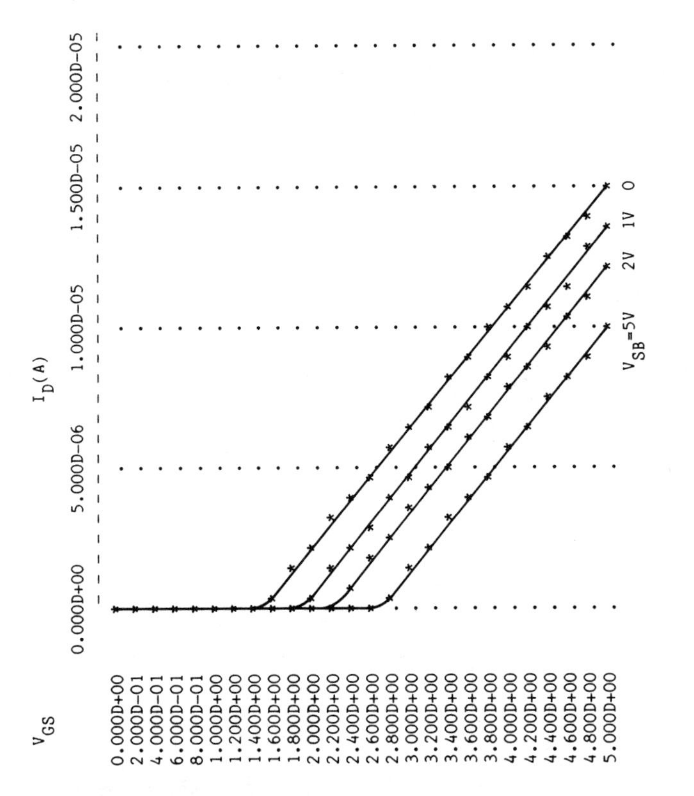

Figure 7E.3 Effect of "backgate bias" V_{SB} on threshold voltage.

The output obtained from the file above is plotted in Fig. 7E.3. It can be seen that application of a "backgate bias" indeed raises the threshold voltage, as predicted by (7.31).

The SPICE input file below can be used to illustrate how application of a backgate bias lowers I_D for given V_{DS} and V_{GS} and reduces $V_{DS,\text{sat}}$ for given V_{GS}.

```
*MOSFET ID-VDS CHARACTERISTICS: THE EFFECT OF VSB
VDS 4 0
VGS 1 0 DC 5
VBS 3 0
VID 4 2 DC 0
MOS1 2 1 0 3 NMOS1 L=10U W=10U
.MODEL NMOS1 NMOS LEVEL=1 NSUB=1E16 TOX=50E-9 VTO=1.4
.WIDTH OUT=80
.DC VDS 0 5 0.2 VBS 0 -5 -5
.PLOT DC I(VID)
.END
```

The output obtained from this file is shown in Fig. 7E.4.

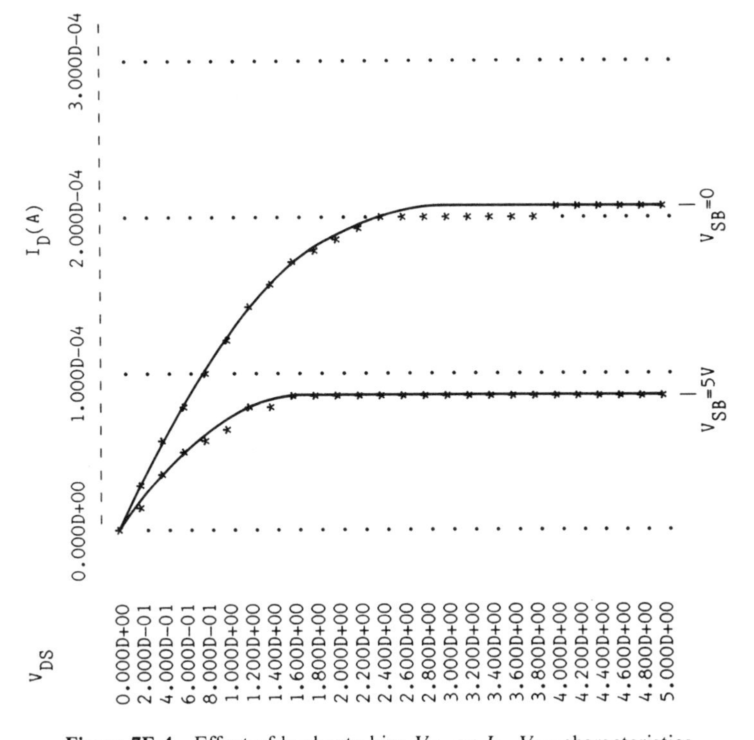

Figure 7E.4 Effect of backgate bias V_{SB} on I_D–V_{DS} characteristics.

7.3.11 Beyond the Bulk Charge Model

In deriving (7.38) we assumed that the potential drop ψ_s across the silicon was pinned at a value of $2\phi_B + V_{CB}$ wherever the surface was strongly inverted. In reality, this is the value of ψ_s at which the surface just enters strong inversion; if the electron charge in the channel is substantial, ψ_s must be greater than this value. From the arguments given in Section 7.2.2, ψ_s will never be much larger than $2\phi_B + V_{CB}$, but the difference might easily amount to a few times the thermal voltage kT/q. We can therefore conclude that the bulk charge model consistently underestimates ψ_s.

If ψ_s is underestimated, so is the magnitude of the depletion region charge Q_B. From (7.8), an underestimate of ψ_s also leads to an overestimate of ψ_{ox} and therefore, from (7.10), an overestimate of the magnitude of the total charge stored in the semiconductor. If $|Q_B|$ is underestimated, and $|Q_s|$ overestimated, then from (7.16) $|Q_n|$ must be overestimated. The bulk charge model therefore consistently overestimates the amount of mobile electron charge in the channel, and thus overestimates the drain current. Fortunately, the error involved is typically no more than 10%.

It is possible to develop an alternative model for current in the long-channel MOSFET, known as the ***charge sheet model***, that is only marginally more complicated than (7.38), but which computes ψ_s at each point in the channel rather than making any assumptions as to the value ψ_s should have. The charge sheet model predicts I_D to an accuracy of about 1% or better for most devices under most operating conditions. Interested readers are directed to Ref. 1 for further details on this model.

7.3.12 Device Breakdown

Increasing V_{DS} with V_{GS} held constant in a MOSFET operating in the saturation regime increases the potential drop across the pinched-off region shown in Fig. 7.14, and thereby raises the electric field in this region. Between the pinch-off point and the drain the current is carried by relatively few electrons moving at very high velocities. If V_{DS} is made sufficiently large, the field in the pinch-off region becomes so high that an electron can gain energy comparable to E_g between collisions. Once this happens, avalanche generation of electrons and holes takes place in the pinch-off region in a process similar to the avalanche breakdown of a reverse-biased *pn* junction discussed in Section 6.5.2. The avalanche process leads to an effective loss of the gate's control over the drain current, and I_D increases very rapidly with increasing V_{DS}. These effects are illustrated in Fig. 7.20. Avalanche-generated holes are swept out of the depletion region and into the substrate, giving rise to a large current at the substrate terminal.

In general, avalanche breakdown in the MOSFET occurs at a V_{DS} value much smaller than the bias that would be required to cause breakdown of the drain–substrate junction in isolation. This comes about because the presence of

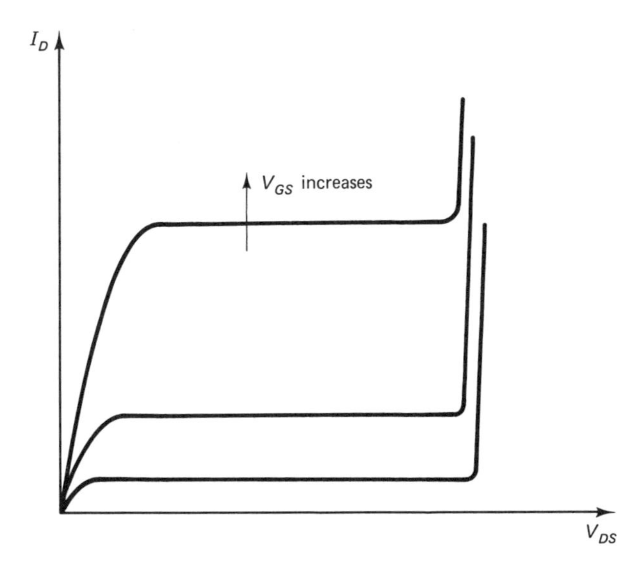

Figure 7.20 Breakdown in the MOSFET caused by avalanche generation of electron–hole pairs in the pinch-off region near the drain.

the gate leads to fields in the pinch-off region that are much higher than those which would be found in the depletion region of an isolated junction subject to a reverse bias equal to V_{DB}.

7.4 FLATBAND VOLTAGE

So far in this chapter we have assumed that the work functions of the metal and semiconductor in the MOSFET are identical and that there is no charge in the oxide layer. In practice, it is unlikely that the metal work function will equal that of the semiconductor, and it is almost certain that a real oxide will contain trapped charge. In a real MOSFET it is therefore likely that there will be some band bending across the semiconductor at zero applied bias. This possibility can be incorporated into the theory of the MOSFET developed above with some relatively minor modifications which will now be considered.

7.4.1 Work Function Difference ϕ_{ms}

Figure 7.21a shows the band diagram at zero applied gate–substrate bias for a MOS capacitor for which the work function of the metal is less than that of the semiconductor. The work function difference results in the inversion of the semiconductor surface even when $V_{GB} = 0$. If V_{GB} is made negative, as suggested in Fig. 7.21b, it is possible to arrive at a condition in which there is no band bending across the semiconductor. This value of V_{GB} is known as the ***flatband voltage*** V_{FB}. From Fig. 7.21b,

$$-V_{FB} + \phi'_m = \phi'_s$$

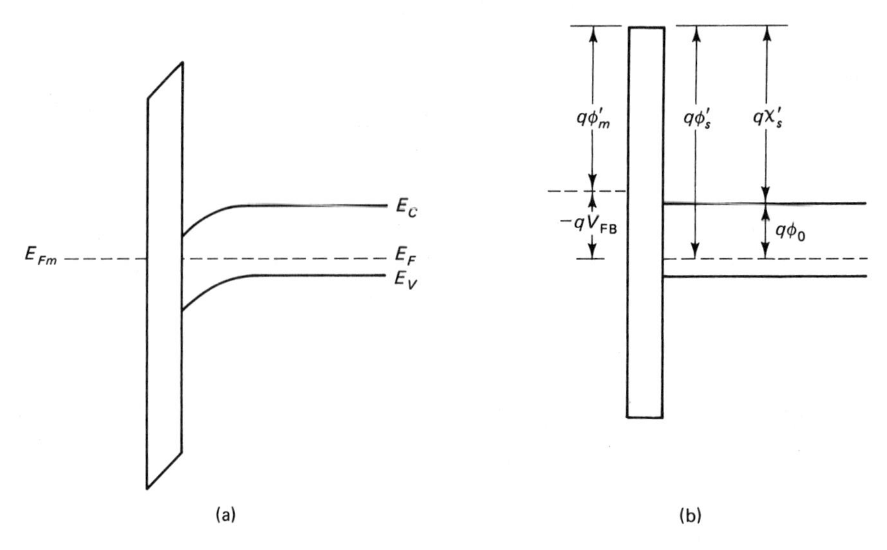

Figure 7.21 Computing V_{FB} for the case of a gate–substrate work function difference: (a) $V_{GB} = 0$; (b) $V_{GB} = V_{FB}$.

so

$$V_{FB} = \phi'_m - \phi'_s = \phi'_m - \chi'_s - \phi_0 \tag{7.57}$$

In most MOSFETs produced commercially today, the gate is made of n^+ polysilicon. The band diagram for a MOS capacitor with an n^+ polysilicon gate formed on a p-type substrate is shown in Fig. 7.22a. Assuming that the Fermi

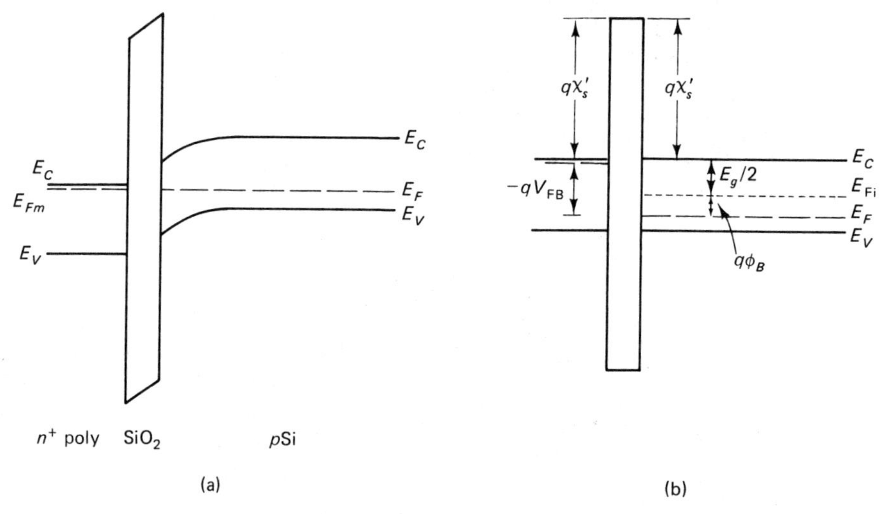

Figure 7.22 Computing V_{FB} for an n^+ polysilicon gate on a p-type substrate: (a) $V_{GB} = 0$; (b) $V_{GB} = V_{FB}$.

energy coincides with the conduction band edge in the heavily doped polysilicon, Fig. 7.22b shows that

$$V_{FB} = -\frac{E_g}{2q} - \phi_B \tag{7.58}$$

In many CMOS processes (see Section 12.2.11) the p-channel transistors also have n^+ polysilicon gates. The band diagram appropriate to this MOS system is shown in Fig. 7.23. Here

$$V_{FB} = -\frac{E_g}{2q} + |\phi_B| \tag{7.59}$$

Finally, in some state-of-the-art CMOS processes the p-channel transistors have p^+ polysilicon gates. Assuming that E_F coincides with E_V in such a gate, the band diagram for the MOS system appears as shown in Fig. 7.24. Here

$$V_{FB} = \frac{E_g}{2q} + |\phi_B| \tag{7.60}$$

7.4.2 Significance of V_{FB}

The analysis of the MOSFET presented to this point has been based on the assumption that there is no charge stored in the semiconductor—in other words, that flatband conditions hold—when $V_{GB} = 0$. If this is not the case, the equations developed in Section 7.3 are still valid provided that the following simple modification is made: everywhere the term V_{GB} appears, it must be replaced by $V_{GB} - V_{FB}$, while the term V_{GS} must be replaced by $V_{GS} - V_{FB}$. This surprisingly

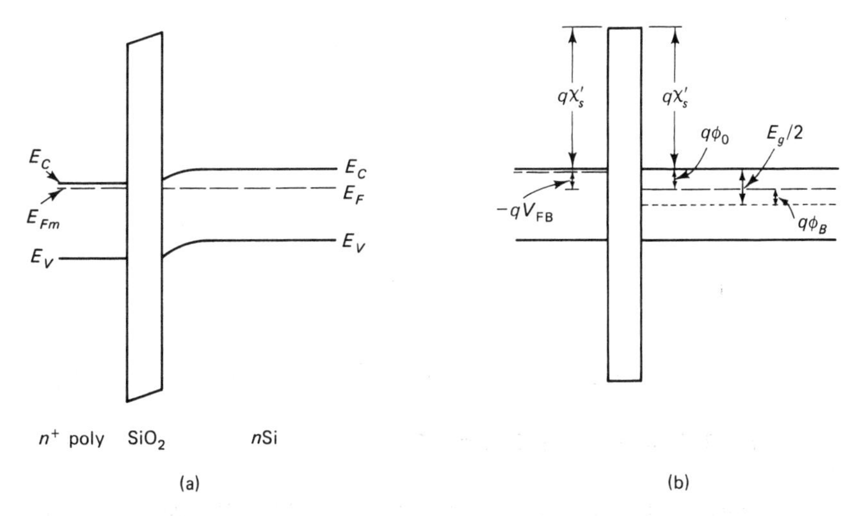

Figure 7.23 Computing V_{FB} for an n^+ polysilicon gate on an n-type substrate: (a) $V_{GB} = 0$; (b) $V_{GB} = V_{FB}$.

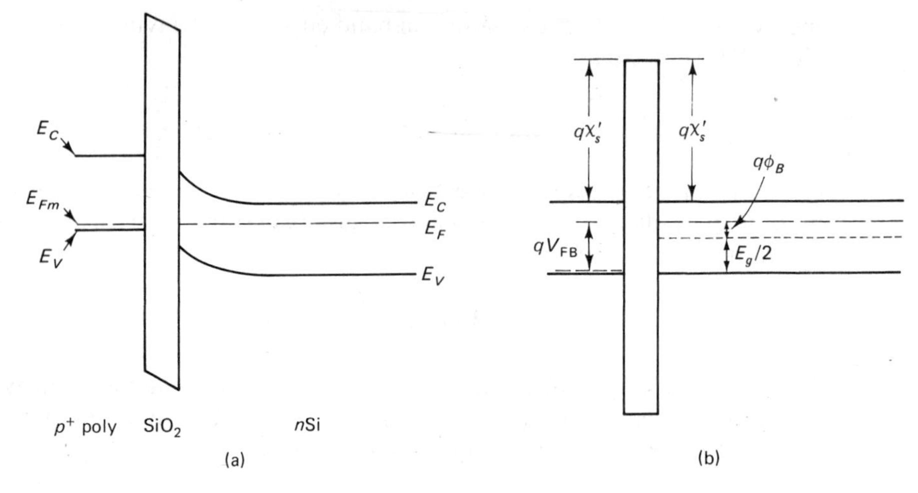

Figure 7.24 Computing V_{FB} for a p^+ polysilicon gate on an n-type substrate: (a) $V_{GB} = 0$; (b) $V_{GB} = V_{FB}$.

simple result can be justified using the superposition theorem of electrostatics. In any electrostatic problem, if two different solutions for the potential and charge distributions satisfying Poisson's equation are known, their sum must also satisfy Poisson's equation. Here one solution to Poisson's equation is for the case $V_{GB} = 0$ while $V_{FB} \neq 0$, and the other solution is for $V_{GB} \neq 0$ while $V_{FB} = 0$.

More specifically, when $V_{FB} \neq 0$, the equations for the threshold voltage, drain current, and saturation voltage in the bulk charge model become

$$V_T = V_{FB} + 2\phi_B + \frac{\sqrt{2\epsilon_s q N_A (2\phi_B + V_{SB})}}{C_{ox}} \tag{7.61}$$

$$I_D = \frac{Z\mu_e C_{ox}}{L}\left\{\left(V_{GS} - V_{FB} - 2\phi_B - \frac{V_{DS}}{2}\right)V_{DS} - \frac{2}{3}\frac{\sqrt{2\epsilon_s q N_A}}{C_{ox}}\left[(2\phi_B + V_{SB} + V_{DS})^{3/2} - (2\phi_B + V_{SB})^{3/2}\right]\right\} \tag{7.62}$$

and

$$V_{DS,\text{sat}} = V_{GS} - V_{FB} - 2\phi_B + \frac{\epsilon_s q N_A}{C_{ox}^2}\left[1 - \sqrt{1 + \frac{2C_{ox}^2(V_{GB} - V_{FB})}{\epsilon_s q N_A}}\right] \tag{7.63}$$

7.4.3 Oxide Fixed Charge Q_f

Whenever silicon dioxide is grown thermally on a silicon substrate, as discussed in Section 12.2.5, a thin layer of positive charge is formed at the silicon–silicon dioxide interface. This charge cannot be altered by the application of a bias to

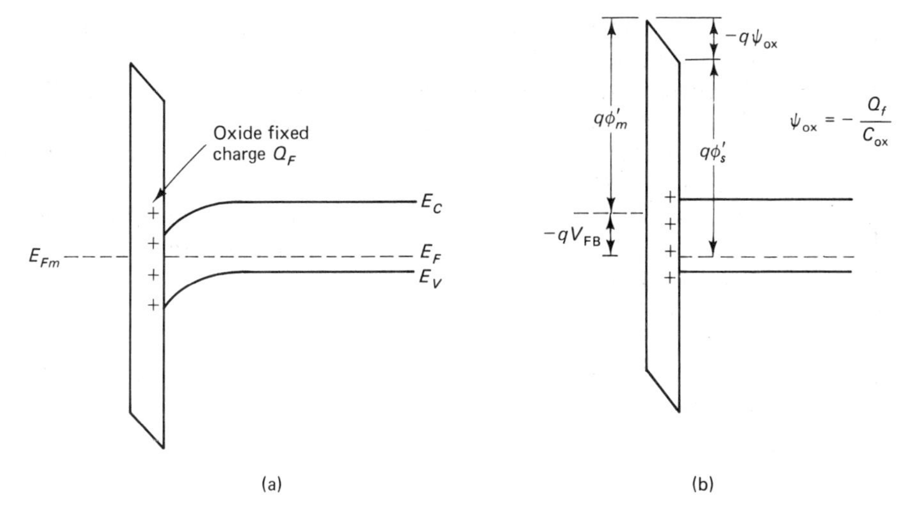

Figure 7.25 Computing V_{FB} when oxide fixed charge Q_f is present: (a) $V_{GB} = 0$; (b) $V_{GB} = V_{FB}$.

the gate, and so has come to be known as the ***oxide fixed charge***, symbolized as Q_f. The magnitude of Q_f depends on the crystallographic orientation of the silicon, and on the ambient (oxidizing or inert) in which the wafer is placed during the last high-temperature treatment used in processing. Q_f can be minimized by annealing in an inert ambient such as argon at a temperature in excess of 900°C. Modern MOS processes are normally designed to hold Q_f to the minimum possible value, which is approximately 2×10^{10} charges/cm^2.

Even if the work functions of the metal and semiconductor are equal, the presence of Q_f will give a nonzero flatband voltage, as shown in Fig. 7.25. The positive charge in the silicon dioxide attracts electrons to the silicon surface, causing the bands in the silicon to bend. To eliminate this band bending, it is necessary to apply a negative voltage to the gate in order to place a charge equal to Q_f in magnitude but opposite in sign on this electrode. In consequence,

$$V_{FB} = -\frac{Q_f}{C_{ox}} \tag{7.64}$$

Figure 7.25b shows how the band diagram changes under flatband conditions. It is important to note that the term "flatband" refers to the bands in the silicon; when charge is present in the oxide, flatband conditions in the silicon can be achieved only when there is a potential drop and hence band bending across the oxide.

Example 7.3 Effect of V_{FB}

This example demonstrates how SPICE can include the effects of gate–substrate work function difference and oxide fixed charge in computing V_T.

Here V_T is obtained by finding the intercept of the I_D-V_{GS} characteristic, as suggested in Section 7.3.6.

The gate work function is specified by the TPG ("type of gate") switch. A TPG value of 1 corresponds to a polysilicon gate of the same doping type as the substrate, TPG = −1 gives a gate of opposite doping type to the substrate, and TPG = 0 gives an aluminum gate.

NSS is the symbol used in SPICE to represent the oxide fixed charge Q_f; a Q_f value of 2×10^{10} charges/cm^2 is assumed here.

The circuit used for the SPICE simulation is that of Fig. 7E.1; the input file is listed below.

```
*MOSFET ID-VGS CHARACTERISTICS: COMPUTATION OF VT
VDS 4 0 DC 0.1
VGS 1 0
VBS 3 0 DC 0
VID 4 2 DC 0
MOS1 2 1 0 3 MOS1 L=10U W=10U
.MODEL MOS1 NMOS LEVEL=2 NSUB=2E15 TOX=100E-9 TPG=1
+NSS=2E10
.WIDTH OUT=80
.DC VGS -1 1 0.1
.PLOT DC I(VID)
.END
```

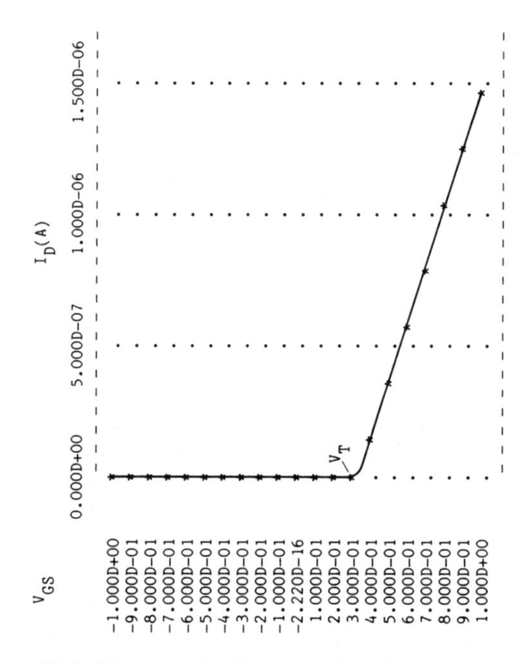

Figure 7E.5 Using the I_D-V_{GS} characteristics to determine V_T.

The $I_D - V_{GS}$ characteristic obtained from SPICE is shown in Fig.7E.5. The intercept of the drain current curve with the V_{GS} axis lies at approximately $V_{GS} = 0.2$ V, which agrees well with the value of V_T computed using (7.61), (7.58), and (7.64).

7.4.4 Mobile Ion Charge Q_m

The first MOSFETs built in the early 1960s tended to have unstable threshold voltages; V_T in these devices would frequently shift by a volt or more over the course of a few minutes when a bias was applied to the gate. This problem ruled out the immediate commercial development of MOS integrated circuits. The source of the instability in V_T remained something of a mystery until 1965, when a variety of careful experiments revealed that the problem was due to contamination of the gate oxide with sodium (Na^+) ions. Extremely low concentrations of Na^+ ions in the oxide were found to give significant instability in V_T. To avoid this contamination problem, it is necessary to take extreme care in minimizing the sodium content of any solutions with which the silicon wafers may come in contact, and also of the components of the high-temperature furnaces in which steps such as oxidation and diffusion are carried out.

Na^+ ions are small, and can move freely through the relatively open structure of SiO_2 in response to an applied gate bias. If the gate is biased positively, the ions move toward the Si–SiO_2 interface, where they shift the flatband voltage in the same manner as Q_f. Application of a negative bias to the gate attracts the Na^+ ions toward the gate, reducing their influence on V_{FB} and V_T. In general, if $\rho(x)$ is the charge density due to Na^+ ions in the oxide, it can be shown (see Problem 7.11) that

$$V_{FB} = -\frac{1}{C_{ox}} \int_0^{t_{ox}} \rho(x) \frac{x}{t_{ox}} \, dx \tag{7.65}$$

where we have temporarily shifted the coordinate origin so that $x = 0$ at the gate electrode.

In modern MOS processes, a variety of techniques is used to hold sodium ion contamination to a low enough level that it has effectively no influence on V_{FB}.

7.4.5 Oxide Trapped Charge Q_{ot}

The bandgap of SiO_2 is roughly 8 eV. If a photon with energy greater than this is incident on an SiO_2 layer, electron–hole pairs are photogenerated in the layer. Electrons are quite mobile in SiO_2, but holes are relatively immobile and are easily trapped. During etching and thin-film deposition steps required in processing, silicon wafers are frequently exposed to "soft" (i.e., low energy) x-rays. As a result, large quantities of electrons and holes are photogenerated in oxide layers, and many of these holes are trapped. The trapped holes constitute a positive charge $\rho(x)$, which shifts V_T in accordance with (7.65). Fortunately, this oxide trapped charge Q_{ot} can be completely removed by annealing the wafers at temperatures in the range 300 to 400°C after exposure to radiation. Q_{ot} is therefore of relatively little importance in completed devices.

If more than one source of V_{FB} is present in a given MOS system, the superposition principle of electrostatics can be used to justify summing the contribution to V_{FB} from each source; that is,

$$V_{FB,\text{total}} = V_{FB,\text{work function}} + V_{FB,\text{fixed charge}} + \cdots \tag{7.66}$$

7.4.6 Interface Trapped Charge Q_{it}

In Chapter 5 we found that electrons inside a crystalline solid can occupy only certain allowed bands of energies. An electron occupying an allowed energy level in the conduction or valence band is free to move throughout the crystal, and should not be viewed as being attached to any one silicon atom. The analysis leading to this result ignored the fact that any real crystal has surfaces. The surface represents a gross disruption of the periodic crystal lattice potential, so at the surface solutions to the Schrödinger equation appropriate deep in the interior of the solid do not apply. In general, it is found that at the surface of a crystal there are many allowed states at energies lying within the gaps that are forbidden in the bulk. Each of these so-called ***surface states*** is associated with a single atom at the surface, and consequently, an electron occupying one of these states is localized—in other words, forced to remain in a restricted region of space centered on that atom.

Surface states are normally able to exchange electrons with one or both of the conduction and valence bands in the semiconductor. For this reason, they are also frequently referred to as ***interface traps***, since they effectively trap free carriers at the silicon–silicon dioxide interface. The charge per unit area stored in the traps is symbolized as $\boldsymbol{Q_{it}}$ while the density of traps per unit area per unit energy in the forbidden gap is symbolized as $\boldsymbol{D_{it}}$. The characteristic time constant for electron exchange can vary over many orders of magnitude for different traps, so reference is frequently made to "fast" and "slow" surface states.

From a chemical bonding viewpoint, surface states can be thought of as being associated with dangling bonds at surface atoms, as suggested in Fig. 7.26. Capture of an electron from the interior of the crystal partly satisfies the bonding requirements of the surface atoms.

In an equilibrium situation, the occupancy of the interface traps is governed by the position of the Fermi energy, just as for any other electron energy level. The charge stored in the traps therefore varies in response to the band bending across the silicon, as shown in Fig. 7.27. For this reason, it is not possible to account for the effect of interface trapped charge on the MOSFET characteristics simply by introducing a flatband voltage shift. When interface traps are present,

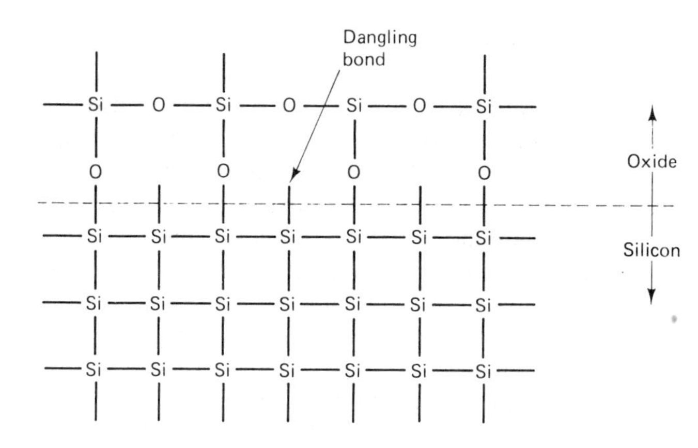

Figure 7.26 Two-dimensional representation of how dangling bonds associated with interface traps arise at the silicon–silicon dioxide interface.

(7.10) must be modified to read

$$\psi_{ox} = -\frac{Q_s}{C_{ox}} - \frac{Q_{it}}{C_{ox}} \tag{7.67}$$

or equivalently,

$$V_{GB} - V_{FB} - \psi_s = -\frac{Q_s}{C_{ox}} - \frac{Q_{it}}{C_{ox}} \tag{7.68}$$

The presence of interface traps increases the change in charge stored in the silicon and at the interface (i.e., the sum $Q_s + Q_{it}$) resulting from a given change in ψ_s. In consequence, the change in the mobile electron charge in the channel

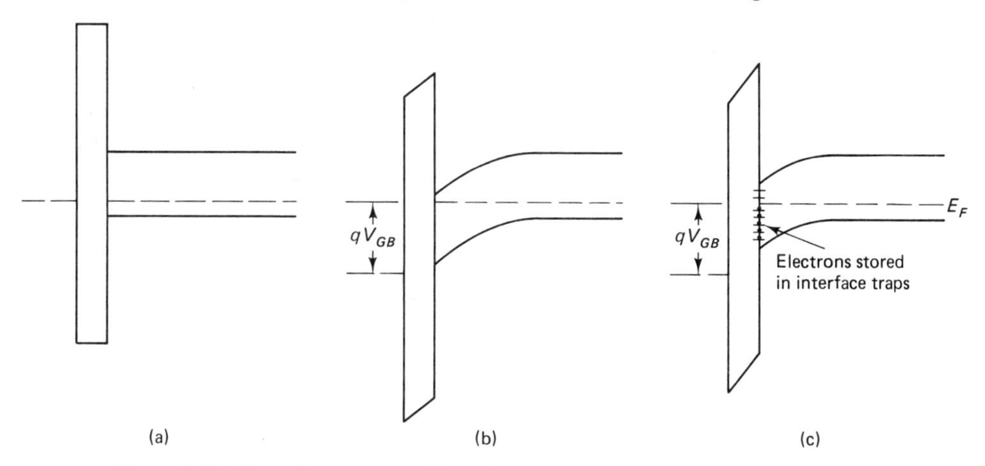

Figure 7.27 How the presence of interface traps reduces the change in ψ_s for a given change in V_{GB}. (a) Band diagram for $V_{GB} = 0$ (V_{FB} is assumed to be 0). (b) Band diagram for $V_{GB} > 0$, no interface traps. (c) Band diagram for $V_{GB} > 0$, with interface traps.

resulting from a given change in gate bias is reduced. Since a carrier captured by an interface trap is confined to a very restricted region of space centered at a single surface atom, charge stored in interface traps makes no contribution to charge flow between the source and drain. For these reasons, interface traps degrade the transconductance of a MOSFET and are highly undesirable.

In the fabrication of silicon integrated circuits, the density of interface traps is minimized by subjecting the wafers to an inert ambient anneal at a temperature close to 1000°C (this procedure is also needed to minimize Q_f), and then annealing in a hydrogen ambient at a temperature of approximately 450°C. The latter step is thought to allow hydrogen atoms to penetrate the gate oxide and satisfy dangling bonds at the silicon surface, as suggested in Fig. 7.28. By following this combined high temperature–low temperature annealing procedure, D_{it} at the middle of the forbidden gap can be reduced below 10^{10} traps cm^{-2} eV^{-1}. At this level, the interface traps have a negligible effect on MOSFET performance.

The various charges that may be present in a thermally grown silicon dioxide layer are illustrated schematically in Fig. 7.29.

7.4.7 Threshold Adjustment by Ion Implantation

In modern MOS integrated circuits, transistor threshold voltages are adjusted by using ion implantation to introduce dopant ions into the silicon under the gate. The implant is usually carried out through the thin gate oxide prior to the deposition of the gate polysilicon; further details are given in Sections 12.2.4 and 12.2.11. The implant energy is kept low so that the implanted ions lie close to the silicon surface.

To determine the effect of the implanted ions on V_T, let us suppose that the MOSFET is biased just below threshold, so that the surface under the gate is depleted. If donor ions (phosphorus or arsenic) were implanted, a layer of extra positive charge would be present near the silicon surface. This would shift V_T in exactly the same manner as the oxide fixed charge Q_f. If ΔV_T denotes the change

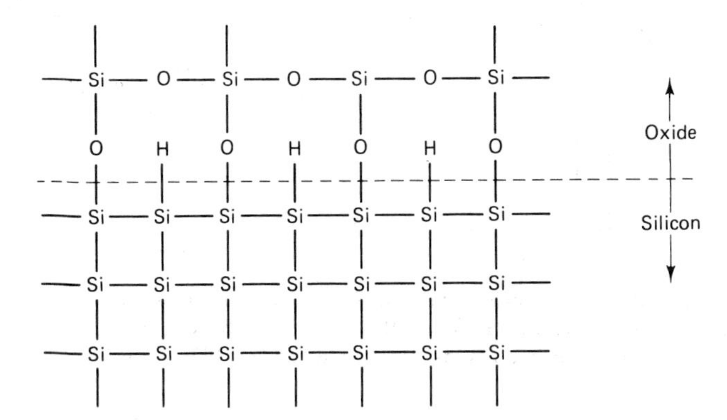

Figure 7.28 How hydrogen can passivate the dangling bonds at the silicon surface shown in Fig. 7.26 and reduce the interface trap density.

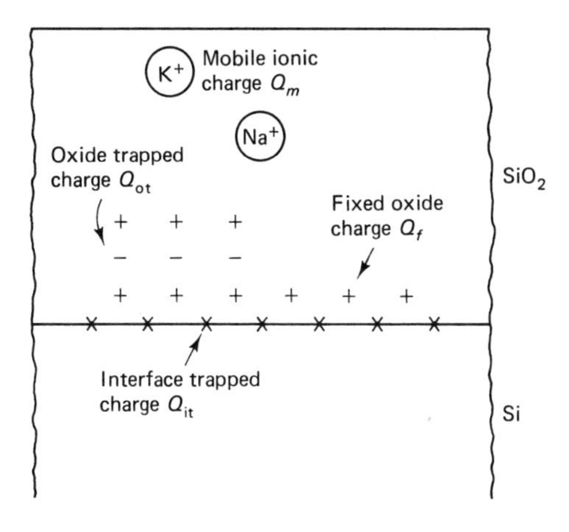

Figure 7.29 Charges associated with thermally grown SiO_2 layers on silicon. (From Deal [5]. Reprinted with permission of the publisher, IEEE, Inc.)

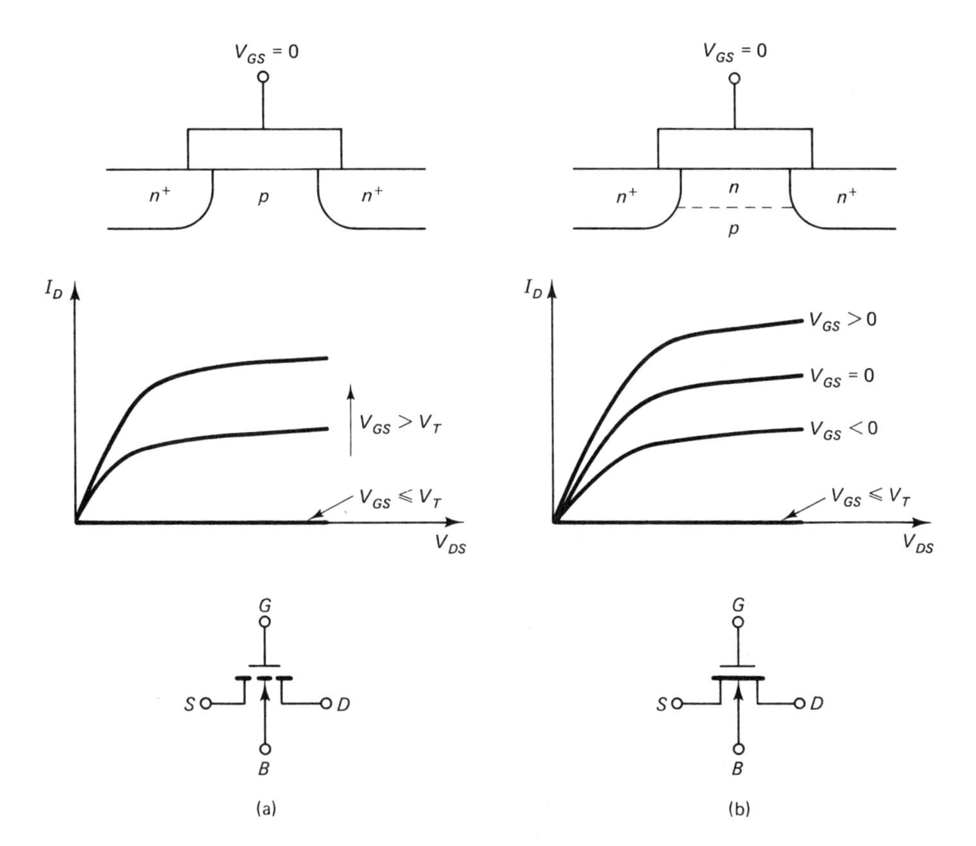

Figure 7.30 Classes of MOSFET: (a) n-channel enhancement; (b) n-channel depletion; (c) p-channel enhancement; (d) p-channel depletion.

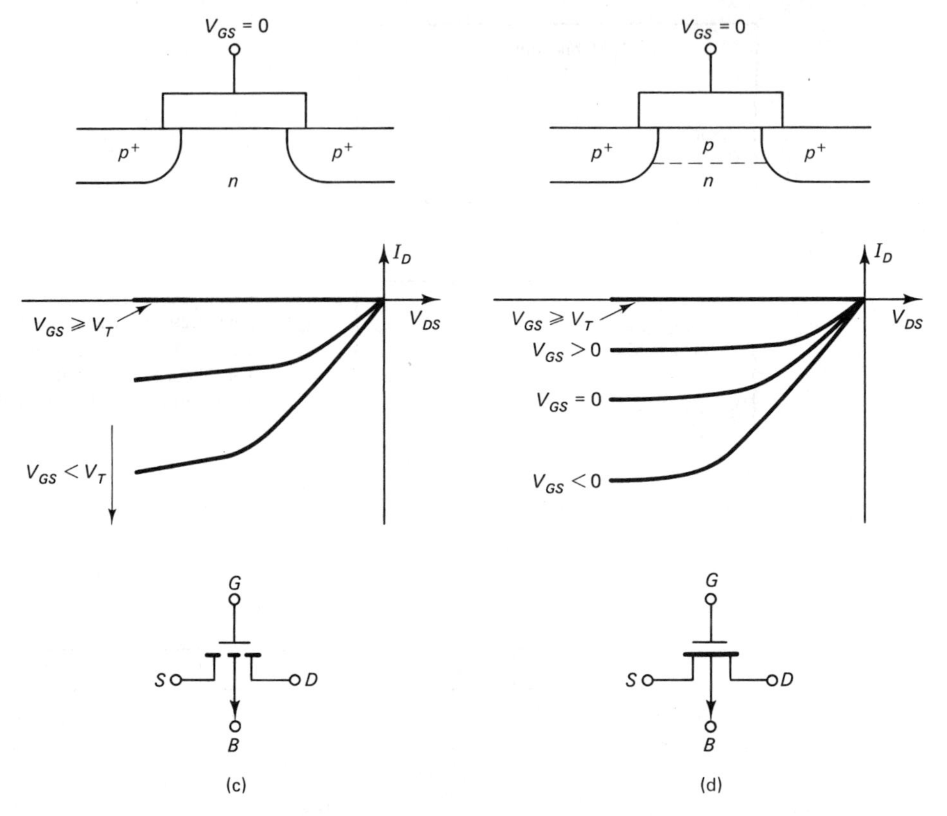

Figure 7.30 *(continued)*

in V_T resulting from the implant, then

$$\Delta V_T = -\frac{qD_I}{C_{ox}} \tag{7.69}$$

where D_I is the implant dose in ions per square centimeter. If acceptor ions (boron) were implanted, V_T would shift by the same amount but in the opposite direction. This can be accounted for by defining D_I to be a negative quantity for negatively charged acceptor ions. In summary, implanting donor ions lowers V_T, while acceptor ions raise V_T.

When ion implantation is used to control V_T, it is possible to set different threshold voltages for different transistors on the same wafer. This is extremely useful in circuit design.

7.4.8 Types of MOSFETs

The threshold voltage of a MOSFET depends on a number of factors: the thickness of the gate oxide, the doping level in the substrate, the work function difference between the gate and the substrate, the level of charges in the oxide, and the possible presence of a threshold adjust implant. Depending on the values of these different parameters, V_T can be either positive or negative. An n-channel MOSFET with a positive threshold voltage is referred to as an ***enhancement***-type device, while ***depletion*** transistors have negative thresholds. Depletion transistors are therefore "on" even when the gate is shorted to the source. Enhancement transistors are usually desired as switches or drivers in logic gates, while depletion transistors are useful primarily as load elements.

So far only n-channel MOSFETs have been considered here. To turn on a p-channel transistor, it is necessary to apply a negative voltage to the gate to attract holes to the surface and form an inversion layer. Enhancement p-channel transistors have negative threshold voltages, while depletion p-channel devices have positive thresholds. Equations (7.61) and (7.62) can be applied to compute the threshold voltage and drain current of a p-channel transistor provided that ϕ_B is computed from (7.6). The saturation voltage $V_{DS,\text{sat}}$ for a p-channel device is given by

$$V_{\text{DS,sat}} = V_{GS} - V_{FB} + 2\phi_B - \frac{\epsilon_s q N_D}{C_{\text{ox}}^2}\left(1 - \sqrt{1 + \frac{2C_{\text{ox}}^2 |V_{GB} - V_{FB}|}{\epsilon_s q N_D}}\right) \tag{7.70}$$

I_D is a negative quantity, in accord with the convention that positive charge flow out of the drain constitutes a negative current.

The I_D–V_{DS} characteristics for the various classes of MOSFET are shown in Fig. 7.30, along with the complete circuit symbol for each type of device. It should be noted, however, that abbreviated symbols are frequently used in diagrams of complex circuits.

7.5 MOSFETS IN INTEGRATED CIRCUITS

A cross section through an inverter fabricated in a typical modern CMOS process (described in Section 12.2.11) is shown in Fig. 7.31, while the circuit diagram for the inverter is shown in Fig. 7.32. The "substrate" for the n-channel transistor is a deep, lightly doped p-type ***well***, while the p-channel transistor is formed in the n-type substrate itself. Both transistors have their sources shorted to their respective substrates, and so are not subject to backgate bias effects.

In any MOS integrated circuit, there is a risk that the metal or polysilicon lines used to interconnect transistors may act as parasitic gates and form conducting channels in the substrate where none were intended. In the structure of Fig. 7.31, the formation of such parasitic channels is avoided by growing a thick ***field oxide*** except in those active regions where it is actually intended that transistors be formed. Increasing the oxide thickness reduces C_{ox} and therefore, from (7.31), raises V_T. To guard further against the formation of parasitic channels, ion implantation is usually used to raise the surface doping level in the field region. This implant is done before the field oxide is grown; boron is implanted into the field regions in the p-type wells, while arsenic or phosphorus is implanted into

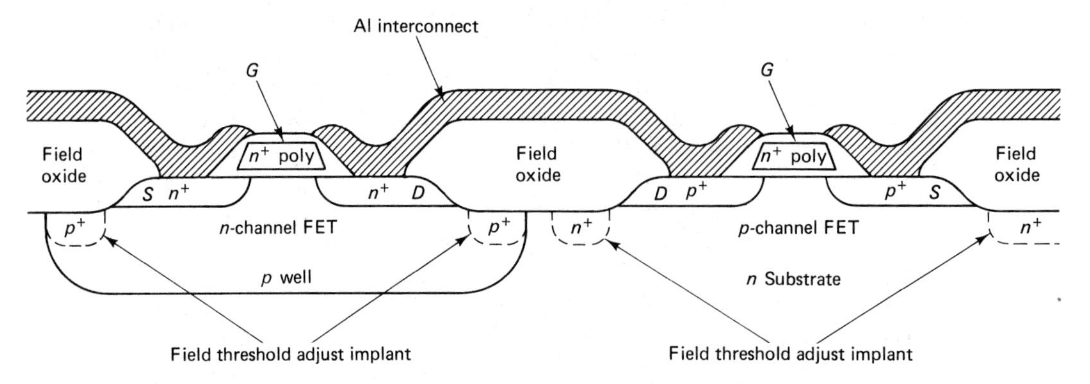

Figure 7.31 Cross section through an inverter fabricated in a p-well CMOS process.

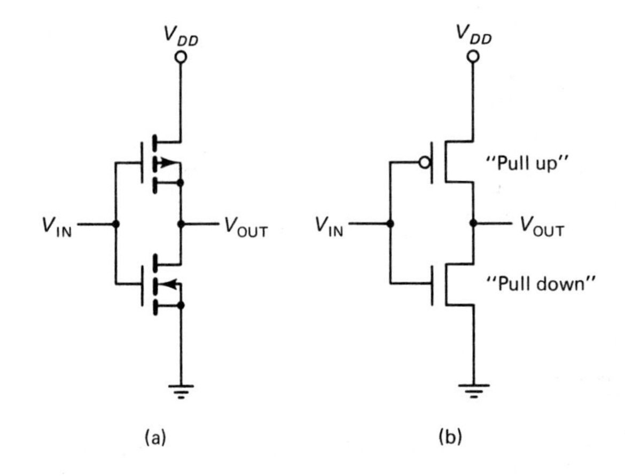

Figure 7.32 Circuit diagram for CMOS inverter of Fig. 7.31. (a) Using full circuit symbols for MOS transistors. (b) Simplified diagram frequently used in IC schematics.

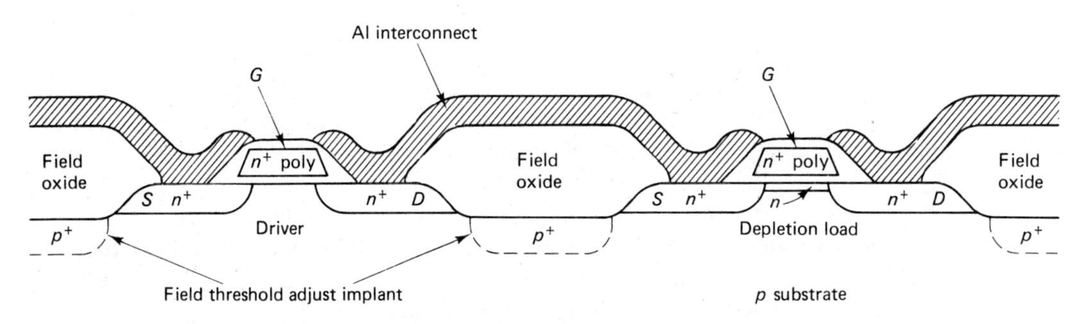

Figure 7.33 Cross section through a depletion-load NMOS inverter.

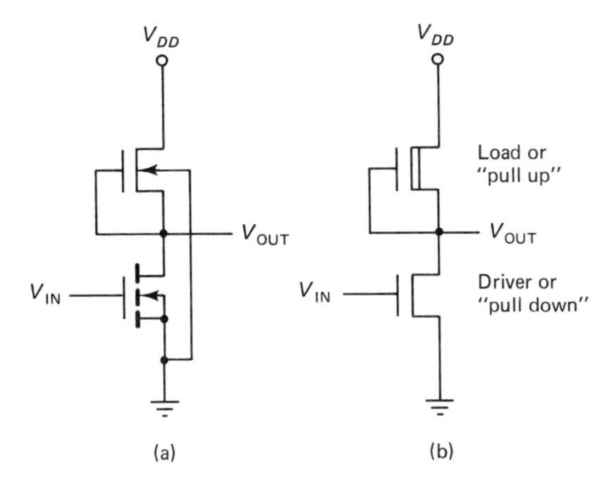

Figure 7.34 Circuit diagram for NMOS inverter of Fig. 7.33. (a) Using full circuit symbols for MOS transistors. (b) Simplified diagram frequently used in IC schematics.

the field regions in the substrate. Equation (7.31) shows that an increase in the substrate doping raises V_T. Whatever measures are taken to control the ***field threshold voltage*** (i.e., V_T for a parasitic transistor formed by an interconnect line running over the field oxide), it is important that it be greater than the largest supply or signal voltage expected in the circuit. Field threshold adjust implants are often referred to as ***channel stop*** implants.

It is worth noting that the field oxide in Fig. 7.31 is partly recessed below the silicon surface. The techniques required to produce this structure are discussed in Section 12.2.6.

Figure 7.33 shows the cross section through a typical NMOS inverter, while Fig. 7.34 shows the corresponding circuit diagram. The load transistor in this circuit is a depletion-type device formed by implanting arsenic or phosphorus into the channel. This transistor is subject to the body effect, since its source rises to a higher potential than the substrate when the output of the inverter goes high. The main impact of the body effect here is to reduce the current available to charge any capacitance that may be loading the inverter when the output goes high.

Example 7.4 Design of a Depletion Load Inverter

Figure 7.34 presents the basic circuit for a depletion load inverter. The SPICE file listed below can be used to generate the transfer characteristic—that is, a plot of V_{OUT} as a function of V_{IN}—for this logic gate. In this file, it has been assumed that the load and driver transistors have equal lengths and widths. The threshold voltage for the driver has been set to 1.0 V, while V_T for the load transistor is -3.0 V.

```
*DEPLETION LOAD NMOS INVERTER: TRANSFER CHARACTERISTICS
VIN 1 0
VDD 3 0 DC 5
MOSLD 3 2 2 0 MOSD L=5U W=5U
```

```
MOSDR 2 1 0 0 MOSE L=5U W=5U
.MODEL MOSE NMOS LEVEL=2 NSUB=5E15 TOX=50E-9 VTO=1.0
.MODEL MOSD NMOS LEVEL=2 NSUB=5E15 TOX=50E-9 VTO=-3.0
.WIDTH OUT=80
.DC VIN 0 5 0.2
.PLOT DC VIN V(2)
.END
```

The transfer characteristic generated by SPICE is shown in Fig. 7E.6. It can be seen that the low output level is roughly 1 V, which is comparable to the threshold of the driver transistor. This is not acceptable, for the output low level of one inverter must be capable of firmly turning off another, identical inverter. To resolve this problem, it is necessary either to increase the effective resistance of the load transistor by reducing Z/L for this device, or decrease the effective resistance of the driver by increasing Z/L for this transistor. The former approach is taken in the SPICE file below.

```
*DEPLETION LOAD NMOS INVERTER
VIN 1 0
VDD 3 0 DC 5
```

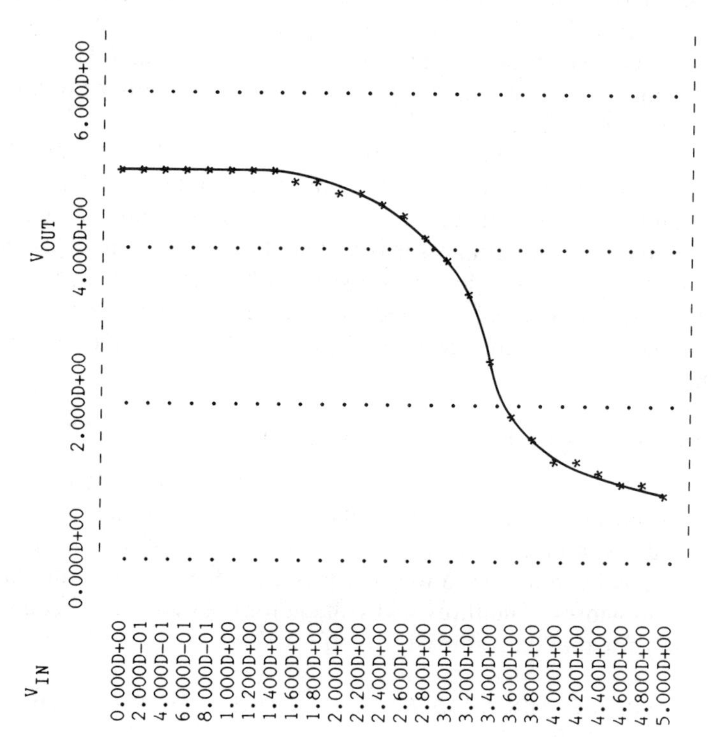

Figure 7E.6 Transfer characteristic of depletion load inverter with $(Z/L)_{\text{driver}} = (Z/L)_{\text{load}}$.

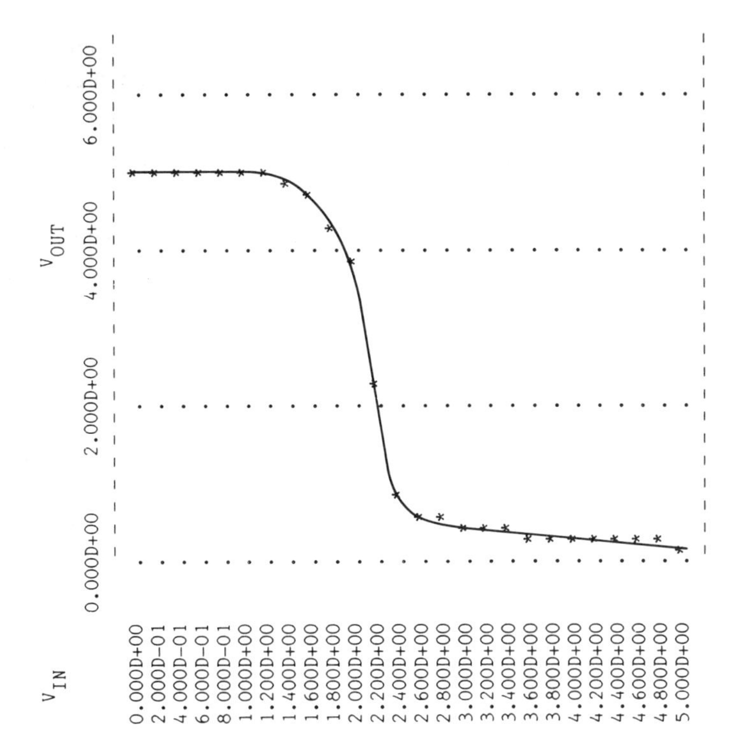

Figure 7E.7 Transfer characteristic for depletion load inverter with $(Z/L)_{\text{driver}} = 4(Z/L)_{\text{load}}$.

```
MOSLD 3 2 2 0 MOSD L=20U W=5U
MOSDR 2 1 0 0 MOSE L=5U W=5U
.MODEL MOSE NMOS LEVEL=2 NSUB=5E15 TOX=50E-9 VTO=1.0
.MODEL MOSD NMOS LEVEL=2 NSUB=5E15 TOX=50E-9 VTO=-3.0
.WIDTH OUT=80
.DC VIN 0 5 0.2
.PLOT DC VIN V(2)
.END
```

The transfer characteristic for the redesigned inverter is shown in Fig. 7E.7. Now that the sizes of the load and driver transistors have been properly matched, the output low level has dropped to 0.2 V. This is sufficient to turn off the driver transistor of the next stage.

7.6 SHORT-CHANNEL EFFECTS

In the early 1970s, a typical channel length for a MOSFET in a commercial integrated circuit was 10 μm. In the late 1980s, channel lengths of 2 to 3 μm are common and 1-μm channels are used in some commercial products. Devices with channel lengths below 0.1 μm have been fabricated on an experimental basis. In another decade, submicron devices will be widely used commercially.

There are two main reasons behind the drive to smaller MOSFET dimensions. The first is fairly obvious: the smaller the size of an individual transistor, the more can be placed in a given area of silicon. This allows integrated circuits of greater functional complexity to be produced, and also allows more circuits to be placed on a given wafer. (The cost of processing a single silicon wafer is roughly independent of the number of transistors it carries.) The second reason is that as shown in Section 7.7, MOSFETs with short channels offer faster switching speed.

As the dimensions of a MOSFET are reduced, there is a tendency for the longitudinal field to grow, since the voltage difference between source and drain is dropped over a smaller distance. Eventually, the longitudinal field becomes comparable to the transverse field, and the gradual channel approximation breaks down. When this happens, the characteristics of the MOSFET begin to depart from those predicted for long-channel devices in Section 7.3, and a variety of so-called ***short-channel effects*** appear. These include a failure of the drain current to saturate at large V_{DS}, a reduction in threshold voltage V_T, increased subthreshold current, and in the extreme, a complete loss of the ability of the gate to control the drain current.

Much of present-day MOSFET design is devoted to producing transistors that are physically small but that still exhibit electrical characteristics similar to those of devices with physically long channels. Qualitatively, to avoid short-channel effects the transverse field should be kept large compared to the longitudinal field. This can be done by reducing the thickness of the gate oxide to bring the gate electrode close to the substrate, and by minimizing the extension of the source and drain depletion regions under the gate. This in turn requires that the depth of the source and drain junctions be minimized. Figure 7.35 illustrates these points by comparing two devices with the same physical channel length, one of which should give effectively long-channel electrical characteristics while the other should show pronounced short-channel effects.

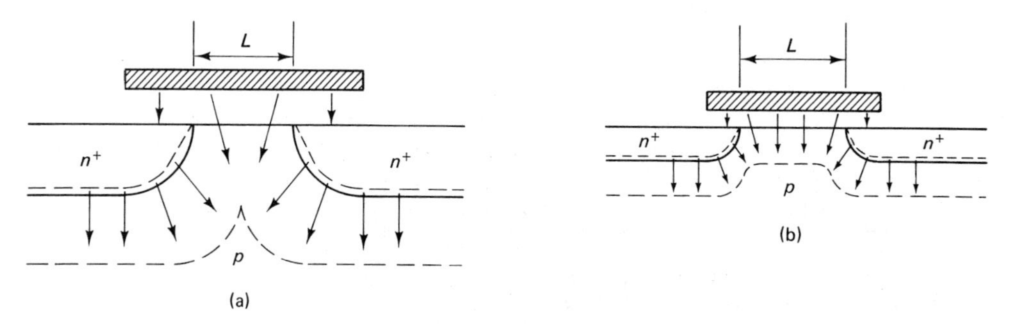

Figure 7.35 Two MOSFETs with identical channel lengths L. Device (a) should show pronounced short-channel effects, while device (b) will have effectively long-channel characteristics. Dashed lines show the boundary of the depletion region, while arrows point in the direction of the electric field.

A useful formula exists for estimating the minimum channel length a MOSFET can have while still exhibiting electrical characteristics resembling those of a long-channel device. The minimum allowable channel length $L_{\min}$ in micrometers is given by [6]

$$L_{\min} = 0.4[r_j t_{\text{ox}}(W_S + W_D)^2]^{1/3} \tag{7.71}$$

where r_j is the depth of the source and drain metallurgical junctions in microns, t_{ox} is the gate oxide thickness in angstroms, and W_S and W_D are the depletion region depths in micrometers under the source and drain, respectively. These quantities are illustrated in Fig. 7.36. In using (7.71), W_S and W_D are to be computed from the one-dimensional depletion approximation (see Section 6.2.1.1).

It should be emphasized that (7.71) is entirely empirical, but it provides a good fit to both experimental data and the predictions of full numerical solutions of the equations of state for the MOSFET.

7.6.1 Channel-Length Modulation in the Saturation Regime

In Section 7.3.4 it was noted that when V_{DS} is raised above $V_{DS,\text{sat}}$, the MOSFET channel pinches off a short distance away from the drain. This effect is illustrated in Fig. 7.37. The gradual channel approximation should still apply between the source and the pinch-off point, so that the current through the device is given by (7.38) with V_{DS} replaced by $V_{DS,\text{sat}}$ and the channel length L replaced by the distance L' between the source and the pinch-off point.

For a MOSFET with a long channel the difference ΔL between L and L' is negligible, so I_D is effectively independent of V_{DS} in saturation. However, for a short-channel device ΔL may be a sizable fraction of the source–drain separation, in which case the drain current will increase indefinitely as V_{DS} is increased beyond $V_{DS,\text{sat}}$. In this case, there is no saturation of the drain current after the channel pinches off.

SPICE offers two possible approaches to determining L' in the pinch-off regime. The first is by specification of the model parameter LAMBDA, in which

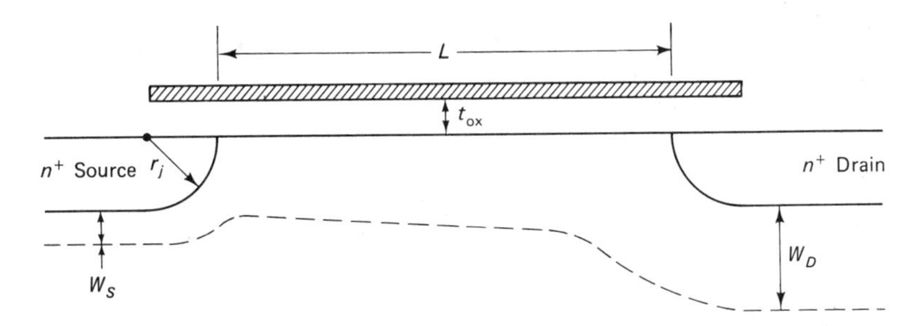

Figure 7.36 Parameters used in empirical formula for onset of short-channel effects in a MOSFET (7.71).

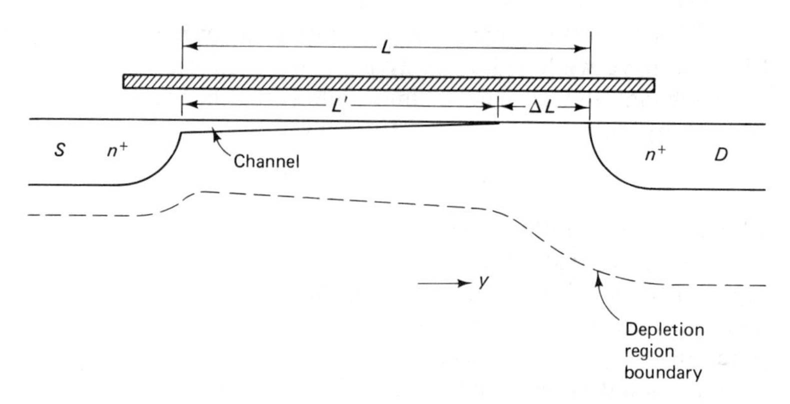

Figure 7.37 Reduction in effective channel length in a short-channel MOSFET.

case ΔL is computed from the simple empirical expression

$$\Delta L = \text{LAMBDA} \cdot V_{DS} \tag{7.72}$$

LAMBDA can be used in either the Level 1 or Level 2 model.

In the Level 2 model, if LAMBDA is not specified, then L is computed by applying one-dimensional depletion approximation electrostatics between the pinch-off point and the drain along the y-direction in Fig. 7.37.

Example 7.5 Channel-Length Modulation

This example illustrates how channel-length modulation results in a failure of I_D to saturate at large V_{DS} in a short-channel MOSFET. The device examined here is the same as that used in Example 7.1, but the channel length has been reduced from 10 μm to 1 μm. The SPICE input file required to generate the I_D–V_{DS} characteristic for $V_{GS} = 2.5$ V is listed below.

```
*MOSFET ID-VDS CHARACTERISTICS: CHANNEL LENGTH MODULATION
VDS 4 0
VGS 1 0 DC 2.5
VBS 3 0 DC 0
VID 4 2 DC 0
MOS1 2 1 0 3 MOS1 L=1U W=10U
.MODEL MOS1 NMOS LEVEL=2 NSUB=1E16 TOX=50E-9
.WIDTH OUT=80
.DC VDS 0 5 0.2
.PLOT DC I(VID)
.END
```

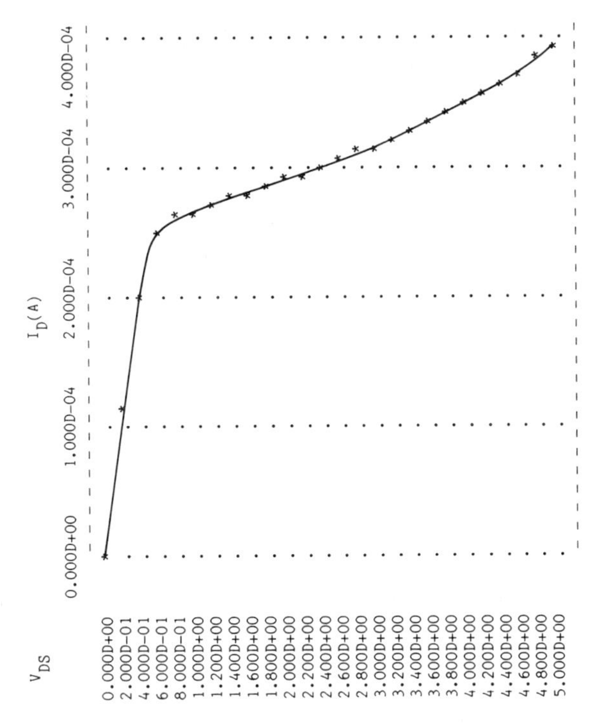

Figure 7E.8 How channel-length modulation results in failure of I_D to saturate in a short-channel device.

The $I_D - V_{DS}$ characteristic for the MOSFET is shown in Fig. 7E.8. It can readily be seen that I_D does not saturate in this short-channel device.

7.6.2 Velocity Saturation

In present-day MOS integrated circuits, the trend is for channel lengths to be scaled down to dimensions approaching 1 μm while the supply voltage is held fixed at 5 V. As a result, the longitudinal electric fields encountered in device channels are increasing. In Section 3.2.1 it was found that at large applied fields the electron velocity drops below the value predicted by the simple expression $v = \mu_e \mathscr{E}$, and that for fields in excess of 10^5 V cm^{-1} the electron velocity saturates at the scatter-limited value v_{sat}.

A very rough estimate of the longitudinal electric field in a MOSFET channel is given by the ratio V_{DS}/L. For $V_{DS} = 5$ V, this simple approximation predicts that a field of 10^5 V cm^{-1} will be reached at a channel length of around 0.5 μm. Since the longitudinal electric field will not be constant along the length of the channel for reasonably large V_{DS}, this calculation shows that velocity saturation rather than pinch-off is likely to limit the drain current in a MOSFET with channel length approaching 1 μm.

Under conditions of velocity saturation, the drain current is given by

$$I_D = -ZQ_n v_{\text{sat}} \tag{7.73}$$

The Level 2 MOSFET model in SPICE allows for velocity saturation provided that the parameter VMAX (equivalent to v_{sat}) is specified in the input file. The value of V_{DS} at which saturation occurs is determined by equating the value of I_D predicted by (7.73) with that predicted by the bulk charge model (7.38). Q_n is computed from (7.34), setting $V_{CB} = V_{DB}$.

A crude approximation for Q_n can be obtained from

$$Q_n \simeq -C_{\text{ox}}(V_{GS} - V_T) \tag{7.74}$$

which gives

$$I_{D,\text{sat}} \simeq ZC_{\text{ox}}(V_{GS} - V_T)v_{\text{sat}} \tag{7.75}$$

7.6.3 Short- and Narrow-Channel Effects on Threshold Voltage

Figure 7.38 shows the cross section through a short-channel MOSFET biased near threshold. The important point to note in this figure is that, contrary to the assumptions of the gradual channel approximation, the electric field lines are not everywhere perpendicular to the silicon surface. In particular, field lines originating in the source and drain penetrate into the depletion region under the gate. It is useful to divide the region under the gate into two parts, one in which field lines from the gate terminate (marked as trapezoid *ABCD* in Fig. 7.38b), and one in which field lines from either the source or drain terminate. Region *ABCD* can therefore be thought of as that part of the device which is under control of the gate.

Overall charge neutrality in the MOSFET requires that the total charge (including both mobile electron charge Q_n' and acceptor ion charge Q_B') in region *ABCD* be balanced by the total charge Q_G' on the gate. (Here the ′ symbol denotes a total charge, as opposed to a charge per unit area.) Thus

$$Q_G' = -(Q_n' + Q_B') \tag{7.76}$$

If the channel length of the MOSFET is reduced while all other dimensions and the bias voltages are held constant, Fig. 7.38 suggests that the average dopant ion charge per area $Q_B = Q_B'/ZL$ stored in the region controlled by the gate will shrink due to the encroachment of the source and drain depletion regions. This suggests that the average electron charge per unit area in the channel must increase. Conversely, a lower gate bias will be required to produce a given channel

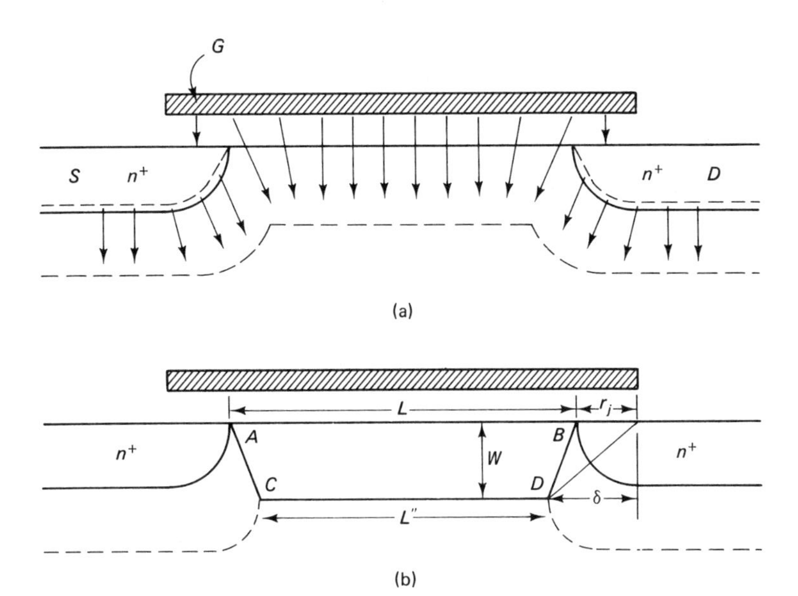

Figure 7.38 (a) Short-channel MOSFET at threshold, showing encroachment of electric field lines (drawn as arrows) from the source and drain into the region under the gate. (b) Geometrical construction used to estimate drop in V_T resulting from field encroachment.

charge, so the threshold voltage of a short-channel MOSFET is reduced compared to that of an otherwise identical long-channel device.

To obtain a rough estimate of the reduction ΔV_T in threshold, we will assume that at the onset of threshold the average potential drop across the silicon is the same for both a long- and a short-channel transistor. Letting $\overline{\Delta Q_B}$ represent the difference in average stored charge per unit area in the depletion region between the long- and short-channel device, we should then have

$$\Delta\overline{\psi}_{\text{ox}} = \frac{\Delta\overline{Q}_B}{C_{\text{ox}}} \tag{7.77}$$

where $\Delta\overline{\psi}_{\text{ox}}$ is the difference in the average potential drop across the oxide between the two devices. If ψ_s is the same for both the long and short transistors, we should have $\Delta V_T = \Delta\overline{\psi}_{\text{ox}}$. Therefore,

$$\Delta V_T = \frac{\Delta\overline{Q}_B}{C_{\text{ox}}} \tag{7.78}$$

From Fig. 7.38, in the short-channel device,

$$Q_B' = -qN_A ZW \frac{L + L''}{2} \tag{7.79}$$

so

$$\overline{Q}_B = -qN_A W \frac{L + L''}{2L} \tag{7.80}$$

In a long-channel device we would have $Q_B = -qN_A W$, so

$$\Delta\overline{Q}_B = \frac{-qN_A W}{2}\left(1 - \frac{L''}{L}\right) \tag{7.81}$$

L'' can be found from Fig. 7.38 by a simple geometric analysis. Assuming that the edge of the source implant or diffusion is cylindrical with radius r_j, we must have

$$L'' = L + 2r_j - 2\delta = L + 2r_j - 2\sqrt{(r_j + W)^2 - W^2} \tag{7.82}$$

Therefore,

$$1 - \frac{L''}{L} = 1 - \frac{L}{L} - \frac{2r_j}{L} + \frac{2r_j}{L}\sqrt{1 + \frac{2W}{r_j}} = \frac{2r_j}{L}\left(\sqrt{1 + \frac{2W}{r_j}} - 1\right) \tag{7.83}$$

and

$$\Delta V_T = -\frac{qN_A W r_j}{C_{ox} L}\left(\sqrt{1 + \frac{2W}{r_j}} - 1\right) \tag{7.84}$$

Similar fringing-field effects can alter the threshold voltage of a ***narrow-channel*** MOSFET—that is, a device in which the channel width is small. Figure 7.39a shows a cross section through such a device taken across the width of the

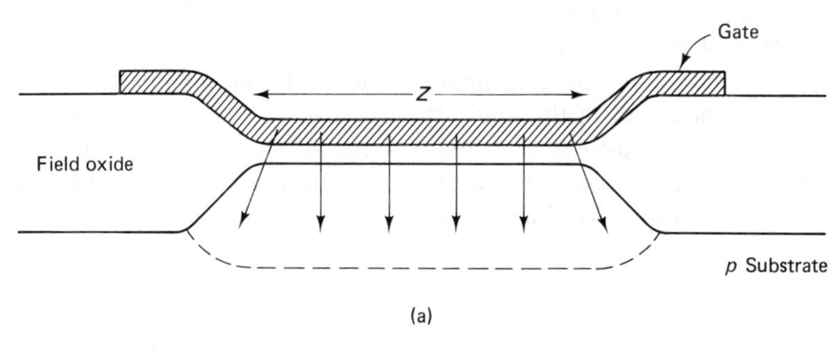

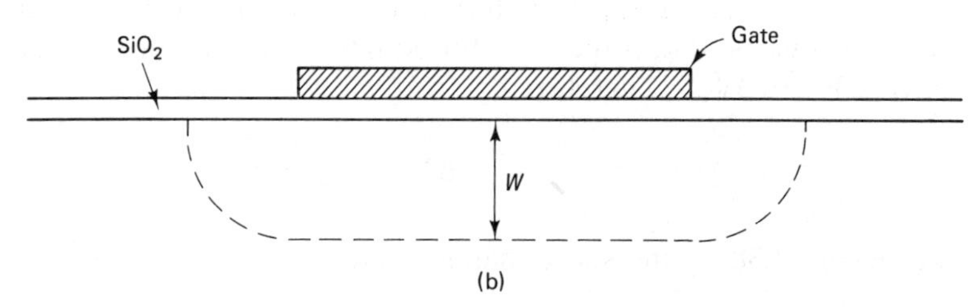

Figure 7.39 (a) Cross section across the width of a small-geometry MOSFET, showing how field lines from gate spread out at edges of channel. (b) Simplified geometry used to compute increase in V_T resulting from narrow-channel effects.

channel. Once again the gradual channel approximation does not apply, for the electric field lines at the edge of the channel are not perpendicular to the silicon surface, but rather spread out laterally. The average charge per unit area stored in the depletion region must therefore be larger than in the case of a wide-channel MOSFET, so from the argument applied to short-channel devices, the threshold voltage should be higher. The difference in V_T should be given approximately by

$$\Delta V_T = -\frac{\Delta \overline{Q}_B}{C_{ox}} \tag{7.85}$$

To compute $\Delta \overline{Q}_B$, we will use the simplified geometry of Fig. 7.39b, which ignores the presence of a recessed field oxide. Assuming that the edges of the depletion region are cylindrical,

$$Q'_B = -ZLqN_AW - 2\left(\frac{\pi W^2 qN_A}{4}\right)L \tag{7.86}$$

so

$$\overline{Q}_B = -qN_AW - \frac{\pi W^2 qN_A}{2Z} \tag{7.87}$$

and

$$\Delta \overline{Q}_B = -\frac{\pi W^2 qN_A}{2Z} \tag{7.88}$$

Therefore,

$$\Delta V_T = \frac{\pi W^2 qN_A}{2ZC_{ox}} \tag{7.89}$$

It is worth stressing that narrow-channel effects tend to raise V_T, while short-channel effects tend to lower it. If both the length and width of a MOSFET are reduced, the short- and narrow-channel effects often tend to cancel, so that V_T remains relatively constant.

It should be noted that this analysis of short- and narrow-channel effects on V_T is approximate, since it deals with average potentials and average values of stored charge, while the drain current is limited by the height of the highest part of the energy barrier faced by electrons trying to leave the source. However, (7.84) and (7.89) are reasonably accurate for devices in which small geometry effects are just beginning to set in.

Example 7.6 Short-Channel Effects on V_T

This example illustrates how short-channel effects reduce the threshold voltage of a MOSFET. The circuit used is, once again, Fig. 7E.1. The device parameters are those of Example 7.3, with the single exception that the channel length for the transistor has been reduced from 10 μm to 1 μm. According to (7.71), this device should exhibit short-channel behavior. The SPICE input file listed below can be used to generate the $I_D - V_{GS}$ characteristic and extract the threshold voltage.

```
*MOSFET ID-VGS CHARACTERISTICS:
SHORT CHANNEL EFFECTS ON × VT
VDS 4 0 DC 0.1
VGS 1 0
VBS 3 0 DC 0
VID 4 2 DC 0
MOS1 2 1 0 3 MOS1 L=1U W=10U
.MODEL MOS1 NMOS LEVEL=2 NSUB=2E15 XJ=2E-6 TOX=100E-9
+ TPG=1 NSS=2E10
.WIDTH OUT=80
.DC VGS -1 1 0.1
.PLOT DC I(VID)
.END
```

The actual $I_D - V_{GS}$ characteristic obtained from SPICE is shown in Fig. 7E.9. From the voltage axis intercept, V_T is seen to be approximately

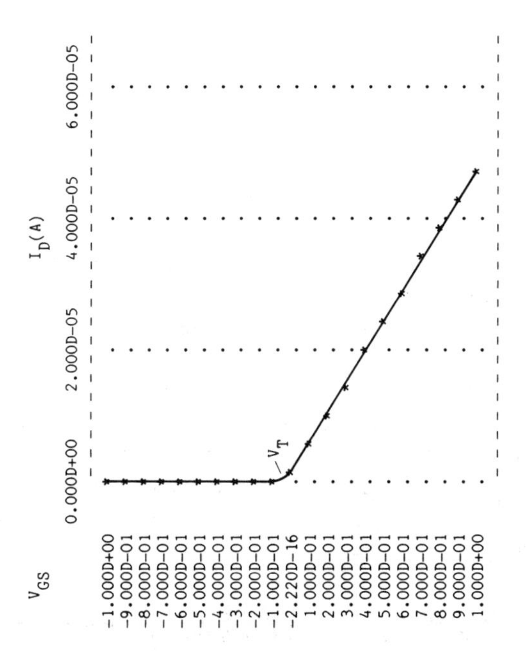

Figure 7E.9 Using $I_D - V_{GS}$ characteristic to determine V_T for short-channel MOSFET.

−0.1 V. This is substantially lower than the V_T value obtained for the corresponding long-channel transistor in Example 7.3.

7.6.4 Punchthrough

Figure 7.40 represents a MOSFET in which a bias has been applied to the gate to produce flatband conditions in the underlying substrate. The right-hand side of this figure shows an energy band diagram drawn along the line AB connecting the source and drain just below the silicon surface when $V_{DS} = 0$. An energy barrier exists between the source and the region under the gate that serves to hold electrons in the source. Applying a positive voltage to the gate lowers this energy barrier, allowing electrons to pour out of the source and form a conducting channel connecting the source and drain, as we have seen. However, for the moment we will assume that V_{GB} is held fixed at V_{FB} so that there is no surface

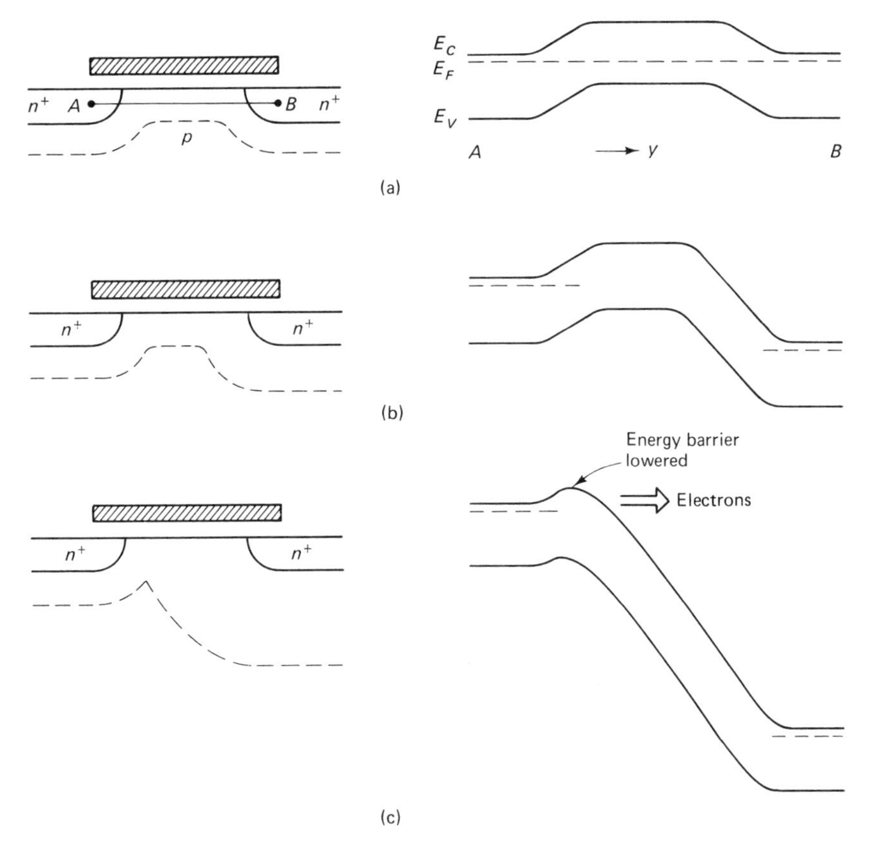

Figure 7.40 How punchthrough occurs in a MOSFET. The energy band diagrams are plotted along line AB below the surface. V_{DS} increases from (a) to (c).

inversion layer. If V_{DS} is increased with V_{SB} held constant, the band diagram appears as shown in Fig. 7.40b. The current at the drain will be just the very small leakage current of the reverse-biased drain–substrate junction.

If V_{DS} is increased still further, a point will eventually be reached at which the potential energy well surrounding the drain overlaps that surrounding the source, as suggested in Fig. 7.40c. Once this happens, the energy barrier holding electrons in the source is lowered, and a large number of electrons will flow from source to drain even though the gate has been biased so as to turn the device off. This situation is qualitatively identical to that of a bipolar transistor in which application of an excessively large collector–base bias has resulted in complete depletion or ***punchthrough*** of the base. Quantitatively, punchthrough in the MOSFET is more difficult to analyze since the problem is inherently two-dimensional. Application of a negative bias to the gate can help suppress punchthrough by forcing electrons away from the silicon surface, but if a large enough V_{DS} is applied, the punchthrough current will still exist irrespective of V_{GS}.

A rough estimate of the bias at which punchthrough sets in can be obtained by determining the value of V_{DS} at which the sum of the depletion widths W_S under the source and W_D under the drain equals the channel length L. W_S and W_D can be found using the one-dimensional depletion approximation (Section 6.2.1.1).

It should be noted that punchthrough is a problem encountered only in small-geometry MOSFETs. If the channel length is greater than approximately 3 μm, avalanche breakdown will occur at the drain end of the channel before the drain depletion region punches through to the source. It should also be pointed out that present versions of the SPICE program make no allowance for punchthrough.

7.6.5 Hot Carrier Effects

For MOSFETs with channel lengths in the 1 μm range used with standard 5 V power supply levels, prolonged operation in the saturation regime leads to a gradual degradation in transconductance and an increase in threshold voltage. Both effects are believed to result from the presence of high-energy (''***hot***'') electrons in the pinched-off region of the channel near the drain. Some of these electrons may gain sufficient energy to enter the silicon dioxide conduction band, where they may be trapped. Negative trapped charge in the gate oxide will shift V_T in the positive direction. Degradation in transconductance is thought to result from the generation of interface traps when part of the energy of a hot electron is used to rearrange bonding at the silicon surface.

7.7 TRANSIENT ANALYSIS

Consider what happens when V_{GS} is suddenly increased in an n-channel MOSFET operating in the triode regime while V_{DS} and V_{SB} are held constant. With a higher V_{GS}, we know that there will be more electrons stored in the channel and a higher drain current under final steady-state conditions. In a typical modern MOSFET, generation of electrons in the depletion region underlying the channel is a relatively slow process, so almost all of the extra electrons will be supplied from the source. During the time these extra electrons are entering the channel, the drain current must be less than its steady-state value, and the source current will be larger than the drain current.

7.7.1 Transit Time

A crude upper bound on the time required for the drain current to respond to the change in V_{GS} can be obtained by noting that the extra electrons supplied from the source cannot move faster than v_{sat}, the scatter-limited velocity in silicon, and that these electrons must travel roughly the length L of the channel. The maximum operating frequency of the MOSFET can therefore not exceed v_{sat}/L. Taking $v_{sat} \simeq 10^7$ cm s^{-1} for electrons in silicon, this simple analysis predicts an absolute maximum operating frequency of approximately 100 GHz or equivalently, a switching time of 10 ps for a 1 μm channel-length device.

7.7.2 Quasi-Static Model

In the SPICE circuit simulator, MOSFET transient response is modeled using the quasi-static approach which was introduced in connection with the *pn* diode in Section 6.6. Rather than attacking the very complicated problem of how charge within the transistor redistributes in response to a rapid change in bias, quasi-static models assume that the terminal voltages change slowly enough that the carrier distributions are always close to their steady-state forms. The quasi-static approximation works quite well for modern MOS integrated circuits, where switching is typically limited by the time taken to charge parasitic capacitances, such as those associated with long interconnect lines, rather than the time required for charge redistribution within a transistor itself. However, it should be kept in mind that this approach may fail for transistors with very little load capacitance.

From the steady-state analysis of the *n*-channel MOSFET in Section 7.3, we know that excess charge is stored in the depletion region, as mobile electrons in the channel, and on the gate. These charges will be symbolized as Q'_B, Q'_n and Q'_G, with the $'$ symbol once again denoting ***total*** stored charge as opposed to charge per unit area. Overall charge neutrality requires that

$$Q'_n + Q'_B + Q'_G = 0 \tag{7.90}$$

Under steady-state conditions, the only appreciable current in the MOSFET results from electron flow from the source to the drain. In transient analysis, this is frequently termed the ***transport current***. In a transient situation, additional currents associated with changes in stored charge in the device may exist not only at the source and drain terminals, but also at the gate and substrate. These current components are referred to as ***charging currents***. Physically, the gate charging current must be associated with a change in Q'_G, the substrate charging current with a change in Q'_B, and the source and drain charging currents with changes in Q'_n. Using the symbol i to represent charging currents, we find that

$$i_G = \frac{dQ'_G}{dt} \tag{7.91}$$

$$i_B = \frac{dQ'_B}{dt} \tag{7.92}$$

and

$$i_S + i_D = \frac{dQ'_n}{dt} \tag{7.93}$$

Equation (7.93) presents something of a problem, since from the viewpoint of circuit simulation it is necessary to find i_S and i_D individually, not just their sum. It is therefore desirable to partition the total channel charge, apportioning a part Q'_{nS} to the source and a part Q'_{nD} to the drain. This is to be done in such a way that

$$i_S = \frac{dQ'_{nS}}{dt} \tag{7.94}$$

and

$$i_D = \frac{dQ'_{nD}}{dt} \tag{7.95}$$

7.7.3 Computing Stored Charge with the Square-Law Model

In order to proceed further with the quasi-static transient model, it is necessary to obtain expressions for Q'_G, Q'_B, Q'_{nS}, and Q'_{nD} as functions of the terminal voltages. This information is normally available from any model used to compute steady-state currents in the MOSFET. As an example, we will now compute Q'_G using the square-law model.

From the gradual channel approximation, the charge per unit area stored on the gate is given by

$$Q_G = C_{ox}(V_{GB} - V_{FB} - \psi_s) \tag{7.96}$$

For operation in the triode regime the square law model assumes that $\psi_s = 2\phi_B + V_{CB}$, where V_{CB} is the channel voltage, so

$$Q_G = C_{ox}(V_{GB} - V_{FB} - 2\phi_B - V_{CB}) \tag{7.97}$$

It is convenient to write (7.97) in the form

$$\begin{aligned}Q_G &= C_{\text{ox}}[V_{GB} - V_{\text{FB}} - 2\phi_B - \sqrt{2\epsilon_s qN_A(2\phi_B + V_{SB})}/C_{\text{ox}} \\ &\quad - V_{CB}] + \sqrt{2\epsilon_s qN_A(2\phi_B + V_{SB})} \\ &= C_{\text{ox}}[V_{GB} - V_T - V_{CB}] + \sqrt{2\epsilon_s qN_A(2\phi_B + V_{SB})}\end{aligned} \tag{7.98}$$

The total charge stored on the gate is obtained by integrating (7.98) along the length of the channel to give

$$Q'_G = C_{\text{ox}}Z \int_{y=0}^{L} (V_{GB} - V_T - V_{CB})\, dy + ZL\sqrt{2\epsilon_s qN_A(2\phi_B + V_{SB})} \tag{7.99}$$

To evaluate the remaining integral in (7.99), we note that $I_D = -Z\mu_e Q_n\, dV_{CB}/dy$, so

$$\begin{aligned}Q'_G &= -C_{\text{ox}}Z \int_{V_{SB}}^{V_{SB} + V_{DS}} (V_{GB} - V_T - V_{CB}) \frac{Z\mu_e Q_n}{I_D}\, dV_{CB} \\ &\quad + ZL\sqrt{2\epsilon_s qN_A(2\phi_B + V_{SB})} \\ &= -C_{\text{ox}}Z \int_0^{V_{DS}} (V_{GS} - V_T - V_{CS}) \frac{Z\mu_e Q_n}{I_D}\, dV_{CS} \\ &\quad + ZL\sqrt{2\epsilon_s qN_A(2\phi_B + V_{SB})}\end{aligned} \tag{7.100}$$

In the triode regime, the square-law model predicts that

$$I_D = \frac{Z\mu_e C_{\text{ox}}}{L}\left(V_{GS} - V_T - \frac{V_{DS}}{2}\right)V_{DS} \tag{7.101}$$

Also, in the square-law model Q_B is computed by assuming that $\psi_s = 2\phi_B + V_{SB}$ along the length of the channel, so Q_n is given by

$$\begin{aligned}Q_n = Q_s - Q_B &= -C_{\text{ox}}(V_{GB} - V_{FB} - 2\phi_B - V_{CB}) + \sqrt{2\epsilon_s qN_A(2\phi_B + V_{SB})} \\ &= -C_{\text{ox}}(V_{GS} - V_T - V_{CS})\end{aligned} \tag{7.102}$$

Therefore,

$$Q'_G = C_{\text{ox}}ZL \int_0^{V_{DS}} \frac{(V_{GS} - V_T - V_{CS})^2\, dV_{CS}}{(V_{GS} - V_T - V_{DS}/2)\, V_{DS}} + ZL\sqrt{2\epsilon_s qN_A(2\phi_B + V_{SB})} \tag{7.103}$$

Evaluating the integral in (7.103) gives

$$Q'_G = \frac{2C_{ox}ZL}{3[2(V_{GS} - V_T) - V_{DS}]V_{DS}} [(V_{GS} - V_T)^3 - (V_{GS} - V_T - V_{DS})^3] + ZL\sqrt{2\epsilon_s qN_A(2\phi_B + V_{SB})} \quad (7.104)$$

It is frequently useful to rewrite (7.104) in terms of V_{GS} and $V_{GD} = V_{GS} - V_{DS}$. With this substitution and much additional algebraic manipulation, (7.104) becomes

$$Q'_G = \frac{2C_{ox}ZL}{3} \frac{(V_{GS} - V_T)^3 - (V_{GD} - V_T)^3}{(V_{GS} - V_T)^2 - (V_{GD} - V_T)^2} + ZL\sqrt{2\epsilon_s qN_A(2\phi_B + V_{SB})} \quad (7.105)$$

Evaluation of the charge Q'_B stored in the depletion region is straightforward when using the square-law model, since Q_B is assumed to be independent of position along the channel. Q'_{nD} and Q'_{nS} can be evaluated following a procedure similar to that used to find Q'_G, although, as noted above, an arbitrary decision must be made as to how the channel charge is to be partitioned between the source and drain. This problem is considered at length in Ref. 4.

7.7.4 Terminal Capacitances

Expressions for Q'_G, Q'_B, Q'_{nD}, and Q'_{nS} such as those derived above are directly useful for large-signal transient analysis with circuit simulators such as SPICE. However, for the purpose of developing an intuitive understanding of the transient response problem, it is preferable to associate a capacitance with each node of the transistor specifying how charge flows into or out of that node as the terminal voltages change. These capacitances can be used to construct an equivalent circuit for the MOSFET that is the basis for analyzing the transient and small-signal ac responses.

To take an example, capacitance C'_{GS} specifies the rate of change of Q'_G with respect to V_S with the voltages at the other terminals (V_G, V_D, and V_B) held constant; that is,

$$C'_{GS} = \frac{\partial Q'_G}{\partial V_S} \quad (7.106)$$

Similarly,

$$C'_{GD} = \frac{\partial Q'_G}{\partial V_D} \quad (7.107)$$

and

$$C'_{GB} = \frac{\partial Q'_G}{\partial V_B} \quad (7.108)$$

These three capacitances are generally the most important in determining the transient response, but others can be defined in an analogous fashion. Since the MOSFET has four terminals, there are in principle $4 \times 4 = 16$ capacitances to consider. It is worth noting that in general the terminal capacitances are ***nonreciprocal***. For example, in saturation changing the voltage on the drain causes only a slight increase in the width of the depletion region under the gate, and so produces virtually no change in Q'_G. However, changing V_G in saturation changes the charge stored in the channel, and so has a major influence on Q'_{nD}. Thus

$$C'_{GD} = \frac{\partial Q'_G}{\partial V_D} \neq C'_{DG} = \frac{\partial Q'_{nD}}{\partial V_G} \tag{7.109}$$

Differentiating the expression for Q'_G given in (7.105) gives, after considerable algebra,

$$C'_{GS} = \frac{2}{3} C_{\text{ox}} ZL \left[1 - \frac{(V_{GD} - V_T)^2}{(V_{GS} - V_T + V_{GD} - V_T)^2} \right] \tag{7.110}$$

and

$$C'_{GD} = \frac{2}{3} C_{\text{ox}} ZL \left[1 - \frac{(V_{GS} - V_T)^2}{(V_{GS} - V_T + V_{GD} - V_T)^2} \right] \tag{7.111}$$

Equations (7.110) and (7.111) apply only in the triode regime.

As noted above, in saturation changing the drain bias has virtually no effect on Q'_G, so C'_{GD} drops to zero in saturation. For similar reasons, when computing C'_{GS} in saturation V_D should be set to its saturation value in (7.110).

When the MOSFET is operating in the subthreshold regime—that is, when the surface under the gate is depleted or accumulated—C'_{GS} and C'_{GD} are both effectively zero. In subthreshold C'_{GB} can be computed using the analysis applied to find the capacitance of the MOS capacitor (see Section 7.9). The formation of an inversion layer in the channel region provides an electrostatic screen between the gate and the substrate, so Q'_G ceases to respond to changes in V_B above threshold. C'_{GB} therefore drops to zero when the MOSFET is turned on.

The bias dependence of C'_{GS}, C'_{GD}, and C'_{GB} is summarized in Fig. 7.41.

7.7.5 Simple Equivalent Circuit

Although highly accurate modeling of the transient response of MOS circuits requires use of all 16 nonreciprocal capacitances mentioned above, in many situations it is possible to obtain acceptable results using a simpler model which incorporates only C'_{GS}, C'_{GD}, and C'_{GB}, and assumes that $C'_{SG} = C'_{GS}$, $C'_{DG} = C'_{GD}$, and $C'_{BG} = C'_{GB}$ (in other words, the simple model assumes the capacitances are reciprocal). This simple model is sometimes called the ***Meyer model***, after its developer [7]. The Meyer model can be represented by the simple equivalent circuit shown in Fig. 7.42.

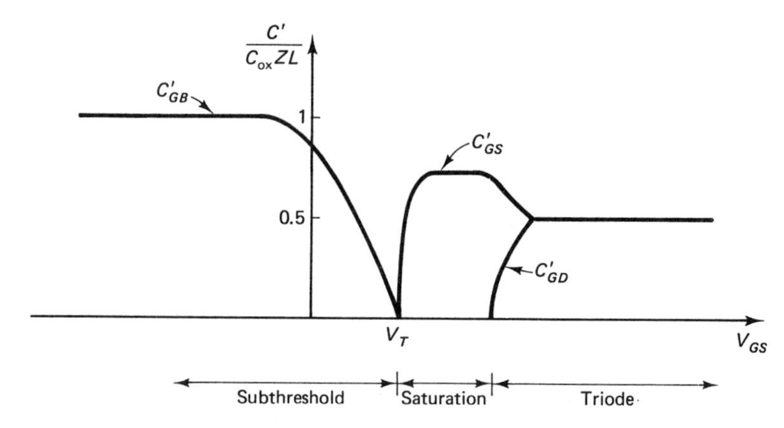

Figure 7.41 Dependence of Meyer model capacitances on V_{GS} for a fixed V_{DS}.

7.7.6 Extrinsic Capacitances

So far our discussion of charge storage and transient analysis in the MOSFET has only considered what might be termed the ***intrinsic*** device. The intrinsic part of the MOSFET can be viewed as a slice beginning at the boundary between the source implant and the channel, and ending at the boundary between the far end of the channel and the drain implant, as illustrated in Fig. 7.43. The properties of the intrinsic part of the device are therefore independent of the size and shape of the source and drain regions, at least as far as the gradual channel approximation applies.

In addition to the intrinsic capacitances discussed above, any real MOSFET has ***extrinsic*** capacitances associated with the inevitable overlap of the gate and the source and drain regions, and with the junction capacitance between the n-type source and drain and the p-type substrate. (Recall that the source and substrate and drain and substrate form n^+p junction diodes.) Both types of extrinsic capacitance are shown in Fig. 7.43. The overlap capacitances C'_{GSO} and C'_{GDO} are effectively bias independent, due to the very heavy doping of the source and drain. These capacitances depend strongly on the process used to fabricate the MOS-

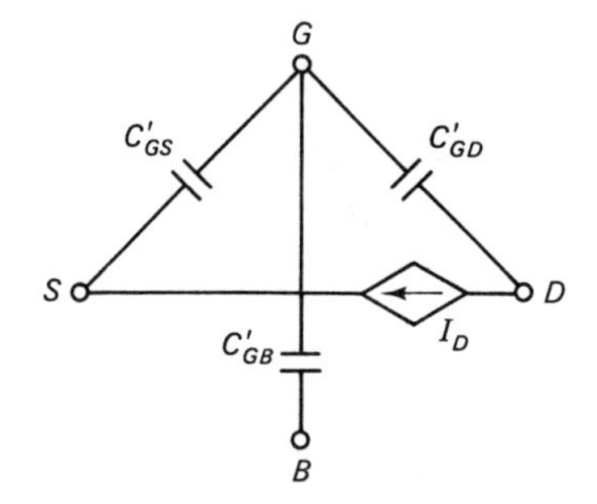

Figure 7.42 Meyer model equivalent circuit for the MOSFET. I_D is the steady-state transport current.

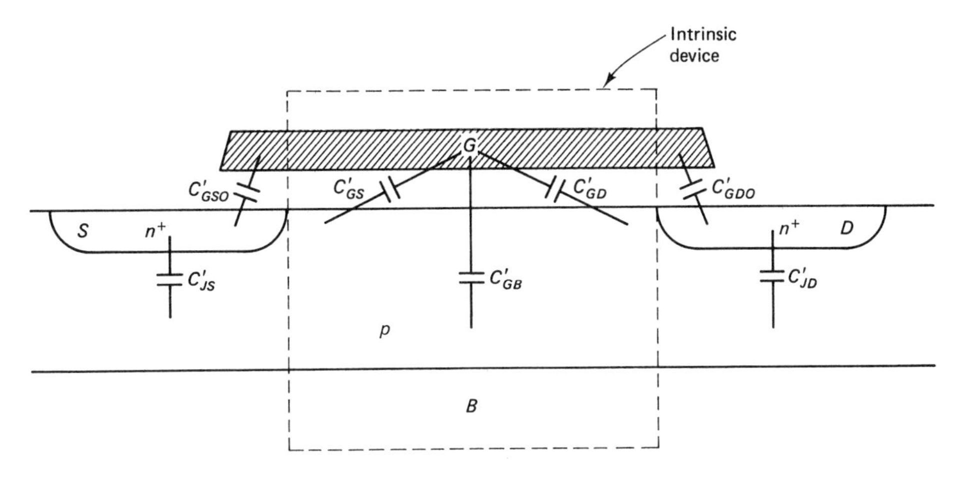

Figure 7.43 Extrinsic and intrinsic capacitances in the MOSFET.

FET, since they result from the lateral diffusion of the impurity used to form the source and drain under the edge of the gate (see Section 12.2.11). The junction capacitances C'_{JS} and C'_{JD} can be computed using the expression for the capacitance of a *pn* junction (Section 6.6.1). A complete equivalent circuit for the MOSFET, including the extrinsic capacitances, is shown in Fig. 7.44. For state-of-the-art short-channel MOSFETs, the extrinsic capacitances are often the more important in determining switching speed.

7.7.7 Transient Modeling with SPICE

The most accurate model for MOSFET transient analysis available in SPICE makes direct use of the expressions for the stored charges Q'_G, Q'_B, Q'_{nS}, and Q'_{nD} as functions of the terminal voltages discussed in Section 7.7.3.

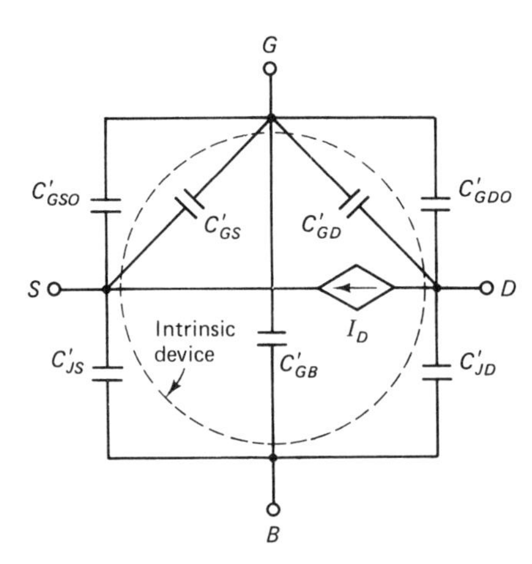

Figure 7.44 Complete Meyer model equivalent circuit for the MOSFET, showing both extrinsic and intrinsic capacitances.

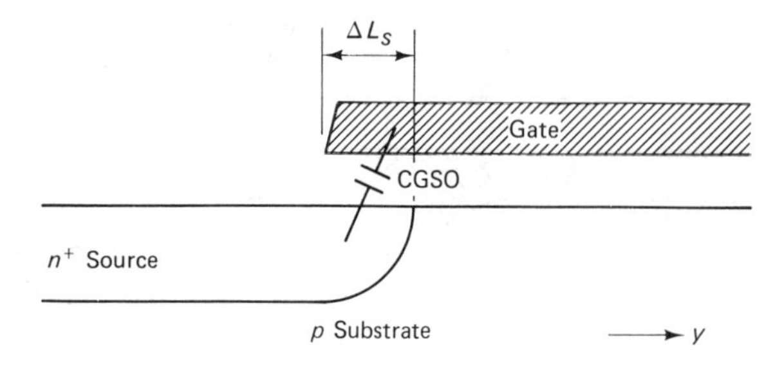

Figure 7.45 Overlap ΔL_S of the gate over the source diffusion or implant which gives rise to the capacitance component CGSO.

This is frequently referred to as the ***Ward and Dutton*** model, after its developers [8]. The Ward and Dutton model is available only as part of the Level 2 MOSFET model, and then only when the model parameter XQC is assigned a value between 0 and 0.5. XQC determines the fraction of the total mobile electron charge Q'_n stored in the channel which is associated with the drain charging current in accord with (7.95); that is,

$$Q'_{nD} = \text{XQC} \cdot Q'_n \qquad (7.112)$$

Ordinarily, XQC is set to a value of 0.5 when the Ward and Dutton model is to be used. In the Level 1 MOSFET model, stored charge is computed using the simple Meyer model equivalent circuit of Fig. 7.42 and the parameter XQC is ignored.

Both the Level 1 and Level 2 MOSFET models allow for the extrinsic capacitances discussed in Section 7.7.6 and shown in the equivalent circuit of Fig. 7.44. The gate–source and gate–drain overlap capacitances are specified through the SPICE model parameters CGSO and CGDO, which are normally identical. CGSO is defined through the equation

$$\text{CGSO} = \frac{\epsilon_{ox}}{t_{ox}} \Delta L_S \qquad (7.113)$$

where ΔL_S is the distance by which the source region extends under the gate, as shown in Fig. 7.45. CGSO is therefore the overlap capacitance per unit channel width; the total overlap capacitance is given by

$$C'_{GSO} = \text{CGSO} \cdot Z \qquad (7.114)$$

Similar considerations apply to CGDO.

The SPICE parameter CGBO specifies the gate–substrate overlap capaci-

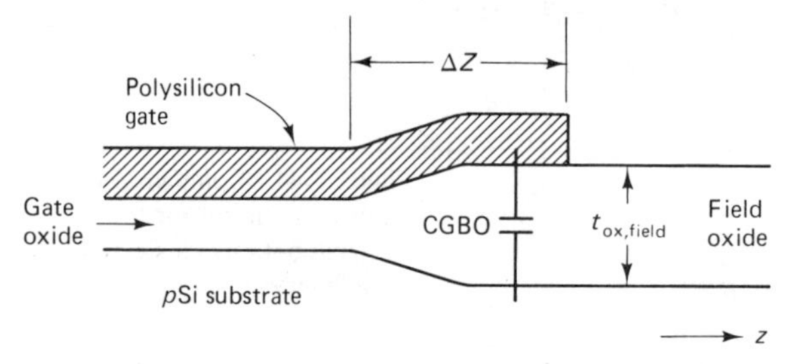

Figure 7.46 How the overlap of the polysilicon gate over the field oxide gives rise to the gate–substrate overlap capacitance CGBO.

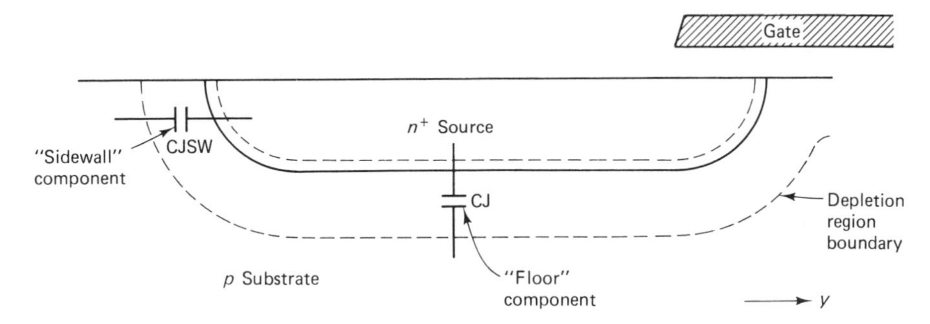

Figure 7.47 Floor and sidewall components of the source or drain junction capacitance.

tance per unit gate length. CGBO is given by

$$\text{CGBO} = \frac{\epsilon_{\text{ox}}}{t_{\text{ox,field}}} \Delta Z \tag{7.115}$$

where $t_{\text{ox,field}}$ is the field oxide thickness and ΔZ is the amount by which the edge of the gate extends over the field oxide, as shown in Fig. 7.46.

SPICE breaks the capacitance of the source–substrate and drain–substrate junctions into two parts, one (CJ) associated with the base or "floor" of the junction and the other (CJSW) with the sidewall, as suggested in Fig. 7.47. CJ specifies the floor capacitance per unit area, so SPICE must also be supplied with the source and drain junction floor areas AS and AD to compute the total capacitance. Similarly, CJSW specifies the sidewall capacitance per unit sidewall perimeter, so SPICE must be told the source and drain perimeters PS and PD to compute the actual sidewall capacitance. (Naturally, PS and PD do not include the part of the perimeter formed by the gate.) Both capacitance components are assumed to depend on bias in accordance with the relation

$$C = C_{j0}\left(1 - \frac{V}{V_{\text{bi}}}\right)^{-m} \tag{7.116}$$

where V_{bi} is the built-in electrostatic potential barrier of the junction, m the so-called "***grading coefficient***," and C_{j0} the junction capacitance at zero applied bias. In Section 6.6.1 it was shown that for a one-dimensional abrupt junction, m should have a value of $\frac{1}{2}$. For a linearly graded junction, m should equal $\frac{1}{3}$. Here m is treated as a semiempirical parameter that can be used to fit the SPICE model to results obtained with experimental circuits. The model parameters are MJSW for the sidewall and MJ for the floor. Since the source–substrate and drain–substrate junctions are normally one-sided and nearly abrupt, the default value of MJ is $\frac{1}{2}$.

Example 7.7 Switching in a CMOS Inverter

A silicon CMOS technology is described in Section 12.2.11. CMOS uses n-channel enhancement and p-channel enhancement transistors. Figure 7E.10

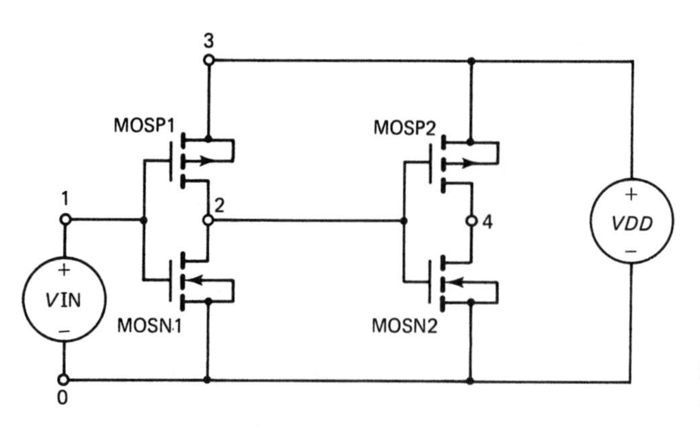

Figure 7E.10 Circuit used to examine transient response of CMOS inverter.

shows a CMOS inverter driving an identical inverter as a load. To obtain symmetric rise and fall times in this logic gate, the Z/L ratio for the p-channel pull-up transistors has been made three times that of the n-channel pull-downs. This compensates for the fact that the mobility of holes is about one-third that of electrons, making the effective resistance of a p-channel transistor about three times higher than that of an n-channel device with the same geometry. The SPICE input file below can be used to examine the transient response of the inverter when its input is switched from a logic "high" level to a "low" level and then back to a "high" level. Only intrinsic capacitances have been included in the device model for this initial simulation. It has been assumed that the inverter is made in a p-well process, so that the substrate doping for the n-channel transistors is substantially higher than that for the p-channel devices (see Section 12.2.11), and that the gate oxide thickness is 50 nm.

```
*TRANSIENT RESPONSE OF CMOS INVERTER
VDD 3 0  DC 5
VIN 1 0  PULSE 0 5 0.5NS 0.05NS 0.05NS 3NS 10NS
MOSP1 2 1 3 3 MOSP L=5U W=15U AS=180P PS=39U
+ AD=180P PD=39U
MOSN1 2 1 0 0 MOSN L=5U W=5U AS=180P PS=39U
+ AD=180P PD=39U
MOSP2 4 2 3 3 MOSP L=5U W=15U AS=180P PS=39U
+ AD=180P PD=39U
MOSN2 4 2 0 0 MOSN L=5U W=5U AS=180P PS=39U
+ AD=180P PD=39U
.MODEL MOSN NMOS LEVEL=2 NSUB=1E16 TOX=50E-9 VTO=1
+ XQC=0.5
.MODEL MOSP PMOS LEVEL=2 NSUB=2E15 TOX=50E-9 VTO=-1
+ XQC=0.5
.WIDTH OUT=80
.TRAN 0.2NS 8NS
.PLOT TRAN V(2)
.END
```

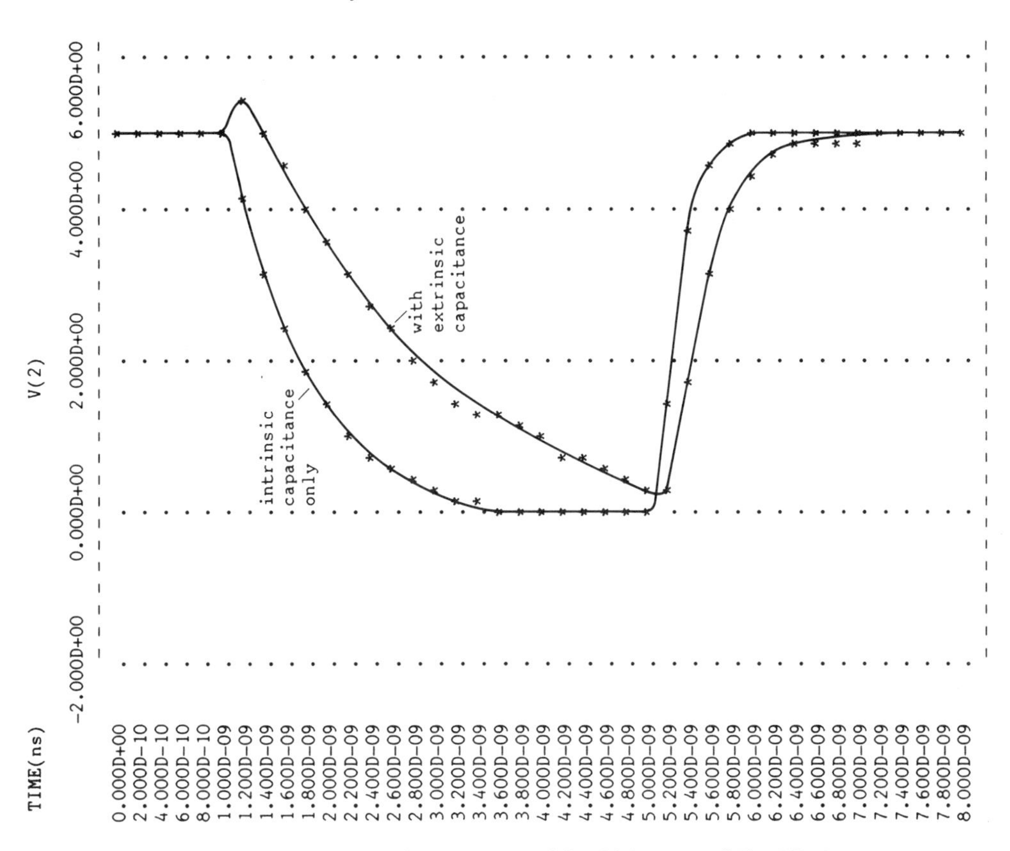

Figure 7E.11 Transient response of CMOS inverter of Fig. 7E.10.

By modifying the .MODEL lines describing the two transistors to read as shown below it is possible to examine the effect of extrinsic capacitance on the inverter response. The capacitance parameters, CJ, CJSW, MJ, and MJSW were computed assuming one-sided junctions and using the values given in the above listing for the area and perimeter of the source and drain junctions. CGSO and CGDO were computed assuming a 0.5 μm overlap of the gate over each of the source and drain regions.

```
.MODEL MOSN NMOS LEVEL=2 NSUB=1E16 TOX=50E-9 VTO=1
+ XQC =0.5 CJ=3.48E-4 CJSW=2.78E-10 CGSO=3.45E-10
+ CGDO=3.45E-10 CGBO=1.72E-10 MJSW=0.5
.MODEL MOSP PMOS LEVEL=2 NSUB=2E15 TOX=50E-9 VTO=-1
+ XQC=0.5 CJ=1.30E-4 CJSW=1.04E-10 CGSO=3.45E-10
+ CGDO=3.45E-10 CGBO=1.72E-10 MJSW=0.5
```

The transient response of the inverter is shown in Fig. 7.E.11, both with and without extrinsic capacitance effects. It can be seen that the extrinsic capacitances substantially slow switching. To identify which of the extrinsic capacitances makes the greatest contribution to the delay, it is instructive to remove these capacitances one at a time from the MOSFET models. When this is done, it is found that the largest delay is associated with the source and drain junction capacitances.

7.8 AC ANALYSIS

7.8.1 Small-Signal, Linearized Model

Choosing the source terminal as a voltage reference, the drain current in a MOSFET can be written generally as

$$I_D = I_D(V_{GS}, V_{DS}, V_{BS}) \tag{7.117}$$

Following the approach established in Section 6.7.1 for the diode, to determine the response of the MOSFET to a small ac signal applied at some given dc operating point we perform a Taylor series expansion of (7.117) and retain only the zeroth- and first-order terms. The resulting expression for the small-signal drain current i_d is

$$i_d = g_m v_{gs} + g_{ds} v_{ds} + g_{mbs} v_{bs} \tag{7.118}$$

where v_{gs}, v_{ds}, and v_{bs} are small-signal ac voltages and

$$g_m = \left. \frac{\partial I_D}{\partial V_{GS}} \right|_{V_{DS}, V_{BS}} \tag{7.119a}$$

is the transconductance,

$$g_d = \left. \frac{\partial I_D}{\partial V_{DS}} \right|_{V_{GS}, V_{BS}} \tag{7.119b}$$

is the channel or drain conductance, and

$$g_{mbs} = \left. \frac{\partial I_D}{\partial V_{BS}} \right|_{V_{GS}, V_{DS}} \tag{7.119c}$$

is the substrate transconductance.

This suggests that for small-signal analysis the MOSFET can be represented by two voltage-controlled current sources and a conductance. Combining these elements with the Meyer model intrinsic capacitances C'_{GS}, C'_{GD}, and C'_{GB} gives the small-signal equivalent circuit shown in Fig. 7.48. It is a straightforward matter to add extrinsic components representing overlap capacitance, series resistance, and junction capacitance to Fig. 7.48, and obtain the complete equivalent circuit for the device.

7.8.2 Small-Signal Analysis with SPICE

The equivalent circuit of Fig. 7.48, augmented by the extrinsic capacitance components mentioned above, is used for small-signal ac analysis when the Level 1 MOSFET model is invoked in SPICE. When the Level 2 MOSFET model is used, the more accurate Ward and Dutton capacitance model mentioned briefly in Section 7.7.7 is available if the parameter XQC is set to a value of 0.5 or less. In both cases the model parameters are identical to those discussed in Section 7.7.7.

7.8.3 Cut Off Frequency f_T

Figure 7.49a shows the circuit for a simple common-source MOSFET amplifier. Amplifiers of this basic type are used in analog MOS integrated circuits, but the load resistor is replaced by another MOSFET (typically, a depletion-type device operated with its gate and drain shorted) since the range of resistor values available in an integrated circuit is very limited. Ordinarily, a dc bias point is chosen so that the MOSFET is operating in the saturation regime, where it most closely resembles an ideal voltage-controlled current source.

The small-signal equivalent circuit corresponding to Fig. 7.49a is shown in Fig. 7.49b. For simplicity, only the intrinsic Meyer model capacitance components have been included in this circuit. It is worth noting that the dc voltage sources are short circuits as far as ac signals are concerned, and that C'_{GD} and C'_{GB} are negligible in saturation. Also, g_{ds} is zero for a long-channel MOSFET in saturation, and so has not been included in the circuit.

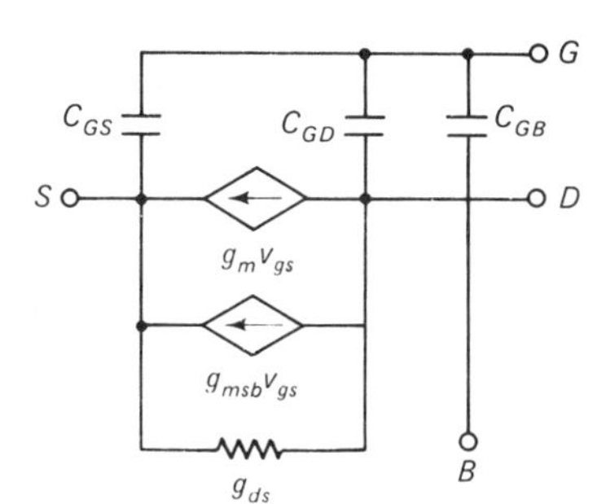

Figure 7.48 Small-signal ac equivalent circuit for the intrinsic MOSFET.

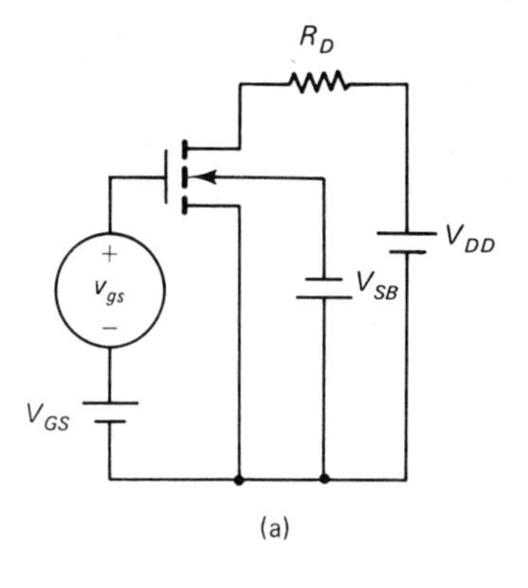

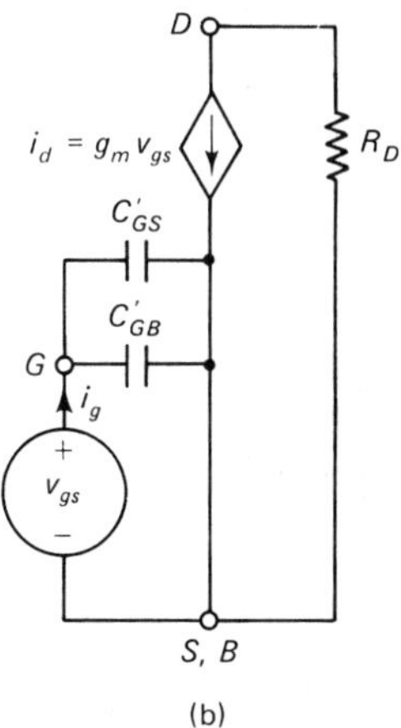

Figure 7.49 (a) Circuit diagram for common-source MOSFET amplifier. (b) Small-signal equivalent circuit.

From Fig. 7.49b we have

$$i_d = g_m v_{gs} \tag{7.120}$$

and

$$i_g = j\omega(C'_{GS} + C'_{GB})v_{gs} \tag{7.121}$$

The current gain a_i of the amplifier is therefore given by

$$a_i = \frac{g_m}{\omega(C'_{GB} + C'_{GS})} \tag{7.122}$$

As the frequency of the small-signal ac source increases, the current gain falls. Defining the cutoff frequency f_T as the frequency at which $a_i = 1$, we have

$$f_T = \frac{g_m}{2\pi(C'_{GB} + C'_{GS})} \tag{7.123}$$

This equation is similar in form to that derived for the cutoff frequency of the

BJT (11.74). Using the square-law model, the transconductance in saturation is given by

$$g_{m,\text{sat}} = \frac{Z}{L}\mu_e C_{\text{ox}}(V_{GS} - V_T) \tag{7.124}$$

while from (7.110), C'_{GS} in saturation is given by

$$C'_{GS} = \frac{2}{3}C_{\text{ox}}ZL \tag{7.125}$$

Substituting these expressions in (7.123) gives

$$f_T = \frac{3\mu_e(V_{GS} - V_T)}{4\pi L^2} \tag{7.126}$$

Equation (7.126) shows once again how important it is to reduce the gate length if good high-frequency performance is required, both in large-signal digital and small-signal analog circuits.

7.9 MOS CAPACITOR

When the MOS capacitor was introduced in Section 7.2.2, it was noted that the main use of this device is as a test structure in processing. In any production line for MOS integrated circuits, extra silicon wafers that do not carry a circuit pattern will be included in any oxide growth step. These ***monitor wafers*** will then have aluminum deposited on their front and back surfaces, and this metal will be patterned following the techniques outlined in Section 12.2.8 to form MOS capacitors, with typical diameters of a few hundred micrometers. By measuring the capacitance–voltage (C–V) characteristics of these capacitors, it is possible to obtain a great deal of information concerning the quality of the oxide grown. This information includes the flatband and threshold voltages and the level of mobile ion contamination. In this section we outline the basic concepts used in recording and analyzing MOS capacitor characteristics. A much more thorough discussion of the tests that can be conducted with MOS capacitors is given in Ref. 9.

7.9.1 Capacitance–Voltage Characteristics

In process control applications MOS capacitor C–V characteristics are normally measured in response to an ac small signal with a frequency of 1 MHz and an amplitude of a few millivolts while a slowly varying large-signal "dc" bias is applied between the gate and the substrate. A typical measurement setup is shown schematically in Fig. 7.50. This section is concerned with explaining the shape of the C–V characteristics obtained in such a measurement.

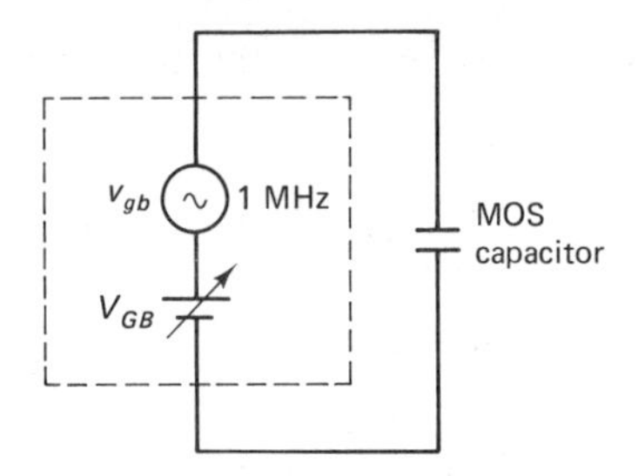

Figure 7.50 Schematic of test equipment used to measure the C–V characteristics of the MOS capacitor.

Since the MOS capacitor must be electrically neutral overall, any charge Q_s stored on the substrate must be balanced by an equal and opposite charge Q_G stored on the gate. The small-signal capacitance per unit gate area measured by the capacitance meter in Fig. 7.50 is given by

$$C = \frac{dQ_G}{dV_{GB}} = -\frac{dQ_s}{dV_{GB}} \tag{7.127}$$

Therefore,

$$\frac{1}{C} = -\frac{dV_{GB}}{dQ_s} = -\frac{d\psi_s}{dQ_s} - \frac{d\psi_{\text{ox}}}{dQ_s} = \frac{1}{C_s} + \frac{1}{C_{\text{ox}}} \tag{7.128}$$

where the semiconductor component C_s of the total capacitance is defined by

$$C_s = -\frac{dQ_s}{d\psi_s} \tag{7.129}$$

C_s is a measure of how charge stored in the semiconductor changes with changes in the band bending ψ_s across the semiconductor. Equation (7.128) suggests that the MOS capacitor can be thought of as consisting of the oxide capacitance C_{ox} in series with the semiconductor capacitance C_s, as shown in Fig. 7.51.

It is possible to solve exactly for Q_s as a function of ψ_s in the MOS capacitor, and thereby obtain exact expressions for C_s and the total capacitance C [9]. However, the exact solution is rather involved, so a much simpler but still relatively accurate approach to determining C_s will be taken here.

When the capacitor is in accumulation, the hole concentration at the silicon surface depends exponentially on ψ_s as specified in (7.23). This implies that a small change in ψ_s will produce a very large change in Q_s, so that C_s approaches infinity. In this case $C = C_{\text{ox}}$. Physically, this result can be explained by noting that the high density of charge carriers in the accumulation layer makes the semiconductor electrode behave almost like a metal plate. In accumulation, then, the MOS capacitor is electrically equivalent to a capacitor with two metal plates.

When the silicon surface is depleted or weakly inverted, Q_s results almost

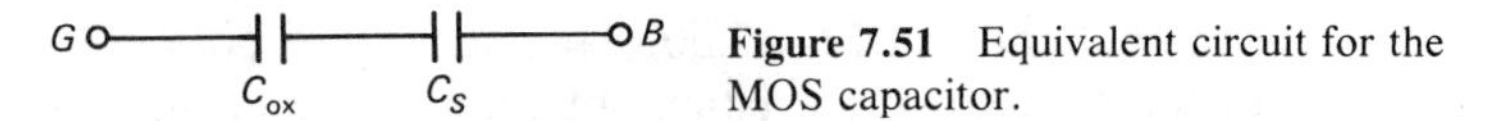

Figure 7.51 Equivalent circuit for the MOS capacitor.

entirely from dopant ion charge in the depletion region. In this case, using (7.17) to determine the depletion width W,

$$Q_s = -\sqrt{2\epsilon_s q N_A \psi_s} \tag{7.130}$$

so

$$C_s = \sqrt{\frac{\epsilon_s q N_A}{2\psi_s}} = \frac{\epsilon_s}{W} \tag{7.131}$$

It is worth noting that this is just the junction capacitance one would expect for a *pn* diode with depletion width W. With C_s specified by (7.131), the total capacitance is given by

$$\frac{1}{C} = \frac{1}{C_{ox}} + \frac{W}{\epsilon_s} \tag{7.132}$$

As V_{GB} is increased the depletion region widens and the capacitance falls. Eventually, the strong inversion condition $V_{GB} = V_T$ is reached, and at this point the depletion region stops expanding. The minimum capacitance is given by

$$\frac{1}{C_{\min}} = \frac{1}{C_{ox}} + \frac{W_{\max}}{\epsilon_s} \tag{7.133}$$

where $W_{\max}$ is given by (7.20).

It might be expected that once the semiconductor surface becomes strongly inverted C_s would again become extremely large. After all, from (7.13) the surface electron concentration has the same exponential dependence on ψ_s as the surface hole concentration, so a small change in ψ_s should produce a very large change in the electron concentration, and hence in Q_s. This is in fact the case provided that ψ_s and V_{GB} change slowly. However, electrons in the surface inversion layer cannot respond to rapid changes in V_{GB}, and in particular cannot respond to the 1-MHz frequency usually used to measure C. This comes about because in the MOS capacitor electrons can only be supplied to the surface inversion layer from thermal generation processes in the depletion region. Unless the minority carrier lifetime is abnormally short, these generation processes are rather slow. For example, in a MOS capacitor fabricated using present processing technology on a high-quality silicon substrate, it normally takes many seconds for the surface inversion layer to form after the capacitor is biased into what should be a strong inversion condition. (The situation is quite different in a MOSFET, where electrons can be drawn into the inversion layer from the n^+ source and drain regions.) For this reason, in strong inversion the small-signal capacitance of the MOS capacitor saturates at the value $C_{\min}$ given in (7.133).

Putting together the results above, we expect the complete C–V_{GB} characteristic of the MOS capacitor to appear as shown in Fig. 7.52. C is largest in accumulation, drops as V_{GB} increases and the width of the depletion region grows, and saturates at $C_{\min}$ in strong inversion.

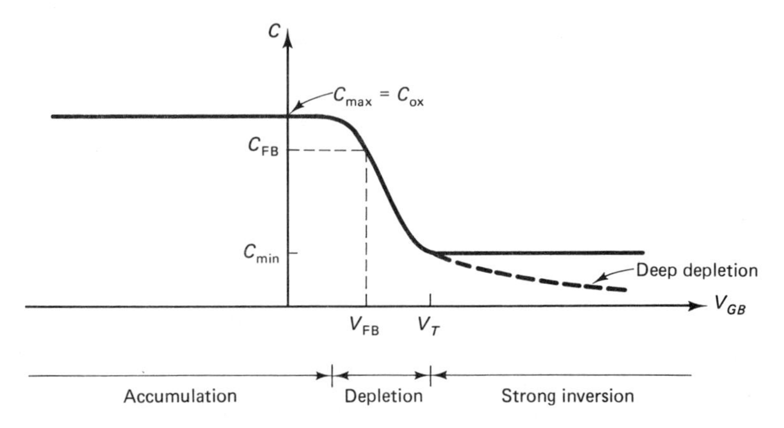

Figure 7.52 MOS capacitor capacitance–voltage characteristics.

If V_{GB} is increased very rapidly starting from a condition of accumulation, there may not be time for an inversion layer to form at the silicon surface. In this case the depletion region width will grow indefinitely as V_{GB} is increased, and the capacitance will fall in accordance with (7.132). This possibility is illustrated by the dashed line in Fig. 7.52. When W is made greater than W_{max} in this kind of transient situation, the capacitor is said to be in a condition of ***deep depletion.***

7.9.2 Analyzing C–V_{GB} Curves

As noted above, a great deal of information concerning an MOS system can be obtained from a single MOS capacitor C–V_{GB} curve. For example, the thickness of the gate oxide can be computed from (7.11) by noting that the maximum capacitance measured in accumulation is just C_{ox}. If C_{min} is then obtained from the strong inversion region of the characteristic, the maximum steady-state depletion region width W_{max} can be found from (7.20). Using (7.20) and (7.5), the substrate doping N_A can be found once W_{max} is known. It is not possible to solve (7.20) and (7.5) explicitly for N_A in terms of W_{max}, so iteration must be used in this calculation. However, ϕ_B is not a very strong function of N_A, so convergence is rapid.

The threshold voltage of the capacitor can be obtained by simply determining the value of V_{GB} at which the capacitance first reaches its minimum value, as shown in Fig. 7.52. Information on the amount of mobile contamination present in the oxide is found by applying a ***bias-temperature stress test*** to the capacitor. In this test a C–V characteristic is first recorded at room temperature. A fairly large positive bias is then applied to the gate of the capacitor, and the device is heated to a temperature of 200°C for a few minutes. At this elevated temperature the mobility of sodium ions in the oxide is greatly enhanced, and any sodium present in the SiO_2 layer is rapidly swept to the silicon–silicon dioxide interface, where, according to (7.65), it has the maximum effect on V_T. The capacitor is then allowed to cool to room temperature, and a second C–V characteristic is recorded. If there is substantial mobile ion contamination, the second characteristic will be displaced relative to the first as shown in Fig. 7.53. The spread between the C–V curves obtained after the positive and negative stress tests can be used to estimate the level of ionic contamination; assuming that all the mobile charge lies at the silicon–silicon dioxide interface after the positive stress and at the gate–silicon dioxide interface after the negative stress,

$$Q_m \simeq \frac{\Delta V}{C_{ox}} \tag{7.134}$$

If the silicon–silicon dioxide interface contains a large number of interface traps, the MOS capacitor C–V characteristic will be altered as shown in Fig. 7.54. In general, interface traps "smear out" the characteristic, increasing the voltage spread between the accumulation and inversion regimes. This distortion of the characteristic comes about primarily because the presence of the interface traps increases the change in V_{GB} required to produce a given change in ψ_s, as discussed in Section 7.4.6. The interface traps themselves do not normally contribute to the capacitance at frequencies as high as 1 MHz. It is possible to infer the density of interface traps from the degree of "smearing out" of the C–V characteristic; details are given in Ref. 9.

Another important quantity that can be readily extracted from a MOS capacitor C–V curve is the flatband voltage V_{FB}. To do this, it is necessary to know the flatband capacitance C_{FB}. At flatband, there is effectively no depletion region in the semiconductor, so it is not possible to use (7.131) to compute the semiconductor contribution to the capacitance. Instead, the functional dependence of Q_s on ψ_s must be determined directly by solving Poisson's equation. For a uniformly doped p-type substrate, this equation (3.29), states

$$\frac{d^2\psi}{dx^2} = -\frac{q}{\epsilon_s}[p(x) - N_A - n(x)] \simeq -\frac{q}{\epsilon_s}[p(x) - N_A] \tag{7.135}$$

Here

$$p(x) = N_V e^{-[E_F - E_V(x)]/kT} = N_V e^{-[E_F - E_V(x_B)]/kT} e^{-q\psi(x)/kT} = N_A e^{-q\psi(x)/kT} \tag{7.136}$$

taking $\psi = 0$ at the back contact where $p = N_A$.

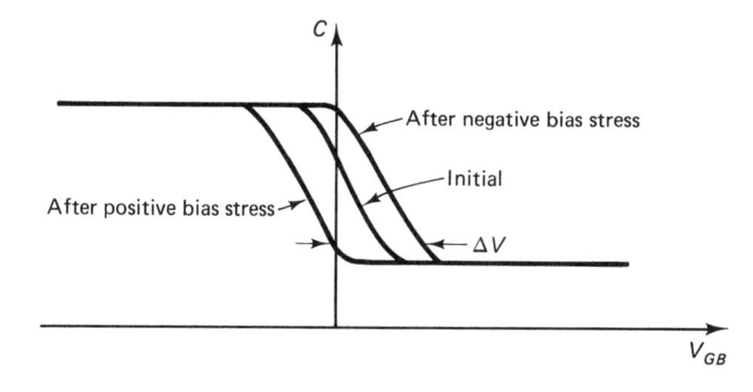

Figure 7.53 Effect of bias-temperature stress test on MOS capacitor characteristics.

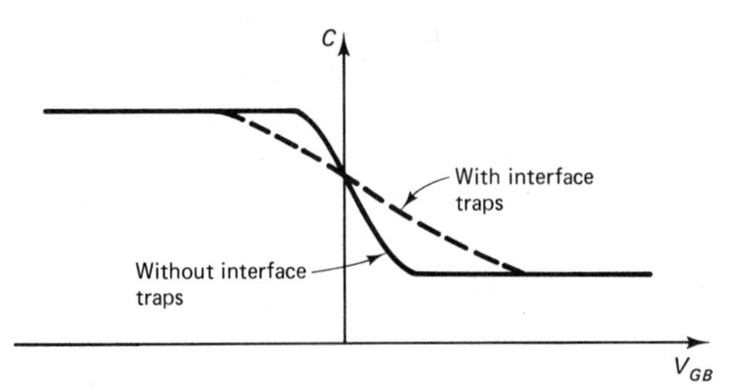

Figure 7.54 Effect of interface traps on MOS capacitor characteristics.

Near flatband, $\psi(x)$ should be small, so we can expand the right-hand side of (7.136) in a Taylor series:

$$e^{-q\psi/kT} = 1 - \frac{q\psi}{kT} \cdots \tag{7.137}$$

Therefore,

$$\frac{d^2\psi}{dx^2} = \frac{q^2 N_A \psi}{\epsilon_s kT} \tag{7.138}$$

Equation (7.138) has the solution

$$\psi(x) = \psi_s e^{-x/L_D} \tag{7.139}$$

The characteristic decay length L_D appearing in the exponential is known as the ***Debye length***, and is given by

$$L_D = \sqrt{\frac{\epsilon_s kT}{q^2 N_A}} \tag{7.140}$$

From (7.139), the electric field $\mathscr{E}_s$ at the semiconductor surface is given by

$$\mathscr{E}_s = -\left.\frac{d\psi}{dx}\right|_{x=0} = \frac{\psi_s}{L_D} \tag{7.141}$$

From Gauss's law,

$$Q_s = -\mathscr{E}_s \epsilon_s = -\frac{\epsilon_s \psi_s}{L_D} \tag{7.142}$$

so

$$C_{S,FB} = -\frac{dQ_s}{d\psi_s} = \frac{\epsilon_s}{L_D} \tag{7.143}$$

and finally, the capacitance of the complete MOS structure at flatband is given by

$$\frac{1}{C_{FB}} = \frac{1}{C_{ox}} + \frac{\epsilon_s}{L_D} \tag{7.144}$$

Once the substrate doping N_A has been found, it is therefore possible to compute C_{FB} and determine V_{FB} from the experimental C–V curve, as suggested in Fig. 7.52.

7.10 CHAPTER SUMMARY

This chapter began with a brief discussion of the MOS capacitor, in which the concepts of ***accumulation***, ***depletion***, and ***inversion*** of a semiconductor surface were introduced. These concepts were then applied to analyze the MOSFET itself. First an expression for the ***threshold voltage*** V_T of the MOSFET was found. The ***gradual channel approximation***, which views the MOSFET as a string of MOS capacitors stretching from source to drain, was then used to develop the ***bulk charge model*** for the drain current I_D as a function of biases V_{GS}, V_{DS}, and V_{SB}. The simpler but considerably less accurate ***square-law model*** was introduced as an approximate form of the bulk charge model. The problem of pinch-off at the drain end of the channel, which results if V_{DS} is increased above some critical value $V_{DS,\text{sat}}$ was examined, and was found to lead to ***saturation*** of the drain current in a long-channel MOSFET. The ***transconductance*** g_m was introduced as an important figure of merit for any field-effect transistor.

A variety of factors that can lead to nonzero flatband voltage in MOSFETs and MOS capacitors was next considered. The contribution of metal-semiconductor ***work function difference*** ϕ_{ms}, ***oxide fixed charge*** Q_f, ***mobile ion charge*** Q_m, and ***oxide trapped charge*** Q_{ot} to V_{FB} were all analyzed. The degradation of MOSFET transconductance resulting from ***interface trapped charge*** Q_{it} stored in ***fast surface states*** was also discussed.

Short- and ***narrow-channel effects*** encountered when the channel length and width are reduced in a MOSFET were examined in a largely qualitative fashion. A very useful empirical expression (7.71) predicting the minimum physical channel length a device can have while still exhibiting electrically long-channel characteristics was presented. The problems of ***channel-length modulation*** in saturation, threshold voltage reduction, ***punchthrough***, and ***hot carrier effects*** in short-channel devices were also discussed.

A simple approach to computing the charge stored in the MOSFET as a function of the terminal voltages for use in quasi-static transient analysis was outlined. By differentiating the expressions for the stored charge with respect to the terminal voltages, a simple equivalent circuit model for the MOSFET consisting of capacitive elements connecting the gate to the source, drain, and substrate was developed. In addition to these inherent or ***intrinsic*** capacitances, a complete transient analysis model must include ***extrinsic*** capacitances associated with the source–substrate and drain–substrate *pn* junctions, and with the inevitable overlap of the gate electrode and the source and drain. In Example 7.7 it was seen that these extrinsic capacitances can dominate the switching speed of a typical logic gate in a MOS IC.

The chapter concluded with a brief discussion of the use of the MOS capacitor as a diagnostic probe for extracting information such as gate oxide thickness, substrate doping level, flatband voltage, and threshold voltage. MOS capacitors are used almost universally in this way as ***process monitors*** in the fabrication of MOS integrated circuits.

7.11 REFERENCES

1. J. R. Brews, "Physics of the MOS transistor," in *Applied Solid State Science*, Suppl. 2A, D. Khang, ed., Academic Press, Inc., New York, 1981.
2. A. Vladimirescu, K. Zhang, A. R. Newton, and D. O. Pederson, "SPICE version 2G.6 user's guide," Department of Electrical Engineering and Computer Science, University of California at Berkeley, 1981.
3. A. G. Sabnis and J. T. Clemens, "Characterization of the electron mobility in the inverted ⟨100⟩ Si surface," *Technical Digest of the IEEE International Electron Devices Meeting*, p. 18, 1979.
4. A. Vladimirescu and S. Lui, "The simulation of MOS integrated circuits using SPICE 2," Memo. no. UCB/ERL/M80/7, Electronics Research Laboratory, University of California at Berkeley, 1980.
5. B. E. Deal, "Standardized terminology for oxide charges associated with thermally oxidised silicon," *IEEE Transactions on Electron Devices*, vol. ED-27, p. 606, 1980.
6. J. R. Brews, W. Fichtner, E. H. Nicollian, and S. M. Sze, "Generalized guide for MOSFET miniaturization," *IEEE Electron Device Letters*, vol. EDL-1, p. 2, 1980.
7. J. E. Meyer, "MOS models and circuit simulation," *RCA Review*, vol. 32, p. 42, 1971.
8. D. E. Ward and R. W. Dutton, "A charge-oriented model for MOS transistor capacitances," *IEEE Journal of Solid-State Circuits*, vol. SC-13, p. 703, 1978.
9. E. H. Nicollian and J. R. Brews, *MOS Physics and Technology*, John Wiley & Sons, Inc., New York, 1982.

PROBLEMS

7.1. The I_D–V_{GS} characteristics for a long *n*-channel MOSFET are shown in Fig. 7P.1. Using the square-law model and information available from the graph, compute the drain current I_D when $V_{GS} = 3$ V and $V_{DS} = 5$ V.

7.2. Solve (7.39) using the quadratic formula to derive the expression for $V_{DS,\text{sat}}$ given in (7.40).

7.3. The analysis of the MOSFET presented in Section 7.4 was based on the gradual channel approximation—that is, the assumption that the electric field $\mathscr{E}_x$ perpendicular to the silicon surface in the channel is much greater than the field $\mathscr{E}_y$ parallel

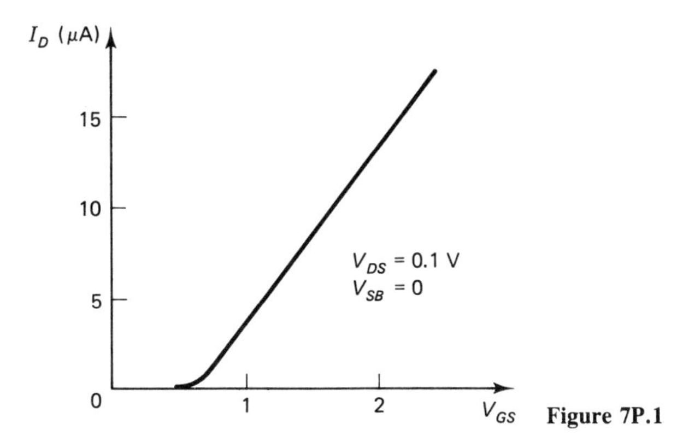

Figure 7P.1

to the surface. This question tests the accuracy of this assumption. If we assume that $\psi_s = 2\phi_B + V_{CB}(y)$ all along the silicon surface, then

$$|\mathscr{E}_y| = \frac{d\psi_s}{dy} = \frac{dV_{CB}}{dy}$$

But

$$I_D = \left| Z\mu_e Q_n \frac{dV_{CB}}{dy} \right|$$

so

$$|\mathscr{E}_y| = \frac{I_D}{Z\mu_e |Q_n|}$$

where Q_n is given by (7.34). $\mathscr{E}_x$ can be determined from the total charge Q_s stored in the silicon by applying Gauss's law:

$$\mathscr{E}_x = \frac{|Q_s|}{\epsilon_s} = \frac{C_{ox}(V_{GB} - V_{FB} - 2\phi_B - V_{CB}(y))}{\epsilon_s}$$

Using these results, compare the magnitudes of $\mathscr{E}_x$ and $\mathscr{E}_y$ for a "typical" MOSFET. Take $L = 5$ μm, $V_{FB} = 0$, $N_A = 4 \times 10^{15}$ cm^{-3}, $t_{ox} = 50$ nm, $V_{SB} = 0$, $V_{GS} = 5$ V, and $V_{DS} = V_{DS,sat}$. Consider channel voltages $V_{CB}(y)$ ranging from zero up to $V_{DS,sat}$. Note how the gradual channel approximation is undermined when the channel starts to pinch off.

7.4. The "body effect"—that is, the increase in V_T resulting from application of a source–substrate bias V_{SB}—is more pronounced for transistors with thick gate oxides and heavy substrate doping. To verify this, use the SPICE input file given in Example 7.2 to determine the change in V_T resulting from a change in V_{SB} from 0 V to 5 V for the following n-channel transistors:

(a) A device with $N_A = 10^{15}$ cm^{-3} and $t_{ox} = 50$ nm.
(b) A device with $N_A = 10^{16}$ cm^{-3} and $t_{ox} = 50$ nm.
(c) A device with $N_A = 10^{15}$ cm^{-3} and $t_{ox} = 100$ nm.

7.5. A MOSFET has been formed on a p-type substrate with doping concentration $N_A = 4 \times 10^{15}$ cm^{-3}. The gate oxide thickness is 50 nm, the gate is made of n^+

polysilicon, the oxide fixed charge $Q_f/q = 2 \times 10^{10}$ cm^{-2}, and a threshold adjust implant of 2×10^{11} boron ions/cm^2 has been done through the gate oxide. (For the purposes of this problem, assume that all the boron ions are electrically active and are concentrated at the Si–SiO_2 interface.) Take $Z/L = 1$ and $\mu_e = 800$ cm^{2}V^{-1}s^{-1}.

(a) Compute the threshold voltage for the MOSFET.

(b) Compute $V_{DS,\text{sat}}$ and the saturation drain current for the MOSFET using (1) the bulk charge model and (2) the square-law model. Take $V_{GS} = 5$ V. Comment on the accuracy of the square-law model.

7.6. Figure 7.31 shows a cross section through an inverter fabricated using a typical p-well CMOS process. Both the n- and p-channel transistors have a gate oxide thickness of 50 nm and n^+ polysilicon gates. The oxide fixed charge $Q_f/q = 2 \times 10^{10}$ cm^{-2}. The n-type substrate has a doping level $N_D = 2 \times 10^{15}$ cm^{-3}, while the acceptor concentration at the surface of the p-well is $N_A = 10^{16}$ cm^{-3} (for the purposes of this problem, the p-well can be taken as uniformly doped). In the following, assume that $V_{SB} = 0$; this is usually the case for transistors in CMOS circuits.

It is desired to shift the threshold voltages of the n- and p-channel transistors using ion implantation so that they are equal in magnitude, that is, so that $V_{Tn} = -V_{Tp}$. To simplify processing, a "blanket" implant is used, with all regions of the wafer receiving the same dose. Should boron or phosphorus be implanted? What implant dose is required?

7.7. A MOSFET with a gate oxide thickness of 40 nm is to be fabricated on a p-type substrate of doping $N_A = 10^{16}$ cm^{-3}. The source/drain regions have depth 0.4 μm and $N_D = 10^{19}$ cm^{-3}. The device is to be operated with $V_{SB} = 0$, but V_{DS} could be as large as 5 V. From (7.71), estimate the minimum channel length $L_{\min}$ this MOSFET can have while still exhibiting electrically long channel characteristics. A smaller value of $L_{\min}$ could be obtained if the substrate doping were increased. Suggest some undesirable side effects that such a move might have.

7.8. In Section 7.6.2 it was noted that the saturation of electron velocity can limit the drain current in a MOSFET with a channel length of 1-μm or less operating from a 5-V supply. Examine this effect by using SPICE to generate the I_D–V_{DS} characteristic of an n-channel MOSFET with $L = 1$ μm formed on a substrate of doping $N_A = 10^{16}$ cm^{-3} with a 50-nm gate oxide. Set $V_{GS} = 5$ V and sweep V_{DS} from 0 to 5 V, using the Level 2 model. First run a simulation in which VMAX is not specified, and then repeat the simulation with VMAX $= 8 \times 10^6$ cm s^{-1}, which corresponds roughly to the scatter-limited velocity for electrons in a MOSFET channel. The SPICE input file of Example 7.1 may be useful as a model.

7.9. Investigate how the transient response of the CMOS inverter examined in Example 7.7 changes as the capacitances associated with the source and drain floor and sidewall and with the overlap of the gate over the source and drain are removed from the device model. Comment on the importance of minimizing the source and drain junction capacitance in order to obtain fast switching in a CMOS circuit.

7.10. To make a CMOS inverter as small as possible, the n- and p-channel transistors should both be minimum-geometry devices—in other words, the length and width of these transistors should be set equal to the minimum feature size that can be resolved in a polysilicon film in the integrated circuit process in question (see Section 12.2.11). However, if this is done, the rise time of the inverter will be substantially greater than the fall time, due to the fact that the mobility of holes is about one-

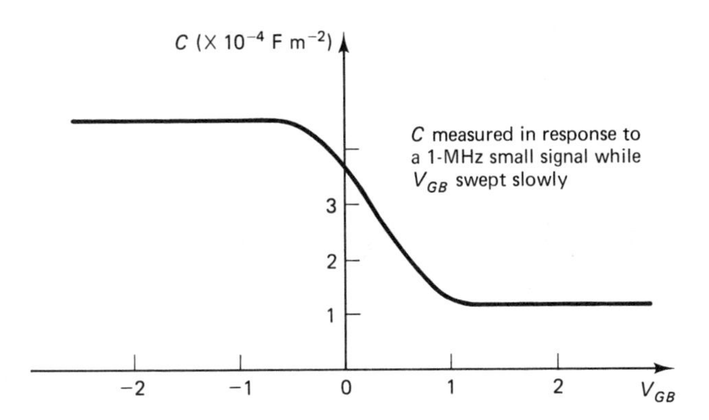

Figure 7P.2

third that of electrons. To verify this, run the SPICE input file given in Example 7.7, but reduce the width of the p-channel transistor from 15 μm to 5 μm.

7.11. The gate oxide layer in a MOS capacitor contains a charge density $\rho(x)$ due to contamination with Na^+ ions. Show that the contribution to the flatband voltage resulting from this charge is given by (7.65).

7.12. The C–V curve for a MOS capacitor formed on a uniformly doped substrate is shown in Fig. 7P.2.

(a) What is the thickness of the gate oxide in the capacitor?

(b) Is the substrate doped with donor or acceptor impurities? Explain how you can tell.

(c) What is the doping concentration in the substrate?

(d) What is the flatband voltage V_{FB} for the capacitor?

(e) Assuming that the oxide fixed charge $Q_f/q = 2 \times 10^{10}$ cm^{-2} and that all other oxide charges are negligible, is the gate of the capacitor made of n^+ or p^+ polysilicon?

7.13. A MOS capacitor has the structure shown in Fig. 7P.3. What bias V_{GB} must be applied to the capacitor to give a small-signal capacitance $C = 10^{-4}$ F m^{-2} measured at 1 MHz? Carry out your calculation assuming that there is no inversion layer at the silicon surface, but check this assumption when you have finished. Assume that $V_{FB} = 0$.

7.14. Starting from (7.132), show that when a MOS capacitor is operated with its surface in depletion or weak inversion, the small-signal capacitance is given as a function of gate voltage by

$$C = \frac{C_{ox}}{\sqrt{1 + 2C_{ox}^2(V_{GB} - V_{FB})/\epsilon_s q N_A}}$$

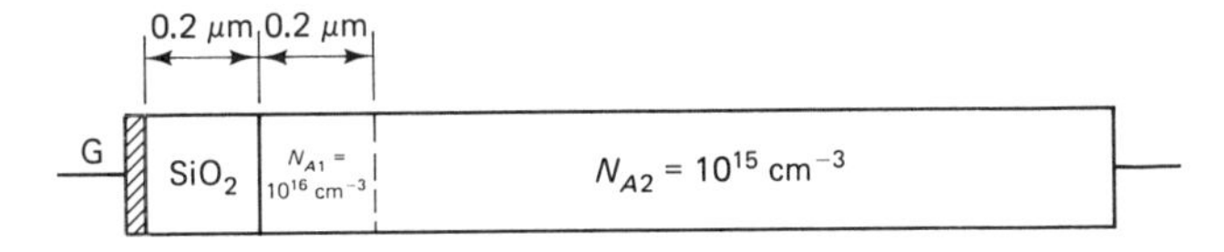

Figure 7P.3

8 Metal-Semiconductor Junction Diodes

8.1 INTRODUCTION

Metal-semiconductor junctions are both the simplest and oldest of semiconductor devices. In the early part of this century these junctions were widely used in "crystal" radios. The crystal in question was typically a piece of semiconducting lead sulfide contacted with a "cat's whisker"—a hard wire drawn to a fine point. The metal-semiconductor contact formed in this way could be made to exhibit rectifying or diode-like properties, and was used to demodulate AM radio signals.

Metal-semiconductor junctions have long been known by the alternative name of ***Schottky barriers*** or ***Schottky diodes*** in recognition of the German physicist Walter Schottky, who did much to develop the theory of these devices and investigate techniques for their fabrication in the 1930s.

8.2 STRUCTURE

The basic structure of a Schottky diode is shown in Fig. 8.1. The device can be formed very simply by the vacuum evaporation of a metal film onto a clean semiconductor surface, using the techniques discussed in Section 12.2.8. In forming Schottky diodes on silicon, it is now quite common to deposit a metal such as platinum which will react with silicon to form a ***silicide***. In the case of platinum, the reaction is

$$\mathrm{Pt} + \mathrm{Si} \longrightarrow \mathrm{PtSi}$$

This reaction proceeds quite rapidly at a temperature of approximately 500°C.

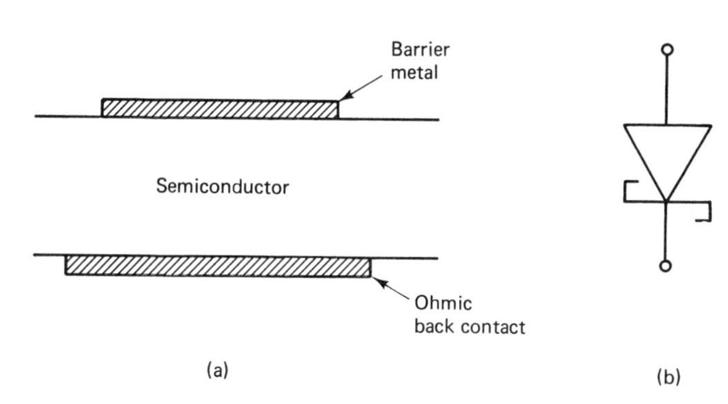

Figure 8.1 (a) Basic structure of Schottky diode. (b) Circuit symbol for a Schottky diode. The correspondence between (a) and (b) is correct for an *n*-type semiconductor. The diode symbol would be reversed for a *p*-type semiconductor.

Other examples of widely used silicides include Pd_2Si, $NiSi_2$, and $TiSi_2$. These compounds have the energy band structure characteristic of a metal and exhibit high electrical conductivity. The advantage of using a silicide in making a Schottky barrier is that the diode properties become insensitive to the cleaning procedure given the semiconductor surface prior to metal deposition, since a thin surface layer of silicon is consumed by the reaction that forms the silicide.

8.3 BARRIER FORMATION

Figure 8.2a shows the electron energy levels for a metal and an *n*-type semiconductor separated by vacuum. Although it is possible to form Schottky barriers on *p*-type semiconductors, these generally have low barrier heights and are consequently of limited interest. Only Schottky barriers with *n*-type substrates will be considered in the following. In the device of Fig. 8.2a an ohmic contact has been made to the semiconductor and connected to the metal so that the Fermi energy is the same in both materials. Here the work function of the semiconductor is less than that of the metal, so an electron has higher potential energy just outside the metal than just outside the semiconductor. An electric field therefore exists in the region between the metal and the semiconductor. This field points away from the semiconductor. When the metal and semiconductor are brought closer together as shown in Fig. 8.2b, the field becomes stronger.

To support the electric field in the gap between the metal and the semiconductor, there must be net charge present on the surfaces of both these solids. The charge on the semiconductor takes the form of uncompensated donor ions in a surface depletion region. The upward bending of the conduction band edge seen in Fig. 8.2b is associated with the depletion region; in fact, the band diagram closely resembles that of the *n*-type side of a p^+n junction diode. This region of band bending can be thought of as a potential energy barrier holding electrons in the bulk semiconductor.

If the metal and semiconductor are brought closer together, the field in the gap between them becomes more intense, and the band bending at the surface of

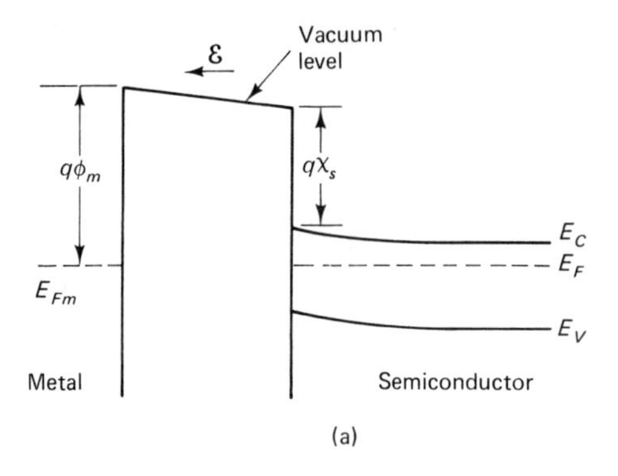

(a)

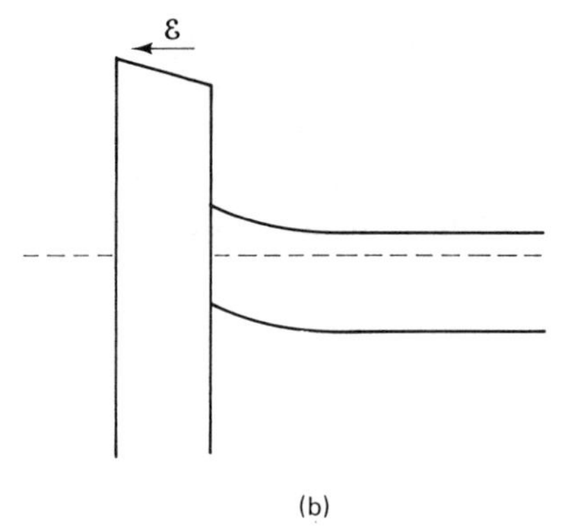

(b)

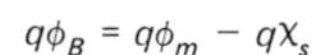

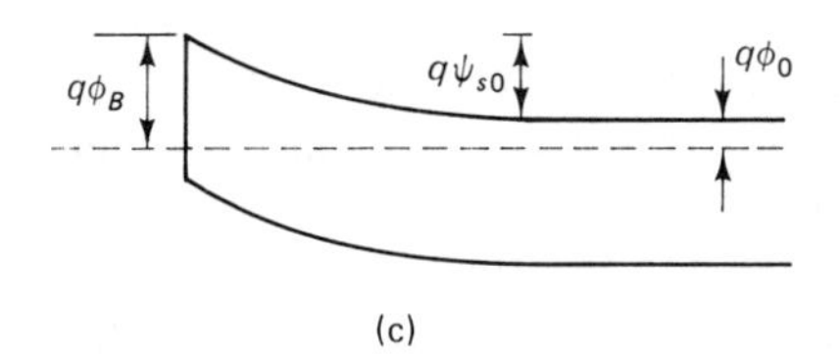

(c)

Figure 8.2 Formation of a Schottky barrier. As the metal and semiconductor are brought closer together, the band bending across the semiconductor increases.

the semiconductor more pronounced. When the two materials come into contact, the band diagram appears as shown in Fig. 8.2c.

One of the most important characteristics of a Schottky barrier is the ***barrier height*** $q\phi_B$, which is defined as the energy difference between the conduction band edge and the Fermi level at the semiconductor surface. From Fig. 8.2c, we expect that

$$\phi_B = \phi_m - \chi_s \tag{8.1}$$

where $q\phi_m$ is the metal work function and $q\chi_s$ the semiconductor electron affinity (see Section 7.2.1).

If the work function of the metal is very high, the bands may bend to such an extent that near the surface of the semiconductor the hole concentration becomes greater than the donor concentration, so that the surface is strongly inverted. This is a rather unlikely but not impossible situation; usually, the semiconductor surface in a Schottky diode is only weakly inverted.

Experiment shows that the picture of Schottky barrier formation presented above is somewhat oversimplified. Although it is found that metals with high work function such as platinum and gold tend to give high barrier heights while reactive, low-work function metals such as aluminum give low barriers, ϕ_B is not proportional to ϕ_m as (8.1) predicts. This is shown in Fig. 8.3, which plots ϕ_B as a function of ϕ_m for various metals on silicon and GaAs substrates. Even for a particular choice of barrier metal, the barrier height may depend quite strongly on the cleaning treatment the semiconductor surface receives prior to metal deposition. The values of ϕ_B given in Fig. 8.3 are for surfaces that were chemically cleaned before metal deposition; for Schottky barriers formed on atomically clean surfaces produced by cleaving crystals under ultrahigh-vacuum conditions, ϕ_B has even less dependence on ϕ_m.

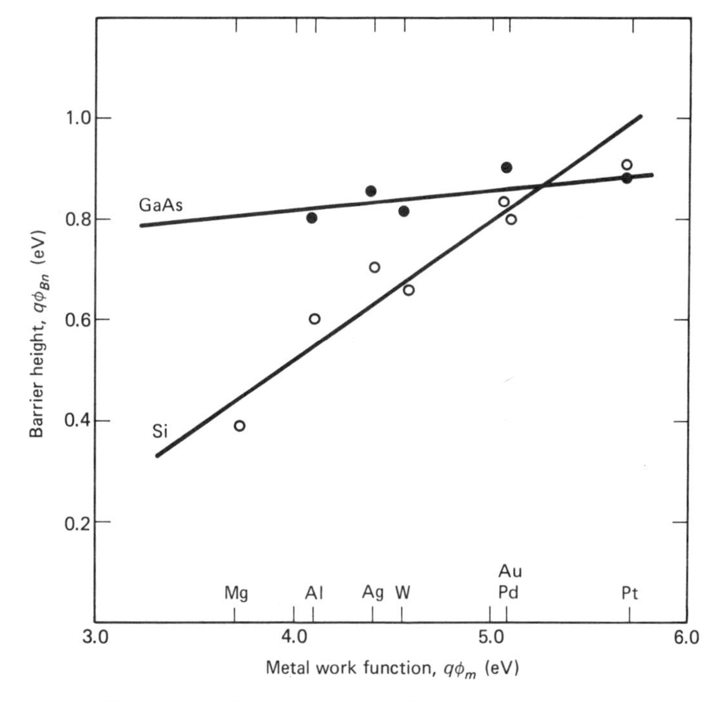

Figure 8.3 Schottky barrier height $q\phi_B$ as a function of metal work function $q\phi_m$ for diodes formed on silicon (open circle) and GaAs (filled circle) substrates. (From Cowley and Sze [1]. Reprinted with permission of the publishers, American Institute of Physics.)

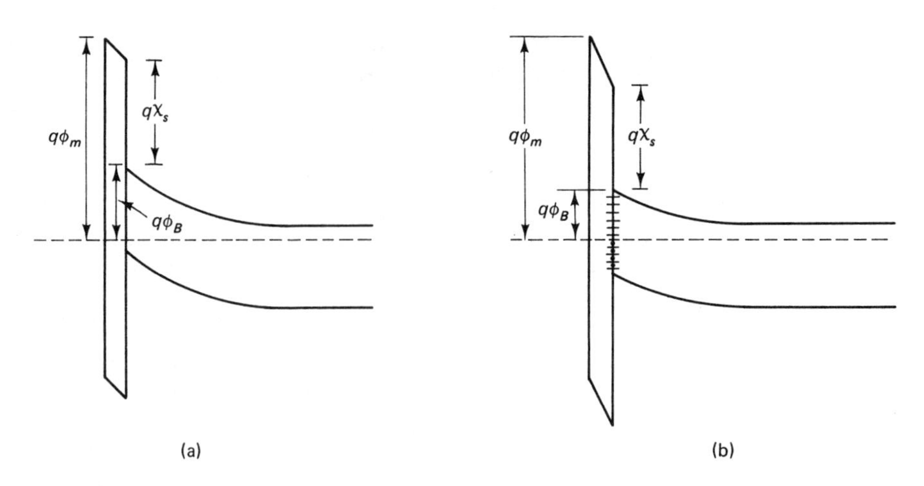

Figure 8.4 Band diagram for Schottky diode with thin interfacial oxide layer (a) without and (b) with interface trap states. The presence of interface traps reduces the band bending across the semiconductor and the Schottky barrier height.

The usual explanation put forward for ϕ_B not being proportional to ϕ_m is that charge stored in interface traps at the semiconductor surface (see Section 7.4.6) influences the barrier height. In this explanation the Schottky diode is viewed as being much like an MOS capacitor, with a very thin layer of oxide separating the metal and the semiconductor. Charge stored in interface traps screens the semiconductor from the influence of the metal work function, reducing both the band bending across the silicon and the Schottky barrier height, as illustrated in Fig. 8.4. This picture is not unreasonable for a Schottky diode formed by depositing metal on a semiconductor surface cleaned chemically and then exposed to the atmosphere, for most semiconductors grow a "native" oxide layer typically 1 to 2 nm thick within seconds of exposure to air. The interface trap density in such a structure is likely to be many orders of magnitude higher than that of a silicon MOS capacitor or MOSFET subjected to careful annealing procedures. The model breaks down, however, for atomically clean surfaces produced under ultrahigh-vacuum conditions, where it is known that there is no oxide separating the semiconductor and the metal. There is as yet no entirely satisfactory model explaining the relationship between ϕ_B and ϕ_m for such systems.

For the silicide Schottky barriers commonly used in silicon integrated circuits, it makes little sense to relate the barrier height to the work function of the silicide measured in isolation. Instead, the barrier heights for various silicide contacts are usually tabulated directly. The barrier heights for some of the more important silicides are given in Table 8.1. The high barrier height obtained with platinum silicide, which is widely used in integrated circuit fabrication, is worth noting.

TABLE 8.1 BARRIER HEIGHTS FOR SOME COMMONLY USED SILICIDE SCHOTTKY BARRIERS ON *n*-TYPE SILICON

Silicide	$q\phi_B$ (eV)
NiSi	0.66
PdSi	0.74
PtSi	0.89
$TaSi_2$	0.59
$TiSi_2$	0.60

8.4 DC ANALYSIS

8.4.1 Thermionic Emission Current

A theory predicting the steady-state current–voltage characteristic of the Schottky diode can be developed fairly quickly by applying the concepts introduced in modeling the *pn* junction in Chapter 6. Just as there is electron flow across the depletion region of a *pn* junction even in equilibrium, we expect that there will be electron flow between the metal and semiconductor in the Schottky diode even when there is no bias applied. On the metal side of the junction, an electron may occasionally acquire enough thermal energy to overcome the barrier $q\phi_B$ and, if it is moving in the correct direction, enter the conduction band of the semiconductor. Conversely, on the semiconductor side of the junction we know that there will be a few electrons in the conduction band near the metal-semiconductor interface. In fact, from (4.9) we can compute the surface electron concentration n_s as

$$n_s = N_C e^{-q\phi_B/kT} \tag{8.2}$$

since from Fig. 8.2c the separation between E_F and the conduction band edge at the surface is clearly $q\phi_B$. To a first approximation we might imagine that all electrons in the conduction band move with the same thermal velocity v_{th}. This is, of course, not really the case, since the electron velocity will generally increase with increasing energy above the conduction band edge, but this approximation will give us an acceptably accurate result for current in the diode. A more exact derivation of the electron flow equation which does not incorporate this assumption is given in Section 8.4.2. To determine the rate of electron flow into the metal, we need to find the average electron velocity component $\overline{v_m}$ directed toward the metal. Assuming that the electron velocities are randomly distributed in direction, from Fig. 8.5 $\overline{v_m}$ is given by

$$\overline{v}_m = \int_{\theta=0}^{\pi/2} \frac{(2\pi v_{\text{th}} \sin\theta)(v_{\text{th}}\cos\theta)v_{\text{th}}\, d\theta}{4\pi v_{\text{th}}^2} = \frac{v_{\text{th}}}{2}\int_0^{\pi/2} \sin\theta\cos\theta\, d\theta = \frac{v_{\text{th}}}{4} \tag{8.3}$$

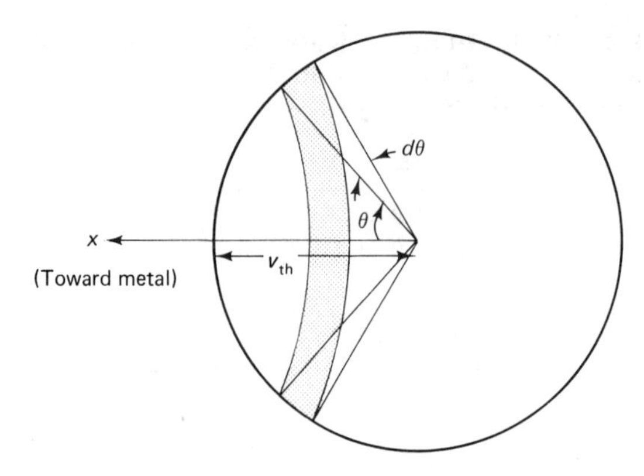

Figure 8.5 Computing the average electron velocity directed toward the metal. If electron velocities are randomly distributed in direction, the fraction of electrons with velocities in the range θ to θ + dθ is given by

$$\frac{2\pi v_{th} \sin\theta\, v_{th}\, d\theta}{4\pi v_{th}^2} = \frac{\sin\theta\, d\theta}{2}$$

These electrons have a velocity component $v_{th} \cos\theta$ directed toward the metal.

The current carried by electrons flowing from the conduction band to the metal is then given by

$$J_{C\to M} = qn_s\bar{v}_m = \frac{qv_{th}}{4} N_C e^{-q\phi_B/kT} \tag{8.4}$$

In equilibrium, this current must be balanced by an equal and opposite flow of electrons from the metal to the semiconductor conduction band.

Figure 8.6 shows how the energy band diagram of the Schottky diode changes when a bias is applied between the semiconductor and the metal. Applying a negative voltage to the semiconductor raises the Fermi energy on the semiconductor side of the junction, reducing the band bending across the semiconductor.

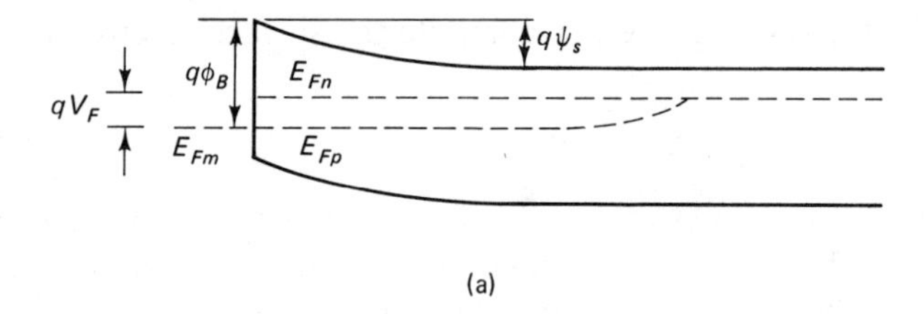

(a)

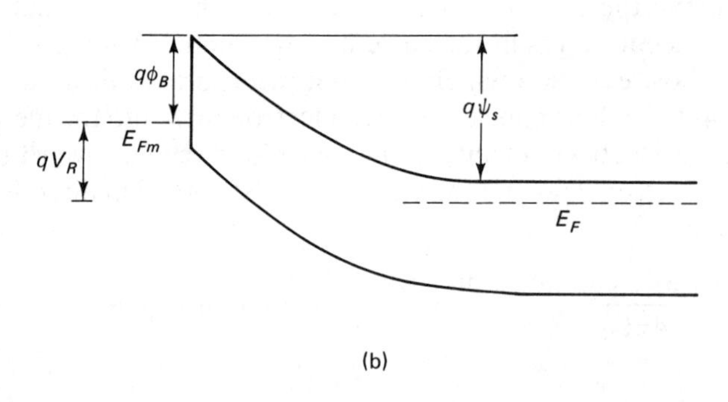

(b)

Figure 8.6 Schottky diode band diagrams for (a) forward and (b) reverse bias. *Note:* With respect to the semiconductor, V_F is a positive voltage and V_R is a negative voltage.

We expect that this reduced band bending will result in an increased electron concentration at the semiconductor surface. If there are more electrons at the surface, it seems reasonable that there will be an enhanced flow of electrons from the semiconductor into the metal. Therefore, when a negative voltage is applied to the semiconductor, there should be a large current in the diode. Conversely, application of a positive voltage to the semiconductor reduces the electron concentration at the surface, effectively shutting down the flow of electrons from the conduction band to the metal. In this situation the energy barrier faced by an electron attempting to cross from the metal to the semiconductor conduction band remains unchanged at $q\phi_B$, so the electron flow from the metal to the semiconductor should remain fixed at its equilibrium value.

In summary, then, for a Schottky diode formed on an n-type semiconductor, a large current exists when a negative bias is applied to the semiconductor, while a relatively small leakage current exists if the semiconductor is made positive relative to the metal. This situation is analogous to that of a p^+n diode, with the p^+ layer replaced by the metal. A positive voltage on the metal relative to the semiconductor constitutes forward bias, while a negative voltage on the metal places the diode in reverse bias.

The description of electron flow in the Schottky diode given above can be made quantitative with relatively little extra work. In reverse bias we expect the electron flow from the metal to the semiconductor to remain at its equilibrium value of $(qv_{\text{th}}N_C/4)e^{-q\phi_B/kT}$, while there is effectively no electron flow from the semiconductor to the metal. The reverse leakage current in the Schottky diode should therefore be given by

$$J_{M\to C} = \frac{-qv_{\text{th}}N_C}{4}e^{-q\phi_B/kT} \tag{8.5}$$

(Here the usual convention that reverse current is negative has been followed.) When a forward bias V is applied to the diode, Fig. 8.6 shows that the band bending across the semiconductor is reduced by an amount qV. In the analysis of the pn junction, a condition of quasi-equilibrium was assumed to hold across the depletion region. More specifically, it was assumed that E_{Fn} and E_{Fp} were effectively flat across the depletion region. If the same assumption is made regarding E_{Fn} in the Schottky diode, application of a forward bias V brings the conduction band edge closer to E_{Fn} at the semiconductor surface by an amount qV compared to equilibrium. This in turn implies that the surface electron concentration is increased by a factor $e^{qV/kT}$, so that the electron flow from the conduction band to the metal should now be given by

$$J_{C\to M} = \frac{qv_{\text{th}}N_C}{4}e^{-q\phi_B/kT}e^{qV/kT} \tag{8.6}$$

Accounting for the electron flow from the metal to the conduction band which should still have the same magnitude as in equilibrium, we arrive at the following expression for the net electron flow between the conduction band and the metal:

$$J_n = J_{C \to M} + J_{M \to C} = \frac{q v_{\text{th}} N_C}{4} e^{-q\phi_B/kT}(e^{qV/kT} - 1) \tag{8.7}$$

A moment's consideration leads us to conclude this equation applies in reverse bias as well as forward bias.

From statistical thermodynamics, we expect that the electron's thermal energy should be proportional to kT. Since classically kinetic energy is $mv^2/2$, we expect that $v_{\text{th}} \propto (kT)^{1/2}$ [see (3.1)]. Also, from Section 4.2, we know that $N_C \propto T^{3/2}$. Thus the product $v_{\text{th}} N_C$ should be proportional to T^2. In consequence, (8.7) is frequently rewritten in the form

$$J_n = J_{0n}(e^{qV/kT} - 1) \tag{8.8a}$$

where the ***saturation current density*** J_{0n} is given by

$$J_{0n} = A^* T^2 e^{-q\phi_B/kT} \tag{8.8b}$$

The factor A^* is termed the ***effective Richardson constant*** and is independent of temperature. The value of A^* depends on the particular semiconductor under investigation, and also varies with crystal orientation. For n-type silicon, A^* is approximately 250 A cm^{-2} K^{-1}, while for n-type GaAs A^* is roughly 8.1 A cm^{-2} K^{-1}.

At this point a quick calculation of the current densities expected in a Schottky diode is useful. In the manufacture of silicon integrated circuits, Schottky barriers are very commonly made using platinum silicide as a barrier metal. From Table 8.1, the barrier height $q\phi_B$ for such a diode is approximately 0.89 eV, so from (8.8b) the saturation current density J_{0n} is about 3×10^{-8} A cm^{-2}. For comparison purposes, the saturation current density of the typical n^+p-junction diode with a substrate doping of 10^{17} cm^{-3} analyzed in Example 6.1 is about 3×10^{-12} A cm^{-2}—four orders of magnitude smaller. Platinum silicide forms a junction with a relatively high barrier height; for metals with lower work function the difference in saturation current density would be even more extreme. We can therefore conclude that at a given forward bias a typical Schottky diode carries a far higher current density than a typical pn-junction diode. Expressed in somewhat different terms, it is frequently said that the ***turn-on*** or ***cut-in*** voltage of a Schottky diode is typically about 250 mV less than that of a pn junction. For example, the current density in the Schottky diode described above reaches a value of 1 A cm^{-2} at a forward bias of 450 mV, while a 700-mV forward bias must be applied to the pn junction to give the same current. Naturally, the high saturation current of the Schottky diode also translates into a higher reverse leakage than is found in the pn junction.

It should be noted that the assumption of quasi-equilibrium for carriers in the depletion region is much harder to justify for the case of a Schottky diode than for a pn junction. It is true that as long as the electron flow across the depletion region is not excessively large, we can expect E_{Fn} will be constant across this region just as in a *pn* diode. The problem arises in the immediate vicinity of the semiconductor surface where, in a forward-biased Schottky diode, we expect that there will be a large flow of electrons from the conduction band to the metal, but almost no return electron flow from the metal to the conduction band. This hardly represents a near-equilibrium situation, for it seems clear that the semiconductor surface will become depleted of electrons with velocities directed toward the metal. This difficulty with the derivation of (8.8) can be avoided if, on moving back from the semiconductor surface, the energy of the conduction band edge drops by an amount large compared to the thermal energy kT in a distance comparable to the mean free path λ that electrons travel between collisions (in silicon, λ is approximately 10 nm). This situation is illustrated in Fig. 8.7. In this case, the electron velocity distribution should be close to its equilibrium form to within a distance λ of the interface, since only a few energetic electrons will be able to traverse the remaining distance and enter the metal. Equation (8.4) still applies, but only to those electrons with sufficient energy to overcome the peak of the barrier. A process of this kind in which only a few very energetic electrons from a large pool are capable of passing over a barrier is frequently referred to as ***thermionic emission***; hence the electron flow from the conduction band to the metal is often termed a ***thermionic emission current***.

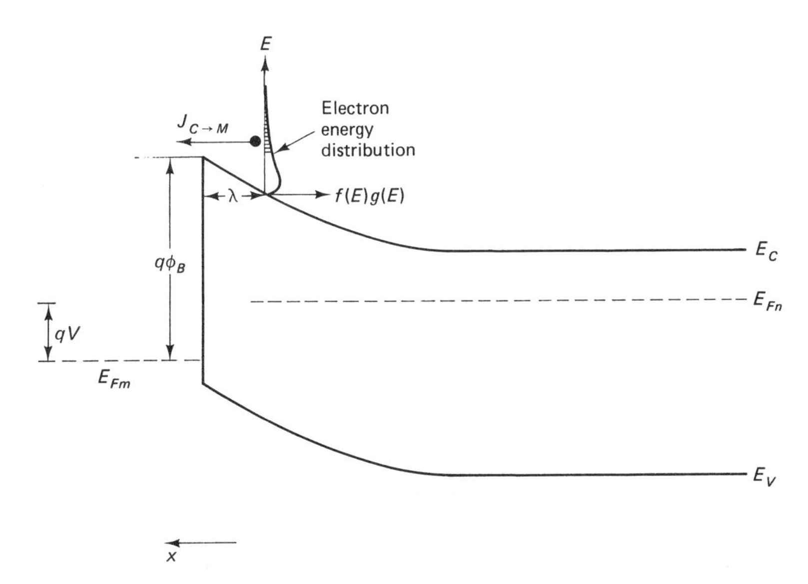

Figure 8.7 Band diagram for forward-biased Schottky diode illustrating the thermionic emission of electrons over the peak of the barrier into the metal. The electrons are assumed to have an equilibrium energy distribution to within a mean free path λ of the interface. Only a few energetic electrons can penetrate the tip of the barrier from this point and enter the metal.

8.4.2 A More Accurate Expression for the Thermionic Emission Current

In developing equation (8.8), it was assumed that every electron contributing to the flow from the semiconductor to the metal moves with the same velocity v_{th}. In this section the derivation of (8.8) is repeated without this assumption. The result obtained is basically the same—in particular, (8.8) still holds—so this section can be skipped without loss of continuity.

In Section 5.4 it was found that an electron's velocity is related to its energy level in the semiconductor band structure by

$$\mathbf{v} = \frac{1}{\hbar}\left(\frac{\partial E}{\partial k_x}, \frac{\partial E}{\partial k_y}, \frac{\partial E}{\partial k_z}\right) \tag{8.9}$$

For simplicity, we will assume here that the x-axis is normal to the metal-semiconductor interface, so that the velocity component of interest in computing the electron flow into the metal is given by

$$v_x = \frac{1}{\hbar}\frac{\partial E}{\partial k_x} \tag{8.10}$$

In Section 5.7 it was noted that in many semiconductors the energy–wave vector relationship is parabolic near the bottom of the conduction band. In this case

$$E = \frac{\hbar^2}{2m^*}(k_x^2 + k_y^2 + k_z^2) \tag{8.11}$$

and

$$v_x = \frac{\hbar k_x}{m^*} \tag{8.12}$$

where m^* is the effective mass of a conduction band electron.

In Section 5.7 it was found that the density of allowed electron energy states per unit volume of crystal per unit volume of (k_x, k_y, k_z) space is $2/(8\pi)^3$. If we let $f(E)$ denote, as usual, the probability that at state at energy E is occupied by an electron, we have

$$J_{C\to M} = q\int_{k_x=0}^{\infty}\int_{k_y=-\infty}^{\infty}\int_{k_z=-\infty}^{\infty}\frac{\hbar k_x}{m^*}f(E)\frac{2}{(2\pi)^3}\,dk_x\,dk_y\,dk_z \tag{8.13}$$

For electrons having sufficient energy to enter the metal, it should be possible to use a Boltzmann approximation to the state occupancy function $f(E)$. In this case

$$J_{C\to M} = q\int_{k_x=0}^{\infty}\int_{k_y=-\infty}^{\infty}\int_{k_z=-\infty}^{\infty}\frac{\hbar k_x}{m^*}e^{-[E_c+(\hbar^2/2m^*)(k_x^2+k_y^2+k_z^2)-E_F]/kT}$$

$$\times\frac{2}{(2\pi)^3}\,dk_x\,dk_y\,dk_z$$

$$= \frac{2qe^{-(E_C - E_F)/kT}}{(2\pi)^3} \int_0^{\infty} \frac{\hbar k_x}{m^*} e^{-\hbar^2 k_x^2/2m^*kT} dk_x \int_{-\infty}^{\infty} e^{-\hbar^2 k_y^2/2m^*kT} dk_y$$

$$\times \int_{-\infty}^{\infty} e^{-\hbar^2 k_z^2/2m^*kT} dk_z \tag{8.14}$$

The leftmost integral appearing in (8.14) can be evaluated by inspection. From tables of definite integrals, we find that

$$\int_{-\infty}^{\infty} e^{-\hbar^2 k_y^2/2m^*kT} dk_y = \frac{\sqrt{2\pi m^* kT}}{\hbar} \tag{8.15}$$

Therefore,

$$J_{C \to M} = \frac{q4\pi m^* k^2 T^2}{h^3} e^{-q\phi_B/kT} e^{qV/kT} \tag{8.16}$$

Subtracting the electron flow from the metal to the semiconductor conduction band from $J_{C \to M}$ leads to the following equation for the thermionic emission current:

$$J_n = \frac{q4\pi m^* k^2 T^2}{h^3} e^{-q\phi_B/kT}(e^{qV/kT} - 1) \tag{8.17}$$

This has the same form as equation (8.8), with the effective Richardson constant related to the electron effective mass by

$$A^* = \frac{q4\pi m^* k^2}{h^3} \tag{8.18}$$

8.4.3 Minority Carrier Injection

So far in our analysis of the Schottky diode we have made no allowance for electron flow between the metal and the valence band in the semiconductor. This is surely an oversight, for the bending of the bands in Fig. 8.6 suggests that there will be many holes present at the semiconductor surface. Since holes are really just vacant states in the valence band, there is plenty of opportunity for electron exchange between the valence band and the metal. When an electron leaves the valence band and enters the metal, a hole is injected into the semiconductor.

In a forward-biased p^+n-junction diode, we expect that the hole quasi-Fermi energy level E_{Fp} will be nearly flat across the depletion region at a value matching that of E_{Fp} on the p^+ side of the junction. Similarly, E_{Fp} should be effectively flat across the depletion region of the Schottky barrier, and should align with E_{Fm}, the Fermi level in the metal. This situation is illustrated in Fig. 8.6a. The reason for this alignment of Fermi levels is that the exchange of electrons between the valence band and the metal is a relatively easy process, while the subsequent diffusion of holes into the bulk semiconductor and their recombination with electrons is relatively slow. In other words, the metal can easily supply holes to the semiconductor, while the bottleneck for the motion of these injected holes is the process of diffusion and recombination in the substrate. Making this argument more quantitative, we expect that the hole current injected into the semiconductor in a forward-biased Schottky diode should be exactly the same as that in a p^+n-junction diode subjected to the same bias. That is, just inside the semiconductor we should have

$$J_p = \frac{qD_h n_i^2}{L_h N_D}(e^{qV/kT} - 1) \tag{8.19}$$

As usual, the total current in the diode is given by the sum of J_n and J_p at one plane.

It is frequently stated that Schottky diodes are majority carrier devices, in other words, that the forward-bias current in these devices is dominated by electron emission from the semiconductor into the metal rather than by hole injection into the semiconductor. This statement may seem paradoxical in view of the conclusion drawn above that the hole current in a forward-biased Schottky diode is the same as that which would exist in a p^+n-junction diode formed on an identical substrate. The apparent paradox can be resolved by noting that, as shown in Section 8.4.1, the electron current in the Schottky diode is orders of magnitude higher than that which would exist in the p^+n junction. In consequence, the minority carrier injection ratio—specifically, the ratio of minority carrier hole current crossing the metal-semiconductor interface to the total current—is very low in the Schottky diode. This has important consequences for the switching speed of the diode, as we shall see shortly.

8.5 JUNCTION BREAKDOWN

Reverse-bias breakdown in Schottky diodes formed on lightly doped substrates occurs through avalanching in high-field regions, just as in a *pn*-junction diode. However, Schottky diodes tend to have lower breakdown voltages than *pn* junctions unless special precautions are taken in their fabrication. In a junction diode, avalanching is likely to occur first near the edge of the junction where the electric field is highest. The deeper the metallurgical junction, the more rounded its edges will generally be and the higher the breakdown voltage. (As usual in electrostatic problems, sharp corners tend to be associated with high electric fields.) In a Schottky diode, the metal-semiconductor interface effectively replaces the metallurgical junction. In consequence, the Schottky diode has essentially zero junction depth, and so has very sharp edges and a correspondingly low breakdown voltage.

One way of improving the breakdown properties of the Schottky diode is to form a p^+ guard ring around the edge of the diode by diffusion or ion implan-

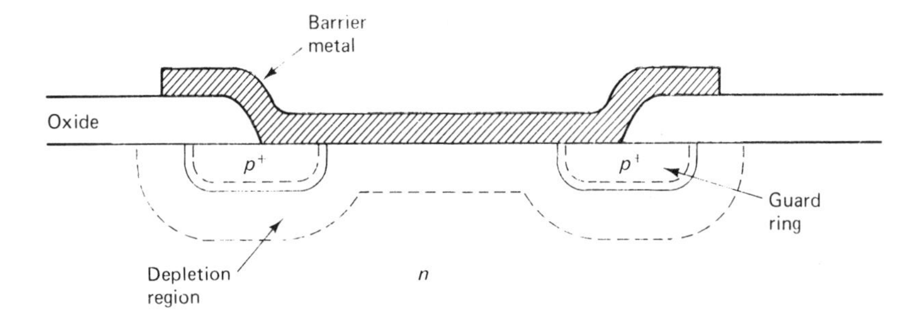

Figure 8.8 Structure of Schottky diode with p^+ guard ring. Note the gentle rounding of the depletion region edge.

tation, as illustrated in Fig. 8.8. This provides a much more rounded corner to the edge of the junction, reducing the electric field in that region.

For Schottky diodes formed on more heavily doped substrates, breakdown may occur through tunneling. Again, the situation is quite comparable to that of Zener breakdown in a *pn* junction. If the substrate is so heavily doped that the depletion region width is less than about 5 nm at a reverse bias too small to produce avalanche breakdown, electrons will be able to tunnel easily between states near E_{Fm} in the metal and the semiconductor conduction band, as illustrated in Fig. 8.9. This effect is frequently termed ***field emission***. An intermediate situation may also be encountered in which the depletion region is too wide to be penetrated by electrons with energies near E_{Fm}, but in which energetic electrons may be able to tunnel through the peak of the barrier. This is also shown in Fig. 8.9. This process is referred to as ***thermionic field emission*** since, although it involves tunneling, electrons must acquire a substantial amount of thermal energy to participate in it. The thermionic field emission current is therefore strongly temperature dependent, while a pure field emission current has relatively little temperature dependence.

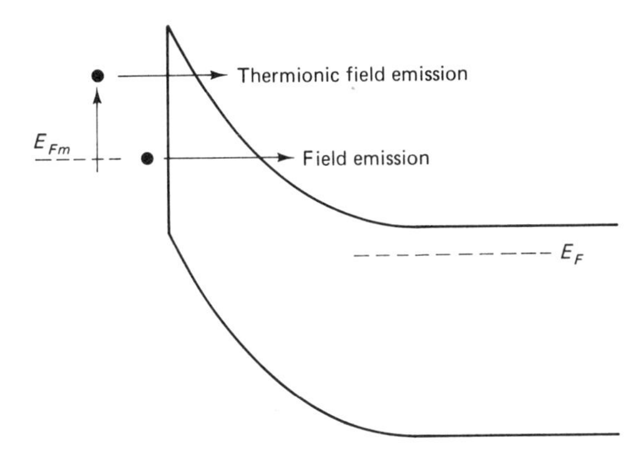

Figure 8.9 Tunneling through the depletion region in a Schottky diode can give rise to field emission or thermionic field emission currents.

8.6 OHMIC CONTACTS

Ohmic contacts constitute a special class of metal-semiconductor junctions. A good ohmic contact has a linear current–voltage characteristic, so that it does not rectify an applied signal, and also possesses a very low resistance. Noting that the resistance of a contact should scale in inverse proportion to area, an appropriate measure of contact quality is the product of resistance and area. This product is known as the ***contact resistivity*** ρ_c. A contact intended for use in a modern VLSI integrated circuit must have a resistivity of no more than 10^{-6} $\Omega\cdot\text{cm}^2$.

In principle, an ohmic contact can be formed by depositing a low-work-function metal on an n-type substrate or a high-work-function metal on a p-type substrate to produce a Schottky barrier with a very low height. In Problem 8.2 it is found that the barrier height $q\phi_B$ must be no more than 0.15 eV to give a contact resistivity of 10^{-6} $\Omega\cdot\text{cm}^2$. In practice, the fact that ϕ_B is not as sensitive to ϕ_m as simple theory predicts makes it difficult if not impossible to obtain a barrier this low. For this reason, the usual approach taken in forming an ohmic contact is to dope the semiconductor surface very heavily. When a metal-semiconductor junction is subsequently formed on the heavily doped region, electrons are able to tunnel through the very narrow depletion region in either forward or reverse bias, giving a very low contact resistivity.

8.7 TRANSIENT ANALYSIS

Since there is very little injection, and hence storage, of minority carriers in the base of a Schottky diode, the transient response is dominated almost entirely by the junction capacitance. In fact, in modeling transient response it is convenient simply to view the Schottky diode as a pn junction with no minority carrier charge storage effects. Just as for any other junction, the capacitance associated with the depletion region in the Schottky diode is given by

$$C = \frac{\epsilon_s}{W} \tag{8.20}$$

where W is the depletion width. As usual, W is related to the band bending ψ_s across the silicon by

$$W = \sqrt{\frac{2\epsilon_s \psi_s}{qN_D}} \tag{8.21}$$

From Fig. 8.6, $\psi_s = \psi_{s0} - V$, where ψ_{s0} is the value of ψ_s in equilibrium. (ψ_{s0} is directly analogous to the built-in potential V_{bi} defined for a pn junction.) From

Fig. 8.2c, we see that

$$\psi_{s0} = \phi_B - \phi_0 \tag{8.22}$$

where $\phi_0 = kT/q \ln (N_C/N_D)$. ($\phi_0$ is the separation between the conduction band edge and the Fermi level deep in the substrate.)

The standard versions of SPICE contain no special model for the Schottky diode. However, a Schottky diode can be described quite accurately using the *pn* junction model by setting the minority carrier storage time TT to zero, specifying a value of saturation current IS appropriate to the diode's barrier height and area, setting the built-in voltage VJ to ψ_{s0}, and specifying an appropriate value for the zero-bias junction capacitance. The diode ideality factor N would ordinarily be set to 1 and the grading coefficient M to 0.5.

In Example 6.4 the reverse recovery transient for a *pn*-junction diode was simulated. It was found that charge continued to flow in the diode for several hundred nanoseconds after the driving voltage was switched from the forward to the reverse direction. Example 8.1 repeats this problem, but with parameters chosen to represent a Schottky diode with a PtSi barrier. In this case it is found that current through the diode shuts down almost immediately after the drive voltage changes polarity.

Example 8.1 Schottky Diode Reverse Recovery Transient

Figure 8E.1 shows the circuit used to study the reverse recovery transient of a *pn*-junction diode in Example 6.4. In this example, the simulation of the reverse recovery transient is repeated with the junction diode replaced by a PtSi Schottky diode with a barrier height of 0.89 eV. To simulate the Schottky diode the minority carrier storage time TT is set to zero in the SPICE file, and the built-in potential VJ and zero-bias capacitance CJO set

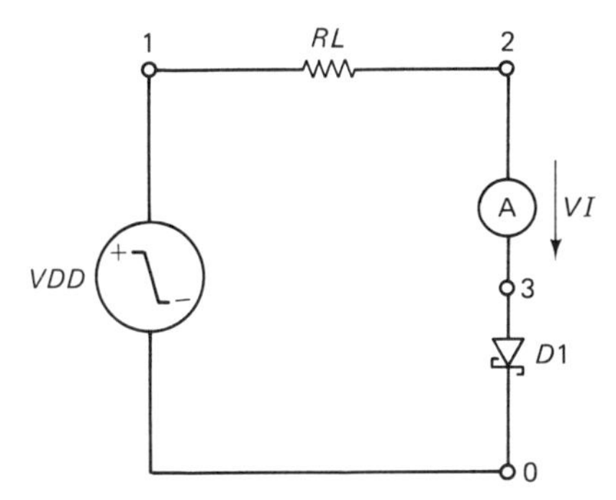

Figure 8E.1 Circuit used to simulate reverse recovery transient of a Schottky diode.

to values appropriate to the diode's barrier height. The actual SPICE input file is listed below.

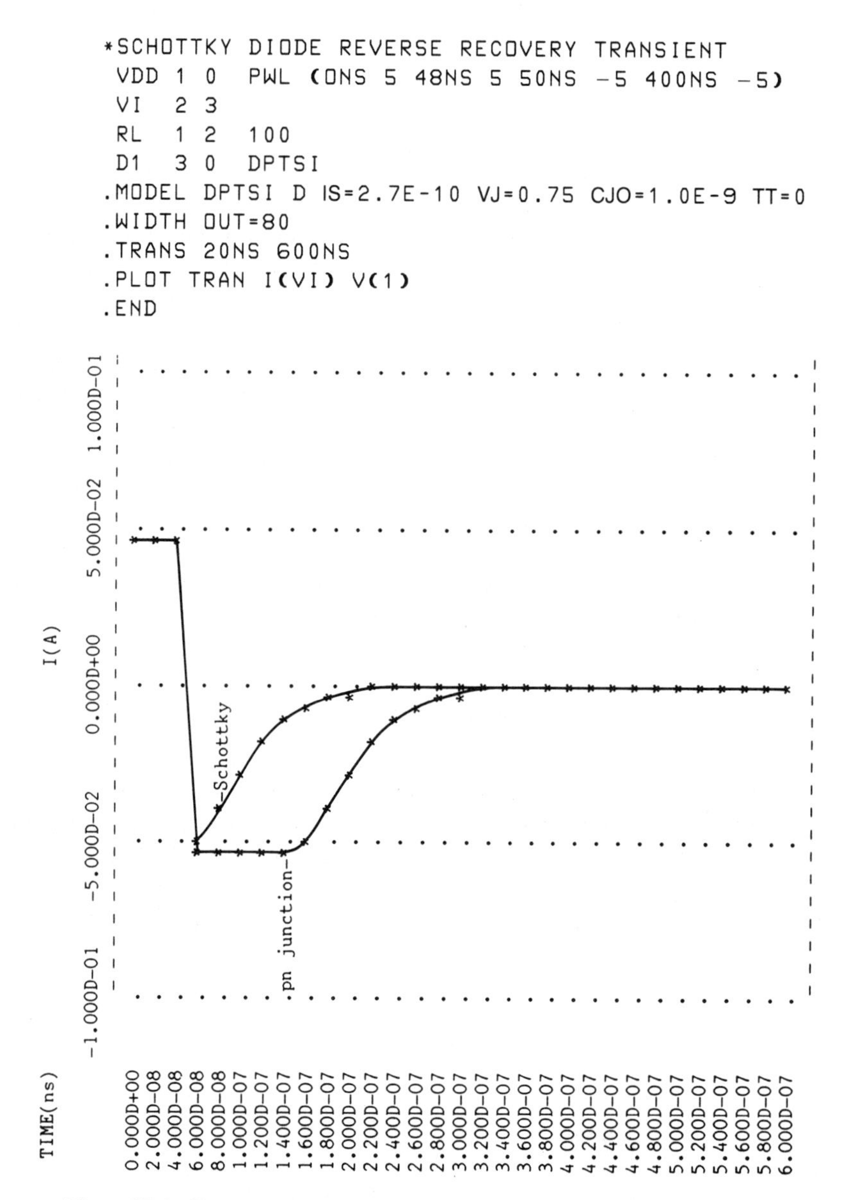

```
*SCHOTTKY DIODE REVERSE RECOVERY TRANSIENT
 VDD 1 0  PWL (0NS 5 48NS 5 50NS -5 400NS -5)
 VI  2 3
 RL  1 2  100
 D1  3 0  DPTSI
.MODEL DPTSI D IS=2.7E-10 VJ=0.75 CJO=1.0E-9 TT=0
.WIDTH OUT=80
.TRANS 20NS 600NS
.PLOT TRAN I(VI) V(1)
.END
```


Figure 8E.2 Reverse recovery transients of Schottky and *pn*-junction diodes.

Figure 8E.2 shows the reverse recovery transient for the Schottky diode and for the junction diode of Example 6.4. As expected, the current in the Schottky diode tends to zero as soon as the drive voltage is switched; there is no delay time associated with the storage of minority carriers in the diode base.

8.8 AC ANALYSIS

Minority carrier charge storage effects in the Schottky diode can generally be ignored from the viewpoint of small-signal ac response as well as large-signal transient analysis. The small-signal equivalent circuit of the diode therefore simply contains the junction capacitance in parallel with a resistor representing the differential junction conductance, as shown in Fig. 8.10.

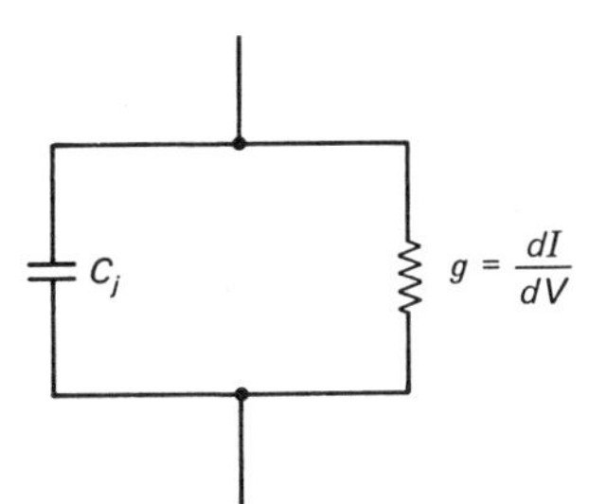

Figure 8.10 Small-signal equivalent circuit for a Schottky diode. C_j is the capacitance associated with the junction depletion region.

8.9 APPLICATIONS OF SCHOTTKY DIODES

Today it might be argued that the most important application of Schottky barriers is in forming the gates for metal-semiconductor field-effect transistors (MESFETs) used in high-speed GaAs integrated circuits. These devices are considered in detail in Chapter 9. However, Schottky diodes have other important and more traditional applications. In Example 11.5, the effect of minority carrier storage in the base on the switching speed of a simple bipolar transistor inverter circuit with a resistive load is examined. It is found that if the device is driven into saturation when the input is high, a considerable time elapses before the collector current drops and the output voltage rises when the input is subsequently switched low. This minority carrier storage delay is equally important in more sophisticated bipolar logic circuits such as those used in the transistor-transistor logic (*TTL*) family. One effective way of avoiding minority carrier storage in the base is to shunt the base–collector junction with a Schottky diode, as shown in the circuit of Fig. 8.11. Now when the transistor is driven into saturation (base–collector junction

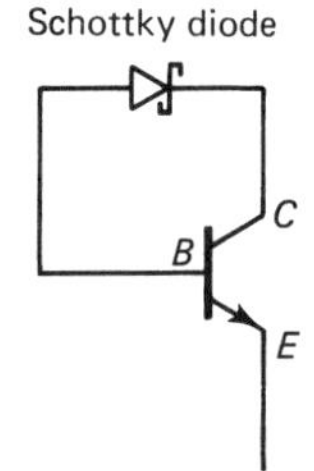

Figure 8.11 Bipolar transistor with Schottky diode clamp between the collector and the base.

forward biased) by applying a large input voltage to the inverter, the Schottky diode with its low turn-on voltage begins to conduct before the base–collector *pn* junction. This virtually eliminates minority carrier injection from the collector into the base, greatly reducing the minority carrier charge stored in the base. This technique is used in the low-power, fast-switching variant of TTL known as ***Schottky TTL*** or ***LSTTL***. The advantages of using a Schottky-clamped switching transistor are illustrated in Example 8.2. This example will be better appreciated after studying the ***RTL*** (resistor-transistor logic) inverter considered in Example 11.5.

Fabrication of a Schottky-clamped bipolar transistor is relatively easy. Usually, all that is necessary is to allow the Schottky barrier metal to overlap both the lightly doped *n*-type epitaxial collector and the more heavily doped *p*-type base. The junction between the metal and the heavily doped base is likely to have near-ohmic characteristics.

Example 8.2 Schottky Diode Clamped BJT Inverter

Figure 8E.3 shows the simple resistor-transitor-logic (RTL) inverter of Example 11.5 modified to include a Schottky clamp diode between the base and collector. We will take the area of the clamp diode as 5 μm × 5 μm, which is one-fourth of the area of the base–emitter junction. The saturation current IS of the diode is therefore $(5 \times 10^{-4}\text{ cm})^2(3 \times 10^{-8}\text{ A cm}^{-2}) = 7.5 \times 10^{-15}$ A. The SPICE file needed to analyze the turn-off transient of this circuit is listed below. It is instructive to run two simulations, one with the Schottky clamp diode, and the other with the clamp diode removed.

```
*BJT SWITCHING CHARACTERISTICS
*TURN-OFF TRANSIENT
*WITH SCHOTTKY CLAMP
```

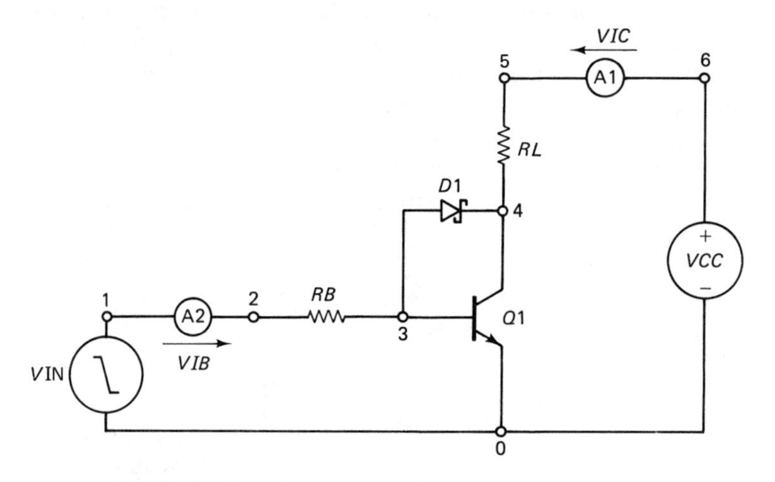

Figure 8E.3 Circuit used to study switching transient of RTL inverter.

```
VCC 6 0 DC 3
VIC 6 5
VIB 1 2
VIN 1 0 PWL(0NS 3 5NS 3 6NS 0 100NS 0)
RL  5 4 2K
RB  3 2 1K
D1  3 4 DPTSI
Q1  4 3 0 Q172
.MODEL Q172 NPN BF=122 BR=0.2 IS=2.65E-17
+CJE=9.37E-14 VJE=0.94 TF=1.4E-9
+CJC=1.09E-14 VJC=0.7 TR=1E-8
.MODEL DPTSI D IS=7.5E-15 VJ=0.75 CJO=2.6E-14
.WIDTH OUT=80
.TRAN 0.5NS 15NS
.PLOT TRAN V(1) V(4) (0,4)
.END
```

```
V(1)
0.000D+00  1.000D+00  2.000D+00  3.000D+00  4.000D+00

TIME(ns)
0.000D+00
5.000D-10
1.000D-09
1.500D-09
2.000D-09
2.500D-09
3.000D-09
3.500D-09
4.000D-09
4.500D-09
5.000D-09
5.500D-09
6.000D-09
6.500D-09
7.000D-09
7.500D-09
8.000D-09
8.500D-09
9.000D-09
9.500D-09
1.000D-08
1.050D-08
1.100D-08
1.150D-08
1.200D-08
1.250D-08
1.300D-08
1.350D-08
1.400D-08
1.450D-08
1.500D-08
```

Figure 8E.4 Switching transient of RTL inverter with and without Schottky diode clamp.

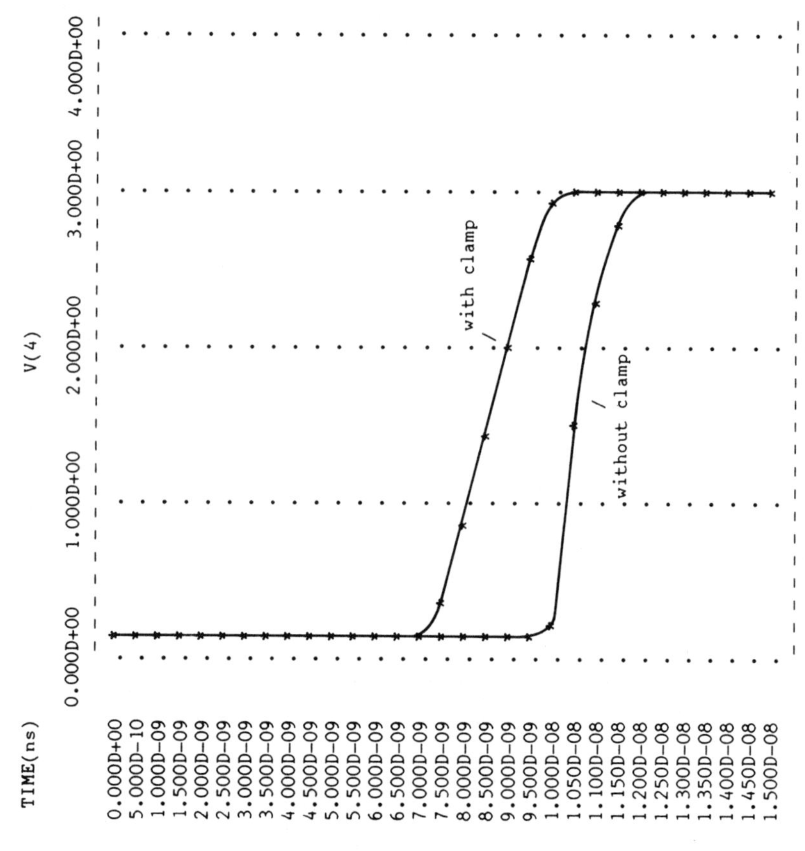

Figure 8E.4 (*continued*)

The inverter input and output waveforms obtained in the two transient simulations are shown in Fig. 8E.4. It is seen that the clamp diode greatly reduces the delay time that must elapse before the excess minority charge is removed from the saturated transistor and the inverter output can respond to the change in input signal.

Because of their fast response, Schottky diodes are also widely used in microwave applications. Here care must be taken to minimize parasitic resistances and capacitances surrounding the diode. For example, it is common to form microwave diodes in lightly doped epitaxial layers grown on much more heavily doped substrates. The light doping of the surface layer provides a wide depletion region and correspondingly small junction capacitance, while the heavily doped substrate minimizes the parasitic resistance of the diode.

8.10 CHAPTER SUMMARY

The first major topic considered in this chapter was the relation between the height of the ***built-in potential barrier*** $q\phi_B$ formed at a metal-semiconductor contact and the difference between the work function $q\phi_m$ of the metal and the electron affinity $q\chi_s$ of the semiconductor. An expression for the ***thermionic emission current*** resulting from electron emission from the semiconductor into the metal when a forward bias is applied to the diode was then derived. Consideration was next given to the ***minority carrier injection ratio*** of the Schottky diode—that is, the ratio of the minority carrier hole current injected into the substrate to the majority carrier thermionic emission current. This ratio was found to be very small for typical Schottky diodes. It was also noted that at a given forward bias the current in a Schottky diode is generally much higher than that in a *pn*-junction diode formed on an identical substrate. It is therefore often said that the ***turn-on*** voltage of a Schottky diode is lower than that of a *pn* junction. Schottky diodes also tend to have high reverse leakage current and frequently exhibit poor reverse breakdown characteristics due to avalanching in the high-field region found at the edge of the junction. The latter problem can be alleviated by surrounding the Schottky barrier with a diffused or implanted ***guard ring***, at the cost of increasing the junction capacitance.

Schottky diodes were found to lack the minority carrier storage delay of their *pn*-junction counterparts, and therefore to be capable of providing very fast switching. This leads to the application of metal-semiconductor junctions as ***clamp diodes*** preventing saturation in many bipolar logic families. The small-signal ac response of a Schottky diode was found to be dominated by the capacitance associated with the depletion region.

8.11 REFERENCE

1. A. M. Cowley and S. M. Sze, "Surface states and barrier height of metal-semiconductor systems," *Journal of Applied Physics,* vol. 36, p. 3212, 1965.

PROBLEMS

8.1. Consider a PtSi Schottky diode formed on an *n*-type silicon substrate of doping $N_D = 10^{16}\ \text{cm}^{-3}$. From Table 8.1, the barrier height ϕ_B of such a diode is 0.89 V. What forward bias V must be applied to the diode to give a thermionic emission current density of $10^3\ \text{A cm}^{-2}$? What is the minority carrier hole current density injected into the substrate at this bias level? What is the minority carrier injection ratio—that is, the ratio of the hole current to the electron thermionic emission current?

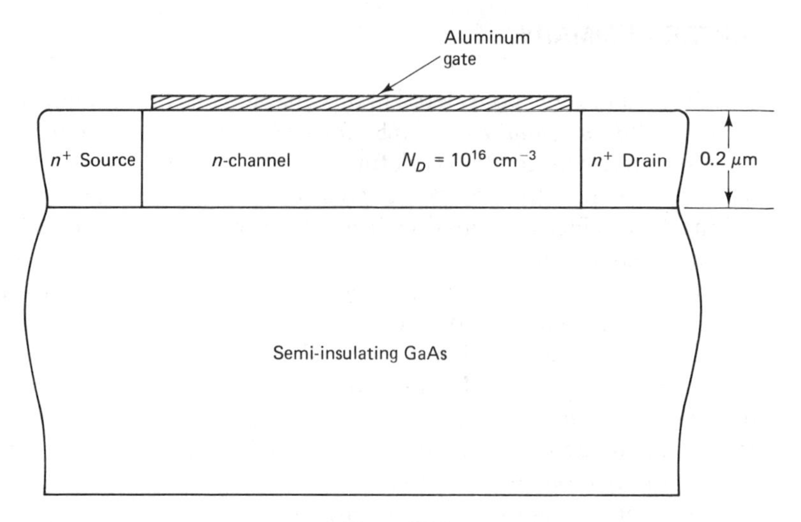

Figure 8P.1

8.2. It is in principle possible to form an ohmic contact to lightly doped silicon using a Schottky diode with a very low barrier height ϕ_B. From (8.8), find an expression for the contact resistivity

$$\rho_c = \left[\left(\frac{dJ}{dV}\right)_{V=0}\right]^{-1}$$

for such a junction. What value of ϕ_B is needed to give $\rho_c = 10^{-6}\ \Omega\cdot\text{cm}^2$?

8.3. A cross section through the channel region of an enhancement-type GaAs MESFET (as discussed in Chapter 9) is shown in Fig. 8P.1. The n-type channel region is 0.2 μm deep and has a doping level $N_D = 10^{16}$ cm^{-3}. This channel region overlies a semi-insulating GaAs substrate, which can be viewed as a perfect insulator for the purposes of this problem. The barrier height of the Schottky diode formed by the aluminum gate is $\phi_B = 0.8$ V.

(a) Show that the channel region is completely depleted when there is no bias between the gate and the channel. There can therefore be no charge flow along the channel when $V_{GS} = 0$.

(b) Determine the forward bias that must be applied to the gate-channel diode to reduce the width of the depletion region to 0.15 μm. What is the diode current at this bias level?

8.4. In Section 8.3.1 the saturation current densities of "typical" Schottky barrier and pn-junction diodes were quoted as 3×10^{-8} A cm^{-2} and 3×10^{-12} A cm^{-2}, respectively. Use these numbers in a SPICE file to generate forward I–V data for diodes of area 10^{-2} cm^{-2}. Taking the turn-on voltage for the junction diode to be 0.5 V, ascertain the turn-on voltage for the Schottky diode.

9
Metal-Semiconductor Field-Effect Transistors

9.1 STRUCTURE

A typical GaAs metal-semiconductor field-effect transistor (MESFET), such as might result from the fabrication sequence described in Section 12.3. is shown in Fig. 9.1. The final device structure is similar to that of a MOSFET inasmuch as it possesses implanted source and drain regions which are self-aligned with the gate. However, the gate in a MESFET is not insulated from the semiconductor. Instead, the metallic gate is in intimate contact with the semiconductor, so forming a Schottky barrier at the interface. Other features to note in Fig. 9.1 are the small thickness of the active region between the source and drain and the employment of a semi-insulating substrate.

Although MESFETs can be made in silicon, such devices are not nearly as

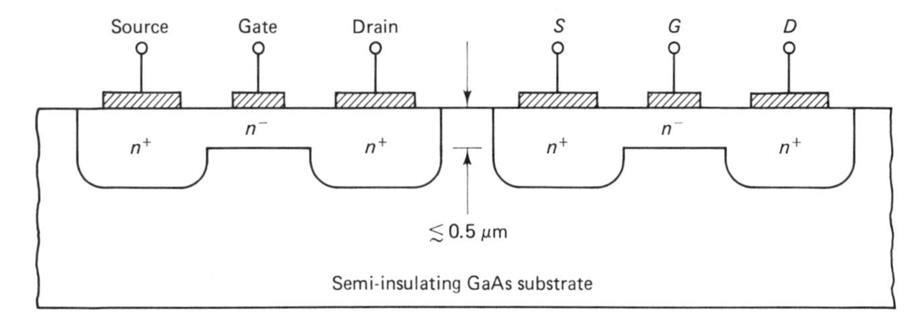

Figure 9.1 GaAs MESFETs isolated by the common semi-insulating substrate.

common commercially, or under such intense study developmentally, as gallium arsenide MESFETs. There are at least three reasons for this. First, Si MOSFETs are so well established and capable of such high performance that there is little incentive to develop another silicon FET structure, particularly one involving MESFETs, which, as this chapter reveals, do not readily yield enhancement devices. Second, in the absence of a GaAs MOSFET technology, the MESFET offers the opportunity of realizing a FET device in GaAs, with all the attendant advantages that the high mobility of this material might bring about. Finally, because of its high bandgap and the role played by certain impurities and defects, GaAs can be grown to yield single-crystal wafers with resistivities approaching 10^7 Ω·cm. Such material is called ***semi-insulating GaAs***. It is of importance to GaAs integrated circuit technology, as it provides excellent electrical isolation between neighboring devices.

9.2 PRINCIPLES OF OPERATION

As is explained in the discussion of FET principles in Section 7.1, FETs are devices in which the flow of majority carriers between the source and drain is controlled by a voltage applied to the gate. In a MESFET the gate voltage modulates the width of the depletion region of the metal-semiconductor Schottky barrier, and hence alters the cross-sectional area, and therefore the resistance, of the channel through which the mobile charges flow. For GaAs the mobility of electrons is so much greater than that of holes that only *n*-type devices are of interest. Figure 9.2 illustrates the situation for a device in which the depletion region does not reach through to the substrate at zero gate–source bias. Such a MESFET is called a ***normally-on*** or ***depletion*** device. When a drain–source potential difference is applied, the voltage is dropped along the *n*-type region, rendering it positive with respect to the gate. This polarity of voltages biases the Schottky barrier diode in the reverse direction, so there is virtually no charge flow between the *n* region and the gate. The electrons drawn from the source are thus constricted to a narrow channel in their passage to the drain (see Fig. 9.2b).

The presence of a steep potential barrier at the *n*-layer/semi-insulating boundary indicates that the surface region of the substrate is effectively doped strongly *p*-type. This comes about because the band bending alters the occupancy of the deep acceptor levels that are present in the type of semi-insulating GaAs material which we consider here. This material is doped with chromium, which acts as an acceptor with energy level very close to the midgap energy. Generally, the acceptor states are mostly empty, so the acceptors are neutral. Another midgap energy level present in semi-insulating material, and thought to be stoichiometry related, is the electron trap known as EL2. This is a donor and is generally full (i.e., neutral). The combination of deep acceptor and donor levels strongly compensates the GaAs, so producing material of high resistivity. The presence of an *n* layer on top of the semi-insulating substrate causes these deep levels to

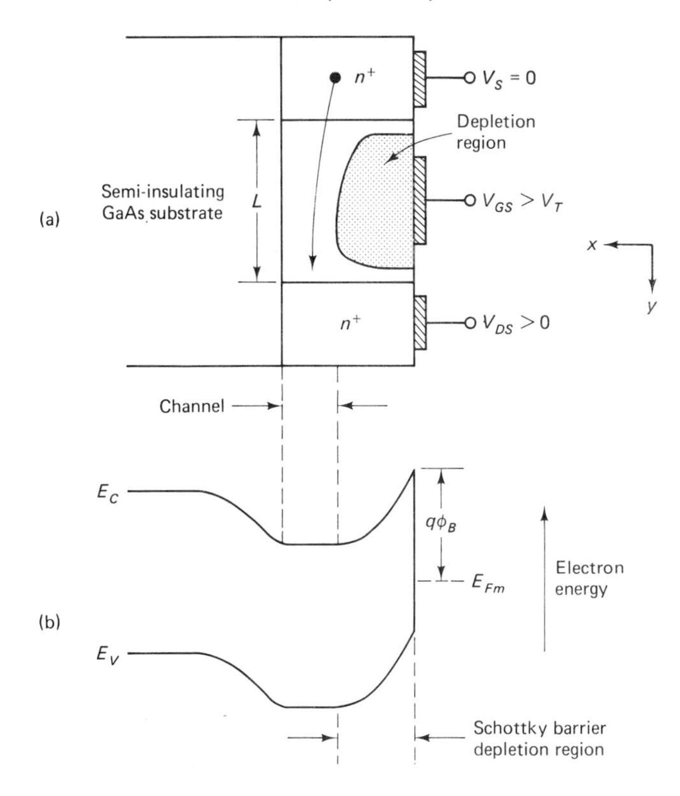

Figure 9.2 Delineation of the channel in a MESFET.

dip below the Fermi energy close to the interface (see Fig. 9.3). The EL2 level remains filled (neutral), but now the acceptor level becomes filled close to the interface. Thus a local negative space charge appears at the surface of the substrate. This negative charge balances the positive charge of the ionized donors on the other side of the interface, so forming the potential barrier that defines the left-hand side of the channel depicted in Fig. 9.2b.

In Fig. 9.2a the depletion region is shown to thicken toward the drain. This is a consequence of the increasingly positive nature of the channel voltage [i.e., $V(y)$ in the channel varies from $V(0) = 0$ to $V(L) = V_{DS}$]. For small values of V_{DS} this thickening is not significant and the relationship between the drain current I_D and V_{DS} is linear, as in a resistor. However, as V_{DS} increases and the widening of the depletion region at the drain end of the channel becomes pronounced, the channel resistance increases and the I_D–V_{DS} relationship becomes less than linear.

Another cause of deviation from linearity is the onset of ***velocity saturation*** of electrons in the channel. The field at which this occurs is lower than in the silicon case (see Fig. 3.4) and is attained at quite low values of V_{DS} in GaAs MESFETs because they are invariably short (i.e., $L < 2\ \mu\text{m}$). The saturation of the velocity of the carriers leads to a saturation of the drain current. Current saturation also occurs by this mechanism in short-channel silicon MOSFETs (see Section 7.6.2).

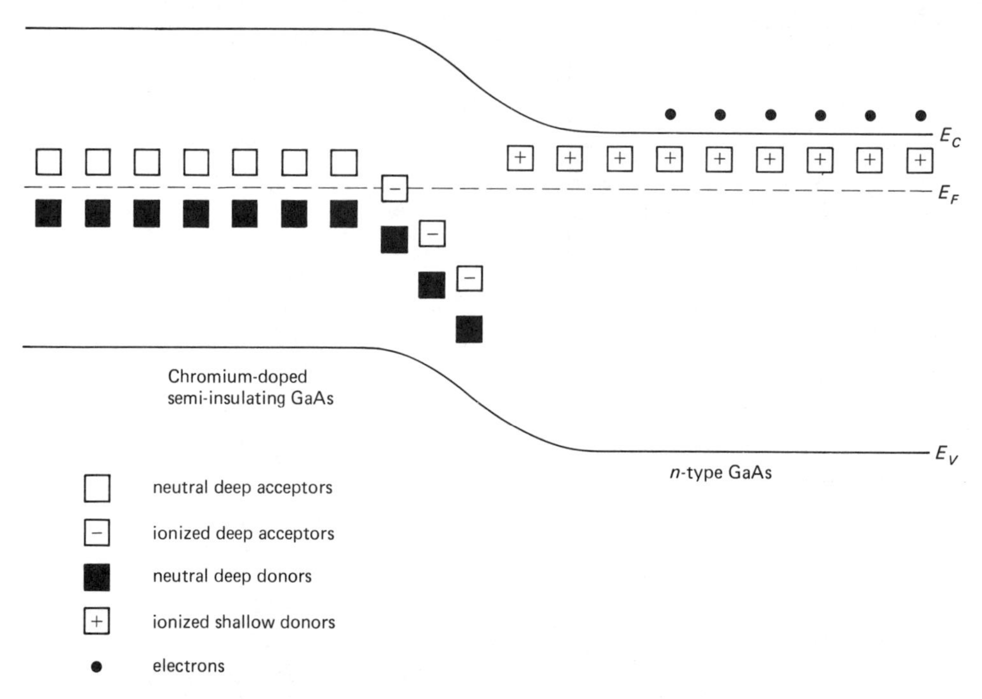

Figure 9.3 Energy band diagram of the *n*-layer/chromium-doped semi-insulating interface in GaAs.

Even though the mechanism of current saturation in MESFETs is different from that in long-channel MOSFETs, the I–V curves for these two devices are of the same form. Figure 9.4a shows the characteristics for a depletion MESFET of the type we have been describing. The gate voltage at which the depletion region reaches through to the substrate, so closing the conducting channel between source and drain, is called the ***threshold voltage***. From Fig. 9.5 it can be appreciated that the threshold voltage is negative in a depletion device.

Figure 9.4b shows the characteristics of a different type of MESFET, one that is called an ***enhancement*** or ***normally-off*** device. In this device the channel is completely pinched off at zero drain–source bias (see Fig. 9.5). To open up a channel in an enhancement device requires a reduction of the depletion region thickness. This necessitates the application of a forward bias to the gate-semiconductor Schottky barrier diode. Thus the threshold voltage is positive in an enhancement device. There is a limit to the amount of forward bias that can be applied; otherwise, there will be a significant flow of carriers from the source to the gate, and transistor action will be lost. The limiting voltage is that of the turn-on voltage for a GaAs Schottky barrier diode (i.e., about 0.75 V).

As was the case for MOSFETs (Chapter 7), there is no need to consider enhancement and depletion MESFET devices separately. Their principles of operation are identical and the two types of device are distinguished by their thresh-

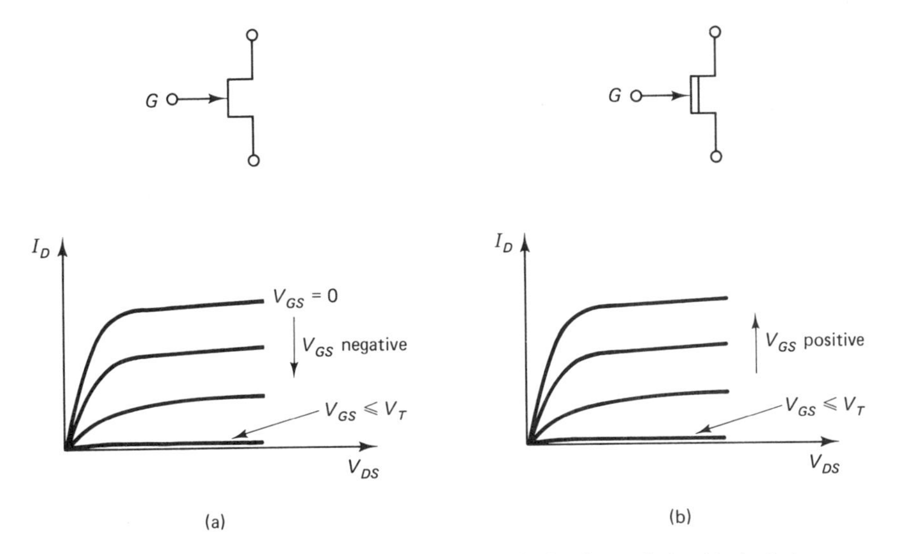

Figure 9.4 MESFET drain characteristics and circuit symbols: (a) depletion MESFET; (b) enhancement MESFET.

old voltage V_T. To compute V_T we need to find the gate voltage at which the depletion region width W just reaches across the entire thickness a of the semiconductor at $V_{DS} = 0$ V. The reference potential for the gate voltage is taken to be 0 V, as exists at the source electrode. From Section 8.7 we have

$$W = \sqrt{\frac{2\epsilon_s}{q}\frac{1}{N_D}(V_{\text{bi}} - V_{GS})} \tag{9.1}$$

where V_{bi} is the built-in potential of the Schottky barrier (referred to as ψ_{s0} in Chapter 8).

When $W = a$, then $V_{GS} = V_T$; therefore,

$$V_T = V_{\text{bi}} - V_P \tag{9.2}$$

where

$$V_P = \frac{a^2 q N_D}{2\epsilon_s} \tag{9.3}$$

V_P is called the pinch-off voltage. It is the potential difference across the depletion region at the threshold condition (see Fig. 9.5).

In the discussion above, we have assumed that the doping density in the channel is uniform. We maintain this assumption throughout this chapter, as it greatly simplifies the analysis of the MESFET. It is a good approximation for MESFETs with epitaxially deposited n layers, but less so for devices in which the active layer is formed by ion implantation into the semi-insulating substrate.

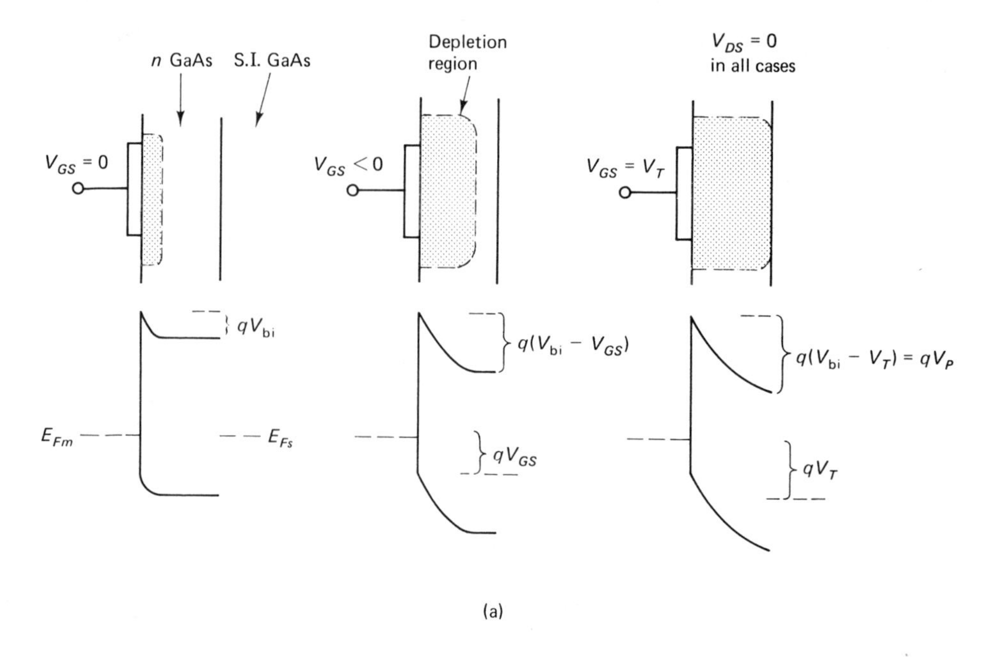

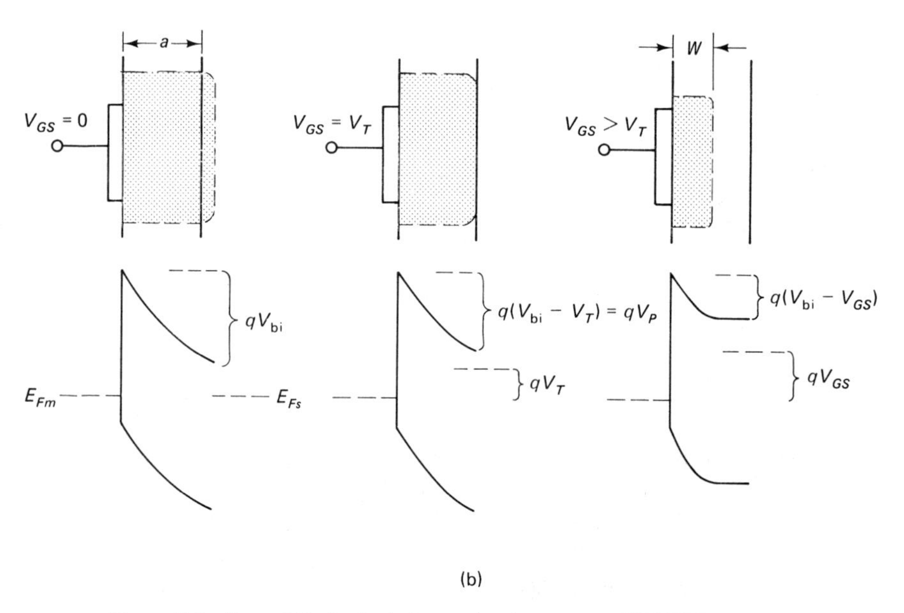

Figure 9.5 V_T and V_P in depletion and enhancement MESFETs: (a) depletion (normally-on) MESFET; (b) enhancement (normally-off) MESFET.

9.3 DC ANALYSIS

9.3.1 Resistive Regime

Consider the depletion MESFET shown in Fig. 9.6. The drain–source voltage is dropped along the channel, raising the potential with respect to the source at some point y to $V(y)$. The potential difference across the depletion region at point y is thus $V_{GS} - V(y)$, and the depletion region width is, extending (9.1), given by

$$W(y) = \sqrt{\frac{2\epsilon_s}{qN_D}\{V_{bi} - [V_{GS} - V(y)]\}} \tag{9.4}$$

We could express $V(y)$ in terms of quasi-Fermi levels as we did in Section 7.3.1 for the MOSFET and proceed to an expression for the drain current. However, here we will take a slightly different approach to arrive at what would be the same answer. The assumptions of the gradual channel approximation, stated explicitly in the beginning of Section 7.3, are implied in the treatment that follows.

We consider the element dy in Fig. 9.6 to have a resistance given by

$$dR(y) = \frac{dy}{A(y)\sigma} \tag{9.5}$$

where A is the cross-sectional area of the channel and σ the conductivity of the semiconductor. Also, we have

$$dR(y) = \frac{1}{I_D}\,dV(y) \tag{9.6}$$

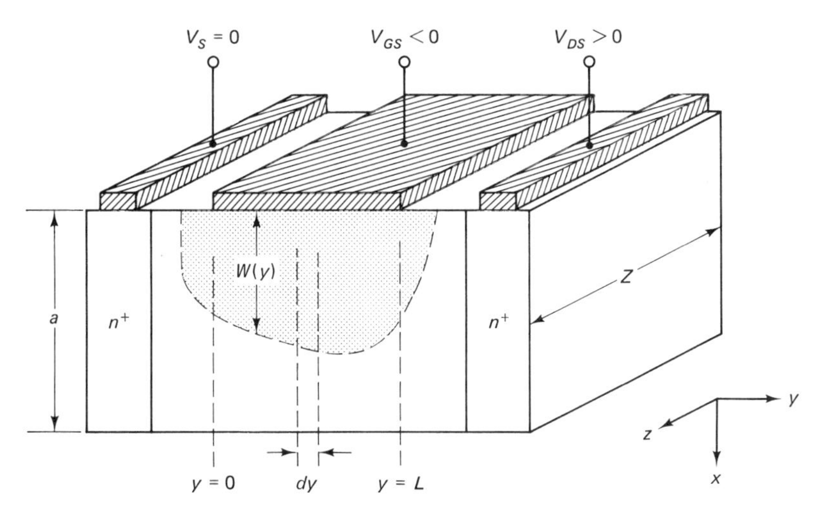

Figure 9.6 Schematic of MESFET operating in the resistive regime. The depletion region is shown shaded.

Combining (9.5) and (9.6), and substituting for A and σ, we obtain

$$dV(y) = \frac{I_D\, dy}{q\mu N_D Z[a - W(y)]} \tag{9.7}$$

where μ is the low-field mobility and is assumed to be constant. Substituting for W from (9.4) and integrating from source to drain, we obtain

$$I_D = G_0 \left\{ V_{DS} - \frac{2}{3V_P^{1/2}} [(V_{DS} + V_{bi} - V_{GS})^{3/2} - (V_{bi} - V_{GS})^{3/2}] \right\} \tag{9.8}$$

where G_0 is the channel conductance when there is no depletion region, that is,

$$G_0 = q \frac{Z}{L} \mu N_D a \tag{9.9}$$

To clarify the fact that (9.8) describes the I–V relation for a resistive element, namely the channel, we expand the first bracketed term as a binomial series and retain only the first-order terms, yielding

$$I_D = G_0 \left(1 - \sqrt{\frac{V_{bi} - V_{GS}}{V_P}} \right) V_{DS} \tag{9.10}$$

This approximation implies that V_{DS} is small compared to $| V_{bi} - V_{GS} |$. Under these circumstances the expression confirms that the channel is resistive, with a resistance that is dependent on V_{GS}.

9.3.2 Saturation Regime

As discussed in Section 9.2, current saturation in GaAs MESFETs arises due to velocity saturation of the electrons traversing the channel from source to drain. To incorporate this phenomenon in a rudimentary model, we first greatly simplify the velocity–field relationship for electrons in GaAs (see Fig. 9.7); that is, we assume that velocity saturation occurs abruptly at some field $\mathscr{E}_{sat}$. This field strength will be attained first at the drain end of the gate. Let us label the channel–source voltage at this point and under these circumstances as $V_{DS,sat}$ and the corresponding width of the depletion region as W_{sat} (see Fig. 9.8). From (9.4) we have

$$V_{DS,sat} = V_{GS} - V_{bi} + \frac{W_{sat}^2 q N_D}{2\epsilon_s} \tag{9.11}$$

The current at this point, where the channel has a thickness $(a - W_{sat})$, is

$$I_{Dsat} = N_D q v_{sat} (a - W_{sat}) Z \tag{9.12}$$

where the electron concentration in the channel has been taken to equal N_D.

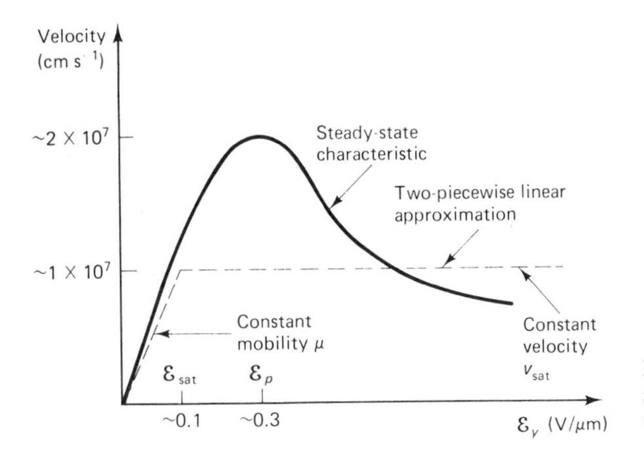

Figure 9.7 Piecewise linear approximation to the steady-state velocity–field characteristic for GaAs.

To compute the drain current in saturation and link up the two portions of our model, we need to equate (9.12) and (9.8), when the latter is evaluated at $V_{DS,\text{sat}}$, and solve for the unknown W_{sat}. Unfortunately, the resulting expression is a cubic for which there is no closed-form solution. A numerical solution is possible, of course, but an analytical solution would be preferable in order to illustrate the relationship between $(a - W_{\text{sat}})$ and the saturation velocity and other device and material properties. An approximate analytical solution can be obtained by using a less rigorous expression for $V_{DS,\text{sat}}$ than that given in (9.11). To derive such an expression for $V_{DS,\text{sat}}$ we first obtain an approximate equation for I_D by expanding the three-halves power term involving V_{DS} in (9.8) and retaining terms up to and including the second order. Physically, this implies that $V_{DS} < |V_{\text{bi}} - V_{GS}|$. The result is

$$I_D = G_0 \left\{ V_{DS} \left[1 - \left(\frac{V_{\text{bi}} - V_{GS}}{V_P} \right)^{1/2} \right] - \frac{V_{DS}^2}{4} \frac{1}{[V_P(V_{\text{bi}} - V_{GS})]^{1/2}} \right\} \tag{9.13}$$

This expression is valid up to the point at which the predicted current reaches a maximum. We associate this value with $I_{D\text{sat}}$, and the corresponding drain–

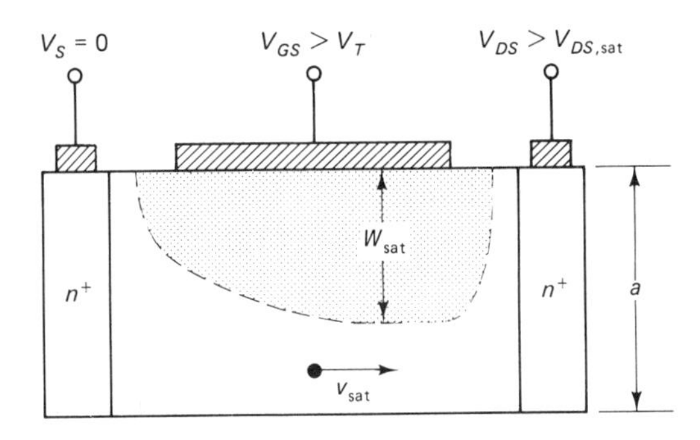

Figure 9.8 Schematic of MESFET operating in the saturation regime.

source voltage with $V_{DS,\text{sat}}$. The results yielded by (9.13) are

$$V_{DS,\text{sat}} = 2V_P \left(\frac{V_{\text{bi}} - V_{GS}}{V_P}\right)^{1/2} \left[1 - \left(\frac{V_{\text{bi}} - V_{GS}}{V_P}\right)^{1/2}\right] \tag{9.14}$$

and

$$I_{D\text{sat}} = G_0 V_P \left(\frac{V_{\text{bi}} - V_{GS}}{V_P}\right)^{1/2} \left[1 - \left(\frac{V_{\text{bi}} - V_{GS}}{V_P}\right)^{1/2}\right]^2 \tag{9.15}$$

Equating this to $I_{D\text{sat}}$ given by (9.12) leads to

$$a - W_{\text{sat}} = \frac{\mu}{L v_{\text{sat}}} \frac{a^3 q N_D}{2\epsilon_s} \frac{V_{\text{bi}} - V_{GS}}{V_{\text{bi}} - V_T} \left[1 - \left(\frac{V_{\text{bi}} - V_{GS}}{V_{\text{bi}} - V_T}\right)^{1/2}\right]^2 \tag{9.16}$$

This expression brings out the fact that a large value for the low-field mobility, a short-channel length, and a low saturation velocity (all properties exhibited by GaAs MESFETs) contribute to increasing the channel thickness and, accordingly, reducing the depletion region thickness at which current saturation occurs. In other words, these properties ensure that the current saturates before the channel pinches off (see Fig. 9.8).

9.3.2.1 Domain formation. We now examine one of the implications of representing the velocity–field relationship for electrons in GaAs more accurately than was done in the analyses of the previous sections. Specifically, we consider Fig. 9.9, which shows velocity saturation occurring at a point y_1 in the channel. Recalling Figs. 3.4 and 9.7, note that attainment of v_{sat} means that the field in the region of velocity saturation is in excess of the field at peak velocity. It follows that on either side of y_1, where the channel widens and the longitudinal field decreases, the carriers will be moving with velocity greater than v_{sat} (see Fig. 9.9b). This causes the electrons to bunch up on entering the narrow part of the channel, and to spread out on leaving it. This arrangement of an accumulation region and a neighboring, partially depleted region constitutes a dipole (see Fig. 9.9c). Any further increase in drain voltage, beyond that necessary to cause velocity saturation, is dropped across this ***stationary domain*** of space charge. Thus the voltage $V(y_2)$ immediately to the left of the domain does not change as V_{DS} increases. As the potential difference $[V(y_2) - V_{\text{source}}]$ provides the driving force for the channel drift current, it follows that this current saturates.

9.3.2.2 Channel-length modulation. In practice, $I_{D\text{sat}}$ is not precisely constant as V_{DS} increases beyond $V_{DS,\text{sat}}$. This is a consequence of the dipole domain in Fig. 9.9 having to grow as it absorbs more potential difference. This shortens the channel to the left of the domain without changing the voltage drop across it. Thus the field in this part of the channel increases, leading to a slight increase in current. Channel shortening also occurs in MOSFETs and, as we noted

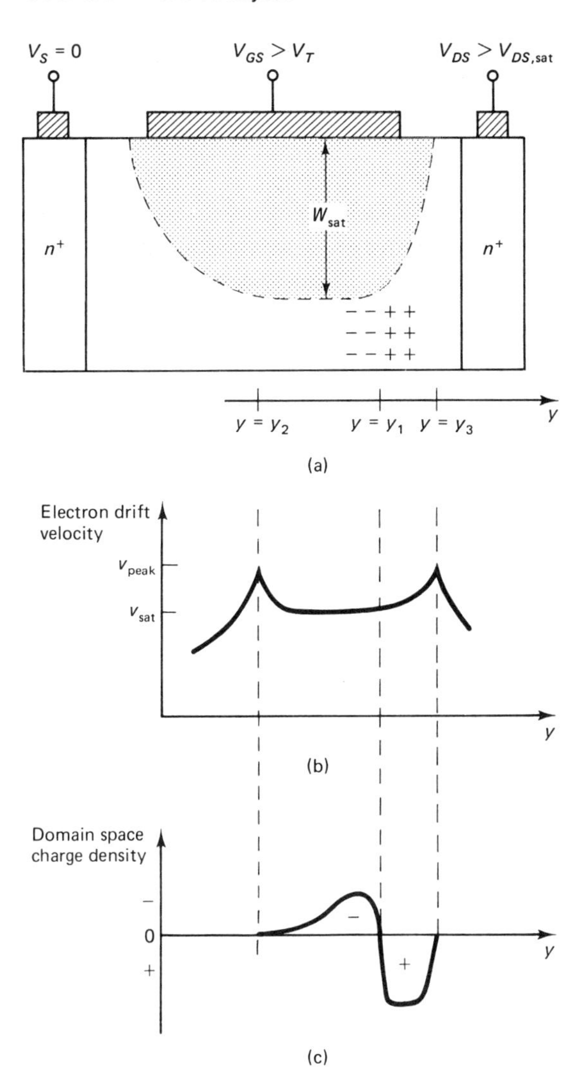

Figure 9.9 Stationary domain formation at the drain end of the channel in the saturation mode.

in Section 7.6.1, can be conveniently accounted for by an empirical factor λ. Thus

$$I_{D\mathrm{sat}} = I^*_{D\mathrm{Sat}}(1 + \lambda V_{DS}) \tag{9.17}$$

where $I^*_{D\mathrm{Sat}}$ is the saturation current predicted using the equations of Section 9.3.2.

9.3.2.3 Velocity overshoot. In gallium arsenide there is a satellite conduction band with a minimum in energy about 0.3 eV higher than the minimum of the lowest conduction band (see Fig. 5.8). In the upper band the effective mass is high, so the mobility is low. In the lower band the lower effective mass yields a high mobility. The potential energy and momentum changes required to effect transfer of electrons from the lower to the upper valley can occur when electrons that have gained kinetic energy from an applied electric field suffer collisions. The resultant change in population of the bands gives the velocity–field curve for n-type GaAs its distinctive form (see Fig. 3.4). The field at which the peak equilibrium velocity is attained is about 0.3 V/μm. On application of fields higher than this, it is possible to accelerate the electrons to velocities higher than the peak equilibrium value, but only for the short time it takes before the changes in potential energy and momentum due to collisions are sufficient to allow transfer to the upper valley. This transient phenomenon is called ***velocity overshoot***. It is often confused with ***ballistic transport***. In true instances of the latter, no collisions with atoms occur, so the velocity increases indefinitely with time, as described by classical Newtonian mechanics. Collisions can occur during velocity overshoot, but as long as the electrons remain in the lower conduction band valley, their high mobility can be maintained out to fields higher than 0.3 V/μm—hence the higher-than-equilibrium velocities.

Detailed calculations indicate that at a field of 1 V/μm, electrons can reach velocities close to 5×10^7 cm/s for about 1 ps before transfer to the upper valley, and consequent retardation, takes place [1]. These figures suggest that to exploit this phenomenon, and realize velocity overshoot throughout the channel of a GaAs MESFET, would require a device length of no more than 0.5 μm. This is one of the reasons for the present considerable research activity in devices of very short channel length.

9.4 DC CIRCUIT MODELS

The dc equivalent circuit for the MESFET is shown in Fig. 9.10. The resistors account for contact resistances and, in the case of R_S and R_D, the resistances of the undepleted n regions between the ends of the channel and the source and drain, respectively. The diodes represent the gate–source and the gate–drain Schottky barrier junctions. The current source can be viewed as a voltage-controlled resistor when $V_{DS} < V_{DS,\text{sat}}$, and as a voltage-controlled constant-current source when $V_{DS} > V_{DS,\text{sat}}$. This interpretation of the current source is the same

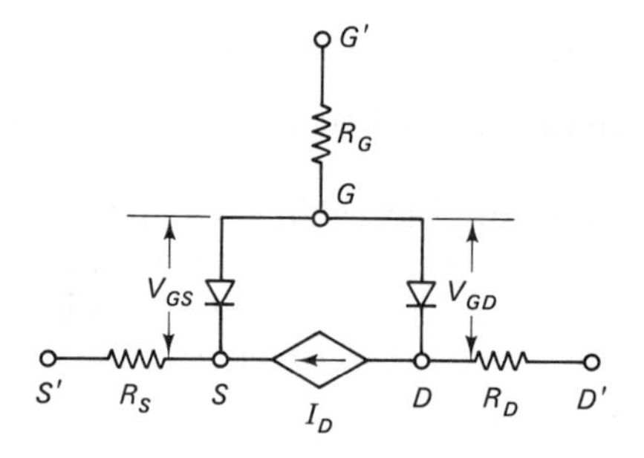

Figure 9.10 Dc equivalent circuit for the MESFET.

as for the MOSFET (see Section 7.3.5). By using the value of $V_{DS,\mathrm{sat}}$ computed from (9.14) as a flag, it should be possible to implement the pair of equations (9.8) and (9.15) in a program to enable the computer-aided analysis of circuits containing GaAs MESFETs. An alternative would be to use a single equation for I_D which describes the drain I–V characteristic over its entire operating range. Such an approach has been developed in the SPICE model described in the following subsection.

9.4.1 SPICE Model

The versions of SPICE available from the University of California at Berkeley, where SPICE was, and continues to be developed, do not contain the subroutines necessary for the analysis of GaAs MESFETs. However, other institutions have developed code that can be patched into SPICE in order to allow simulation of GaAs MESFET circuits. We describe here the most complete model available at present in the public domain. It is from the University of New York at Stony Brook [2, 3].

The dc form of the equivalent circuit used in this SPICE "patch" is the same as shown in Fig. 9.10, except for the omission of the gate–drain Schottky barrier diode. When compared to the gate–source Schottky barrier diode, the gate–drain diode is either more reverse biased in a depletion MESFET, or less forward biased in an enhancement device. In each case the gate–drain diode contributes little to the gate current, so justifying its omission from the equivalent circuit.

The development of a single expression to describe the drain I–V relation over the entire operating range begins with (9.8). The difficulty of computing a value for $V_{DS,\mathrm{sat}}$ is overcome by proposing that current saturation occurs when the channel is pinched-off at the drain end (i.e., when W_{sat} in Fig. 9.8 equals the n-region thickness a). This is akin to the situation in long-channel silicon MOSFETs. We ignore for the moment the fact that such a mechanism in GaAs MESFETs is contradictory to our earlier assertions of current saturation due to velocity saturation. By doing so we arrive at a closed-form solution for $I_{D\mathrm{sat}}$. In the pinch-off case, (9.11) becomes

$$V_{DS,\mathrm{sat}}(\text{pinch-off}) = V_{GS} - V_{\mathrm{bi}} + \frac{a^2 q N_D}{2\epsilon_s} = V_{GS} - V_{\mathrm{bi}} + V_P \qquad (9.18)$$

When this is substituted into (9.8), we obtain

$$I_{D\mathrm{sat}} = \frac{G_0 V_P}{3}\left[1 - 3\left(\frac{V_{\mathrm{bi}} - V_{GS}}{V_P}\right) + 2\left(\frac{V_{\mathrm{bi}} - V_{GS}}{V_P}\right)^{3/2}\right] \qquad (9.19)$$

Eliminating V_{bi} by using (9.2) yields

$$I_{D\mathrm{sat}} = \frac{G_0 V_P}{3}\left\{-2 + 3\left(\frac{V_{GS} - V_T}{V_P}\right) + 2\left[1 - \left(\frac{V_{GS} - V_T}{V_P}\right)\right]^{3/2}\right\} \qquad (9.20)$$

Further simplification of this equation is possible by expanding the three-halves power term as a binomial series and retaining terms up to the second order. Doing this and substituting for G_0 from (9.9) yields

$$I_{D\text{sat}} = \text{BETA}(V_{GS} - V_T)^2 \tag{9.21}$$

where BETA is a SPICE model parameter given by

$$\text{BETA} = \frac{Z}{L}\frac{\mu\epsilon_s}{2a} \tag{9.22}$$

where μ is the low-field mobility.

Our derivation of (9.21) is an example of a situation where the end justifies the means. The end is the attractive one of an equation with exactly the same form as the simple equation for current saturation in a MOSFET (7.50). It neatly describes the often-observed square-law dependence of the current on the excess of the gate voltage over the threshold voltage. The means to this end necessitated use of equations based on the unphysical notion of saturation by pinch-off, and a binomial expansion of $(V_{GS} - V_T)/V_P$. The latter implies that (9.21) is valid only for $V_{GS} \approx V_T$ or $(V_{GS} - V_T) \ll V_P$. However, its familiar form is so appealing that, for device-modeling purposes within the context of a circuit analysis program, it is attractive to retain this equation and introduce an empirical factor that renders the equation capable of giving a good fit to experimental data over a wide range of gate voltages. Further, by carefully choosing the empirical factor, it is possible to fit experimental data *not only* in the saturation regime, *but also* in the resistive regime. The factor is (tanh ALPHA V_{DS}) and the full expression for I_D, taking channel shortening into account, is

$$I_D = \text{BETA}(V_{GS} - \text{VTO})^2 \tanh(\text{ALPHA}\, V_{DS})(1 + \text{LAMBDA}\, V_{DS}) \tag{9.23}$$

In addition to BETA, ALPHA, LAMBDA and the threshold voltage VTO, the SPICE dc model parameter set includes the resistive elements RG, RS, and RD and the Schottky diode parameters IS (the saturation current), N (the ideality factor), and VBI (the built-in potential).

Example 9.1 MESFET Drain *I–V* Characteristics: Effect of Device Thickness

For this example we consider two devices that are identical in all physical respects except for the device thickness *a* (see Table 9E.1). This table lists all the parameters needed to compute the drain characteristics via (9.23). The circuit for the SPICE simulation of MESFET drain characteristics in the common-source connection is shown in Fig. 9E.1. The SPICE input listing is given below.

```
*MESFET COMMON SOURCE DC CHARACTERISTICS
*EFFECT OF DEVICE THICKNESS
VDS    11 0
```

```
VID1   11 1
VGS1    2 0 DC -1.5
VID2   11 3
VGS2    4 0 DC -1
VID3   11 5
VGS3    6 0 DC -0.5
VID4   11 7
VGS4    8 0 DC 0
VID5   11 9
VGS5   10 0 DC 0.5
*GaAs node order; drain, gate, source
B1  1  2  0  FET0.5
B2  3  4  0  FET0.5
B3  5  6  0  FET0.5
B4  7  8  0  FET0.5
B5  9 10  0  FET0.5
.MODEL FET0.5 GASFET(VTO=-1.02, VBI=0.7, ALPHA=2.0,
+BETA=5.8E-3, LAMBDA=0.05, IS=2.8E-14)
.MODEL FET0.25 GASFET(VTO=0.27, VBI=0.7, ALPHA=2.0,
+BETA=11.6E-3, LAMBDA=0.05, IS=2.8E-14)
.DC VDS 0 3 0.1
.WIDTH OUT=80
.PLOT DC I(VID1) I(VID2) I(VID3) I(VID4) I(VID5) (0,8E-4)
.END
```

TABLE 9E.1 DEVICE AND MODEL PARAMETERS USED IN EXAMPLE 9.1

Parameter	Value	Comment
a	0.5 μm	Depletion device
	0.25 μm	Enhancement device
Z	100 μm	
L	1 μm	
N_D	1×10^{16} cm^{-3}	
μ	5000 cm^2 V^{-1} s^{-1}	From Fig. 3.5
ϕ_B	0.8 V	For Al/GaAs, from Fig. 8.3
V_P	1.72 V	For a = 0.5 μm, from (9.3)
	0.43 V	For a = 0.25 μm, from (9.3)
VBI	0.70 V	From (8.22)
VTO	−1.02 V	For a = 0.5 μm, from (9.2)
	+0.27 V	For a = 0.25 μm, from (9.2)
IS	2.8×10^{-14} A	From (8.8b)
BETA	5.8×10^{-3} A V^{-2}	For a = 0.5 μm, from (9.22)
	11.6×10^{-3} A V^{-2}	For a = 0.25 μm, from (9.22)
ALPHA	2.0 V^{-1}	Empirical, illustrative value
LAMBDA	0.05	Empirical, illustrative value
ϵ_s	$13.1 \times 8.85 \times 10^{-14}$ F cm^{-1}	
N_C	4.2×10^{17} cm^{-3}	
kT/q	0.0259 V	At 300 K

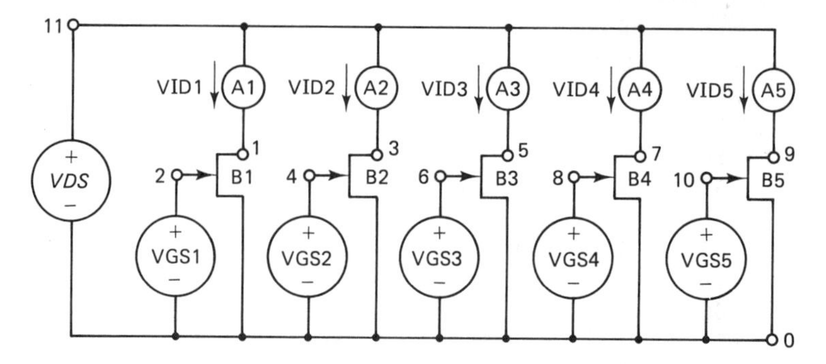

Figure 9E.1 Circuit for the SPICE simulation of MESFET common-source characteristics.

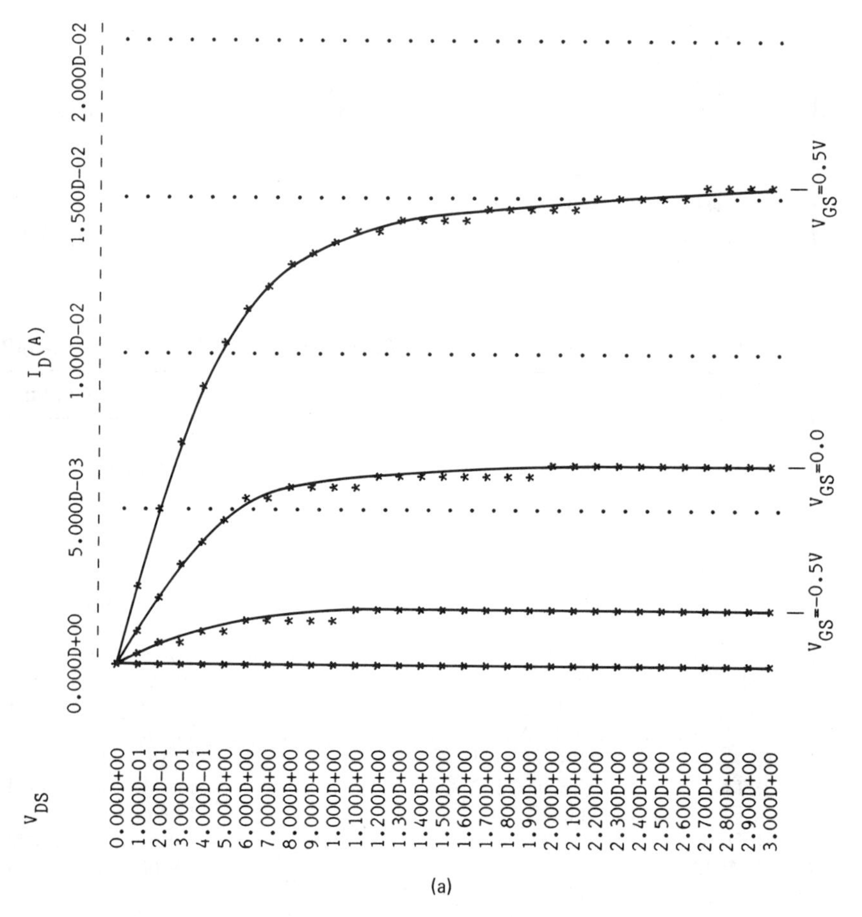

Figure 9E.2 Computed drain characteristics using the data from Table 9E.1. (a) Depletion device with $a = 0.5\ \mu\text{m}$ and $V_T = -1.02$ V.

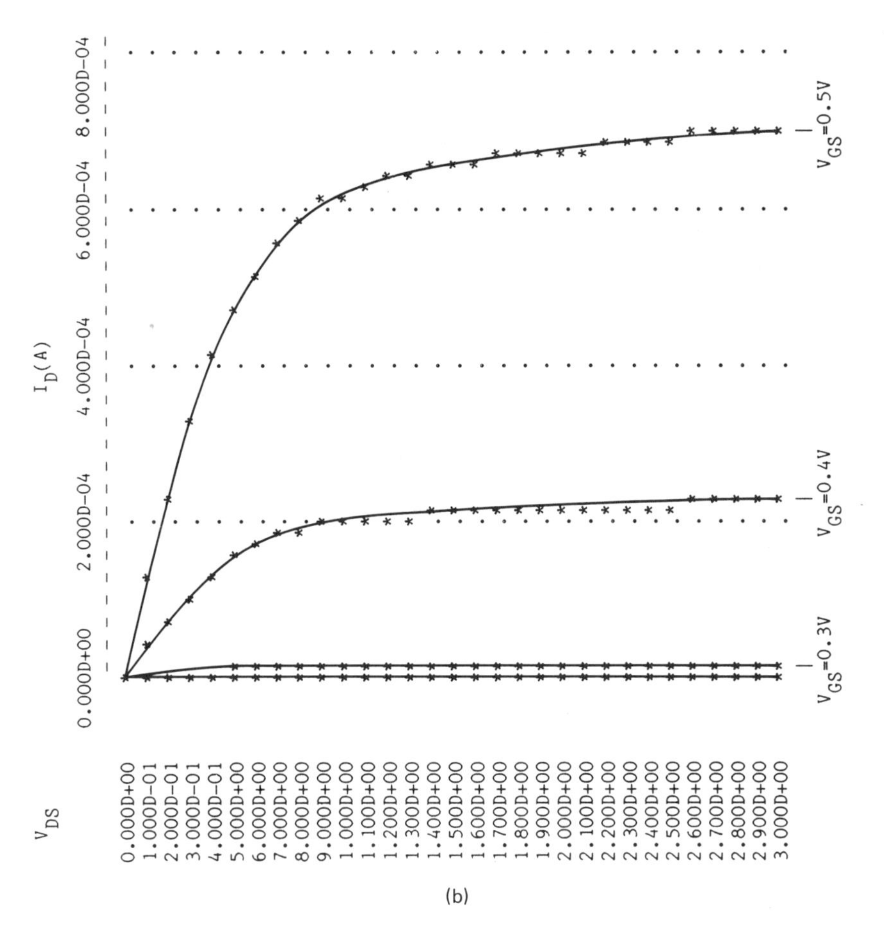

Figure 9E.2 *(continued)* (b) Enhancement device with $a = 0.25\ \mu\text{m}$ and $V_T = 0.27$ V.

The dc characteristics for the two devices are shown in Fig. 9E.2. Note the following:

(a) Neither device conducts until $V_{GS} > V_T$.

(b) For a given V_{GS}, the current in the enhancement device is significantly less than in the depletion device. This is because of the narrower channel of the enhancement device.

(c) The fact that the device type can be changed from depletion to enhancement by altering the thickness of the n layer is the basis for one of the approaches currently being followed toward a feasible technology for enhancement–depletion GaAs FET logic. This form of logic is very appealing, as it would give GaAs an NMOS-like capability

9.5 TRANSIENT ANALYSIS

The transient performance of a device is determined by the manner in which charges in the device redistribute in response to a change in applied voltage or current. In a MESFET there is no minority carrier charge storage, as occurs in a BJT, nor is there an inversion layer charge, as is present in a MOSFET. The charges of consequence are the ionized donors in the depletion region under the gate, and the charges in the stationary domain at the drain end of the gate. The formation of a domain depends on the doping density–thickness product ($N_D a$) [4]. For MESFETs in which $N_D a < 10^{12}\ \text{cm}^{-2}$, the amount of charge accumulation in domains is negligible [4]. We assume that this condition holds for the MESFETs analyzed in this book.

Denoting the total depletion charge (in coulombs) by Q'_B and the total gate charge by Q'_G, charge neutrality demands that

$$Q'_B + Q'_G = 0 \tag{9.24}$$

In device modeling of transient phenomena it is convenient to analyze an equivalent circuit of the device and represent charge storage via capacitative elements. We follow this approach in the transient analyses of all the devices discussed in this book. The only difficulty in doing so for the case of the MESFET is in determining how to partition the gate charge between the capacitances associated with the gate–source and the gate–drain electrodes. We elect to do this by defining the capacitances as follows [5]:

$$C_{GS} = -\left.\frac{\partial Q'_G}{\partial V_S}\right|_{V_G, V_D \text{ const}} \tag{9.25}$$

and

$$C_{GD} = -\left.\frac{\partial Q'_G}{\partial V_D}\right|_{V_G, V_S \text{ const}} \tag{9.26}$$

There are two operating regimes to consider: when the MESFET is ON (conducting channel present) and when it is OFF.

9.5.1 Capacitances in the ON Condition

We base our calculation on the simplified depletion layer charge distribution shown in Fig. 9.11. The charge is divided into three regions. Region 1 comprises a rectangle and a triangle, enabling the charge directly under the gate to be written as

$$\begin{aligned} Q'_{B1} &= qN_D Z[W_S L + \tfrac{1}{2}(W_D - W_S)L] \\ &= \left(\frac{qN_D\epsilon_s}{2}\right)^{1/2}[(V_{\text{bi}} - V_G + V_S)^{1/2} + (V_{\text{bi}} - V_G + V_D)^{1/2}]ZL \end{aligned} \tag{9.27}$$

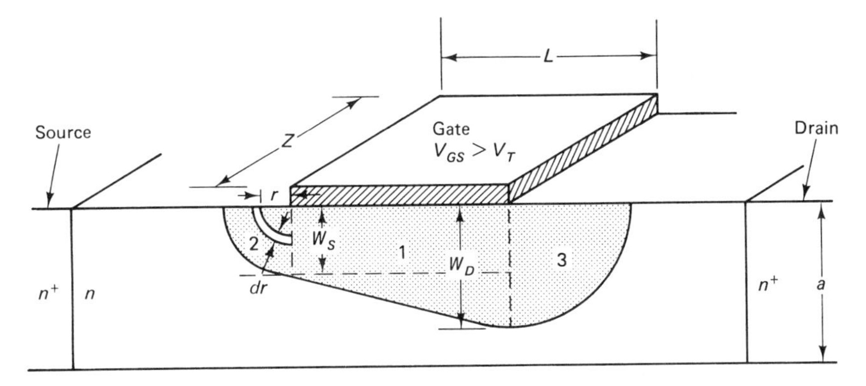

Figure 9.11 Simplified depletion region charge distribution for a MESFET in the ON condition.

where use has been made of (9.1) to relate the depletion layer thicknesses W_S and W_D to voltages.

Regions 2 and 3 are treated as quarter arcs, allowing Q'_{B2}, for example, to be written as

$$Q'_{B2} = \int_{r=0}^{r=W_S} qN_D Z \frac{2\pi r}{4}\, dr$$

$$= \frac{Z\pi\epsilon_s}{2}(V_{bi} - V_G + V_S) \tag{9.28}$$

Similarly, for Q'_{B3} we have

$$Q'_{B3} = \frac{Z\pi\epsilon_s}{2}(V_{bi} - V_G + V_D) \tag{9.29}$$

Summing (9.27), (9.28), and (9.29), and performing the appropriate differentiations leads to

$$C_{GS} = \frac{Z\pi\epsilon_s}{2} + \frac{ZL}{2\sqrt{2}}\left(\frac{qN_D\epsilon_s}{V_{bi} - V_{GS}}\right)^{1/2} \tag{9.30}$$

and

$$C_{GD} = \frac{Z\pi\epsilon_s}{2} + \frac{ZL}{2\sqrt{2}}\left(\frac{qN_D\epsilon_s}{V_{bi} - V_{GD}}\right)^{1/2} \tag{9.31}$$

9.5.2 Capacitances in the OFF Condition

When V_{GS} is more negative than the threshold voltage, there is no conducting channel in a MESFET (see Fig. 9.5). Under these circumstances the depletion region in the n layer can be approximated as shown in Fig. 9.12. For simplicity,

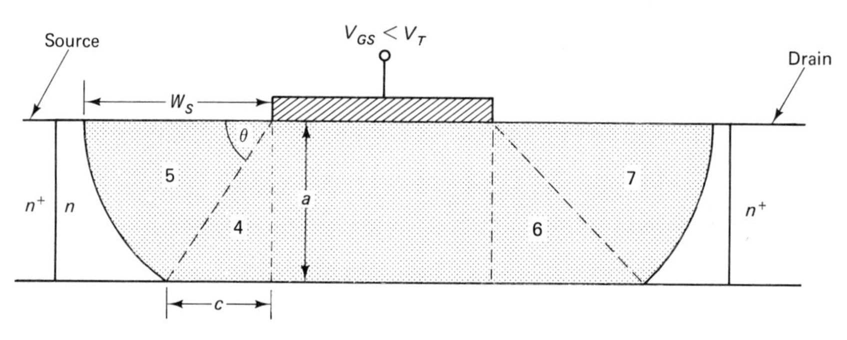

Figure 9.12 Simplified depletion region charge distribution for a MESFET in the OFF condition.

we ignore the depletion region charge in the substrate at the boundary with the n layer (see Fig. 9.3). This means that the only capacitances of significance are those associated with the portions of the depletion region which are not directly under the gate. With reference to Fig. 9.12, we have

$$Q'_{B4} = \frac{1}{2} acZqN_D$$

$$= \frac{1}{2} aZqN_D \sqrt{\frac{2\epsilon_s}{q}\frac{1}{N_D}(V_{\text{bi}} - V_G + V_S) - a^2} \tag{9.32}$$

and

$$Q'_{B5} = qN_D \frac{\theta}{\pi/2} Z \frac{\pi}{2}\frac{W_S^2}{2}$$

$$= Z\theta\epsilon_s(V_{\text{bi}} - V_G + V_S) \tag{9.33}$$

Similar expressions, with V_S replaced by V_D, apply to regions 6 and 7 beyond the drain edge of the gate. Performing the appropriate summations and differentiations leads to

$$C_{GS} = \epsilon_s Z \tan^{-1}\sqrt{\frac{V_{\text{bi}} - V_T}{V_T - V_{GS}}} \tag{9.34}$$

and

$$C_{GD} = \epsilon_s Z \tan^{-1}\sqrt{\frac{V_{\text{bi}} - V_T}{V_T - V_{GD}}} \tag{9.35}$$

9.5.3 Transit-Time Effects

In computing the capacitances above, it has been assumed that the terminal voltages change slowly enough that the depletion layer charge distribution remains close to its steady-state form (i.e., the quasi-static approach to transient modeling, discussed in Section 6.6., is valid). For example, consider the case of a change in V_G when V_S and V_D are held constant. At the instant $t = t_1$ when $V_G = V_G(t_1)$, the depletion layer width is taken to be given by (9.1) evaluated at $V_G(t_1)$. In fact, it takes some time for the depletion layer charge to respond to the gate voltage change. For example, in a depletion MESFET with $V_T < V_G < 0$, as V_G changes toward zero volts, electrons are drawn from the source to nullify some of the ionized donor charge and so reduce the depletion layer width. The delay involved in this process is called the ***transit time***. An upper bound to the transit time is the time taken for electrons to traverse the full length of the channel in order to change the depletion layer width at the drain end of the gate. This leads to a definition of transit time as

$$\tau = \frac{L}{v_{\text{drift}}} \tag{9.36}$$

In GaAs the scatter-limited velocity for electrons is about 10^7 cm·s^{-1}; thus for a MESFET of channel length 1 μm, the transit time is about 10 ps. The commercial viability of GaAs devices depends on their being able to operate at high speeds. It is likely, therefore, that successful GaAs devices will be so small that the time required to charge the gate capacitance will be comparable to the transit time, rather than in excess of it, as is usually the case in Si MOSFETs. It follows that the transit time needs to be taken into account when modeling GaAs MESFETs.

9.5.4 SPICE Model

The SPICE model described here [2, 3] includes transit-time effects when a value for the model parameter TAU is specified. TAU is not related to any component in the equivalent circuit but is used within the SPICE program in such a manner that a drain current, for example, when computed for the instant $t = t_1$, is actually considered to be the current at $t = t_1 - \tau$.

The large-signal equivalent circuit used in SPICE is shown in Fig. 9.13. It is the dc circuit of Fig. 9.10, augmented by the capacitors C_{GS} and C_{GD}. In the OFF condition these capacitances are computed from (9.34) and (9.35). The model parameters needed for their specification are FC ($\equiv \epsilon_s Z$) and VBI ($\equiv V_{\text{bi}}$). The capacitances in the ON condition are computed from (9.30) and (9.31). The additional model parameters needed for their specification in SPICE are CGSO and CGD, the zero-bias values for the portions of C_{GS} and C_{GD}, respectively, due to

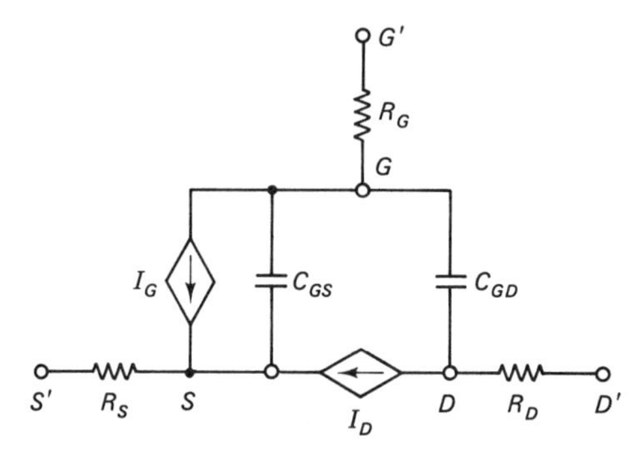

Figure 9.13 Large-signal equivalent circuit for the transient analysis of the MESFET.

the depletion charge directly under the gate. For example, inspection of (9.30) reveals

$$\text{CGSO} = \frac{ZL}{2\sqrt{2}}\left(\frac{qN_D\epsilon_s}{V_{bi}}\right)^{1/2} \tag{9.37}$$

Additional, empirical expressions for the gate capacitances are used in SPICE for cases when either V_{GS} or V_{GD} is within about one-tenth of a volt of V_T. This is to force a smooth transition between the capacitances computed in the ON and OFF conditions. No additional model parameters are needed to compute these capacitances in the transition region.

Finally, note that the current source I_G represents the current in the gate–source Schottky barrier diode, as determined by (8.8a). It is specified by the model parameters IS and N (see Section 8.7).

Example 9.2 Switching Characteristics of MESFETs: A Depletion FET Inverter

DFET logic is a GaAs logic family that uses only depletion-mode MESFETs. The basic form of the simplest gate in this family, the inverter, is shown in Fig. 9E.3.

DFETs have $V_T < 0$, so the top transistor in the inverter, for which $V_{GS} = 0$, is always ON. It behaves, therefore, like a resistor and is often called the ***load*** transistor. The other transistor, DFET2, is called the ***drive***, or ***pull-down***, transistor. When DFET2 is on ($V_{IN} > V_T$) the output voltage is determined by the resistance ratio of the two transistors. To illustrate the operation of this circuit, we use the device parameters given for this example in Refs. 3 and 6. A load capacitance at the output terminal of 6 fF is assumed. This is of the same order of magnitude as the zero-bias parasitic capacitances of the transistors (see SPICE input listing below).

```
*MESFET SWITCHING CHARACTERISTICS
*UNBUFFERED INVERTER
VDD 1 0 DC 3.5
```

```
VIG 3 0 PULSE(0 -1 50PS 50PS 50PS 250PS 1S)
CL  2 0 6FF
B1  1 2 2 GF1
B2  2 3 0 GF2
.MODEL GF1 GASFET(VTO=-2.5, VBI=0.5, ALPHA=1.5,
+BETA=32.5E-6, CGSO=3FF, CGD=0.5FF)
.MODEL GF2 GASFET(VTO=-2.5, VBI=0.5, ALPHA=1.5,
+BETA=65E-6, CGSO=6FF, CGD=1FF, TAU=10PS)
.TRAN 10PS 500PS
.WIDTH OUT=80
.PLOT TRAN V(2) V(3) (-1,4)
.END
```

The capacitances of the load transistor are one-half that of the drive transistor. This is a consequence of making the latter twice the width of the former in order to give the load transistor greater resistance. The results of the SPICE simulation are shown in Fig. 9E.4. Note the following:

(a) At the input voltage level of -1 V (logic low), DFET2 is not completely OFF ($V_T = -2.5$ V), so the output high signal does not reach the full value of *VDD* (3.5 V in this case).

(b) The rise time of the output signal is relatively slow and is little affected by the transit time of electrons in the drive transistor. This is because the dominant time factor is that associated with the charging of the load capacitance via the relatively high resistance of the load transistor.

(c) The fall time of the output signal is relatively fast and does show some dependence on TAU. This indicates that a significant discharge of the output node capacitance takes place through the channel of the drive transistor in a time comparable to the transit time (10 ps in this case).

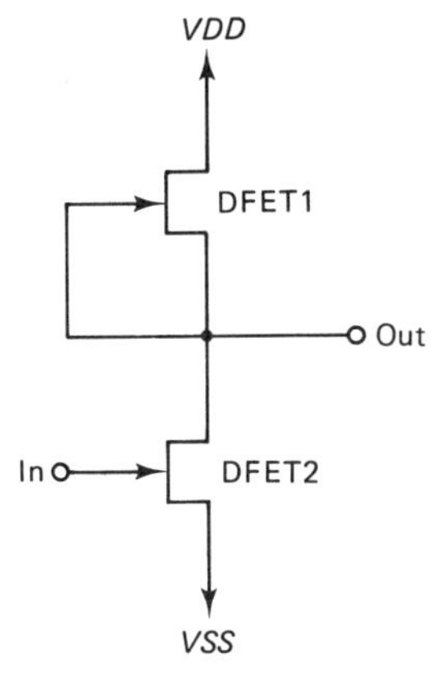

Figure 9E.3 Basic inverter circuit using depletion-mode MESFETs.

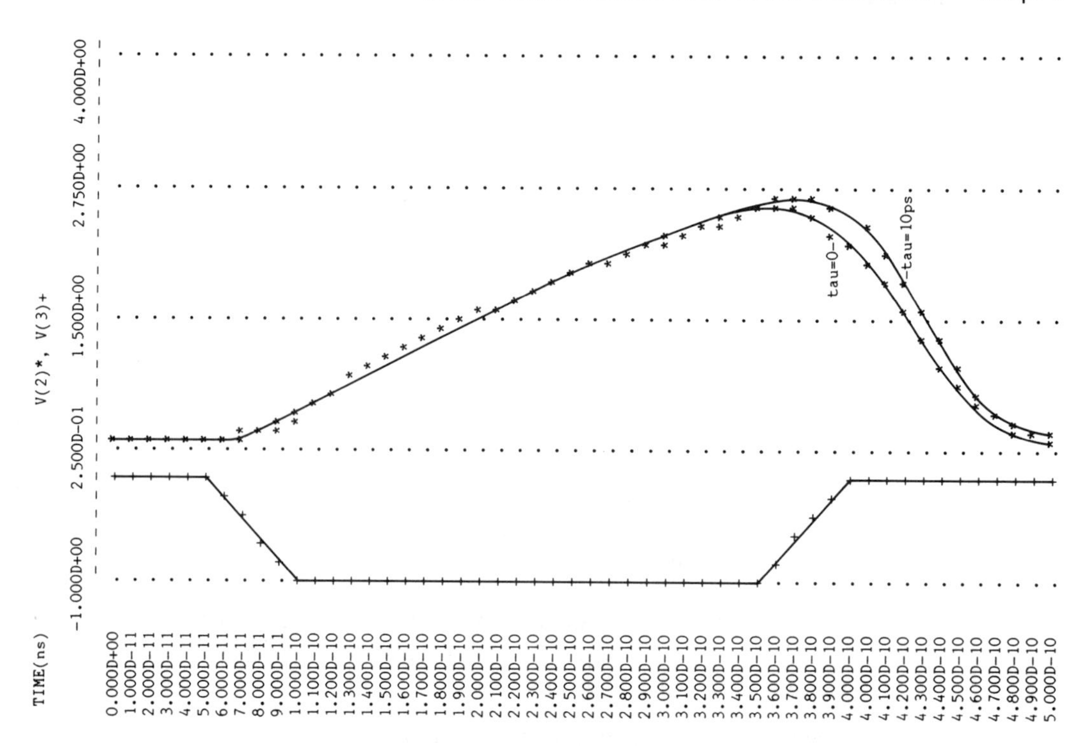

Figure 9E.4 Computed results for the inverter of Fig. 9E.3, using two different values for the transient time in the drive transistor.

(d) The output voltage swing in this example is from +0.37 V to +2.6 V. Such a waveform cannot be fed directly to another inverter stage, as the forward conduction of the gate–source Schottky barrier diode would be prohibitive when V_{GS} exceeded about +0.75 V. For this reason DFET logic uses a level shifting, or buffering, stage between gates. This extra circuitry ensures that the voltage swing at the inputs to all gates is the same (i.e., 0 to −1 V in this example). The full name for this logic family is, therefore, buffered DFET logic. The usual acronym is ***BFL***.

9.6 AC ANALYSIS

The MESFET is a three-terminal device, so the drain current is a function of two voltage variables. Choosing the source terminal as the reference, we have

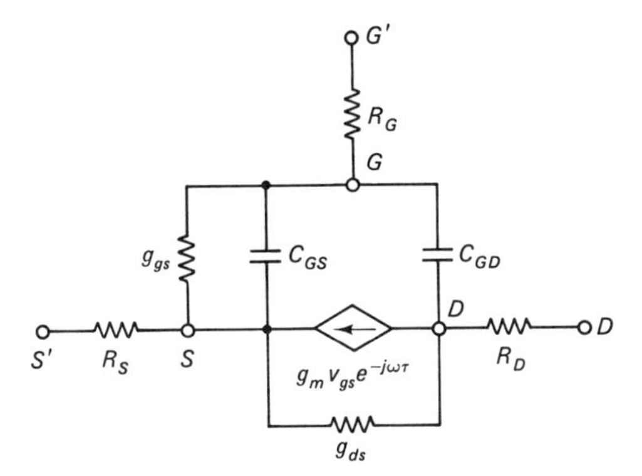

Figure 9.14 Small-signal ac, linearized equivalent circuit for the MESFET.

$$I_D = I_D(V_{GS}, V_{DS}) \tag{9.38}$$

Following the pattern established in Section 6.7.1 for the diode and in Section 7.8.1 for the MOSFET, we perform a Taylor series expansion of (9.38) and retain only the zeroth- and first-order terms. The resulting expression for the small-signal ac current at the drain is

$$i_d = \left.\frac{\partial I_D}{\partial V_{GS}}\right|_{V_{DS}} v_{gs}e^{-j\omega\tau} + \left.\frac{\partial I_D}{\partial V_{DS}}\right|_{V_{GS}} v_{ds} \tag{9.39}$$

where v_{gs} and v_{gd} are small-signal ac voltages applied to the device when it is at an operating point determined by the dc voltages V_{GS} and V_{GD}. The exponential term in (9.39) represents the phase shift between the gate voltage and the drain current, which arises due to the transit-time phenomenon described in Section 9.5.3. Equation (9.39) can be written in terms of the transconductance g_m and the channel, or output, conductance g_{ds}, that is,

$$i_d = g_m v_{gs} e^{-j\omega\tau} + g_{ds} v_{ds} \tag{9.40}$$

These conductances combine with the two gate capacitors, three parasitic resistors, and the Schottky barrier diode small-signal conductance g_{gs} to produce the MESFET small-signal equivalent circuit shown in Fig. 9.14.

9.6.1 SPICE Model

The equivalent circuit of Fig. 9.14 is used in the SPICE model of Ref. 3 for carrying out small-signal analyses of GaAs MESFETs. The required model parameters are those described in Sections 9.4.1 and 9.5.4.

Example 9.3 Small-Signal AC Analysis of MESFETs: A Single-Stage RF Amplifier

The power-amplifying properties of GaAs MESFETs at radio-frequency (RF)/microwave frequencies can be illustrated by analyzing the circuit of Figure 9E.5. The circuit shows the dc biasing arrangement for the transistor, together with 50-Ω resistors to represent the characteristic impedances of the transmission lines which might be employed as the input and output leads to the amplifier. A typical signal frequency would be 4 GHz. The dc drain voltage *VD* is applied to the MESFET via inductor *L*2. This component exhibits a resistance of 0 Ω at dc, but presents a very high impedance to the small RF signal that is to be amplified. The drain dc current passes through the source resistor *RS* and so biases the source electrode to a voltage of ($I_D RS$). This arrangement allows the dc operating point to be set by using only one power supply. It has the other desirable property of providing transient protection. For example, if *VD* were to rise, I_D would follow suit and raise the source voltage. This would increase the reverse bias of the gate–source Schottky barrier and so drive the transistor toward the harmless state of cutoff. The inductor *L*1 holds the gate at a dc voltage of 0 V. The capacitors *C*1, *C*2, and *C*3 block the dc voltages but provide low-impedance paths for the RF signal. A typical value of capacitance is 50 pF, which gives an impedance of less than 1 Ω at 4 GHz. Capacitor *C*4 provides RF bypassing of the source resistor.

For this example we employ the depletion MESFET of Example 9E.1 and analyze the circuit of Fig. 9E.5 using SPICE with the input file shown below. The parasitic capacitances for the MESFET have been computed

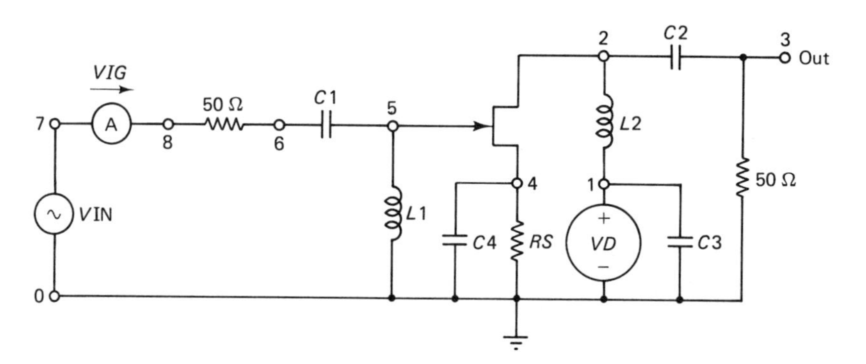

Figure 9E.5 Circuit for the MESFET single-stage RF amplifier used in Example 9.3.

from (9.30) and (9.31), neglecting the effect of charges at the periphery of the gate.

```
*GASFET SMALL-SIGNAL RF AMPLIFIER
*POWER GAIN AT 4GHZ
VD   1 0 DC 2
VIN  7 0 AC 0.01
VIG  7 8
RIN  8 6 50
ROUT 3 0 50
RS   4 0 30
C1   6 5 50PF
C2   2 3 50PF
C3   1 0 50PF
C4   4 0 50PF
L1   5 0 400NH
L2   2 1 400NH
B1   2 5 4 FET0.5
.MODEL FET0.5 GASFET (VTO=-1.02, VBI=0.7,
+ALPHA=2.0, BETA=5.8E-3, +LAMBDA=0.05,
+IS=2.8E-14, CGSO=18FF, CGD=18FF, TAU=10PS)
.AC LIN 1 4000MEG 4000MEG
.WIDTH OUT=80
.PRINT AC VM(3) IM(VIG)
.END
```

The operating point for $RS = 30\ \Omega$ and $VD = 2$ V is $I_D = 4.84$ mA and $V_{GS} = -0.145$ V. Under these conditions SPICE gives, for an input sine wave of amplitude 10 mV and frequency 4 GHz, an output signal of magnitude 5.65 mV. The output power delivered to the 50 Ω load is, therefore, 6.38 μW. The ac input current is predicted to be 4.7 μA, indicating an input power level of 0.47 μW. Thus the power gain is 13.6 or, in decibels, 11.3 dB. This modest gain is a result of not making any attempt to match the MESFET parameters to that of the 50 Ω characteristic impedance of the connecting circuitry. Details of how proper matching is achieved can be found in textbooks on microwave transistor design [7].

9.6.2 Cutoff Frequency

The cutoff frequency for a transistor can be expressed as

$$f_T = \frac{1}{2\pi\tau} \tag{9.41}$$

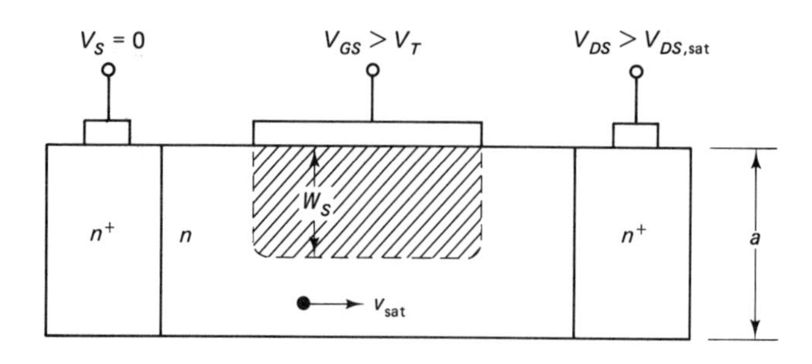

Figure 9.15 Simplified representation of the depletion charge distribution in the MESFET for the purpose of estimating the cutoff frequency.

where τ is a characteristic time. For ultimate performance τ is the transit time which, for FETs, is the transit time of majority carriers along the channel. For long-channel silicon MOSFETs the transit time is short compared to the time required to charge the intrinsic capacitances of the device, so a time greater than the transit time must be used in estimating a realistic value for f_T [see (7.126)]. However, for GaAs MESFETs, as we have intimated in Section 9.5.3, the transit time may well be an appropriate parameter to use in (9.41).

We can demonstrate this by considering a MESFET in which the carriers move with saturated velocity throughout the full length of the channel (see Fig. 9.15). In this case, the saturation current from (9.12) is given by

$$I_D = qN_D Z v_{\text{sat}} \left[a - \sqrt{\frac{2\epsilon_s}{q} \frac{1}{N_D} (V_{\text{bi}} - V_{GS})} \right] \tag{9.42}$$

Approximating the depletion charge by that due to ionized donors directly under the gate, we have

$$-Q'_G = Q'_B = qN_D ZL \sqrt{\frac{2\epsilon_s}{q} \frac{1}{N_D} (V_{\text{bi}} - V_G + V_S)} \tag{9.43}$$

Noting that $g_m = \partial I_D / \partial V_{GS}$ and that $C_{GS} = -\partial Q'_G / \partial V_S$, and realizing that (7.123) (with $C_{GB} = 0$) is appropriate to the equivalent circuit of Figure 9.14, it transpires that

$$f_T = \frac{1}{2\pi} \frac{v_{\text{sat}}}{L} \tag{9.44}$$

which is the sought-for result as L/v_{sat} is the transit time under the gate [see (9.36)].

Equation (9.44) highlights the need to reduce the channel length in order to obtain devices with high operating frequencies. It also explains the interest in devices operating in the overshoot mode, where the drift velocity can exceed the scatter-limited value (see Section 9.3.2.3).

9.7 GaAs MESFETs VERSUS Si MOSFETs

Gallium arsenide is superior to silicon when it comes to considerations of ***very high speed*** FET circuitry. This is principally because the low-field mobility for electrons in GaAs is about five times higher than that for electrons in silicon. However, detractors of GaAs point out that this mobility advantage is lost in devices that operate under conditions where the electron velocity is saturated throughout the length of the channel. This is because there is little difference between the scatter-limited velocities for electrons in the two semiconductors (see Fig. 3.4). In reality, the most usual operating condition is likely to be one in which velocity saturation occurs in only part of the channel. Furthermore, velocity saturation does not necessarily occur all the time. In digital logic applications, for example, FETs switch between ON and OFF states, thus spending a fraction of a cycle in the low-field regime and a fraction of a cycle in the saturation regime. We can conclude, therefore, that the low-field mobility is an important factor in determining the transit time, and hence the speed of operation of a FET, but its effect is mitigated by velocity saturation.

Another metric of FET performance is the ***transconductance***. Taking (7.50) and (9.21), as derived from simple models of saturation by channel pinch-off in MOSFETs and MESFETs, respectively, we have

$$g_m = \frac{Z}{L} \mu \frac{\epsilon}{X} (V_{GS} - V_T) \tag{9.45}$$

where ϵ is the permittivity of either the oxide (MOSFET case) or the semiconductor (MESFET case), and X is either the oxide thickness (t_{ox} for a MOSFET) or the active layer thickness (a for a MESFET). The mobility parameter in (9.45) is the low-field value. Taking μ_e (GaAs) $= 5 \times \mu_e$ (Si), along with ϵ (GaAs) $=$ $3 \times \epsilon$ (SiO_2) and a (GaAs) $= 4 \times t_{ox}$ (SiO_2), indicates that, other things being equal, g_m (GaAs) $\approx 4 \times g_m$ (Si). Obviously, this is a very rough calculation, especially as velocity saturation effects are not embodied in (9.45). Nevertheless, the scale of the predicted transconductance difference is very close to that reported in some comparisons of actual devices [8].

The transconductance advantage enjoyed by GaAs FETs can be viewed in another way. Equal transconductances imply lower values of $(V_{GS} - V_T)$ for GaAs than for Si devices. In digital switching circuits, $(V_{GS} - V_T)$ is usually comparable to the logic voltage swing, V_{sw}. It is desirable to make V_{sw} small in order to reduce the ***dynamic switching energy***. This is the energy that needs to be supplied to charge a capacitance C to V_{sw}; it is given by $0.5CV_{sw}^2$. Thus GaAs FETs have favorable properties as regards dynamic power dissipation.

The situation as regards ***static power dissipation*** is presently not as advantageous for GaAs. This is because most readily available GaAs logic families still employ only depletion (normally-on) transistors. Some improvement in static power dissipation in GaAs FET circuits can be expected when production of both depletion and enhancement (normally-off) devices can be accomplished in a commercially feasible technology. However, even then, this NMOS-like technology cannot be expected to match Si CMOS technology on static power dissipation grounds. To be fair to GaAs, though, it should be noted that the competition in very high speed circuits is not necessarily Si CMOS, but more likely Si bipolar, in the form of emitter-coupled logic (ECL), and this is also a "power-hungry" technology (see Problem 11.13).

To realize ***extremely high speed performance***, all FETs must be small. Equation (7.126) highlights this fact by showing that the cutoff frequency is inversely proportional to the square of the channel length. For very short devices the phenomenon of velocity overshoot can perhaps be exploited to produce an increase in operating speed beyond that due merely to the increase in transconductance and decrease in capacitance embodied in (7.126). In the case of GaAs FETs the relevant channel length is about 0.5 μm, and the electron velocity can be as high as 5×10^7 cm/s for an applied field of 1 V/μm (see Section 9.3.2.3). Velocity overshoot can also occur in silicon, but the nature of the band structure and of the collision mechanisms restricts the extent of the effect. Calculations indicate that at a field of 2 V/μm, electrons in silicon can reach a velocity of about 1.5×10^7 cm/s for times of about 0.3 ps [1]. Thus channel lengths of about 0.05 μm would be needed for velocity overshoot to be important in Si MOSFETs.

A final basis for comparison of Si and GaAs FET circuits is with regard to the substrate on which the transistors are based. The availability of high-quality ***semi-insulating GaAs substrates*** is a key point here. Such a substrate provides excellent isolation between devices and eliminates most of the parasitic capacitance at the junction of the substrate and source–drain regions. Thus denser, faster circuits are possible when the semiconducting substrate is replaced by a semi-insulating one. This fact has not been lost on silicon researchers, and several techniques for the fabrication of MOSFETs in thin silicon layers grown on insulating films are presently being studied vigorously.

9.8 CHAPTER SUMMARY

In this chapter detailed descriptions of the principles of operation of the metal-semiconductor field-effect transistor, and of its electrical performance under both dc and ac conditions, have been presented. The reader should now have an appreciation of the following topics:

1. ***Structure. MESFETs*** are presently the only FETs commercially made in GaAs. They comprise a ***source*** and a ***drain*** region, as in a MOSFET, with the conductivity in the ***channel*** between these two regions being controlled by modulating the width of the depletion region under a ***Schottky barrier gate*** electrode. The high mobility of electrons in GaAs, and the fact that devices can be fabricated with ***very short channel lengths***, makes the devices of use in very high frequency and very fast switching applications.

2. *Modes of operation. Enhancement*- and ***depletion***-type devices are possible and are characterized by a channel that is closed or open, respectively, at zero gate–source bias. For gate–source biases above the ***threshold voltage*** the channel conducts. For low values of drain–source voltage the channel is ***resistive*** in nature. When V_{DS} exceeds $(V_{GS} - V_T)$, the drain current ***saturates***; saturation can occur due to either ***pinching-off*** the channel at the drain end, or, more likely in short-channel GaAs devices, to ***velocity saturation*** of the mobile carriers in the channel. At high ***longitudinal fields*** in the channel the ***negative differential mobility*** of GaAs can lead to ***stationary domain*** formation, yet another phenomenon that causes drain current saturation. In devices with very short channel lengths (less than 0.5 μm), the application of longitudinal fields in excess of 0.3 V/μm can lead to ***velocity overshoot***.

3. *MESFET modeling.* The dc equivalent circuit of the MESFET models the channel as a voltage-dependent current source and the gate–source and gate–drain regions as Schottky barrier diodes. The inclusion of capacitors to represent ***gate-channel*** and diode capacitances renders the model suitable for ***large-signal ac analyses***, as is illustrated by the SPICE simulation of a rudimentary ***inverter*** circuit. When parasitic capacitances and resistances are very small, the ***transit time*** of charge along the channel may be important in determining the speed of operation of MESFETs and, therefore, may need to be included in the device model. In ***small-signal ac*** situations the diodes are replaced by their small-signal equivalent components and the dependence of the channel current on V_{GS} is accounted for by a voltage-controlled current source. This model is used in this chapter in a SPICE simulation of the MESFET as an ***RF amplifier***.

4. *FET comparisons.* GaAs MESFETs have a number of advantages over Si MOSFETs as regards use in very high speed circuitry, namely, higher mobility, higher transconductance, lower dynamic switching energy, the existence of crystalline semi-insulating substrates, and the possibility of capitalizing on the velocity overshoot phenomenon.

9.9 REFERENCES

1. J. G. Ruch, "Electron dynamics in short channel field-effect transistors," *IEEE Transactions on Electron Devices,* vol. ED-19, pp. 652–654, 1972.
2. S. E. Sussman-Fort, S. Narasimhan, and K. Mayaram, "A complete GaAs MESFET computer model for SPICE," *IEEE Transactions on Microwave Theory and Techniques,* vol. MTT-32, pp. 471–473, 1984.
3. S. E. Sussman-Fort, J. C. Hantgan, and F. L. Huang, "A SPICE model for enhancement- and depletion-mode GaAs FETs," *IEEE Transactions on Microwave Theory and Techniques,* vol. MTT-34, pp. 1115–1118, 1986.
4. K. Yamaguchi, S. Asai, and H. Kodera, "Two-dimensional analysis of stability criteria of GaAs FETs," *IEEE Transactions on Electron Devices,* vol. ED-23, pp. 1283–1290, 1976.

5. T. Takada, K. Yokoyama, M. Ida, and T. Sudo, "A MESFET variable-capacitance model for GaAs integrated circuit simulation," *IEEE Transactions on Microwave Theory and Techniques,* vol. MTT-30, pp. 719–723, 1982.
6. W. R. Curtice, "A MESFET model for use in the design of GaAs integrated circuits," *IEEE Transactions on Microwave Theory and Techniques,* vol. MTT-28, pp. 448–456, 1980.
7. R. S. Pengelly, *Microwave Field-Effect Transistors—Theory, Design and Applications,* Chap. 5, Research Studies Press, Chichester, West Sussex, England, 1982.
8. B. M. Welch, R. C. Eden, and F. S. Lee, "GaAs digital integrated circuit technology," in *Gallium Arsenide: Materials, Devices and Circuits,* M. J. Howes and D. V. Morgan, eds., p. 523, John Wiley & Sons, Inc., New York, 1985.

PROBLEMS

9.1. The barrier height of a metal/*n*-GaAs MESFET is 0.95 V. The GaAs *n*-layer is 0.5 μm thick and is doped with a donor concentration of 10^{16} cm^{-3}.
(a) Determine whether the MESFET is depletion type or enhancement type.
(b) Compute the pinch-off voltage and the threshold voltage for this device.

9.2. Under certain operating conditions the MESFET can be viewed as a voltage-controlled resistor. For the depletion device used in Example 9.1:
(a) Confirm that $V_{DS} = 0.1$ V is a suitable operating point for observing voltage-controlled resistance behavior over the range of V_{GS} from -0.5 to $+0.5$ V.
(b) Compute the channel resistance for $V_{GS} = -0.5$, 0, and $+0.5$ V.

9.3. Consider a MESFET in which the electrons exhibit ballistic behavior. In this mode of operation all the potential energy gained by the electron from the drain–source longitudinal field is converted into kinetic energy.
(a) Estimate the drain–source voltage required to produce an electron velocity equal to twice that of the peak equilibrium value.
(b) If this velocity could be sustained over the entire 0.2-μm gate length of a MESFET, what would be the theoretical cutoff frequency of the transistor?

9.4. Two sets of equations are used in the text to describe current saturation due to velocity saturation of electrons in the channel:

Set A:	(9.13) for I_D	(9.14) for $V_{DS,\text{sat}}$	(9.15) for $I_{D\text{sat}}$
Set B:	(9.8) for I_D	(9.11) for $V_{DS,\text{sat}}$	(9.12) for $I_{D\text{sat}}$

Compare the predictions of these two sets of equations by using data for the depletion device listed in Table 9E.1. Specifically, for $V_{GS} = 0$ V:
(a) Compute $V_{DS,\text{sat}}$ from (9.14), evaluate I_D up to $V_D = V_{DS,\text{sat}}$ from (9.13), and then use (9.15) for $I_{D\text{sat}}$. Plot the results as an *I–V* characteristic over the range of V_{DS} from 0 to 5 V.
(b) Plot the corresponding characteristic (on the same graph) resulting from solution of the equations of set B. An iterative method for their solution is suggested. For example, use (9.16) to guess W_{sat}, use this value in (9.12) to guess $I_{D\text{sat}}$ and also

in (9.11) to guess $V_{DS,\text{sat}}$; use this value of $V_{DS,\text{sat}}$ in (9.8) for another approach to guessing $I_{D\text{sat}}$. Compare the two values of saturation current, if they differ markedly, use the second value in (9.12) to get a new value of W_{sat}; iterate until acceptable convergence is obtained.

(c) Comment on any differences between the two predictions vis-à-vis the assumptions made in deriving the equations of set A.

9.5. Another model of current saturation is the channel pinch-off model, which is described by

$$(9.8) \text{ for } I_D \qquad (9.18) \text{ for } V_{DS,\text{sat}} \qquad (9.19) \text{ for } I_{D,\text{sat}}$$

Use these equations to generate an I–V characteristic to compare with those calculated in Problem 9.4. Comment on the agreement of the predictions of the three models.

9.6. **(a)** Derive an expression for the transconductance of a MESFET operating under such conditions that the electron velocity at all points in the channel can be assumed to be equal to the saturation velocity.

(b) Compute $g_{m\text{sat}}$, using the equation derived in part (a), for the depletion device listed in Table 9E.1 and with $V_{GS} = 0$ V.

9.7. The separation of the source and gate electrodes in a MESFET gives rise to a parasitic series resistance between the source terminal and the beginning of the channel (see Fig. 9P.1).

(a) If this separation is 2 μm for the device considered in Problem 9.6, estimate the source series resistance.

(b) Show that this resistance reduces the transconductance computed in Problem 9.6 by about 15%.

9.8. Consider the driver transistor (DFET2) in the unbuffered inverter of Example 9.2. For this device $Z = 10$ μm, $L = 1$ μm, $N_D = 3 \times 10^{16}$ cm^{-3}, $\mu = 5000$ cm^2 V^{-1} s^{-1} and the active layer thickness $a = 0.25$ μm. Take the separation of gate and source electrodes to be 1 μm.

(a) Estimate the value of the parasitic source resistance due to this separation.

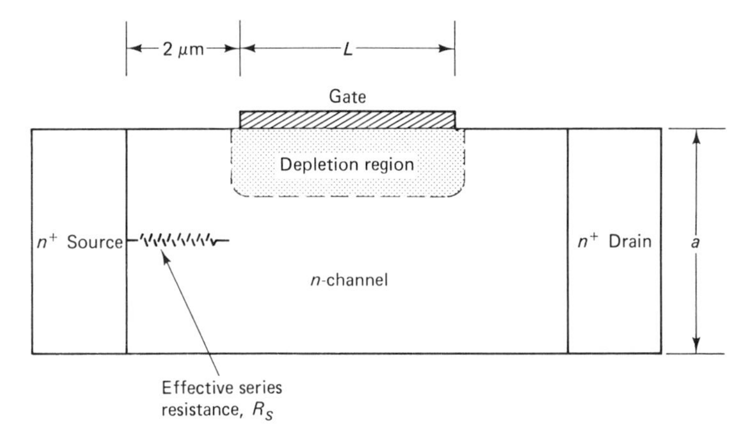

Figure 9P.1

(b) Include this resistance as RS in the input listing of Example 9.2 and run SPICE.
(c) Compare the output response with that shown in Fig. 9E.4 for the case of RS = 0. Comment on the role played by RS in causing the differences between the two responses.

9.9. One way of adjusting the threshold voltage in MESFETs is by changing the doping density in the active region of the device. Investigate this phenomenon for the case of a device in which $a = 0.25$ μm, $Z = 100$ μm, $L = 1$ μm, and the Schottky barrier height is 0.8 eV. Specifically, consider doping densities in the range 10^{15} to 10^{16} cm^{-3}.
(a) Compute V_T.
(b) Plot I_D–V_{DS} for $V_{GS} = 0.5$ V using SPICE. Take Example 9.1 as a guide.

10 Junction Field-Effect Transistors

10.1 STRUCTURE

Junction field-effect transistors (JFETs) are the earliest of the family of field-effect devices. The basic JFET structure is shown in Fig. 10.1 and the circuit symbols are shown in Fig. 10.2. In an n-channel device the n layer is either implanted into, or epitaxially deposited upon, a p-type substrate. This step is followed by diffusions or implantations of n^+ regions to facilitate ohmic contact to the source and drain electrodes, and of a p^+ region to form a junction diode. By reverse biasing this diode, the depletion layer on the n side of the junction can be made to penetrate deep into the n region, so controlling the effective thickness of the n channel through which electrons pass on their way from the source to the drain. The p^+n-junction diode performs, therefore, exactly the same function in a JFET as the metal-semiconductor Schottky barrier diode does in a MESFET.

The reverse current through a reverse-biased pn-junction diode is small (see Example 6.2), but not as small as that through a metal-insulator-semiconductor diode. As this current represents unwanted leakage current in a transistor, silicon JFETs are not nearly so commonly used as silicon MOSFETs. The relative simplicity of fabrication, and suitability for very large scale integration, are other reasons for the popularity of MOSFETs over JFETs.

In view of the minor importance of JFETs and the fact that their operation is, in principle, so very similar to that of the MESFET described in Chapter 9, the ensuing treatment of the JFET is rather brief. The discussion is devoted to silicon devices because, owing to the fact that the state of the art in making pn

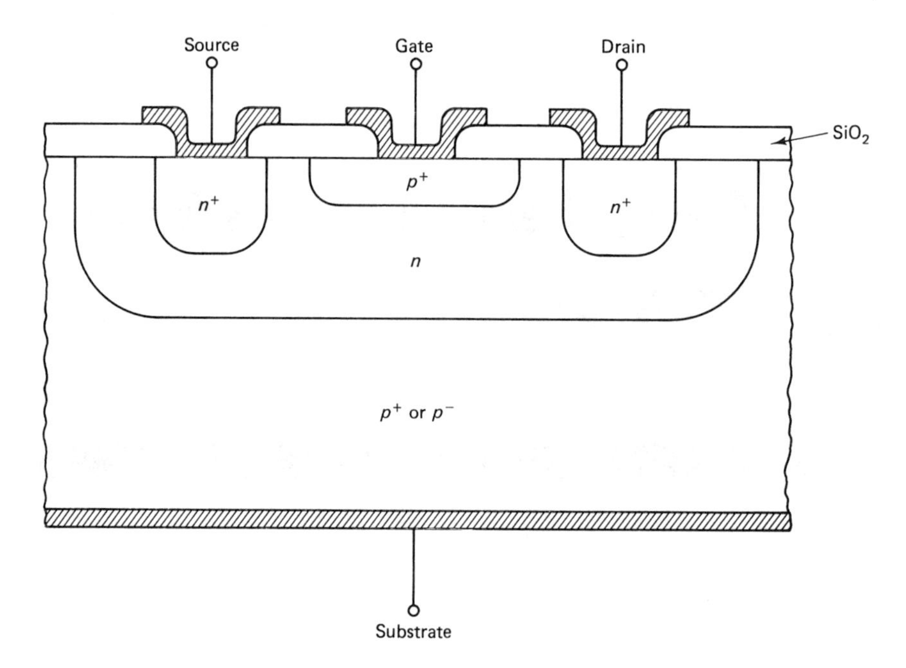

Figure 10.1 Basic structure of an *n*-channel silicon JFET. By changing all the doping types a *p*-channel device can be realized.

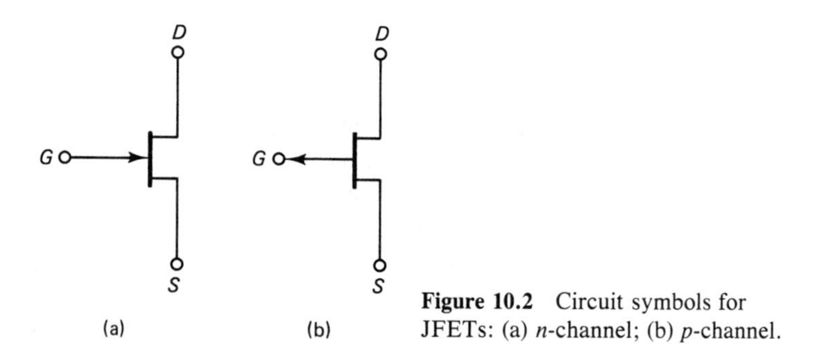

Figure 10.2 Circuit symbols for JFETs: (a) *n*-channel; (b) *p*-channel.

junctions in GaAs integrated circuits is not nearly as advanced as that of making GaAs Schottky barriers, JFETs based on GaAs are presently very uncommon.

10.2 PRINCIPLES OF OPERATION

Figure 10.3 shows two *n*-channel JFETs biased so as to operate in the resistive regime. In the case of a lightly doped *p* substrate connected to ground, the principal modulation of the thickness of the conducting *n* channel between the source and

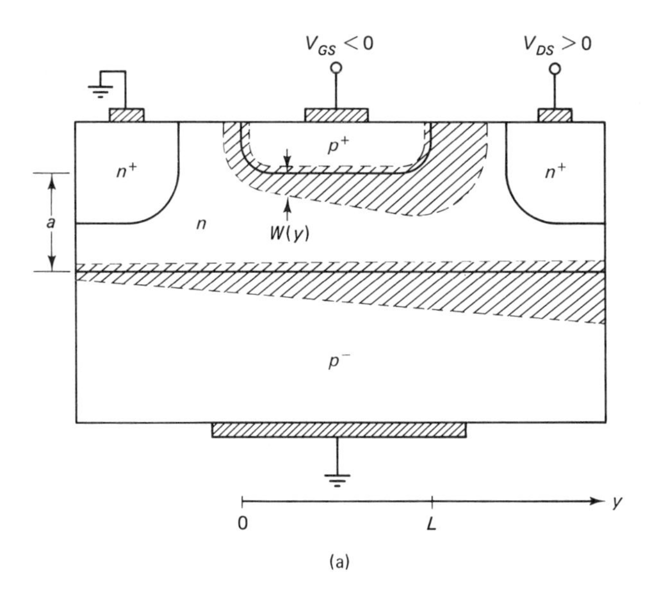

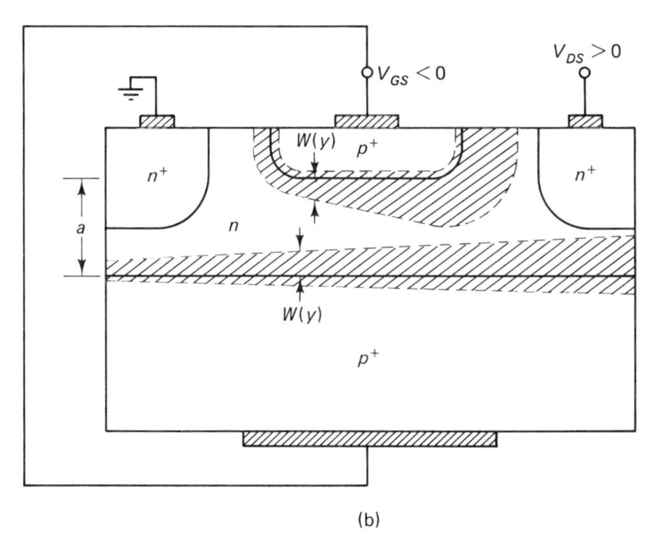

Figure 10.3 JFETs biased in the resistive regime. The depletion regions are shown as shaded for (a) a p^- substrate connected to the source, and (b) a p^+ substrate connected to the gate.

drain occurs via the depletion region under the gate p^+n diode. If we ignore the small contribution to channel narrowing of the depletion region on the n side of the p^-n substrate diode, the thickness of the channel is given by

$$a - W(y) = a - \sqrt{\frac{2\epsilon_s}{q}\frac{1}{N_D}[V_{\text{bi}} - V_{GS} + V(y)]} \tag{10.1}$$

This equation follows directly from (9.4), as used for the MESFET, which is not surprising, as the silicon p^- substrate used here plays a similar role to that of the semi-insulating GaAs substrate employed in the MESFETs described in Chapter 9.

In the case of a heavily doped p-type substrate (Fig. 10.3b), the n-side depletion region thickness of the p^+n substrate diode cannot be ignored. It is usual in JFETs with such substrates to connect together the substrate and top gate electrodes; the thickness of the channel then becomes

$$a - 2W(y) = a - 2\sqrt{\frac{2\epsilon_s}{q}\frac{1}{N_D}[V_{bi} - V_{GS} + V(y)]} \tag{10.2}$$

Thus a given gate–source voltage leads to the JFET with the p^+ substrate having a narrower conducting channel than that of the JFET with the lightly doped substrate. In principle, however, the operation of the two devices is identical. In the remainder of this chapter we elect to describe the device shown in Fig. 10.3a, as many of the equations appropriate to its characterization are identical to those developed in Chapter 9.

Note that enhancement-type JFETs are not very practical because of the process tolerances involved in precise control of the depth of the gate p^+ region, so that the zero-bias depletion layer just penetrates the entire thickness of the n region.

10.3 DC ANALYSIS

Two approaches to deriving expressions for the drain I–V characteristic of silicon JFETs are described here. The first assumes that the electrons in the channel have a constant mobility and that current saturation occurs when the channel is pinched off at the drain. The second method takes into account the field dependence of the electron mobility, and attributes current saturation to a saturating of the carrier velocity.

10.3.1 Constant Mobility, Channel Pinch-Off Model

This model, in common with some of the models we have discussed for MOSFETs and MESFETs, divides the I–V characteristic into two parts: a resistive regime and a saturation regime. For small values of V_{DS} the depletion region thickness can be taken as constant over the length of the device. The channel can be regarded, therefore, as a resistor whose resistance is determined by V_{GS}. The I–V characteristic in this regime is described by (9.10). This equation is a reduced case of the more general equation (9.8), which takes into account the variation of the depletion layer thickness due to the voltage drop along the channel arising from V_{DS}. This equation is repeated here for convenience:

$$I_D = G_0\left\{V_{DS} - \frac{2}{3V_P^{1/2}}[(V_{DS} + V_{bi} - V_{GS})^{3/2} - (V_{bi} - V_{GS})^{3/2}]\right\} \tag{10.3}$$

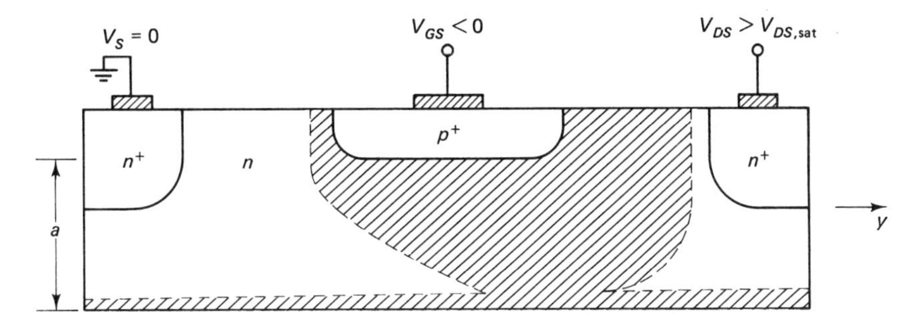

Figure 10.4 JFET in the saturation regime with the channel pinched off.

As V_{DS} increases, the depletion region thickens and eventually, or so the model indicates, the channel becomes pinched off at the drain end where the reverse bias of the gate–channel diode is greatest. The situation is illustrated in Fig. 10.4 and is analogous to that in a MOSFET at pinch-off; that is, the current saturates at a value determined by $V_{DS,\text{sat}}$, the drain–source voltage at which channel pinch-off first occurs. The surplus of V_D over $V_{D\text{sat}}$ is dropped across the region between the drain and the end of the depletion layer, so providing the drift field necessary to collect at the drain the carriers injected into the depletion region from the channel. The magnitude of $V_{DS,\text{sat}}$ is given by (9.18), that is,

$$V_{DS,\text{sat}} = V_{GS} - V_{\text{bi}} + V_P \tag{10.4}$$

From (10.3) and (10.4) the saturation current can be deduced [equation (9.20)], that is,

$$I_{D\text{sat}} = \frac{G_0 V_P}{3}\left[-2 + 3\left(\frac{V_{GS} - V_T}{V_P}\right) + 2\left(1 - \frac{V_{GS} - V_T}{V_P}\right)^{3/2}\right] \tag{10.5}$$

Thus the complete I–V characteristic is of the same form as that of the other FETs described in this book. Equation (10.3) is valid for $V_{DS} < V_{DS,\text{sat}}$, whereas (10.5) holds for larger values of drain voltage. $V_{DS,\text{sat}}$ is readily ascertained from (10.4), as it depends on V_{GS} and known material and device parameters.

10.3.2 Field-Dependent Mobility, Velocity Saturation Model

This model takes into account the implications of the fact that the mobility of electrons in the channel varies with the longitudinal electric field $\mathscr{E}_y$ along the channel length. As the channel narrows, the field must increase in order to maintain a constant current. From Fig. 3.4 it can be appreciated that if the field exceeds about 6×10^4 V cm^{-1}, the velocity of the carriers in the channel will saturate. The current at some point y in the channel can be written as

$$I_D = N_D q v(y) Z[a - W(y)] \tag{10.6}$$

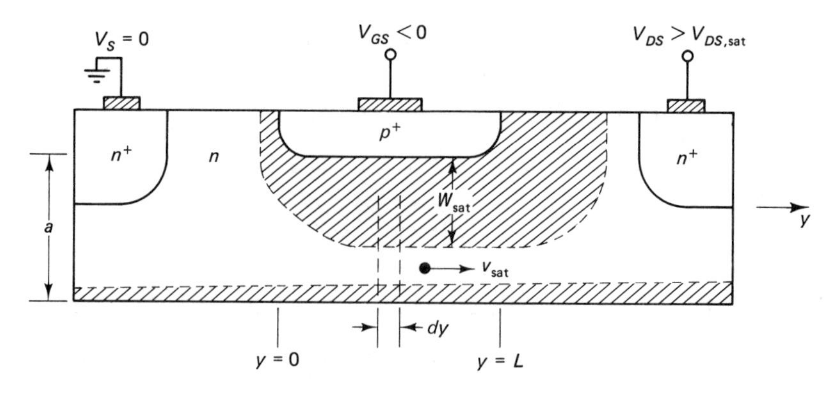

Figure 10.5 JFET with electrons moving with saturated velocity in part of the channel.

where the electron concentration in the channel has been taken to equal N_D. Thus, if $v(y)$ saturates, $W(y)$ must saturate also, and we have the situation shown in Fig. 10.5. To take into account the effects of velocity saturation, we approximate the velocity-field characteristic for silicon (Fig. 3.4) by

$$v = \frac{\mu \, | \mathscr{E}_y |}{1 + \mu \, | \mathscr{E}_y | / v_{sat}} \tag{10.7}$$

where v_{sat} is $1 \times 10^7 \text{ cm}\cdot\text{s}^{-1}$ and μ is the low-field mobility (see Fig. 3.5) as used in the analysis of Section 10.3.1.

Combining (10.1), (10.6), and (10.7), and noting that $| \mathscr{E}_y | = dV(y)/dy$, we have

$$I_D \, dy = \left[N_D q \mu Z a \left(1 - \sqrt{\frac{V_{bi} - V_{GS} + V(y)}{V_P}} \right) \frac{dV(y)}{dy} \right] \Bigg/ \left(1 + \frac{\mu}{v_{sat}} \frac{dV(y)}{dy} \right) \tag{10.8}$$

Integrating along the length of the channel from $y = 0$, $V(0) = V_S = 0$ to $y = L$, $v(L) = V_{DS}$, we arrive at [1]

$$I_D = \frac{G_0\{V_{DS} - (2V_P^{-1/2}/3)[(V_{DS} + V_{bi} - V_{GS})^{3/2} - (V_{bi} - V_{GS})^{3/2}]\}}{1 + \mu V_{DS}/L v_{sat}} \tag{10.9}$$

Thus the effect of the field dependence of the mobility is to reduce the drain current by a factor of $(1 + \mu V_{DS}/L v_{sat})$ with respect to the current in the resistive regime as predicted by the constant mobility model [equation (10.3)]. This reduction can be significant for devices with a large electron mobility and/or short-channel length (see Example 10.1). Equation (10.9) is valid only up to the point at which I_D reaches its maximum value. We can associate this value with the saturation current I_{Dsat}. However, no closed-form solution for $V_{DS,sat}$ the drain voltage at the onset of saturation, can be obtained from this model.

10.4 EQUIVALENT CIRCUITS FOR THE JFET

10.4.1 DC Circuit Model

The dc equivalent circuit model for the JFET is shown in Fig. 10.6. As is to be expected, it is of the same form as that presented for the MESFET in Fig. 9.10.

The circuit of Fig. 10.6 is used in SPICE for the computer-aided analysis of circuits containing JFETs. The resistances r_s and r_d, which represent the parasitic resistances of the n region between the channel and the source and the drain, respectively, are specified in SPICE by the model parameters RS and RD. The two diodes formed by the gate–source and gate–drain p^+n junctions are modeled by a saturation current parameter IS (see Section 6.4.1). The current source I_D is modeled in SPICE by a simpler set of equations than those derived in Sections 10.3.1 and 10.3.2. To derive the equations used in SPICE, we start with (9.2) and (9.13), which result in

$$I_D = G_0\left\{V_{DS}\left[1 - \left(1 + \frac{V_T - V_{GS}}{V_P}\right)^{1/2}\right] - \frac{V_{DS}^2}{4V_P}\left(1 + \frac{V_T - V_{GS}}{V_P}\right)^{-1/2}\right\} \tag{10.10}$$

Expanding the square-root terms as binomial series leads to

$$I_D = G_0\left\{\frac{V_{DS}}{2}\left(\frac{V_{GS} - V_T}{V_P}\right) - \frac{V_{DS}^2}{4V_P}\left[1 - \frac{1}{2}\left(\frac{V_T - V_{GS}}{V_P}\right)\right]\right\} \tag{10.11}$$

Making the expansions implies that $(V_T - V_{GS})/V_P < 1$. Applying this inequality to the last bracketed term in (10.11), and substituting for G_0 from (9.9), gives

$$I_D = \frac{Z\mu\epsilon_s}{2La}[2(V_{GS} - V_T)V_{DS} - V_{DS}^2] \tag{10.12}$$

This equation should appear familiar, as it is of precisely the same form as that used for MOSFETs in the Level 1 version of SPICE [see (7.46)]. It is used in SPICE for values of V_{DS} below $(V_{GS} - V_T)$, where the JFET threshold voltage V_T is given by (9.2). The model parameters needed for its specification are BETA $(\equiv Z\mu\epsilon_s/2La)$ and VTO $(\equiv V_T)$.

The expression for the saturation regime is deduced from (10.12) in exactly

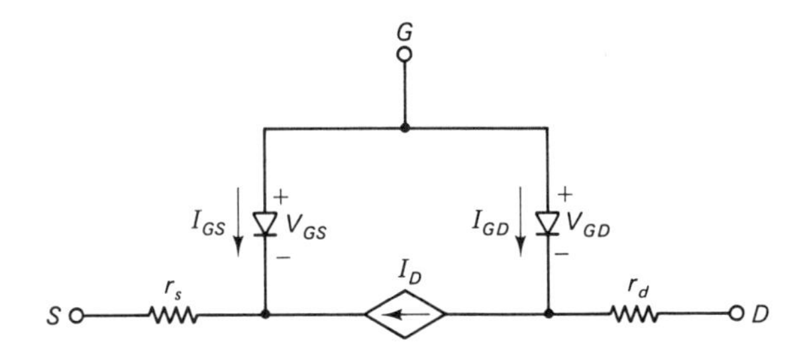

Figure 10.6 Dc equivalent circuit for the JFET.

the same manner as described for the MOSFET [see (7.50)], that is, by evaluating I_D at $V_{DS} = (V_{GS} - V_T)$. Thus

$$I_{D\text{sat}} = \text{BETA}\ (V_{GS} - \text{VTO})^2 \qquad \text{for } V_{DS} > V_{GS} - V_T \tag{10.13}$$

The dc parameter set for SPICE is completed by adding a sixth parameter LAMBDA to account for the shortening of the channel as the pinch-off point moves toward the source when V_{DS} exceeds $(V_{GS} - V_T)$. This parameter appears in a multiplicative term (1 + LAMBDA V_{DS}), which is used in (10.13) in exactly the same manner as for the MESFET [see (9.23)].

Example 10.1 JFET Drain Characteristics: Comparison of DC Models

In this example we compare the predictions of the three JFET models described in this chapter. The device parameters used in this comparison are listed in Table 10E.1. The constant mobility, channel pinch-off model is represented by (10.3) for I_D, (10.4) for $V_{DS,\text{sat}}$, and (10.5) for $I_{D\text{sat}}$. The field-dependent mobility, velocity saturation model is represented by (10.9) for I_D, up to the point at which I_D reaches a maximum. This extremum is determined graphically and is taken to be the value of $I_{D\text{sat}}$. The third model is that used in SPICE, for which I_D is represented by (10.12), $V_{DS,\text{sat}}$ by (10.4), and $I_{D\text{sat}}$ by (10.13).

The results are shown in Fig. 10E.1. Comparing the pinch-off and velocity saturation models, we see that the current predicted by the latter is always less than that computed for the former. This is to be expected, as the velocity saturation model is based on a mobility that becomes progressively smaller, as V_{DS} increases, than that of the constant, low-field mobility used in the pinch-off model. The I–V characteristic predicted by the SPICE model falls in between those of the other two models. Evidently, this is a

TABLE 10E.1 DEVICE AND MODEL PARAMETERS USED IN EXAMPLE 10.1

Parameter	Value	Comment
a	1 μm	Channel depth (maximum)
Z	10 μm	Device width
L	5 μm	Channel length
N_D	5×10^{15} cm^{-3}	Channel doping density
N_A	1×10^{19} cm^{-3}	Gate doping density
μ_e	1298 cm^2 V^{-1} s^{-1}	From (3.8)
V_{bi}	0.87 V	From (6.18)
V_P	3.80 V	From (9.3)
V_T	−2.93 V	From (9.2)
v_{sat}	1×10^7 cm s^{-1}	From Fig. 3.4

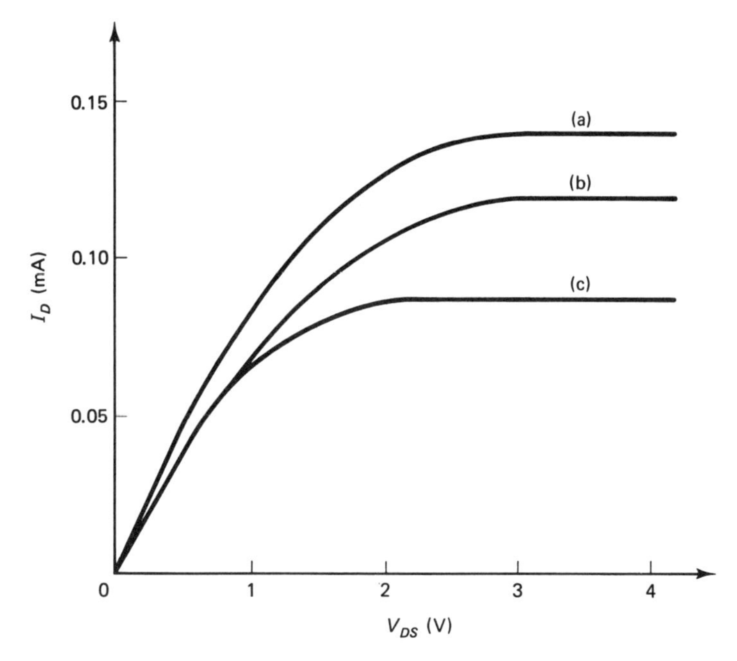

Figure 10E.1 Comparison of I–V characteristics from the three models of the JFET discussed in the text: (a) constant mobility model; (b) SPICE model; (c) field-dependent mobility model. $V_{GS} = 0$ in all cases.

consequence of the approximations (binomial expansions) made in deriving the equations of the SPICE model from those of the constant mobility model.

10.4.2 Large-Signal Model for Transient Analysis

The large-signal equivalent circuit for the JFET shown in Fig. 10.7 is obtained simply by adding to the dc equivalent circuit two capacitive components which represent charge storage in the gate–source and gate–drain p^+n diodes. Since neither of these diodes is normally forward biased, there is no minority carrier storage, so the capacitances involved are just those due to the ionic space charge in the depletion regions (see Section 6.6.1). In SPICE these are modeled in accordance with (6.72) as

$$C_{GS} = \text{CGS}\left(1 - \frac{V_{GS}}{\text{PB}}\right)^{-1/2} \quad \text{and} \quad C_{GD} = \text{CGD}\left(1 - \frac{V_{GD}}{\text{PB}}\right)^{-1/2} \tag{10.14}$$

where the model parameters CGS and CGD are the zero-bias values for the junction capacitances of the respective diodes, and are evaluated from (6.67) at $V = 0$ and with the built-in potential specified by another model parameter PB.

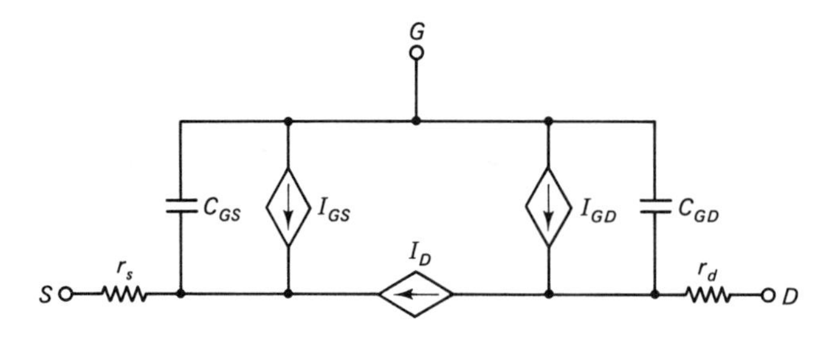

Figure 10.7 Large-signal equivalent circuit for the transient analysis of the JFET.

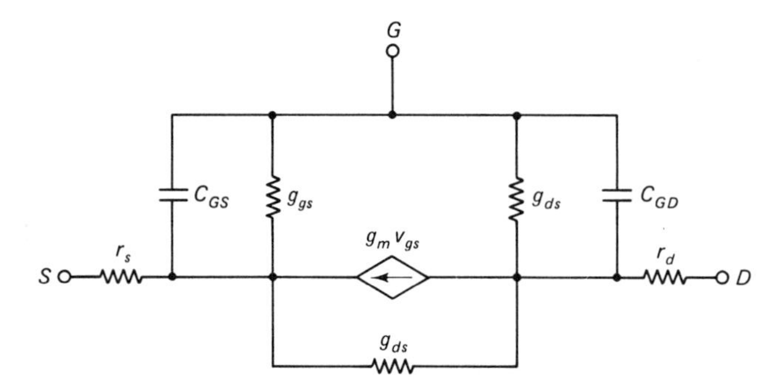

Figure 10.8 Small-signal ac, linearized equivalent circuit for the JFET.

10.4.3 Small-Signal AC Equivalent Circuit

The small-signal, linearized model for the JFET is given in Fig. 10.8. It is derived in the manner followed in Sections 6.7.1, 7.8.1, 9.6, and 11.7.1 for the other devices considered in this book. The conductances g_{gs} $(= dI_G/dV_{GS})$ and g_{gd} $(= dI_G/dV_{GD})$ represent the conductances of the gate–source and gate–drain p^+n diodes, respectively. As these diodes are usually reverse biased, the conductances are very small (see Section 6.4).

Because silicon JFETs are often quite large devices, with channel lengths usually in excess of 5 to 10 μm and channel widths often considerably larger, capacitance charging times tend to dominate the ac response of these devices. For this reason, the transit time of electrons in the channel is of little consequence in JFETs, unlike the situation in small-dimension MESFETs (see Section 9.6).

10.5 CHAPTER SUMMARY

In this chapter brief descriptions of the principles of operation of the junction field-effect transistor, and of its electrical performance under both ac and dc conditions, have been presented. The treatment has been kept short deliberately, for two reasons: (1) the JFET is operationally identical to the MESFET, which is treated in detail in Chapter 9; (2) nowadays the JFET finds only limited use in special applications, such as differential-input amplifiers (see Problem 10.6). The

JFET differs from the MESFET in that a ***pn-junction diode***, rather than a Schottky barrier diode, is used as the ***gate element*** to control the width of the conducting channel between source and drain.

The drain dc I–V characteristics exhibit the usual resistive and saturation regimes as seen in other FETs. The saturation regime can be modeled from the standpoints of either channel pinch-off or velocity saturation. A comparison of the two models is presented in the chapter. The dc model is extended to a large-signal transient model and a small-signal ac model in the same manner as for the MESFET.

10.6 REFERENCE

1. K. Lehovec and R. Zuleeg, "Voltage–current characteristics of GaAs JFETs in the hot electron range," *Solid-State Electronics,* vol. 13, pp. 1415–1426, 1970.

PROBLEMS

10.1. An n-channel JFET of the type shown in Fig. 10.3a has a width of 30 μm, a length of 5 μm, and is made from an n-type silicon wafer of resistivity 20 Ω·cm. If no depletion regions were present, the channel resistance would be 5 kΩ. Calculate $V_{DS,\text{sat}}$ when $V_{GS} = -3$ V. Use the constant-mobility, channel pinch-off model.

10.2. For a JFET of the type shown in Fig. 10.3a, show that the small-signal channel resistance measured before pinch-off at a given V_{GS} equals $1/g_{m\text{sat}}$, where the transconductance $g_{m\text{sat}}$ is measured in saturation at the same V_{GS}. Use the constant-mobility, channel pinch-off model.

10.3. An n-channel JFET of the type shown in Fig. 10.3b has $Z/L = 170$, $a = 1.5$ μm, $N_D = 7.5 \times 10^{15}$ cm^{-3}, and $N_A = 10^{19}$ cm^{-3}. For this transistor, using the constant-mobility, channel pinch-off model:

(a) Calculate the value of the pinch-off voltage.

(b) Find the value at which I_D saturates when $V_{GS} = -1$ V.

10.4. An unlabeled JFET of the type shown in Fig. 10.3b is evaluated using the circuit of Fig. 10P.1 with the results shown in the accompanying table.

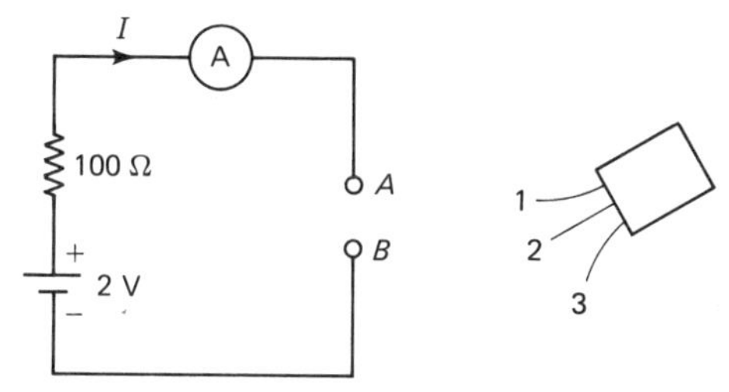

Figure 10P.1

(a) Is this an *n*-channel or a *p*-channel device?
(b) Determine which leads are the source, drain, and gate.
Assume that the voltage drop across the ammeter is negligible and that the bottom gate electrode is connected to the *source*.

Connection		
A	*B*	*I* (A)
1	2	1×10^{-9}
1	3	5×10^{-3}
2	1	1×10^{-2}
2	3	1×10^{-2}
3	1	1×10^{-2}
3	2	1×10^{-9}

10.5. **(a)** Using (10.13) for $I_{D\text{sat}}$, compute the saturation regime transconductance $g_{m\text{sat}}$ for the device described in Table 10E.1. Take $V_{GS} = 0$ V.
(b) For this same JFET, use (10.14) to compute the gate–source and gate–drain capacitances when $V_{GS} = 0$ V and $V_{DS} = 5$ V.
(c) Show that the cutoff frequency for a JFET can be approximated by

$$f_T = \frac{g_m}{2\pi C_{GS}} \tag{10P.1}$$

(d) Use the results of parts (a), (b), and (c) to estimate the cutoff frequency for the JFET specified in Table 10E.1.

10.6. The reverse-biased gate–source *pn* junction gives the JFET a high input impedance.

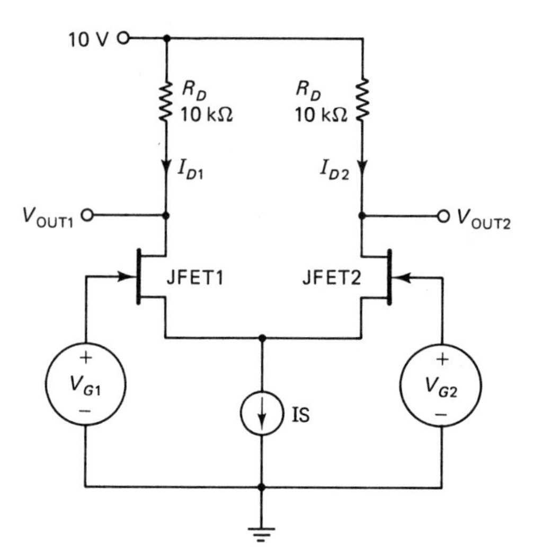

Figure 10P.2

Because of this property, JFETs are widely used in the differential input stage of monolithic op-amps. A typical circuit is shown in Fig. 10P.2. The sum of the currents in the two main branches of the circuit is constrained to the value IS of the current source. IS is usually chosen to be just less than $I_{D\text{sat}}$ of the transistors. Using a pair of identical transistors, with specifications as listed in Table 10E.1, establish the range of differential input voltage ($V_{G1} - V_{G2}$) required to produce a swing in the differential output voltage from ($+R_D$IS) to ($-R_D$IS). Confirm the result using SPICE to generate the transfer characteristic of the circuit.

11 Bipolar Junction Transistors

11.1 STRUCTURE

The bipolar junction transistor (BJT), in its silicon manifestation, is usually formed by two diffusions or ion implantations into a host silicon layer (see Section 12.2.2 and Fig. 11.1). The material of the second diffusion or implantation forms the ***emitter*** region of the BJT and is wholly contained within the ***base***, which is formed by the material first introduced into the host silicon. The base and emitter regions are defined by photolithography and silicon dioxide masking, as described in Chapter 12. The third part of the device, the ***collector*** region, is usually an epitaxial layer grown on a substrate of opposite doping type. Maintaining a reverse bias across the epilayer–substrate *pn* junction helps to isolate the transistor from neighboring devices built on the same substrate.

The device shown in Fig. 11.1 is built within an *n*-type layer and is called an *npn* BJT. By starting with a *p*-type epitaxial layer a *pnp* structure can be fashioned. These devices are examples of ***homojunction*** devices, that is, ones in which the junctions occur between like-semiconductor material (silicon in this case).

BJTs based on gallium arsenide are still in their infancy, primarily because of the immaturity, relative to silicon, of GaAs material and bipolar device processing technologies. An additional problem with GaAs bipolar transistors is the very short minority carrier lifetime, which requires a very thin base region (about 0.1 μm) that is difficult to fabricate. A structure that may find applications in very high speed circuits is shown in Fig. 11.2. The device is actually a ***heterojunction*** structure, as one of the junctions (emitter–base) is made from two different ma-

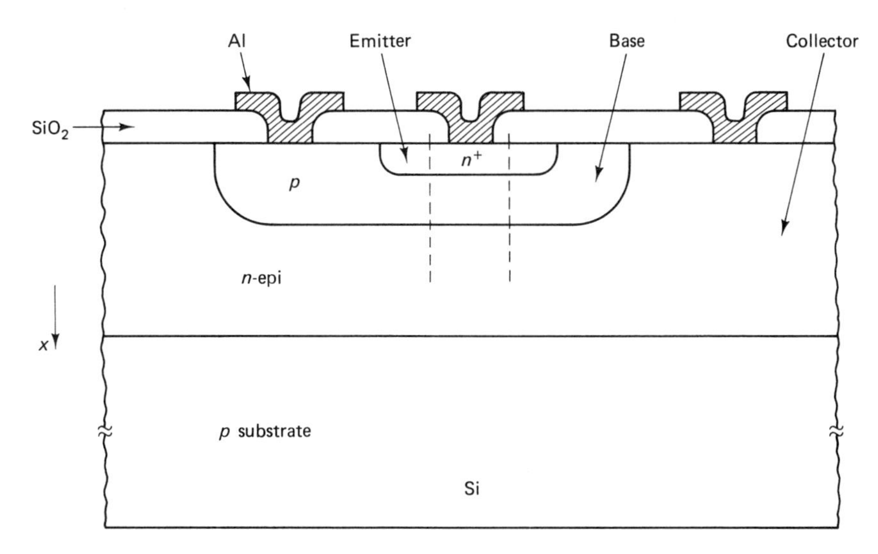

Figure 11.1 Cross section through a silicon bipolar transistor. The region within the dashed lines is elaborated on in Fig. 11.3.

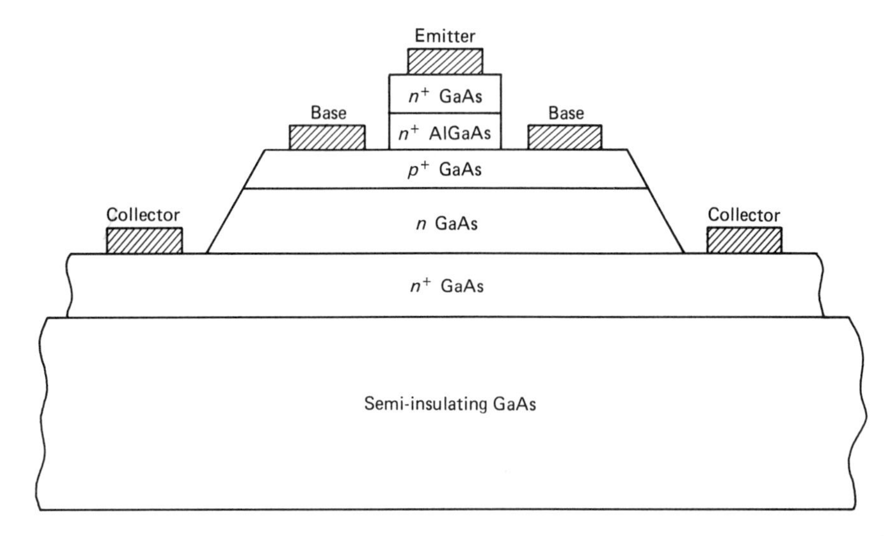

Figure 11.2 Cross section illustrating the basic features of an AlGaAs/GaAs heterojunction bipolar transistor. Note the stack of thin epitaxial layers on top of the semi-insulating substrate.

terials, an n-type layer of aluminum gallium arsenide deposited on top of a p-type GaAs layer.

11.2 PRINCIPLES OF OPERATION

The operation of a BJT can be described conveniently via a one-dimensional representation of the structure shown in Fig. 11.1. The resulting, simplified structure is shown in Fig. 11.3. Not only is this model convenient, but it is also quite accurate, as the charge flows in a BJT are determined by concentration gradients and electric fields that act primarily in the x-direction. From Fig. 11.3 it can be appreciated that physically the BJT comprises two, closely spaced pn junctions that share one region—the base.

As a first step toward understanding the operation of the BJT, let us take the left-hand junction, the emitter–base junction, to be forward biased and the

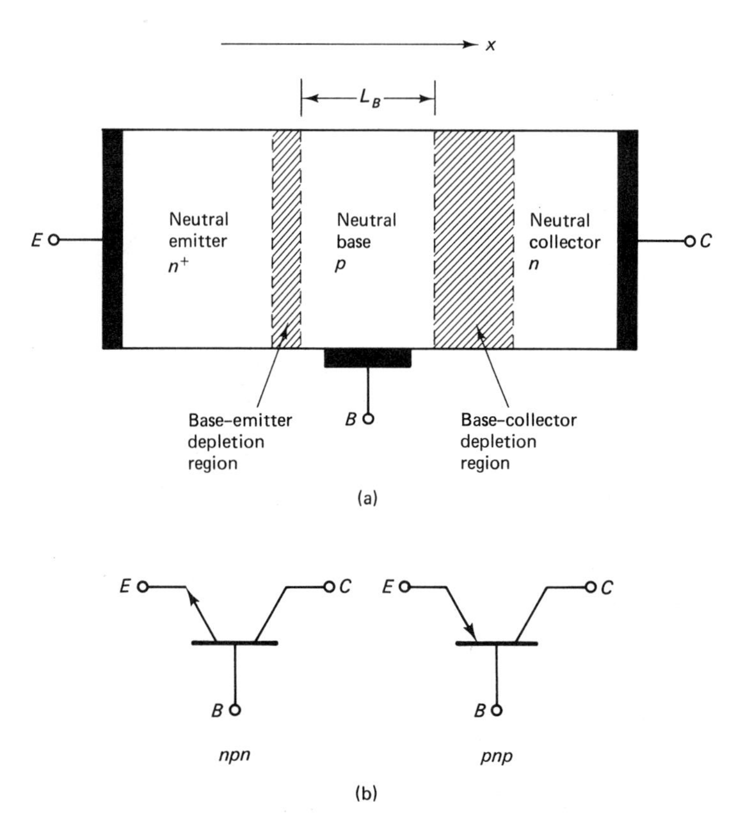

Figure 11.3 (a) One-dimensional representation of a BJT. The section shown is that within the dashed lines of Fig. 11.1. (b) Circuit symbols for BJTs.

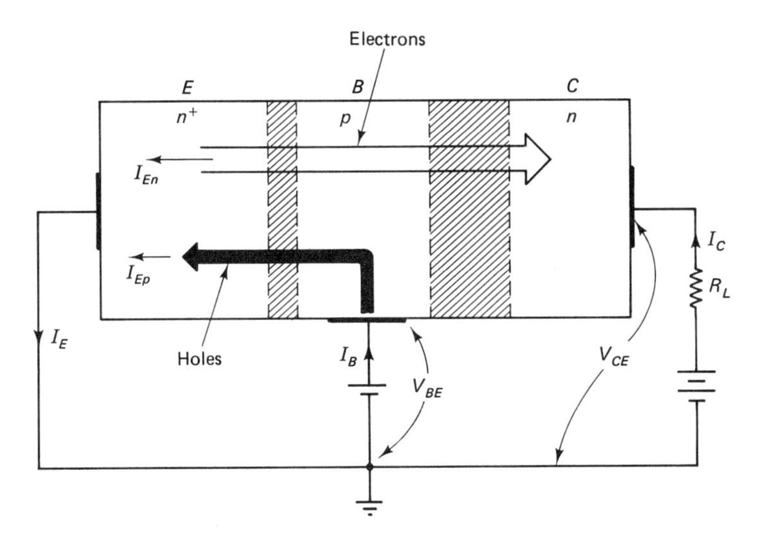

Figure 11.4 Principal charge flows in a BJT biased in the active mode. *Note:* $V_{CE} > V_{BE}$; therefore, the collector–base junction is reverse biased and the emitter–base junction is forward biased.

other junction, the base–collector junction, to be reverse biased. This mode of operation is a very common one in BJT circuits and is called the ***active mode***. The emitter in homojunction devices is deliberately made more heavily doped than the base, so that in the case of an n^+ emitter and a *npn* transistor, the forward-bias current at the emitter–base junction is carried primarily by electrons (see Fig. 11.4). These electrons are emitted into the base. Let us consider the case of a base which is so narrow that no recombination occurs therein. The electrons diffuse, therefore, to the depletion region of the reverse-biased collector–base junction. They appear there as extra minority carriers and are swept into the collector to constitute a drift current that can be measured in the collector lead. In Fig. 11.4 this current is assumed to be far greater than the saturation current which is inherent to a single reverse-biased *pn* junction.

In the situation just described, the only component of base current is that due to holes which are injected into the emitter. If the emitter is longer than several hole diffusion lengths, this component, I_{Ep}, is given by the expression derived for a long *pn*-junction diode in Section 6.3.1, namely (6.41)

$$I_{Ep} = qA \frac{D_h^E}{L_h^E} p_{0n}^E \left[\exp\left(\frac{qV_{EB}}{kT}\right) - 1 \right] \tag{11.1}$$

where p_{0n}^E is the equilibrium hole concentration in the emitter. The corresponding electron component of the emitter current, I_{En}, cannot be given by the direct counterpart to (11.1) because, having just assumed the base to be short, we cannot now take it to be very long. The correct expression for I_{En} is derived later [equation

(11.23)], but it suffices for the present argument to note that

$$I_{En} \propto n_{0p} \tag{11.2}$$

where n_{0p} is the equilibrium electron concentration in the base.

The point of making the emitter more heavily doped than the base is that $p_{0n}^{E} \ll n_{0p}$ and, therefore, $I_{En} \gg I_{Ep}$. The point of making the base narrow is so that $I_C \approx I_{En}$. The point of having a three-terminal device is so that I_{Ep} can appear in one lead (the base) and I_C can appear in a separate lead (the collector). The large change in collector current that can appear in a properly designed BJT in response to a change in base current is, in effect, current amplification, a very useful circuit property. If a resistor is placed in the collector lead, as in Fig. 11.4, a voltage will be dropped across it due to the presence of I_C. As long as this voltage drop does not reduce the voltage at the collector below that needed to maintain the reverse bias at the collector–base junction, the current I_C ($\approx I_{En}$) will not be diminished. Thus a small change in base current (I_{Ep}) is the instigator of a large change in collector voltage. This is a manifestation of ***transistor action***, that is, control by a third terminal of the current at, or the voltage between, two other terminals of the device.

11.3 DC ANALYSIS

11.3.1 Active Regime

Before describing some simple, yet very useful equations that illustrate the relationship between the currents in a BJT operating in the active mode, let us improve the model depicted in Fig. 11.4. To do this we include the collector–base saturation current (I_{CBO}) and the current due to recombination of electrons and holes in the base (I_{rec}). The new model is shown in Fig. 11.5, from which

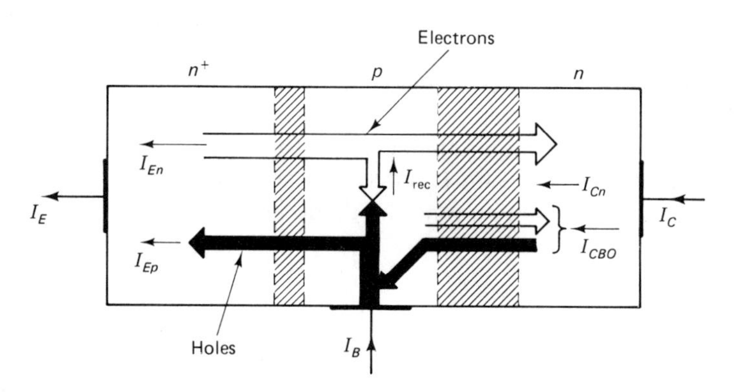

Figure 11.5 As Fig. 11.4 but with extra charge flows due to recombination in the base and generation in the collector–base depletion region.

the terminal currents in the active mode can be seen to be

$$I_E = I_{En} + I_{Ep} \tag{11.3}$$

$$I_C = I_{Cn} + I_{CBO} \tag{11.4}$$

$$I_B = I_{Ep} + I_{\text{rec}} - I_{CBO} \tag{11.5}$$

Whereas these equations are helpful in linking the terminal currents to the internal currents, they do not readily give an appreciation of how "good" a transistor is. For example, how good is the emitter at emitting electrons, and how successful are the electrons in reaching the collector? To provide a ready estimate of these attributes, the following parameters are defined:

$$\text{emitter injection efficiency} = \frac{I_{En}}{I_E} \equiv \gamma \tag{11.6}$$

$$\text{base transport factor} = \frac{I_{Cn}}{I_{En}} \equiv B \tag{11.7}$$

Incorporating these terms into (11.4), we have

$$I_C = \gamma B I_E + I_{CBO} \tag{11.8}$$

This is a very instructive equation because it succinctly states the origins of the collector current in the active mode; that is, the components are a reverse-bias current inherent to a *pn* junction, augmented by some fraction of the emitter current. The better the transistor is at transferring charge from the emitter to the collector, the higher γ and B will be and thus the larger the collector current will be. It is convenient to combine γ and B into a single parameter, the ***dc common-base current gain factor*** $\boldsymbol{\alpha}$, that is,

$$\alpha \equiv \gamma B \tag{11.9}$$

The maximum possible value of α is unity. In practice, values of 0.999 are easily attained.

Equation (11.8) can also be used to explain the subscripts CBO used for the collector/base junction saturation current. C and B refer to the ***c***ollector and ***b***ase, respectively, while O indicates that the third terminal of the device, the emitter in this case, is *o*pen-circuit.

While α is a sufficient parameter to describe the transport properties of a transistor, another parameter is often used to specify its current-amplifying properties. To derive this new parameter, we use (11.8) and Kirchhoff's current law, namely,

$$I_E = I_C + I_B \tag{11.10}$$

Using these two equations to eliminate I_E, we have, in the active mode,

$$I_C = \frac{\alpha}{1 - \alpha} I_B + \frac{I_{CBO}}{1 - \alpha} \tag{11.11}$$

The term $\alpha/(1 - \alpha)$ will be a large number if $\alpha \approx 1$. This term, as it appears in (11.11), neatly illustrates the fact that the collector current is an amplified version of the base current. It is defined as the ***dc common-emitter current gain factor*** and given the symbol $\boldsymbol{\beta}$; thus, in the active mode,

$$I_C = \beta I_B + I_{CEO} \tag{11.12}$$

where $I_{CEO} = I_{CBO}/(1 - \alpha)$ and is so labeled as it represents the current through two terminals, the *c*ollector and the *e*mitter, with the third terminal, the base in this case, *o*pen. Note that because $\alpha \approx 1$, at least for the operating conditions discussed so far, $I_{CEO} \gg I_{CBO}$. The physical reason for this stems from the fact that the collector–emitter voltage is distributed across the two junctions of the transistor. Most of V_{CE} is dropped across the high resistance of the reverse-biased collector–base junction, but some will appear across the emitter–base junction. The polarity is such that the emitter–base diode is forward biased. Therefore, some electrons will be emitted into the base and subsequently collected by the collector, so augmenting I_{CBO}, the inherent reverse-bias current of the collector–base junction.

Equations (11.8) and (11.12) are represented graphically in Figure 11.6 in the form of collector I–V relations for the active mode. To measure the characteristic of Fig. 11.6a necessitates the transistor being connected in such a way that I_E can be varied independently. The connection is shown in the figure and is known as the ***common-base connection***, because the base is common to both the emitter and collector circuits. Correspondingly, the configuration shown in Fig. 11.6b is called the ***common-emitter connection***.

11.3.2 Saturation Regime

To complete the plots shown in Fig. 11.6 we need to consider a different mode of operation of the BJT, namely one in which both the emitter–base and collector–base junctions are forward biased. This mode of operation is called the ***saturation mode***.

Considering the common-base connection first, if the collector–base junction is forward biased, electrons are emitted from the collector into the base. This electronic flow is in opposition to the flow of electrons into the collector from the emitter. Thus the net flow of electrons into the collector is reduced. As the collector–base junction becomes more forward biased, the electron flow from the collector into the base increases exponentially and the collector current eventually reverses (see Fig. 11.7a).

In the common-emitter configuration the forward biasing of the collector–base junction can be brought about by decreasing V_{CE}. All that is needed, in an *npn* BJT, is for the collector voltage to drop below the base voltage. Note that V_{CE} does not have to change polarity for this to occur. Thus the diminution of the collector current with respect to its value in the active mode commences in the first quadrant of the $I-V$ plane (see Fig. 11.7b).

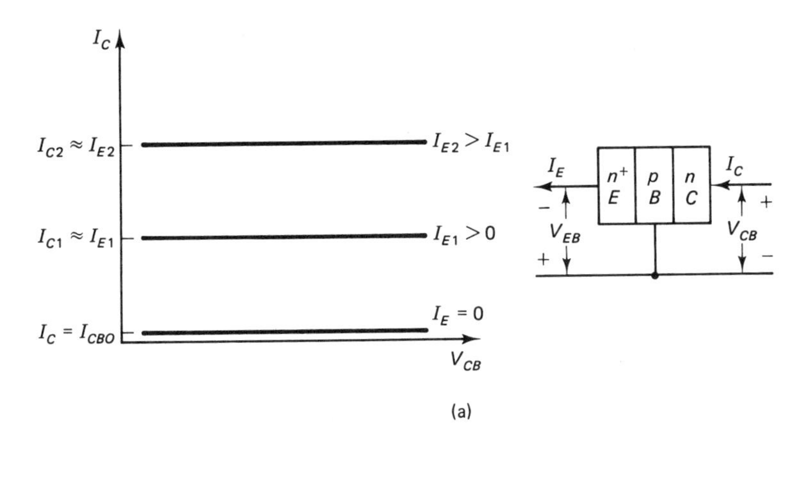

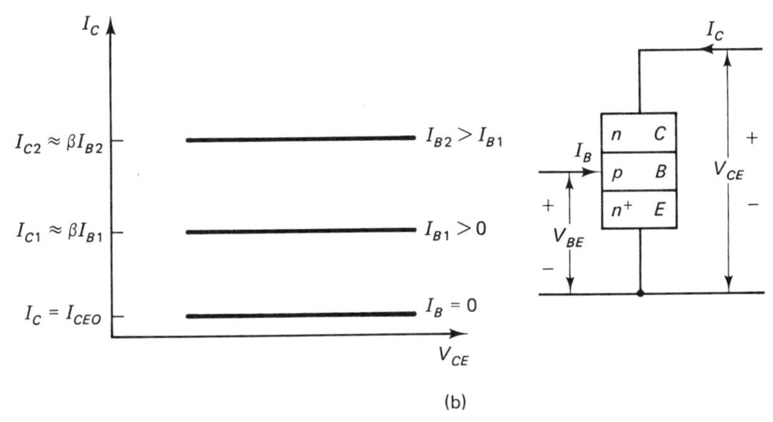

Figure 11.6 Collector I–V characteristics in the active mode for (a) the common-base connection, and (b) the common-emitter connection.

The fact that the individual curves tend to coalesce gives this mode of operation its appelation "saturation." That is, in the common-emitter configuration at low V_{CE}, for example, increasing the base current does not lead to an increase in collector current (i.e., the collector current is saturated).

11.3.3 Cutoff Regime

A third mode of operation of the BJT is when both junctions are reverse biased. This mode is called the ***cutoff*** mode, in recognition of the fact that the emitter–base reverse bias cuts off the supply of electrons from the emitter to the base and, hence, to the collector. As the collector–base junction is also reverse biased, the collector current is very small. The base current is also very small, as it only supplies the holes for the reverse-bias currents across the two junctions.

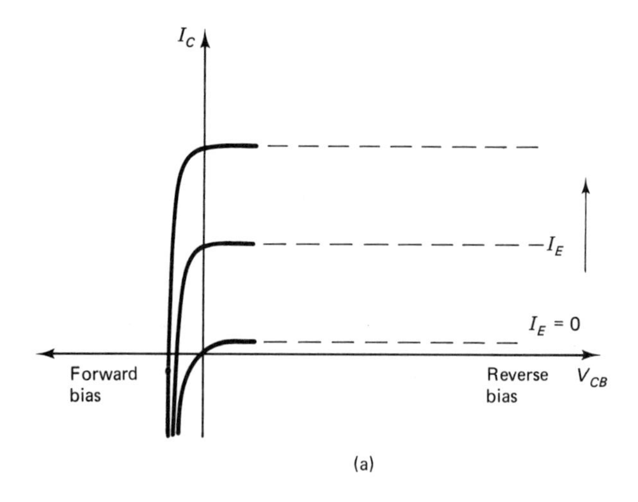

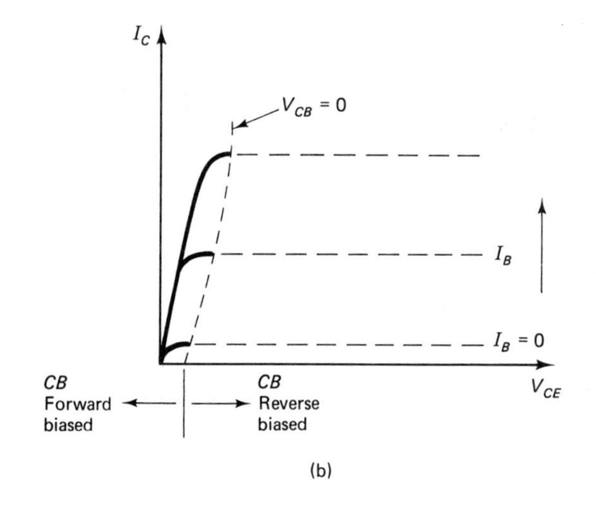

Figure 11.7 Collector I–V characteristics in the saturation mode for (a) the common-base connection, and (b) the common-emitter connection. The dashed curves are for the active region (see Fig. 11.6).

11.3.4 Inverse Mode

The final possible permutation of junction biases is the one in which the emitter–base junction is reverse biased and the collector–base junction is forward biased. This is akin to the normal active mode but with the true emitter acting as a collector, and vice versa, hence the appelation "***inverse*** active." Because the device is not physically symmetrical, owing to the difference in doping densities of the emitter and the collector, the collector currents in the normal active and inverse active modes are not of the same magnitude. The lower doping density of the collector will make it a poorer emitter of electrons, so both α and β will be lower in the inverse active mode.

The normal and inverse modes of operation are sometimes referred to as

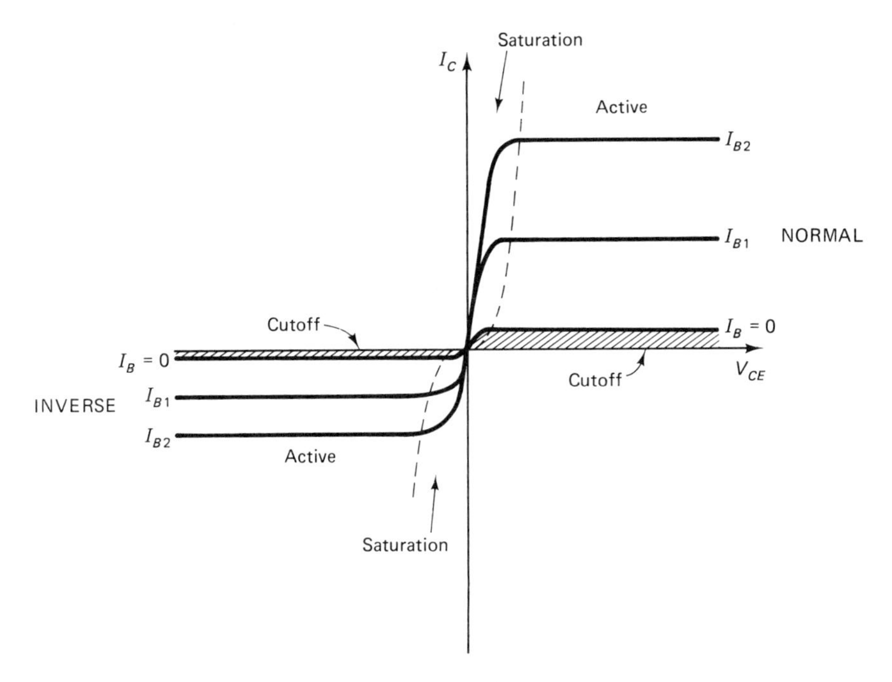

Figure 11.8 Summary of the various modes of operation in the common-emitter configuration.

the ***forward*** and ***reverse*** modes, respectively. We prefer to use "normal" and "inverse" and to reserve "forward" and "reverse" for descriptions of junction bias. However, to be consistent with the notation used in SPICE, we do, later in this chapter, use the subscripts F and R for current gains in the normal and inverse modes, respectively.

With the true collector acting as emitter, saturation and cutoff modes are also possible, as illustrated and summarized in Fig. 11.8.

11.3.5 Detailed Analysis

The preceding four subsections describe the operation of the BJT and also provide some equations that relate the transistor currents in the active mode to readily understandable parameters such as α and β. This subsection provides a more detailed analysis of BJT performance, one that relates the terminal currents and voltages to the basic device properties of doping densities and regional widths, and is valid for all modes of operation. The model of the transistor to be analyzed is that shown in Fig. 11.9.

In using Fig. 11.9, electron and hole currents inside the transistor will be calculated following the rule that positive charge flow in the x-direction constitutes a positive current. However, the final terminal currents I_E, I_B, and I_C will be

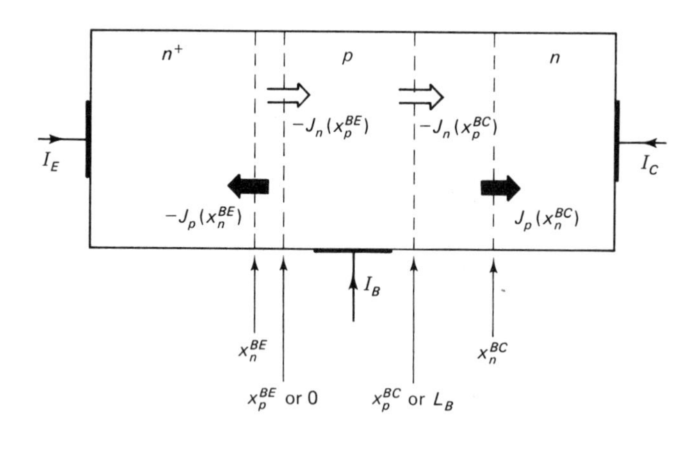

Figure 11.9 Collector and emitter current components in a BJT.

taken as positive when positive charge flows into the corresponding terminal, in accord with long-established convention.

To avoid adding an undue amount of detail to the initial analysis, a number of simplifying assumptions will be made. The modifications to the model that are necessary to remove these assumptions are relatively straightforward, and are considered in Section 11.5. The assumptions are as follows:

1. Recombination and generation processes in the two depletion regions make a negligible contribution to charge flow.
2. The emitter, base, and collector are uniformly doped, and therefore there are no built-in electric fields outside the depletion regions.
3. The emitter and collector regions are wide compared to the minority carrier diffusion lengths in these regions.

From assumption 3 and the results obtained in the analysis of the diode in Section 6.3, it follows immediately that the hole current crossing the emitter–depletion region boundary is given by

$$J_p(x_n^{BE}) = -\frac{qD_h^E}{L_h^E} p_{0n}^E(e^{qV_{BE}/kT} - 1) \tag{11.13}$$

Here the superscript E refers to the emitter. Similarly, the hole current crossing the collector–depletion region boundary is given by

$$J_p(x_n^{BC}) = \frac{qD_h^C}{L_h^C} p_{0n}^C(e^{qV_{BC}/kT} - 1) \tag{11.14}$$

In normal operation, the base–emitter junction is forward biased and the base–collector junction reverse biased, so the current described by (11.13) results from hole injection from the base into the emitter, while that described by (11.14) is the reverse leakage current of the collector–base junction resulting from the thermal generation of holes within a diffusion length of the collector–base depletion region.

The next task in the analysis is to compute the electron current at the two borders of the base. This is done using an extension of the *pn* junction analysis of Section 6.2.2.2. We have assumed that there is no electric field in the uniformly doped base, so the minority carriers in this region move only by diffusion. Under steady-state conditions with no optical generation of carriers, the electron continuity equation then becomes

$$\frac{d^2 n_p}{dx^2} = \frac{n_p - n_{0p}}{L_e} \tag{11.15}$$

For convenience, we will temporarily set $x = 0$ at x_p^{BE} and $x = L_B$ at x_p^{BC}, L_B being the width of the neutral base. One possible form of the general solution to (11.15) was given in (6.31). In the base of a transistor, a more useful general solution is

$$n_p(x) - n_{0p} = A \cosh \frac{x}{L_e} + B \sinh \frac{x}{L_e} \tag{11.16}$$

where A and B are constants that must be chosen to fit the problem at hand, and the hyperbolic functions cosh and sinh u are defined by

$$\cosh u = \frac{e^u + e^{-u}}{2} \tag{11.17a}$$

and

$$\sinh u = \frac{e^u - e^{-u}}{2} \tag{11.17b}$$

It can be seen that (11.16) is really just a linear combination of the functions e^{x/L_e} and e^{-x/L_e}, and so is physically no different from (6.31).

The boundary conditions applicable to (11.16) are

$$n_p(0) = n_{0p} e^{qV_{BE}/kT} \tag{11.18}$$

and

$$n_p(L_B) = n_{0p} e^{qV_{BC}/kT} \tag{11.19}$$

The arguments justifying the use of these boundary conditions are the same as those presented in Section 6.2.2.2. With these boundary conditions, we find that

$$A = n_{0p}(e^{qV_{BE}/kT} - 1) \tag{11.20}$$

and

$$B = \frac{n_{0p}}{\sinh(L_B/L_e)} (e^{qV_{BC}/kT} - 1) - n_{0p} \coth \frac{L_B}{L_e} (e^{qV_{BE}/kT} - 1) \tag{11.21}$$

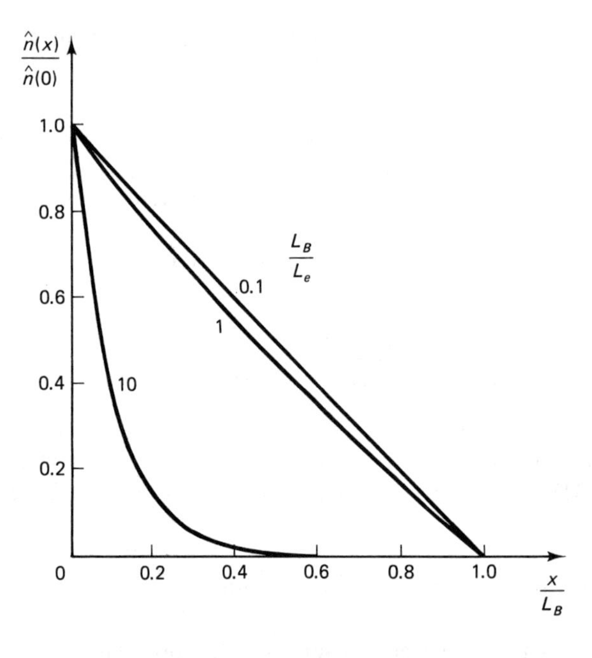

Figure 11.10 The excess electron distribution in the base, as computed from (11.16)–(11.21) for $V_{CB} = 0$.

The resulting electron distribution for normal active operation is shown in Fig. 11.10. The electron diffusion current is given by

$$J_n(x) = qD_e \frac{dn_p}{dx} = qD_e \left(\frac{A}{L_e} \sinh \frac{x}{L_e} + \frac{B}{L_e} \cosh \frac{x}{L_e} \right) \tag{11.22}$$

We therefore have

$$J_n(x_p^{BE}) = \frac{qD_e n_{0p}}{L_e} \left[\frac{e^{qV_{BC}/kT} - 1}{\sinh(L_B/L_e)} - \coth \frac{L_B}{L_e} (e^{qV_{BE}/kT} - 1) \right] \tag{11.23}$$

and using the identity $\cosh^2 u - \sinh^2 u = 1$,

$$\begin{aligned} J_n(x_p^{BC}) &= \frac{qD_e n_{0p}}{L_e} \left[\left(\sinh \frac{L_B}{L_e} - \frac{\cosh^2(L_B/L_e)}{\sinh(L_B/L_e)} \right) (e^{qV_{BE}/kT} - 1) \right. \\ &\quad \left. + \frac{\cosh(L_B/L_e)}{\sinh(L_B/L_e)} (e^{qV_{BC}/kT} - 1) \right] \\ &= \frac{qD_e n_{0p}}{L_e} \left[\coth \frac{L_B}{L_e} (e^{qV_{BC}/kT} - 1) - \frac{e^{qV_{BE}/kT} - 1}{\sinh(L_B/L_e)} \right] \end{aligned} \tag{11.24}$$

In the absence of recombination in the depletion regions,

$$J_n(x_n^{BE}) = J_n(x_p^{BE}) \tag{11.25}$$

and

$$J_n(x_n^{BC}) = J_n(x_p^{BC}) \tag{11.26}$$

We must also have

$$J_E = J_n(x_n^{BE}) + J_p(x_n^{BE}) \tag{11.27}$$

$$J_C = -J_n(x_n^{BC}) - J_p(x_n^{BC}) \tag{11.28}$$

and

$$J_B = -(J_E + J_C) = J_n(x_n^{BC}) - J_n(x_n^{BE}) + J_p(x_n^{BC}) - J_p(x_n^{BE}) \tag{11.29}$$

The actual terminal currents I_E, I_B, and I_C are obtained by multiplying the current densities J_E, J_B, and J_C by the area A_{BE} of the base–emitter junction. Using (11.13), (11.14), (11.23), (11.24), and (11.27)–(11.29), it is found that

$$I_E = -a_{11}(e^{qV_{BE}/kT} - 1) + a_{12}(e^{qV_{BC}/kT} - 1) \tag{11.30}$$

$$I_C = a_{21}(e^{qV_{BE}/kT} - 1) - a_{22}(e^{qV_{BC}/kT} - 1) \tag{11.31}$$

and

$$I_B = (a_{11} - a_{21})(e^{qV_{BE}/kT} - 1) + (a_{22} - a_{12})(e^{qV_{BC}/kT} - 1) \tag{11.32}$$

where a_{11}, a_{12}, and a_{21}, and a_{22} have all been chosen to be positive constants. We find that

$$a_{11} = A_{BE}\left(\frac{qD_h^E p_{0n}^E}{L_h^E} + \frac{qD_e}{L_e} n_{0p} \coth\frac{L_B}{L_e}\right) \tag{11.33}$$

$$a_{12} = a_{21} = A_{BE}\left[\frac{qD_e n_{0p}}{L_e} \frac{1}{\sinh(L_B/L_e)}\right] \tag{11.34}$$

and

$$a_{22} = A_{BE}\left(\frac{qD_h^C p_{0n}}{L_h^C} + \frac{qD_e n_{0p}}{L_e} \coth\frac{L_B}{L_e}\right) \tag{11.35}$$

Equations (11.30)–(11.35) are the celebrated ***Ebers–Moll relations*** for the BJT, named after the two men who first derived them in 1954 [1]. These equations form the basis for the BJT model used in SPICE.

The current gains in the normal and inverse modes, labeled with subscripts F and R, respectively (see Section 11.3.4), are given by

$$\alpha_F = \frac{a_{12}}{a_{11}} \tag{11.36}$$

$$\beta_F = \frac{a_{12}}{a_{11} - a_{12}} \tag{11.37}$$

$$\alpha_R = \frac{a_{12}}{a_{22}} \tag{11.38}$$

and

$$\beta_R = \frac{a_{12}}{a_{22} - a_{12}} \tag{11.39}$$

As noted earlier, in most modern transistors, $L_e \gg L_B$. In this case, (11.33)–(11.35) can be simplified using the expansions

$$\cosh u \simeq 1 + \frac{u^2}{2} \cdots \simeq 1 \tag{11.40a}$$

and

$$\sinh u \simeq u + \frac{u^3}{6} \cdots \simeq u \tag{11.40b}$$

for $u \to 0$. We then have

$$a_{11} = A_{BE}\left(\frac{qD_h^E p_{0n}^E}{L_h^E} + \frac{qD_e n_{0p}}{L_B}\right) \tag{11.41}$$

$$a_{12} = A_{BE}\frac{qD_e n_{0p}}{L_B} \tag{11.42}$$

and

$$a_{22} = A_{BE}\left(\frac{qD_h^C p_{0n}^C}{L_h^C} + \frac{qD_e n_{0p}}{L_B}\right) \tag{11.43}$$

11.4 DC CIRCUIT MODEL

There are a number of possible equivalent-circuit representations of the Ebers–Moll equations [2]. The one we present in Fig. 11.11 is called the ***nonlinear hybrid π version***. By evaluating the terminal currents, and setting $I_S = a_{12}$, it is easy to confirm the equivalence of the circuit and the equations (11.30)–(11.35). The

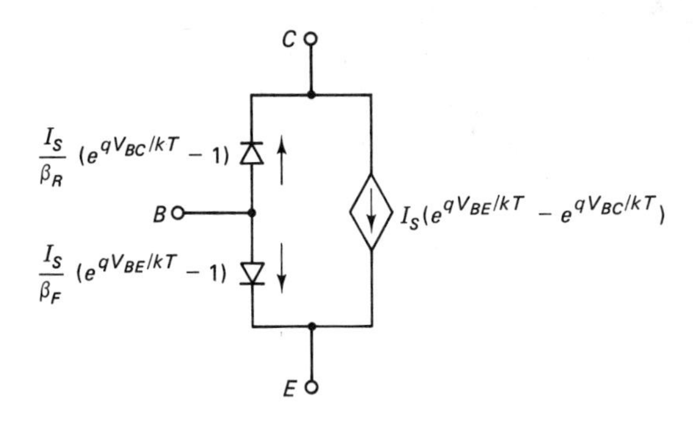

Figure 11.11 Nonlinear hybrid-π version of the Ebers–Moll model. This is the basic model used to describe the BJT in SPICE. The current directions are correct for an *npn* transistor. They must be reversed for a *pnp* device.

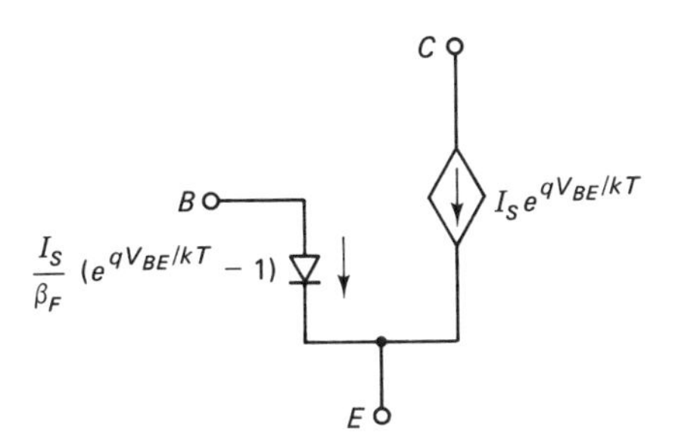

Figure 11.12 Simplified version of the Ebers–Moll model for normal active operation. The reverse leakage current of the collector–base junction has been ignored.

diodes in this circuit do not represent the actual physical junctions of the transistor. Instead, they represent the components of the base current in the BJT. As the base current is usually the controlling signal in the transistor, this representation is very convenient. As an illustration of this, let us ignore the leakage current of the collector–base junction and consider the normal active mode of operation. The equivalent circuit simplifies to Fig. 11.12. This last circuit has an appealing physical interpretation. Looking between the base and emitter terminals at the input to the amplifier, one sees the relatively low impedance of the forward-biased base–emitter diode. The forward bias on the base–emitter junction produces a relatively small current in the base lead, but a far larger one in the emitter and collector, represented by the current source in Fig. 11.12. Since the value of this controlled source depends exponentially on V_{BE}, we would not normally apply an input signal voltage directly to the base to obtain linear amplification, unless the signal were small compared to kT/q. Instead, an input voltage source is converted to a current source by the addition of a large resistance in series with the base lead, leading to the classic common-emitter amplifier configuration of Fig. 11.13.

Linearization of Fig. 11.12 for small-signal operation combined with the addition of internal capacitances which must be considered at high-frequency leads to the hybrid-model of the BJT discussed in Section 11.7.1.

Figure 11.11 provides insight into the operation of the transistor in regimes other than the normal active mode. In the cutoff regime, both collector and emitter

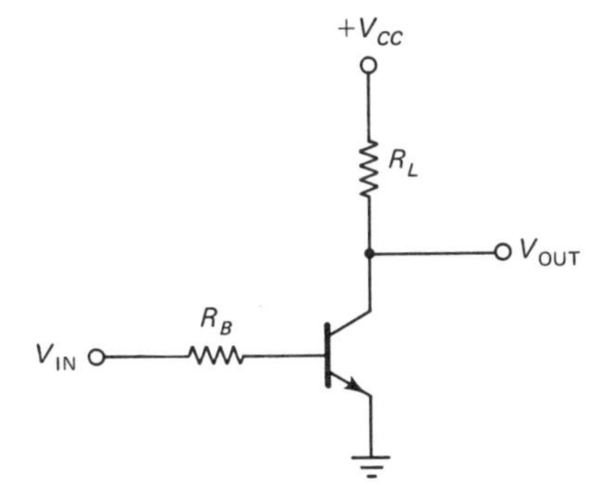

Figure 11.13 Basic common-emitter amplifier. V_{IN} includes both an ac small-signal component and a dc component maintaining the forward bias on the base–emitter junction.

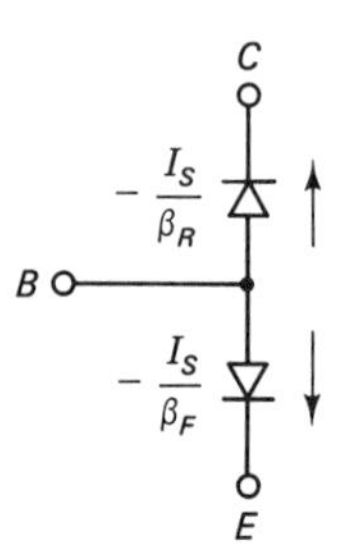

Figure 11.14 Ebers–Moll model for operation in the cutoff mode.

junctions are reverse biased, so the terminal currents are simply the small reverse leakage currents of the two junctions, as suggested in Fig. 11.14. In inverse active operation the base–collector junction is forward biased and the base–emitter reverse biased, leading to the circuit of Fig. 11.15. Here the current in the emitter lead can be viewed as controlled by the bias on the collector–base junction. In saturation, both the base–emitter and base–collector junctions are forward biased, and the full circuit of Fig. 11.11 is needed to describe the transistor. Both junctions inject opposing electron currents into the base and, in fact, if $V_{BE} = V_{BC}$, it is possible to arrive at a situation in which the current source in Fig. 11.11 is reduced to zero. In saturation the base is flooded with excess electrons from both junctions, which can have important consequences for the switching speed of the transistor. This is discussed in more detail in Section 11.6.

11.4.1 Basic SPICE Model

The basic BJT model used in SPICE is simply that represented by the circuit of Fig. 11.11. Three parameters must be supplied to SPICE to use this model: the saturation current IS of the controlled current source, the normal common-emitter current gain BF (β_F), and the inverse current gain BR (β_R). The following SPICE examples provide an introduction to the BJT model and show how the normal active characteristics vary with base current, base doping, and basewidth.

It is important before we proceed with the examples to make a distinction between current gain in a device and current gain in a circuit. The common-emitter current gain in the normal mode, as defined by (11.37), is a constant for a given

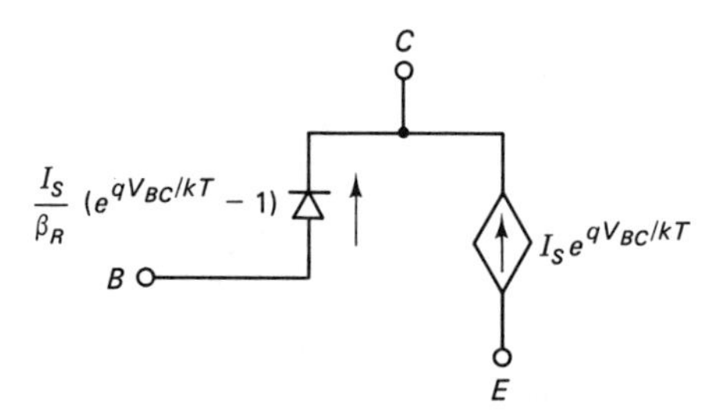

Figure 11.15 Ebers–Moll model for inverse active operation.

BJT. It depends only on the doping densities, basewidth, and minority carrier diffusion properties. As such, β_F is a ***device*** parameter. The actual current gain of a transistor in a circuit depends on the mode of operation of the device. For example, in the common-emitter connection with I_B fixed, the collector current in saturation is less than in the active region. Thus the actual dc gain (I_C/I_B) is reduced. The actual gain is, therefore, a ***circuit*** parameter. It is unfortunate that actual gain and device gain are often described by the same symbol β. To minimize confusion, we reserve β_F and β_R for the common-emitter ***device*** parameters in the normal and inverse modes, respectively. They are identical to the ***device model*** parameters BF and BR used in SPICE. For the actual ***circuit*** current gains we follow SPICE and use BETADC and BETAAC, defined as

$$\text{BETADC} = \frac{I_C}{I_B} \tag{11.44}$$

$$\text{BETAAC} = \left.\frac{\partial I_C}{\partial I_B}\right|_{\text{operating point}} \tag{11.45}$$

Example 11.1 BJT Common-Emitter Characteristics: The Effect of Base Current

To carry out this exercise we need to be able to estimate values for the model parameters BF, BR, and IS from physical data for the device. The relevant equations are (11.33)–(11.39), together with

$$\text{IS} = a_{12} \tag{11E.1}$$

The required physical properties are the doping densities, diffusion lengths, and neutral basewidth. The diffusion lengths are doping density dependent (see Example 6.1 and Problem 3.9).

Table 11E.1 summarizes the numerical data for *npn* transistors of cross-sectional area 10^{-6} cm^2 (10 μm × 10 μm) and various base doping densities and neutral basewidths. In all cases the emitter and collector donor doping densities are taken to be 10^{19} and 10^{15} cm^{-3}, respectively. The hole diffu-

TABLE 11E.1 NUMERICAL VALUES OF THE PHYSICAL AND SPICE PARAMETERS USED IN THE BJT EXAMPLES

N_A (base) (cm^{-3})	L_B (μm)	L_e (μm)	D_e (cm^2 s^{-1})	BF	BR	IS (A)
10^{17}	1	19.0	21.3	354	0.4	5.33×10^{-17}
10^{16}	2	36.9	32.6	597	3	4.04×10^{-16}
10^{17}	2	19.0	21.3	122	0.2	2.65×10^{-17}
10^{18}	2	3.9	6.3	4	0.006	7.54×10^{-19}
10^{17}	5	19.0	21.3	24	0.09	1.05×10^{-17}

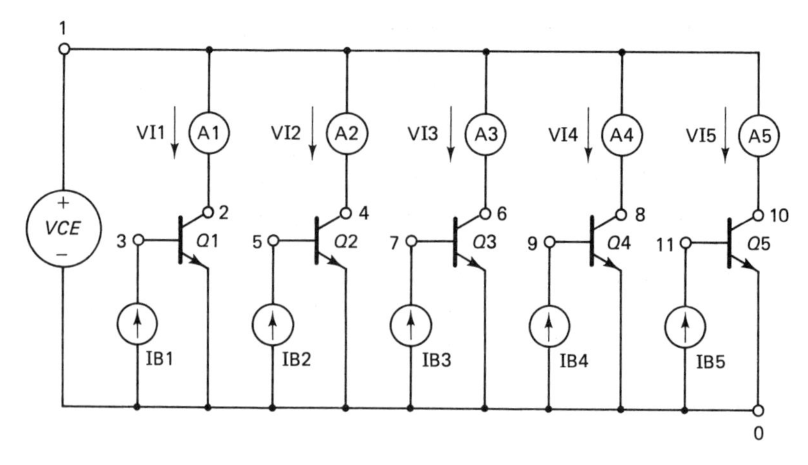

Figure 11E.1 Circuit for the SPICE simulation of collector common-emitter characteristics.

sivities and diffusion lengths for these two regions are 1.7 $cm^2\ s^{-1}$ and 0.6 μm for the emitter, and 12 $cm^2\ s^{-1}$ and 24.2 μm for the collector. The temperature is 300 K.

To examine the effect of base current on the collector current we use the circuit shown in Fig. 11E.1. The transistors in this circuit are identical; the notation in the model line of the listing given below for this circuit labels the transistors as 172, which refers to a base doping density of 10^{17} cm^{-3} and a neutral basewidth of 2 μm.

```
*BJT COMMON-EMITTER CHARACTERISTICS
*EFFECT OF BASE CURRENT
VCE 1 0
VI1 1 2
IB1 0 3  DC 0U
VI2 1 4
IB2 0 5  DC 2.5U
VI3 1 6
IB3 0 7  DC 5U
VI4 1 8
IB4 0 9  DC 7.5U
VI5 1 10
IB5 0 11 DC 10U
Q1  2  3 0  Q172
Q2  4  5 0  Q172
Q3  6  7 0  Q172
Q4  8  9 0  Q172
Q5 10 11 0  Q172
.MODEL Q172 NPN BF=122 BR=0.2 IS=2.65E-17
```

```
.DC VCE 0 2 0.05
.WIDTH OUT=80
.PLOT DC I(VI1) I(VI2) I(VI3) I(VI4) I(VI5) (0,1.5M)
.END
```

Each transistor is driven by a different base current in order to generate a family of I_C curves for the values of collector–emitter voltage *VCE* specified on the control line of the SPICE input listing. The results are plotted in Fig. 11E.2. Note the following:

(a) The characteristic is of the form derived from intuitive arguments in Section 11.3 and displayed in Fig. 11.7b.

(b) When there is no base current, the collector current is essentially zero (i.e., the transistor is cut off).

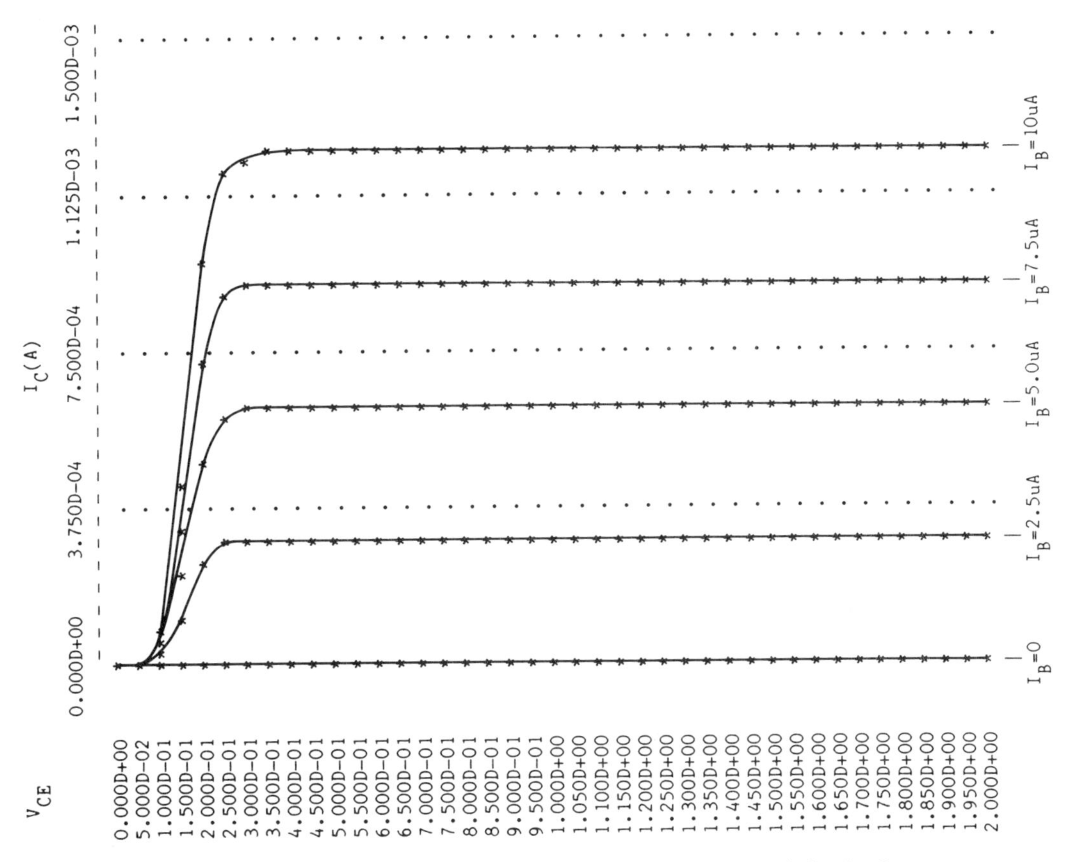

Figure 11E.2 Computed collector common-emitter characteristics for the transistor described in Example 11.1.

(c) In the active mode of operation, I_C is given by BF times I_B, in agreement with (11.12) and our notation for current gains (see the beginning of this section).

Example 11.2 BJT Common-Emitter Characteristics: The Effect of Base Doping Density

In this example the effect of base doping density on the collector I–V characteristic at a fixed base current of 5 μA is investigated. The circuit of Fig. 11E.1 with only three transistors connected in parallel is suitable for this example. The input listing is given below.

```
*BJT COMMON-EMITTER CHARACTERISTICS
*EFFECT OF BASE DOPING
VCE 1 0
VC1 1 2
IB1 0 3  DC 5U
VC2 1 4
IB2 0 5  DC 5U
VC3 1 6
IB3 0 7  DC 5U
Q1  2 3 0   Q172
Q2  4 5 0   Q162
Q3  6 7 0   Q182
.MODEL Q172 NPN BF=122 BR=0.2   IS=2.65E-17
.MODEL Q182 NPN BF=4   BR=0.006 IS=7.54E-19
.MODEL Q162 NPN BF=597 BR=3     IS=4.04E-16
.DC VCE 0 2 0.05
.WIDTH OUT=80
.PLOT DC I(VC3) I(VC2) I(VC1) (0,4E-3)
.END
```

The three transistors have the same basewidth (2 μm) but different doping densities (see Table 11E.1). The results are shown graphically in Fig. 11E.3. Note the following:

(a) The collector current decreases markedly on changing the base doping density over the range 10^{16} to 10^{18} cm^{-3}.

(b) A lower doping density in the base means less "back-injection" current, that is, the hole contribution to the emitter current [$J_p(x_n^{BE})$ in (11.13) and Fig. 11.9] is greatly reduced. Consequently, the emitter injection efficiency is enhanced, leading to an increase in collector current.

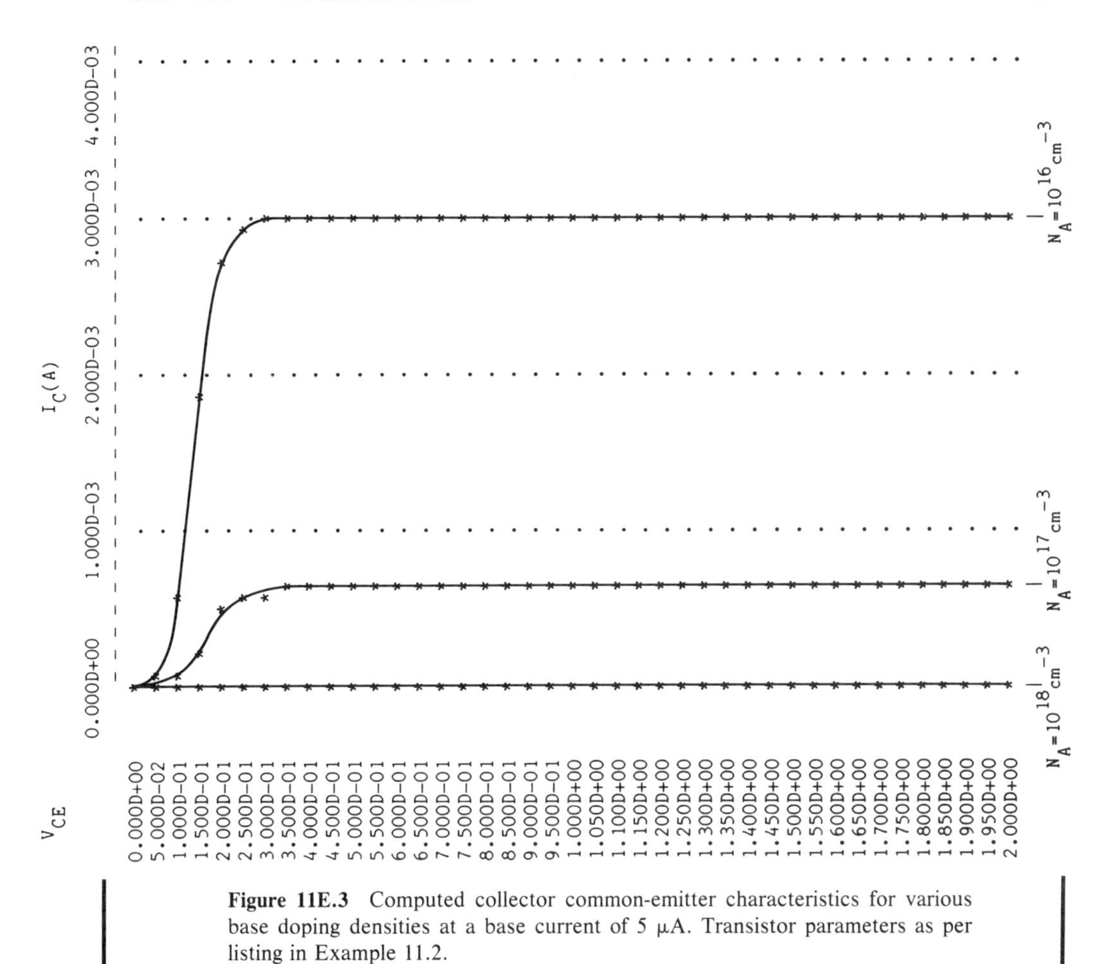

Figure 11E.3 Computed collector common-emitter characteristics for various base doping densities at a base current of 5 μA. Transistor parameters as per listing in Example 11.2.

Example 11.3 BJT Common-Emitter Characteristics: The Effect of Neutral Basewidth

Some of the electrons injected into the base from the emitter recombine with holes before they reach the collector (see Fig. 11.5). The holes are supplied, in effect, from the base lead, so this component of current appears in the base–emitter circuit rather than in the collector–emitter circuit. The smaller the ratio of diffusion length to neutral basewidth, the larger the recombination, as illustrated in Fig. 11.10. The effect on the collector *I–V* characteristics can be observed by simulating the circuit of Fig. 11E.1, using three transistors, each with a different neutral basewidth. The model parameter affected by changes in L_B is IS (see Table 11E.1 and the model lines in the listing given below.)

```
*BJT COMMON-EMITTER CHARACTERISTICS
*EFFECT OF BASEWIDTH
VCE 1 0
VC1 1 2
IB1 0 3   DC 5U
VC2 1 4
IB2 0 5   DC 5U
VC3 1 6
IB3 0 7   DC 5U
Q1   2 3 0    Q172
Q2   4 5 0    Q171
Q3   6 7 0    Q175
```

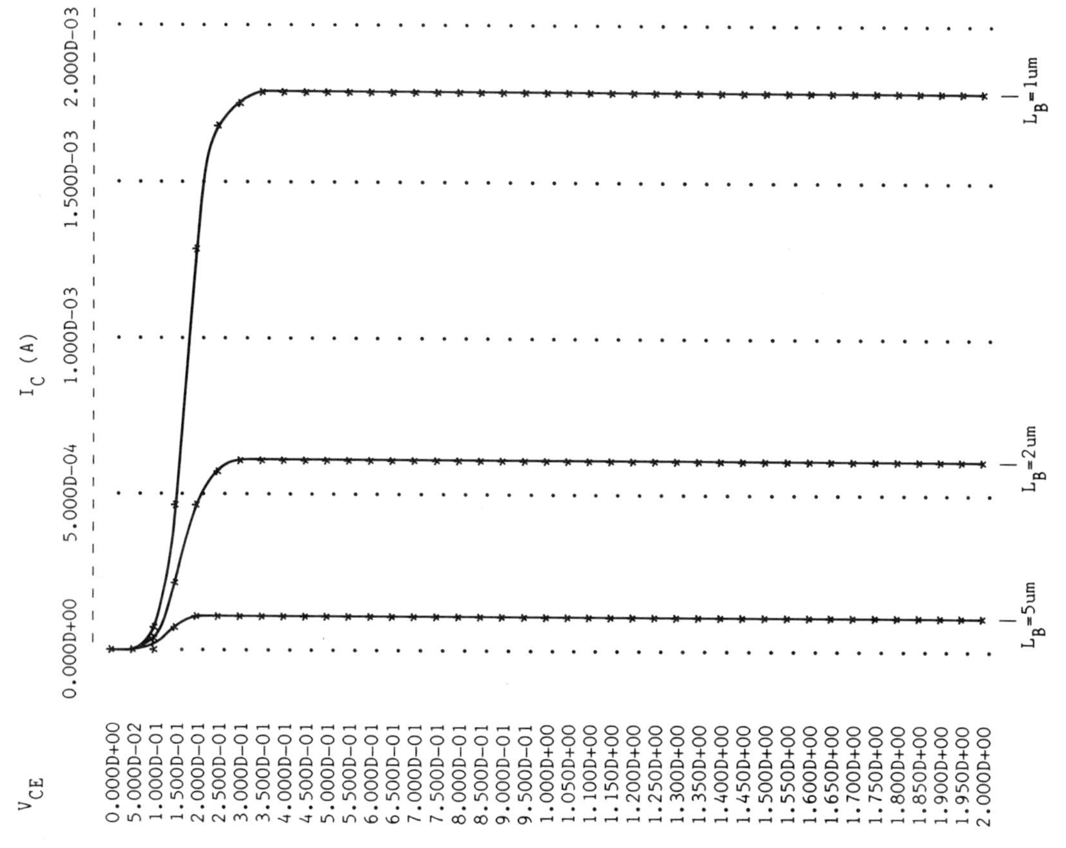

Figure 11E.4 Computed collector common-emitter characteristics for various neutral basewidths at a base current of 5 μA. Transistor parameters as per listing in Example 11.3.

```
.MODEL Q172 NPN BF=122   BR=0.2   IS=2.65E-17
.MODEL Q171 NPN BF=354   BR=0.4   IS=5.33E-17
.MODEL Q175 NPN BF=24    BR=0.1   IS=1.05E-17
.DC VCE 0 2 0.05
.WIDTH OUT=80
.PLOT  DC I(VC1) I(VC2) I(VC3) (0,2E-3)
.END
```

Results for a base doping density of 10^{17} cm^{-3} and basewidths of 1, 2, and 5 μm are shown in Fig. 11E.4. Note the following:

(a) The collector current in the active mode increases as the neutral basewidth is reduced. This is a direct consequence of the improvement in base transport factor (11.7) due to the diminution in volume of the neutral base in which recombination can occur.

(b) There is a strong dependence of I_C in the active mode on the basewidth L_B, as expected from (11.31), (11.34), and (11.35).

11.5 SECONDARY EFFECTS IN REAL BJTS

11.5.1 Basewidth Modulation

In the normal active mode of operation, for example, the depletion region on the base side of the reverse-biased collector–base junction encroaches into the base. This shortening of the neutral base region effectively brings the collector closer to the emitter. This lessens the opportunities for carriers injected from the emitter to recombine in the base before being swept into the collector. Thus the collector current increases. A similar effect was obtained in Example 11.3 by deliberately reducing the width of the base. However, in the case of ***basewidth modulation***, the basewidth is voltage dependent and the effect is manifest in the active region of the transistor characteristic by a finite conductance, called the output conductance (see Fig. 11.16).

The effect is modeled in SPICE in the same manner as channel shortening is modeled in MOSFETs and MESFETs, namely, by including an extra voltage-dependent term in the relevant expression for the device current. For a BJT in the normal active mode the appropriate equation is (11.31), which in SPICE notation, becomes, on modification,

$$I_C = \mathsf{IS}\left[\exp\left(\frac{qV_{BE}}{kT}\right)\right]\left(1 + \frac{V_{CB}}{\mathsf{VAF}}\right) \tag{11.46}$$

where the SPICE parameter IS is equal to a_{12} (11E.1) and the new SPICE parameter VAF is called the normal-mode Early voltage, after J. M. Early, who first described the effect of basewidth modulation in 1952 [3]. Note that to ensure the

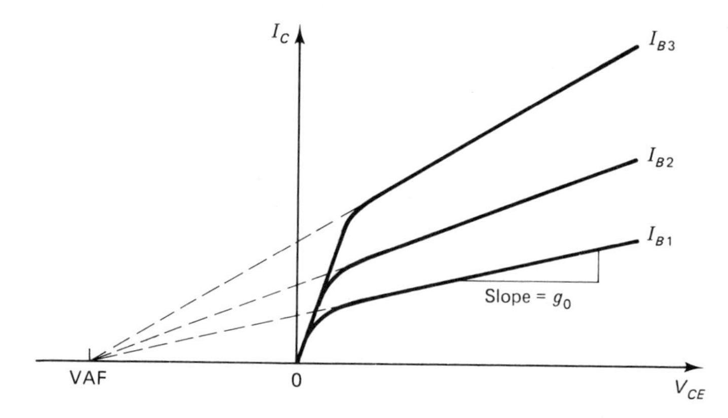

Figure 11.16 Effect of basewidth modulation on the common-emitter collector I–V characteristic. The conductance in the active region becomes finite. The intercept on the voltage axis of the extrapolated curves is the forward Early voltage.

correct direction of change of I_C, V_{CB} and VAF are both taken to be positive quantities. This is also the case for *pnp* transistors.

For cases in which VAF $\gg V_{CB}$, the output conductance is given by

$$g_0 = \frac{dI_C}{dV_{CE}} \approx \frac{dI_C}{dV_{CB}} = \frac{I_C}{\text{VAF}} \tag{11.47}$$

This characterization of the output conductance as being proportional to the collector current is reasonably accurate for many BJTs.

Note from (11.46) that the collector current is reduced to zero when $V_{CB} = -\text{VAF}$. This suggests a means of measuring VAF in a real device (see Fig. 11.16). The corresponding Early voltage used in SPICE to model the effect of basewidth modulation when the BJT is operating in the inverse mode is labeled VAR.

11.5.2 Recombination–Generation in the Depletion Regions

As discussed in detail in Sections 6.3.1 and 6.3.2, recombination and generation of minority carriers in the depletion region give rise to components of current in a *pn* junction. In forward bias, recombination dominates over generation and the resulting current is given by (6.46) and (6.47). In reverse bias, generation prevails over recombination and the current is given by (6.56) and (6.57). Either pair of these two sets of equations yields

$$J_{RG} = \frac{qn_i}{2}\left[\frac{x_Z}{\tau_h} + \frac{x_A}{\tau_e}\right]\left[\exp\left(\frac{qV}{2kT}\right) - 1\right] \tag{11.48}$$

The extra components of base current due to recombination–generation in the

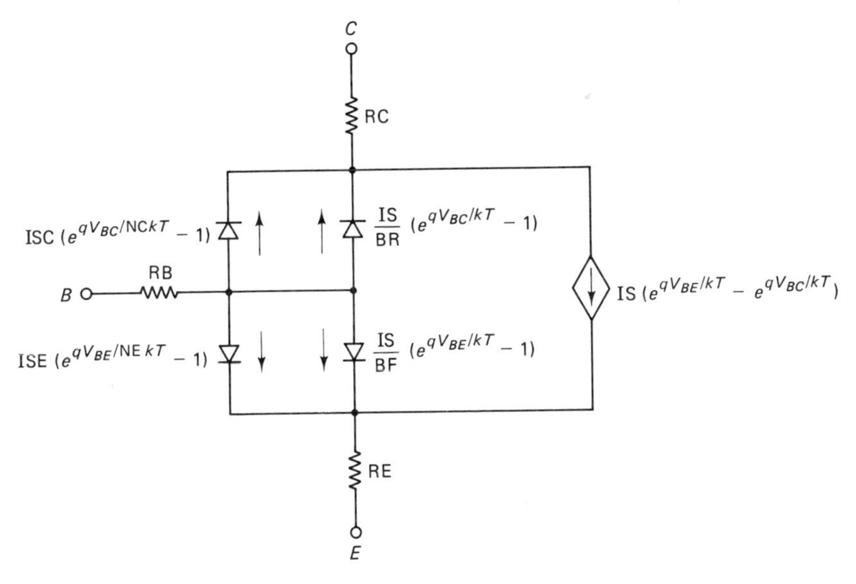

Figure 11.17 Nonlinear hybrid-π equivalent circuit, including the effects of recombination–generation in the depletion regions and series resistance in the neutral regions. All the parameters are listed in SPICE notation.

depletion regions can be modeled in an equivalent circuit by adding diodes as shown in Fig. 11.17.

The parameters appearing in Fig. 11.17 are those employed in SPICE. There is an obvious association between the physical parameters appearing in (11.48) and the model parameters ISE, for the emitter–base junction, and ISC, for the collector–base junction. The other two model parameters used in describing depletion region recombination–generation are NE and NC. The default values for these are 2, in keeping with the physical relationship expressed in (11.48). The four model parameters can also be extracted from the base ln I–V characteristics of a real device (see Fig. 11.18).

11.5.3 High-Level Injection

As discussed in Section 6.5.1 in the context of *pn* junctions, the onset of high-level injection causes the diode I–V characteristic to deviate from the exponential relationship described by the ideal diode equation. In a BJT, high-level injection occurs when the concentration of minority carriers injected from the emitter equals or exceeds the concentration of majority carriers in the base. The effect is shown in Fig. 11.18 and is allowed for in SPICE if the model parameter IKF is specified. This is the "corner" current, that is, the collector current at the intersection of the extrapolated portions of the low-level and high-level ln I–V

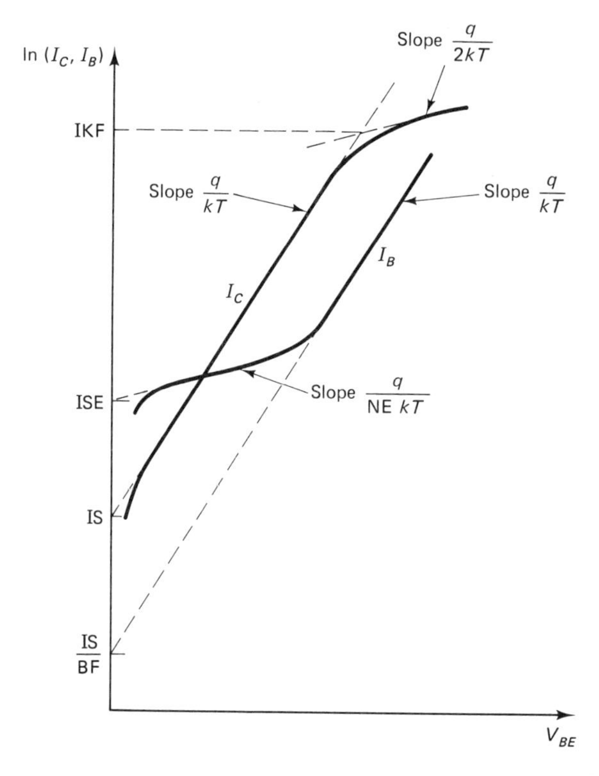

Figure 11.18 Collector and base currents in the active mode, showing the effect on the base current of recombination in the emitter–base depletion region, and the effect on the collector current of high-level injection.

curves. The high-level injection current has the form [4]

$$I_C(\text{high-level injection}) \propto \exp\left(\frac{qV_{BE}}{2kT}\right) \tag{11.49}$$

An equation of this form is incorporated in the SPICE model and is invoked by the specification of a finite value for IKF. The corresponding model parameter applicable to the inverse mode is IKR.

11.5.4 Junction Breakdown

As discussed in Section 6.5.2, reverse biasing of a *pn* junction can lead to avalanche breakdown. Thus breakdown is possible in the collector–base junction of a BJT operating in the normal active mode. The effect of breakdown on the common-emitter collector I–V characteristics is shown in Fig. 11.19. Note that breakdown can occur even when the base is open-circuit. In this case, the electrons that participate in the avalanche originate in the emitter and are injected into the base by the small portion of V_{CE} which is dropped across the emitter–base junction as a forward bias.

Because the doping densities in the collector and the base are generally quite low (seldom greater than 10^{17} cm^{-3}), Zener breakdown is unlikely to occur in

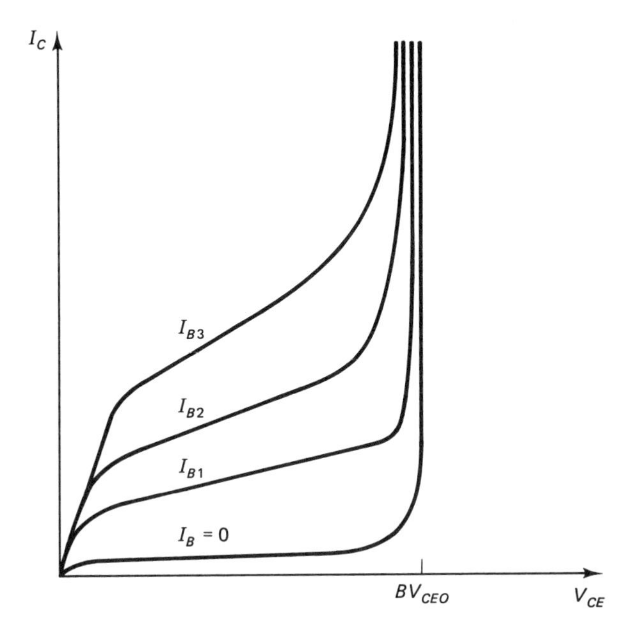

Figure 11.19 Effect of collector–base junction breakdown on the common-emitter collector characteristic. The breakdown voltage when $I_B = 0$ (i.e., $I_C = I_{CEO}$) is called BV_{CEO}.

BJTs. However, another breakdown-like phenomenon called ***punchthrough*** is possible. This occurs when the depletion region on the base side of the collector–base junction reaches all the way across the base and merges with the depletion region at the emitter–base junction. As the energy band diagram of Fig. 11.20 shows, the two n regions of the emitter and the collector become joined by a continuous depletion region. Thus the controlling action of the base on the collector current is lost. Any increase in V_{CE} beyond the punchthrough value reduces the barrier at the emitter–base junction and enables a massive electron charge to flow.

11.5.5 Nonuniform Doping in the Base

In real BJTs, the base region is normally formed by diffusion or ion implantation and so is not uniformly doped. As shown in Fig. 11.21, the boron concentration drops on moving from the base–emitter boundary toward the collector. In consequence, the base region is quasi-neutral and contains a built-in electric field pointing from the collector toward the emitter. (The distinction between truly neutral and quasi-neutral regions, and the presence of built-in electric fields in nonuniformly doped material is discussed in Section 6.2.2.4.) The built-in field opposes the tendency of holes to diffuse from the more heavily doped side of the base, near the emitter, to the lightly doped side near the collector.

The presence of the built-in field actually improves the performance of the transistor in normal active operation, in that it tends to sweep electrons from the emitter to the collector. This reduces the time the average electron spends in the

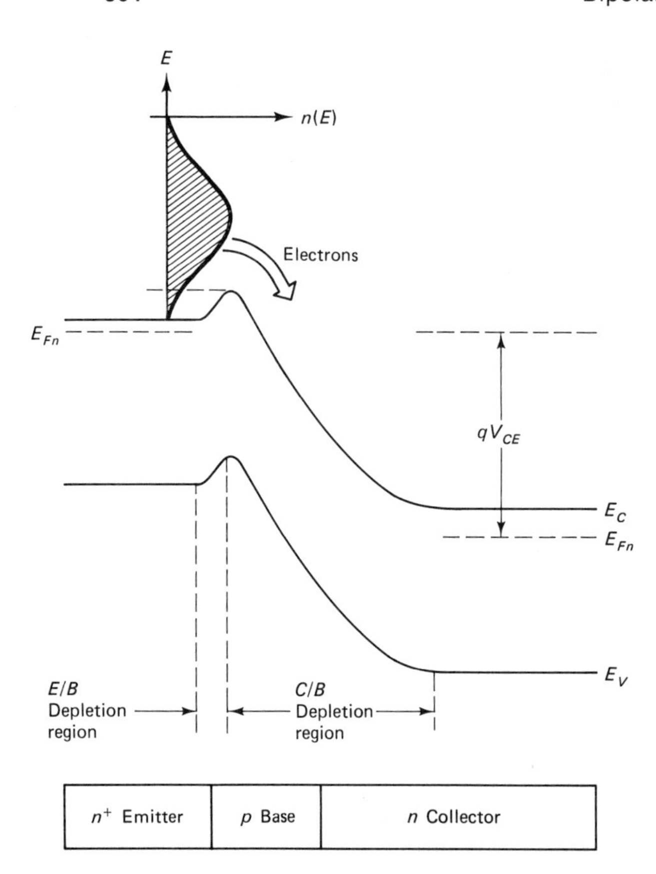

Figure 11.20 Energy band diagram for an *npn* BJT at punchthrough.

base, leaving it with less opportunity to recombine with holes and thereby improving the gain. A reduction in the electron transit time across the base helps improve the switching speed and high-frequency response of the transistor, since it increases the speed with which the collector current can respond to a change in base–emitter bias. (The factors influencing transient response and operating frequency are considered in more detail in Sections 11.6 and 11.7.)

Attempting to incorporate the effects of a built-in field in the base in the analysis of current flow in the BJT given in Section 11.3.5 leads to an extremely difficult mathematical problem which can only be solved using numerical methods. However, the problem can be made tractable if one assumes that almost all the electrons injected from the emitter cross the base without recombining. This is the case with most modern BJTs, where the base is narrow and minority carrier lifetimes are usually quite long.

In the presence of an electric field in the base, the minority carrier electron drift current can no longer be ignored. The electron current is therefore given by

$$J_e = qD_e \frac{dn_p}{dx} + q\mu_e n_p \mathscr{E} \tag{11.50}$$

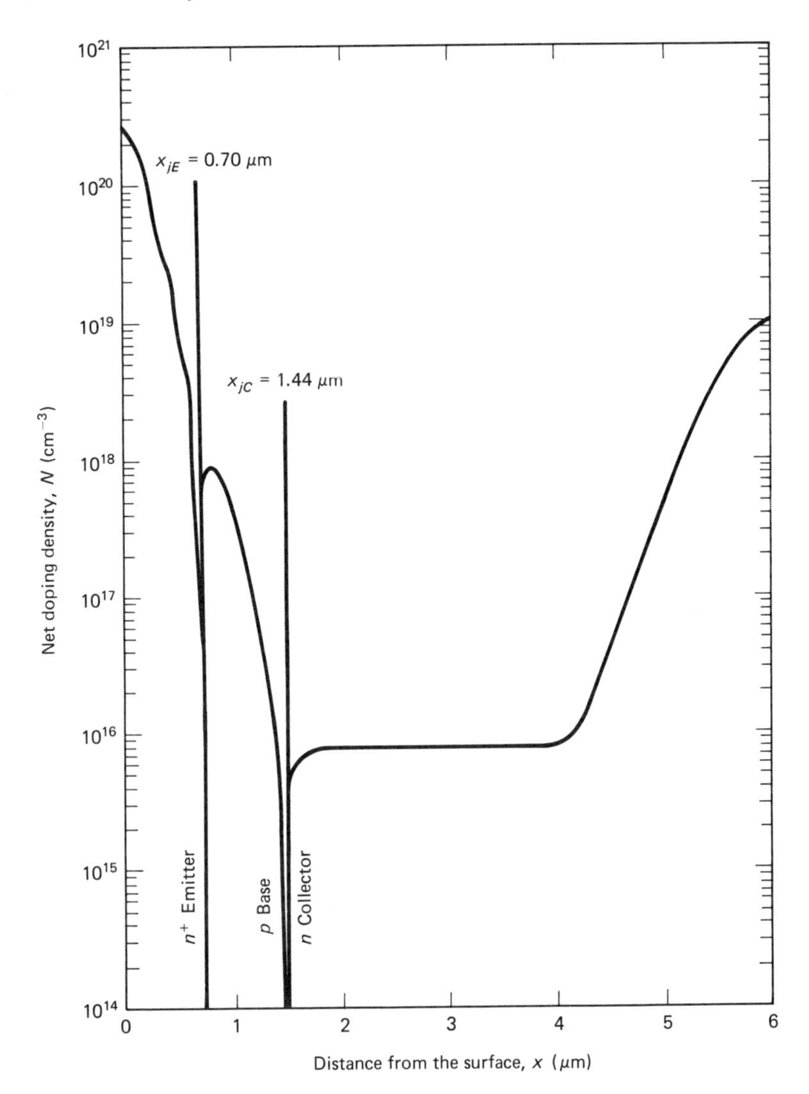

Figure 11.21 Doping profile in a typical *npn* bipolar transistor. Note the non-uniform doping in the emitter and base. (From Colclasser [5]. Reprinted with permission of the publisher, John Wiley & Sons, Inc.)

To make use of (11.50), it is necessary to determine the electric field $\mathscr{E}(x)$. Under low-level injection conditions the minority carriers should make a negligible contribution to the net space charge in the base, so $\mathscr{E}(x)$ should be the same as in equilibrium. In equilibrium, the requirement that there be no hole current gives

$$J_h = q\mu_h p\mathscr{E} - qD_h \frac{dp}{dx} = 0 \tag{11.51}$$

Applying the quasi-neutrality condition $p(x) \simeq N_A(x)$ and using the Einstein relation $D = (kT/q)\mu$ gives

$$\mathscr{E}(x) = \frac{kT}{q}\frac{1}{N_A}\frac{dN_A}{dx} \tag{11.52}$$

Substituting (11.52) into (11.50) and using the Einstein relation again gives

$$J_e = qD_e\left(\frac{dn_p}{dx} + \frac{n_p}{N_A}\frac{dN_A}{dx}\right) \tag{11.53}$$

or

$$\frac{J_e N_A}{D_e} = qN_A\frac{dn_p}{dx} + qn_p\frac{dN_A}{dx} = q\frac{d(n_p N_A)}{dx} \tag{11.54}$$

In steady-state conditions, the assumption of no electron–hole recombination in the base implies that the electron current must be constant across the base. Therefore,

$$\begin{aligned}\int_0^{L_B} J_e\frac{N_A(x)}{D_e(x)}\,dx &= J_e\int_0^{L_B}\frac{N_A(x)}{D_e(x)}\,dx \\ &= q\int_0^{L_B}\frac{d(n_p N_A)}{dx}dx = qn_p(x)N_A(x)\Bigg|_0^{L_B} \\ &= -qn_i^2(e^{qV_{BE}/kT} - 1)\end{aligned} \tag{11.55}$$

Here we have assumed normal active operation with $V_{BC} = 0$, so that $n_p(0) = [n_i^2/N_A(0)]e^{qV_{BE}/kT}$ at the edge of the base–emitter depletion region and $n_p(L_B) = n_{p0}$ at the edge of the base-collector depletion region. The possibility of the electron diffusivity varying with position in the base due to the nonuniform doping profile has also been allowed for.

Rearranging (11.55) and defining $J_C = -J_e$ gives

$$J_C = \frac{qn_i^2}{G_B}(e^{qV_{BE}/kT} - 1) \tag{11.56}$$

where the ***base Gummel number*** G_B is given by

$$G_B = \int_0^{L_B}\frac{N_A(x)}{D_e(x)}\,dx \tag{11.57}$$

("Gummel numbers" are named in honor of H. K. Gummel, who made major contributions to the use of numerical analysis techniques in modeling semiconductor devices in the 1960s.) Base Gummel numbers in the range

10^{10} to 10^{11} s cm^{-4} are common in modern transistors, corresponding to an integrated base doping of from 10^{12} to 10^{13} atoms/cm^2.

The main conclusion to be drawn from (11.56) is that, in the absence of recombination, the collector current is determined by the total or integrated doping in the base; ignoring slight variations in D_e with position, the actual distribution of dopant is of no consequence. A rather surprising corollary is that the same collector current is obtained if a transistor is operated in inverse active rather than normal active mode at a given bias level. (The gain is far lower in inverse active operation, due to the poor injection efficiency of the lightly doped "emitter," which normally serves as the collector of the transistor.)

It is possible to define an ***emitter Gummel number*** G_E by analogy to the base Gummel number. Assuming that recombination in the quasi-neutral base and in the base–emitter depletion region are negligible, the base current is dominated by the back injection of holes into the emitter. In this case the base current can be written as

$$J_B = \frac{qn_i^2}{G_E}\left(e^{qV_{BE}/kT} - 1\right) \tag{11.58}$$

For a long, uniformly doped emitter, we would have [see (11.13)]

$$G_E = \frac{L_h^E N_D^E}{D_h^E} \tag{11.59}$$

Emitter Gummel numbers in the range 10^{13} to 10^{14} s cm^{-4} are common in modern transistors. The current gain of the BJT is determined by the ratio of the emitter and base Gummel numbers:

$$\beta_F = \frac{G_E}{G_B} \tag{11.60}$$

11.5.6 Incorporation of Secondary Effects in SPICE

Of the various phenomena discussed in the preceding five subsections, only basewidth modulation, depletion region recombination/generation, and high-level injection are modeled in SPICE. The required new model parameters are those introduced in Sections 11.5.1 to 11.5.3. In addition, SPICE includes three resistive elements to model series resistance effects at the contacts to, and in the bulk regions of, the three regions of the BJT. These resistances are akin to RS, as used in the diode model described in Section 6.4.1. In total, there are 14 parameters in the SPICE dc model for the BJT. Example 11.4 calls upon most of them in modeling the dependence of the common-emitter current gain BETADC on collector current.

Example 11.4 Common-Emitter Current Gain: The Effect of Depletion Region Recombination and High-Level Injection

The actual dc common-emitter gain of a BJT in a circuit is given by (11.44)

$$\text{BETADC} = \frac{I_C}{I_B}$$

With this definition, inspection of Fig. 11.18 reveals that if depletion region currents are considered and if high-level injection effects are present, BETADC will not be the same for all values of I_C.

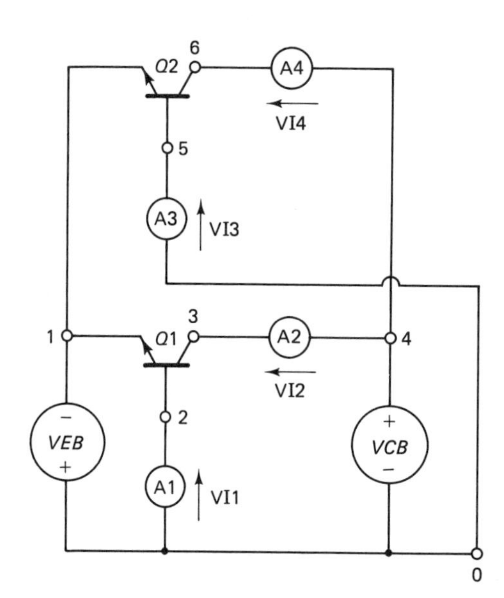

Figure 11E.5 Circuit for the SPICE simulation of the effect on common-emitter current gain of recombination in the base–emitter depletion region and high-level injection.

To investigate this phenomenon, we use SPICE, with the following input listing, to simulate the circuit of Fig. 11E.5.

```
*BJT GAIN VERSUS COLLECTOR CURRENT
* EFFECT OF DEPLETION REGION RECOMBINATION
*AND HIGH LEVEL INJECTION
VEB 0 1
VCB 4 0  DC 2
VI1 0 2
VI2 4 3
VI3 0 5
VI4 4 6
Q1  3 2 1   Q172
Q2  6 5 1   Q172DH
.MODEL Q172    NPN BF=122  BR=0.2  IS=2.65E-17
```

```
.MODEL Q172DH  NPN BF=122  BR=0.2  IS=2.65E-17
+ISE=6.5E-14  NE=2  IKF=5E-3
.DC VEB 0.1 0.9 0.05
.WIDTH OUT=80
.PRINT DC I(VI1) I(VI2) I(VI3) I(VI4)
.END
```

Transistor $Q1$ is our "standard" BJT with a base doping density of 10^{17} cm^{-3} and a neutral basewidth of 2 μm, as used in previous examples in this chapter. $Q2$ includes the effects of both recombination in the emitter–base junction (by using data for the $10^{19}/10^{17}$ diode from Table 6E.1), and high-level injection (by specifying a value for IKF). By keeping VCB at 2 V reverse bias and sweeping VEB up to 0.9 V, we ensure that the transistors remain in the active mode. From the printed results, BETADC for each transistor can be computed at various values of I_C; the results are shown graphically in Fig. 11E.6. Note the following:

(a) The constancy of BETADC, at a value equal to BF, for the transistor

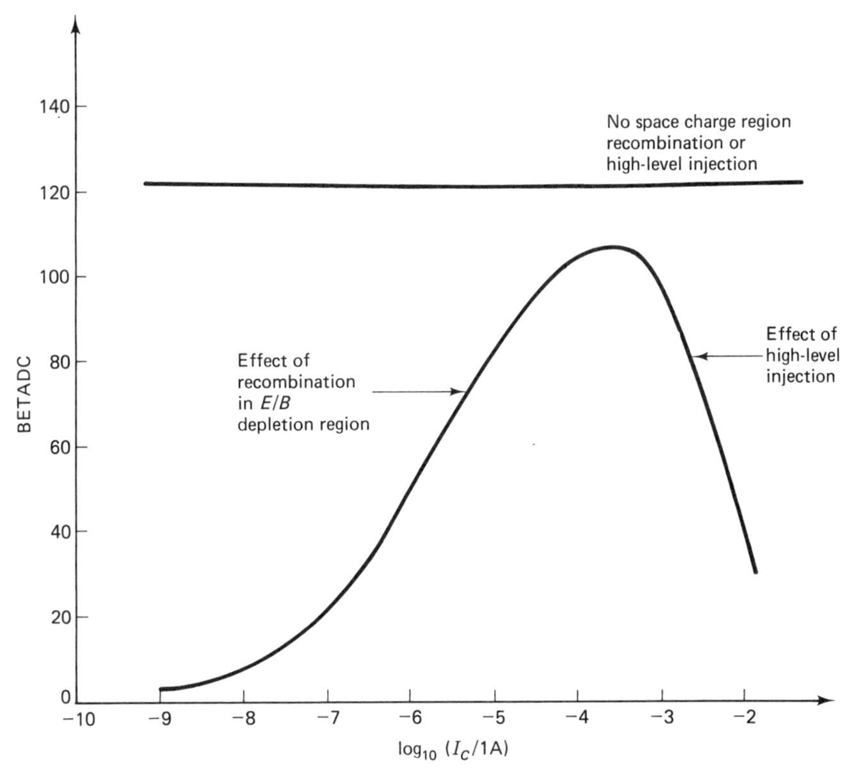

Figure 11E.6 Computed common-emitter current gain as a function of collector current for the transistors described in Example 11.4.

in which secondary effects are not modeled, confirms that it is operating in the active mode.

(b) When recombination in the emitter–base depletion region is significant, BETADC at low collector currents is drastically reduced. This is to be expected from Fig. 11.18. The electrons that recombine in the emitter–base depletion region are, obviously, not injected into the base. Thus the emitter injection efficiency and, consequently, BETADC, are reduced. Instead of contributing to the collector current, these electrons recombine with holes which are, in effect, supplied by the base lead, so the base current is enhanced.

(c) If high-level injection effects are significant, they will show up at high currents. As the collector current is larger than the base current, the former will be affected first (see Fig. 11.18). This leads to the reduction in BETADC at high collector currents that is illustrated in Fig. 11E.6.

(d) The SPICE parameter IKF used to model the high-level injection phenomenon in this example was arbitrarily chosen. A higher value than used here would allow BETADC to attain its maximum value (BF) before degrading at high currents.

11.6 TRANSIENT ANALYSIS

In digital switching applications of BJTs, for example, transient conditions arise when the base current or voltage is suddenly altered in an attempt to change the state of the transistor. The "on" state of a BJT is often taken to be the saturation mode of operation, when the collector current is high and V_{CE} is low (see Fig. 11.8), that is, when the resistance between collector and emitter is low. Conversely, the "off" state is the cutoff mode when the collector current is essentially zero and the collector–emitter resistance is very high.

Consider the case of turning on an *npn* transistor by increasing the forward bias on the base–emitter junction. The transition between the cutoff and saturation states is governed primarily by the time taken to establish a new minority carrier concentration profile in the base. In a long n^+p diode the equivalent time is related directly to the minority carrier lifetime because all the electrons injected from the n^+ region recombine in the p material. However, the situation in a BJT is different because the p region of the base is deliberately made very narrow in order to minimize recombination. Instead of recombining in the base, most of the electrons traverse the base to the collector. The establishment of the new minority carrier profile in the base is, therefore, more likely to be governed by the transit time than by the minority carrier lifetime. A rough relationship between these two times can be obtained as follows.

Imagine that electrons injected from the emitter spend, on average, a time of τ_t in transiting the base. The collector current can be approximated, therefore, by $qnAL_B/\tau_t$, where n is the electron concentration, A the cross-sectional area, and L_B the basewidth. The base current due to recombination in the base of some of the injected electrons with holes can be approximated as $qnAL_B/\tau_e$, where τ_e is the lifetime of the minority carriers (electrons in this case). The ratio of currents can be written as

$$\frac{I_C}{I_B} \approx \frac{qnAL_B/\tau_t}{qnAL_B/\tau_e} = \frac{\tau_e}{\tau_t} \tag{11.61}$$

from which it follows that

$$\tau_t = \frac{\tau_e}{\text{BETADC}} \tag{11.62}$$

This relationship, although obviously not very exact, does indicate that the transit time is shorter than the minority carrier lifetime.

In modeling transient phenomena in the *pn* diode, we followed an equivalent circuit approach and developed equations for the capacitances in a *pn* junction. If we are to extend this approach to BJTs, it follows that there will be four capacitances and that the storage components must be modified, as the storage capacitance, (6.71), is directly proportional to the minority carrier lifetime. The simplest way to avoid overestimating this capacitance in a BJT is to replace τ_e by τ_t. This approach is adopted in SPICE, as described in the next section.

11.6.1 SPICE Model

In SPICE the total capacitance associated with a *pn* diode is given by (6.72). In SPICE notation, this equation for the emitter–base junction, for example, is

$$C_{BE} = \frac{q}{kT}\,\text{TF IS}\exp\left(\frac{qV_{BE}}{kT}\right) + \text{CJE}\left(1 - \frac{V_{BE}}{\text{VJE}}\right)^{-1/2} \tag{11.63}$$

where CJE is the zero-bias capacitance [equivalent to C_{j0} resulting from (6.67)], VJE is the built-in potential [equivalent to V_{bi} in (6.67)], and TF is the transit time (equivalent to τ_t as introduced above). The corresponding parameters for the collector–base capacitance C_{BC} are CJC, VJC, and TR. The last term becomes important in the saturation mode when, under the influence of forward bias at the collector–base junction, minority carriers are injected from the collector into the base. It is a straightforward matter to include the capacitances in the equivalent circuit model for the BJT (see Fig. 11.22).

Example 11.5 provides an opportunity to investigate the transient response of BJTs using SPICE.

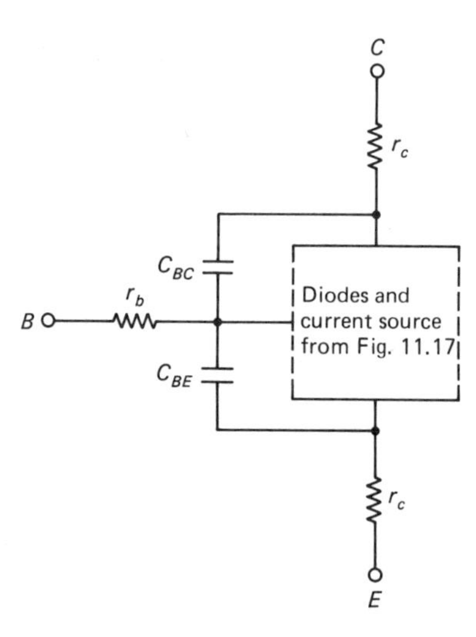

Figure 11.22 Ebers–Moll ac equivalent circuit.

Example 11.5 Switching Characteristics of BJTs: The Turn-Off Transient

To turn off a BJT it is necessary to change the ***operating point*** of the device from either the saturation or active mode (where the collector current is high) to the cutoff mode (where the collector current is essentially zero). This shift in operating points is illustrated in Fig. 11E.7. The three points shown lie on the ***load line***, which is associated with the resistor in the collector lead of the circuit in the inset to the figure. The equation of the load line is

$$I_C = \frac{V_{CC} - V_{CE}}{R_L} \tag{11E.2}$$

The intersection of the load line with the relevant base current curve in the collector characteristic gives the dc operating point of the transistor. The circuit used to simulate the turn-off transient is shown in Fig. 11E.8. This is the same circuit as used in Example 8.2, but without the Schottky diode clamp. The SPICE input listing is as follows:

```
*BJT SWITCHING CHARACTERISTICS
*TURN-OFF TRANSIENT
*SATURATION TO CUT-OFF
VCC 6 0 DC 3
VIC 6 5
VIB 1 2
VIN 1 0 PWL(0NS 3 5NS 3 6NS 0 100NS 0)
RL  5 4 2K
RB  3 2 1K
Q1  4 3 0 Q172
```

```
.MODEL Q172 NPN BF=122 BR=0.2 IS=2.65E-17
+CJE=9.37E-14 VJE=0.94 TF=1.4E-9
+CJC=1.09E-14 VJC=0.7 TR=1E-8
.WIDTH OUT=80
.TRAN 0.5NS 15NS
.PLOT TRAN I(VIB) I(VIC) (-1E-3,3E-3)
.END
```

The transistor under study is the *Q*172 device of previous examples, but further specified on the model line so as to include the effects of junction capacitance and minority carrier transit time. The magnitudes quoted for CJE, VJE, CJC, and VJC follow from (6.12) and (6.67), evaluated at zero bias for the appropriate doping densities. TF, the transit time for the normal mode of operation, is computed from (11.62), taking BETADC = BF and $\tau_e = 1.7 \times 10^{-7}$ s (from Table 6E.1). This gives TF = 1.4 ns. Specification of TR is not so straightforward because the argument presented in Section 11.6, which led to (11.62), implied that conditions in the base alone determined BETADC (i.e., the emitter injection efficiency was essentially unity).

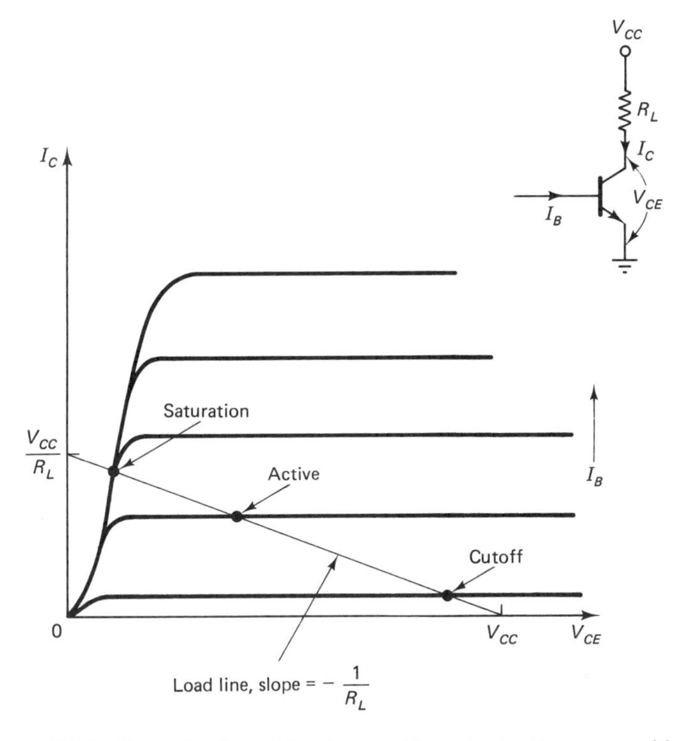

Figure 11E.7 Determination of the dc operating point by the superposition of transistor characteristics and the load line. Three such points, for different base currents, are shown.

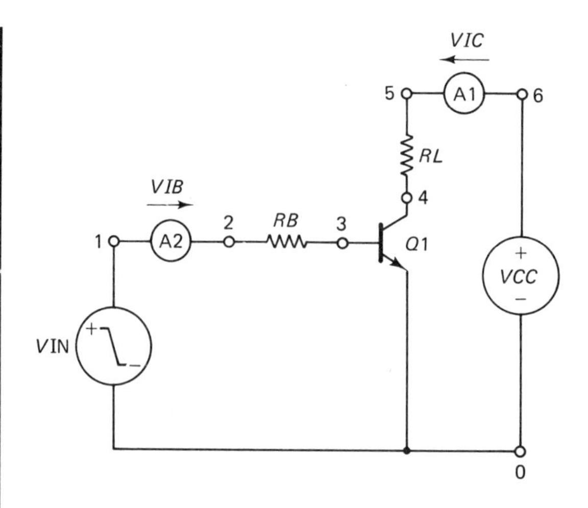

Figure 11E.8 Circuit for the SPICE simulation of BJT transient response.

Although this is a good approximation for the normal mode of operation, it is poor for the inverse mode. This is because the collector doping density is generally lower than that of the base. Thus, using BETADC = BR in (11.62) will seriously overestimate TR. To avoid doing this in this example, we set TR, somewhat arbitrarily, at 10 ns.

The control line in the input listing instructs SPICE to perform a transient analysis and obtain a solution every 0.5 ns over a 15-ns period. SPICE commences the analysis by finding the dc operating point. The information is automatically printed in the output listing and the results for this example are

$$I_B = 2.16 \times 10^{-3}\ \text{A} \qquad V_{BE} = 0.845\ \text{V}$$

$$I_C = 1.47 \times 10^{-3}\ \text{A} \qquad V_{BC} = 0.786\ \text{V}$$

$$\text{BETADC} = 0.7 \qquad V_{CE} = 0.058\ \text{V}$$

The positive value of V_{BC} (indicating forward bias) and the low value of BETADC confirm that the transistor is biased initially in the saturation region.

The results of the transient analysis are shown in Fig. 11E.9. Note the following:

(a) After the base current changes, signifying reverse biasing of the emitter–base junction, there is a discernible delay (about 4 ns) before the collector current starts to fall.

(b) The base current switches to a negative value before attaining the near-zero value expected for reverse-bias operation.

These delays in switching deserve some explanation. The large, negative base current is fueled by the minority carriers, which are a legacy of the prior forward-bias conditions at both the emitter–base and the collector–

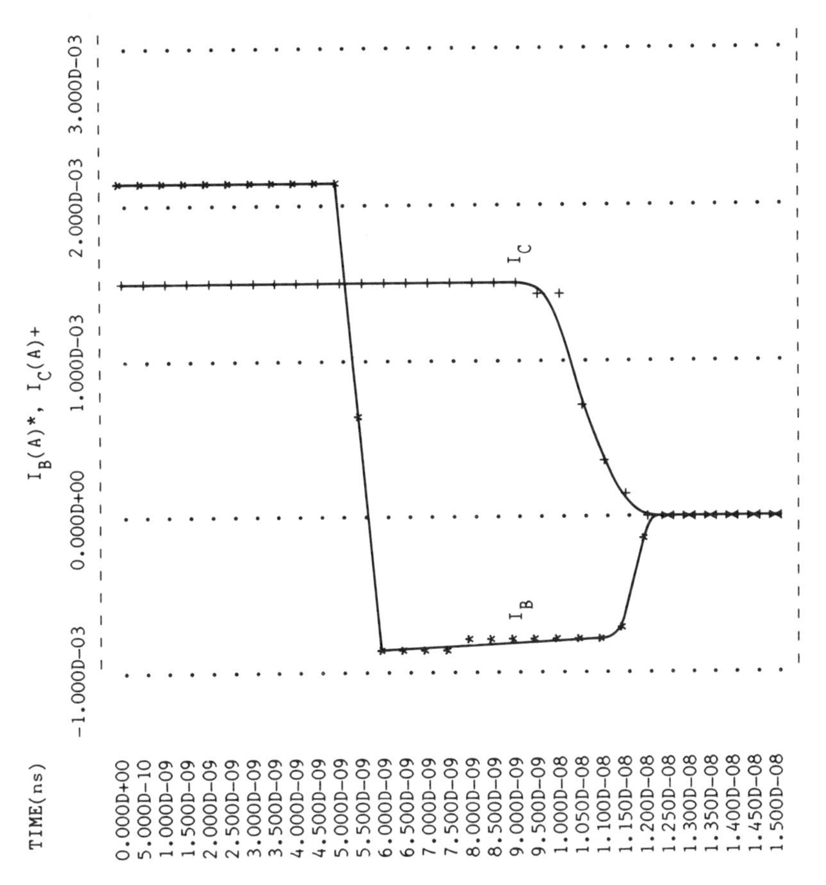

Figure 11E.9 Computed transient response for switching the transistor of Example 11.5 from saturation to cutoff.

base junctions. The stored minority carrier charge is removed either by diffusion to the depletion regions, where extraction is performed by the electric fields, or by recombination in the base with majority carrier holes. V_{BE} depends logarithmically on the excess electron concentration at the depletion region boundary with the base [see (11.18)]. Therefore, V_{BE} changes only slowly as the excess electron concentration diminishes. Consequently, the current driven by V_{BE} through *RB* (and hence recorded as a negative current) remains essentially constant. The base current inclines toward the very small value expected for a reverse-biased junction only when the excess carrier concentration in the base has been reduced to zero. Note the similarity with the *pn*-diode turnoff transient as investigated in Example 6.4. As both junctions in a saturated transistor supply minority carriers to the base, the stored charge is substantial. Figure 11E.10 illustrates the situation and indicates how much charge has to be nullified before the device

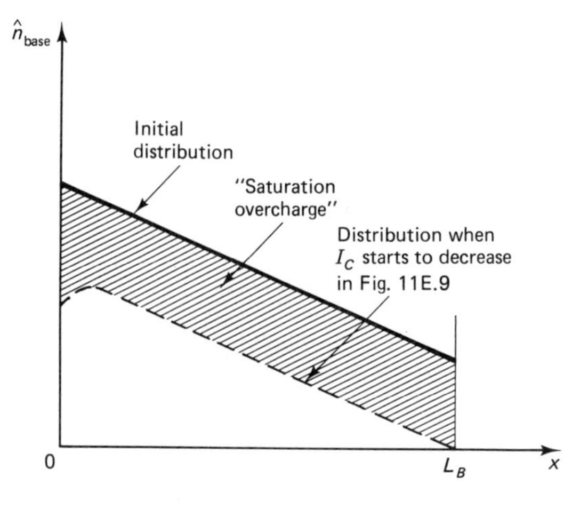

Figure 11E.10 Illustration of the change in excess electron profile in the base on turning off a BJT from the saturation mode. The positive slope close to $x = 0$ for the lower profile indicates extraction of excess electrons by diffusion to the base–emitter depletion region.

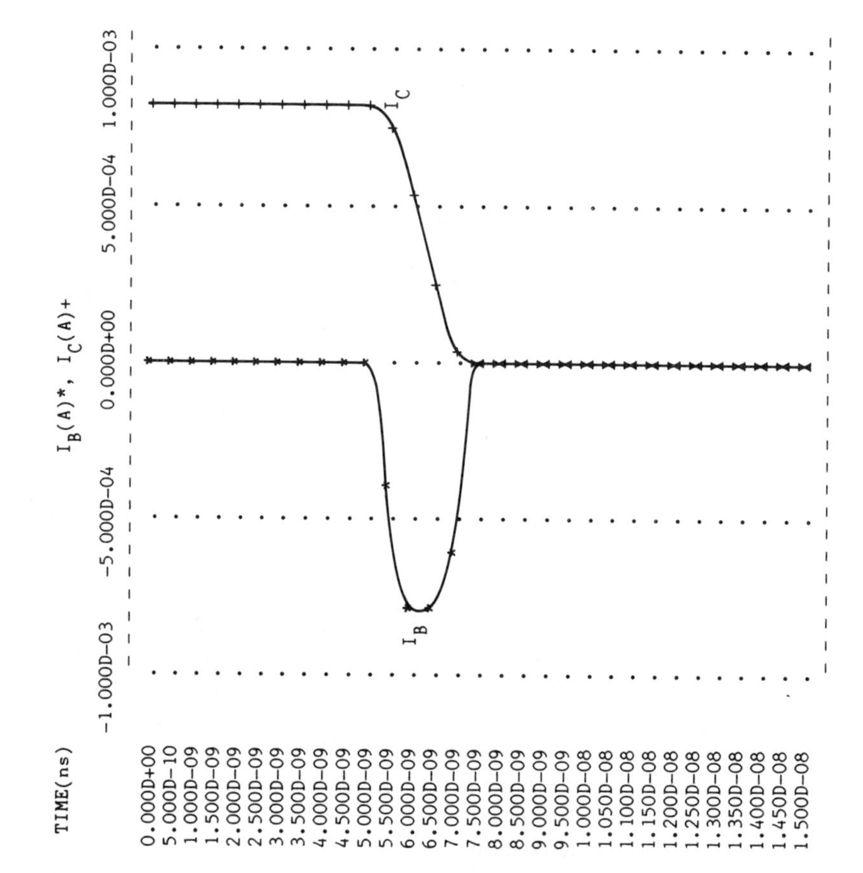

Figure 11E.11 Computed transient response for switching the transistor of Example 11.5 from active to cutoff.

leaves the saturaton mode. While in the saturation mode V_{CE} is small, and therefore I_C stays constant at a value close to VCC/RL [see (11E.2)]. Only when the "saturation overcharge" has been reversed can V_{CE} rise and the collector current commence falling.

It follows from this description that the rapidity of the turn-off transient can be improved by ensuring that the transistor does not enter saturation. This is the principle employed in emitter-coupled logic (ECL) and makes this technology the fastest available in silicon bipolar integrated circuits. We can illustrate the situation by redoing the simulation above, but with an initial base current of such a magnitude that the device is "on" in the active mode rather than in saturation. The change of initial operating point is easily accomplished by changing V_{IN} from 3 V to 0.8 V. The resulting operating point conditions are

$$I_B = 4.88 \times 10^{-6}\text{ A} \qquad V_{BE} = 0.795\text{ V}$$

$$I_C = 5.95 \times 10^{-4}\text{ A} \qquad V_{BC} = -1.015\text{ V}$$

$$\text{BETADC} = 122 \qquad V_{CE} = 1.810\text{ V}$$

The value of BETADC = BF and the negative collector–base voltage confirm that the device is initially in the active mode. The result of turning off the device is shown in Fig. 11E.11. As expected, there is virtually no waiting time before the collector current starts to decrease.

It is possible to prevent a BJT from entering saturation by connecting a Schottky diode between the collector and base terminals (see Example 8.2).

11.7 AC ANALYSIS

11.7.1 Small-Signal, Linearized, Hybrid-π Model

To develop a small-signal model for the BJT, we make the same assumptions as used in the diode case (Section 6.7.1), that is, the ac voltages are considerably smaller than the dc biases that set the operating point of the device, and the frequency of the ac signals is such that quasi-static conditions apply. The existence of two pn junctions in the BJT means that the terminal currents are functions of two voltage variables. Therefore, the appropriate Taylor series expansion for the collector current, when the device is in the common-emitter configuration, for example, is

$$i_C = I_C(V_{BE} + v_{be}, V_{CE} + v_{ce}) = I_C(V_{BE}, V_{CE}) + \left.\frac{\partial I_C}{\partial V_{BE}}\right|_{V_{CE}} v_{be} + \left.\frac{\partial I_C}{\partial V_{CE}}\right|_{V_{BE}} v_{ce} \tag{11.64}$$

where I_C, V_{BE}, and V_{CE} are quasi-static (dc) parameters, v_{be} and v_{ce} are small signal ac voltages, and i_C is the total emitter current, which is given by the sum of the dc and ac currents, that is,

$$i_C = I_C + i_c \tag{11.65}$$

Equating (11.64) and (11.65), we find

$$i_c = \left.\frac{\partial I_C}{\partial V_{BE}}\right|_{V_{CE}} v_{be} + \left.\frac{\partial I_C}{\partial V_{CE}}\right|_{V_{BE}} v_{ce} \tag{11.66}$$

This equation can also be written as

$$i_c = g_m v_{be} + g_O v_{ce} \tag{11.67}$$

where g_m is defined as the ***transconductance*** and g_o is the ***output conductance*** (see Fig. 11.16). Following a similar procedure for the base current yields

$$i_b = \left.\frac{\partial I_B}{\partial V_{BE}}\right|_{V_{CE}} v_{be} + \left.\frac{\partial I_B}{\partial V_{CE}}\right|_{V_{BE}} v_{ce} \tag{11.68}$$

This equation can also be written in a linear fashion in terms of conductances, that is,

$$i_b = g_\pi v_{be} + g_\mu v_{ce} \tag{11.69}$$

where g_π is the ***input conductance*** and g_μ is called the ***reverse feedback conductance***.

Equations (11.67) and (11.69) suggest that the BJT can be represented by four conductors. If we add two capacitors to represent junction and storage capacitance at each of the two *pn* junctions, and three resistors to represent parasitic resistance in each of the three regions of the device, we obtain a small-signal, linearized model which is known as the hybrid-π model (see Fig. 11.23).

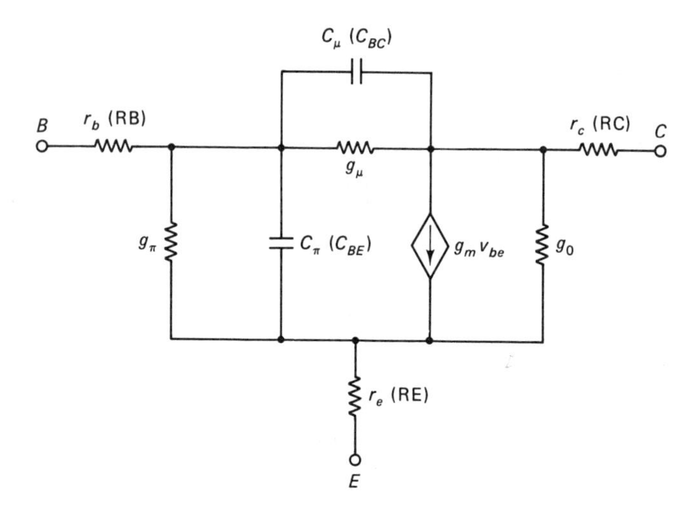

Figure 11.23 Small-signal, linearized hybrid-π equivalent circuit for a BJT. SPICE notation, where different from standard hybrid-π notation, is given in parentheses.

11.7.2 SPICE Model

The hybrid-π model of Fig. 11.23 is used in SPICE for carrying out small-signal analyses of BJTs. The conductor parameters are defined by (11.66)–(11.69) and are computed from the appropriate derivatives, at the desired operating points, of the dc I–V characteristics. The capacitors are the same as used in the transient analysis of the BJT (see Section 11.6.1). The parasitic resistances can be computed from measurements similar to that used in estimating the series resistance of a diode (see Section 6.4.1).

The SPICE output from a small-signal simulation actually prints out values for the hybrid-π parameters, as can be verified by following Example 11.6. Also printed out are values for the actual dc and ac common-emitter currrent gain, as defined in (11.44) and (11.45), and evaluated at specific operating points. Note that

$$BETAAC = \left.\frac{\partial I_C}{\partial I_B}\right|_{V_{CE}} = \frac{g_m}{g_\pi} \tag{11.70}$$

Example 11.6 Small-Signal Analysis of BJTs: A Single-Stage Amplifier

The circuit of Fig. 11E.12 shows a simple, single-stage BJT amplifier. The dc operating point is determined by *RBB* and *VBB*, which set the base current, and by *RL* and *VCC*, which set the load line (see Fig. 11E.7). For high-fidelity amplification it is necessary that the operating point be located in the active region. Otherwise, when the ac signal is applied to the base via the dc-blocking capacitor *C*IN, the transistor might be driven into saturation or cutoff. In both these regimes the collector current is limited, and therefore the linearity between output and input would be lost. With the circuit values

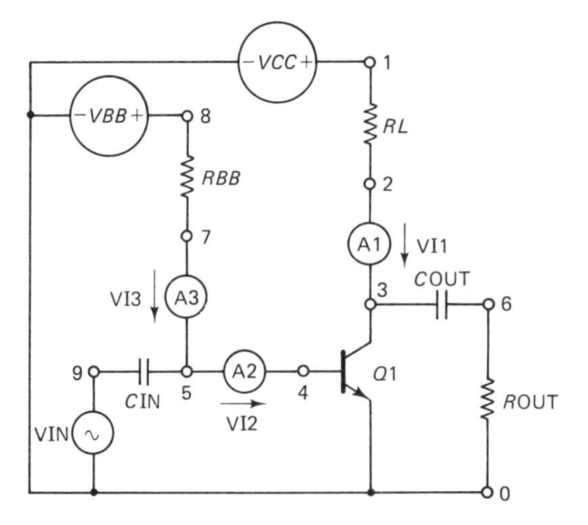

Figure 11E.12 Circuit for the SPICE simulation of BJT small-signal ac performance.

shown in the following listing, the operating point is comfortably in the active region, as can be confirmed from the value of BETADC = BF and the negative value of V_{BC} which the SPICE output listing provides along with other operating-point information, for example,

$$\text{BETADC} = 122 \qquad V_{BE} = 0.81\text{V}$$

$$V_{CE} = 1.21 \text{ V} \qquad V_{BC} = -0.41 \text{ V}$$

```
*BJT SMALL-SIGNAL AMPLIFIER
VCC  1 0 DC 3
VI1  2 3
VI2  5 4
VI3  7 5
VBB  8 0 DC 3
VIN  9 0 AC 0.01
RBB  8 7 300K
RL   1 2 2K
ROUT 6 0 100MEG
CIN  9 5 1U
COUT 3 6 1U
Q1   3 4 0 Q172
.MODEL Q172 NPN BF=122 BR=0.2 IS=2.65E-17
+CJE=9.37E-14 VJE=0.94 TF=1.4E-9
+CJC=1.09E-14 VJC=0.7 TR=1E-8
.WIDTH OUT=80
.AC DEC 1 100 1MEG
.PRINT AC VP(6) VM(6)
.END
```

The control line in the input file specifies an ac analysis over the frequency range 10^2 to 10^6 Hz in steps of 1 decade. The input signal is a sine wave of amplitude 10 mV. The results are shown in Table 11E.2. Note the following:

(a) At frequencies above 1 kHz, the output voltage is constant. This is

TABLE 11E.2 COMPUTED VOLTAGE GAIN AND PHASE DIFFERENCE BETWEEN INPUT AND OUTPUT SIGNALS FOR THE CIRCUIT OF FIG. 11E.12

Frequency (Hz)	Phase difference (deg)	Voltage gain
10^2	155.5	62.8
10^3	177.4	68.9
10^4	179.7	69.0
10^5	180.0	69.0
10^6	180.0	69.0

because at these frequencies the impedances of the capacitors CIN and $COUT$ are very small and do not cause any attenuation of the ac signal.

(b) At the higher frequencies, again because of the negligible effects of CIN and $COUT$, the phase difference between input and output signals is 180°. This antiphase behavior arises because an increase in base–emitter voltage causes an increase in base current, which, in turn, results in an increase in collector current. More I_C means a bigger voltage drop across RL and thus a decrease in collector–emitter voltage.

11.7.3 Cutoff Frequency

Another parameter that appears in the output listing of a small-signal BJT SPICE simulation is the ***cutoff frequency***, f_T. This parameter is often used to specify the high-frequency performance of a transistor. It is properly defined as the frequency at which the common-emitter current gain equals unity when the output terminals of the transistor are short-circuited.

To derive an expression for f_T we start with the hybrid-π circuit of Fig. 11.23 and, first, ignore the parasitic resistors and assume that $1/g_\mu = \infty$. This last assumption is commensurate with a reverse bias at the collector–base junction and is therefore valid for the BJT operating in the normal active mode. The stipulation of a short circuit at the output removes the influence of g_o and effectively connects the two capacitors in parallel. The effect of all these conditions is to reduce the equivalent circuit to that shown in Fig. 11.24. Inspection of this circuit reveals that

$$i_c = g_m v_{be} \tag{11.71}$$

and

$$i_b = [g_\pi + j\omega(C_{BE} + C_{BC})]v_{be} \tag{11.72}$$

Thus the magnitude of the current gain is

$$\frac{i_c}{i_b} = \frac{g_m}{\sqrt{g_\pi^2 + \omega^2(C_{BE} + C_{BC})^2}} \tag{11.73}$$

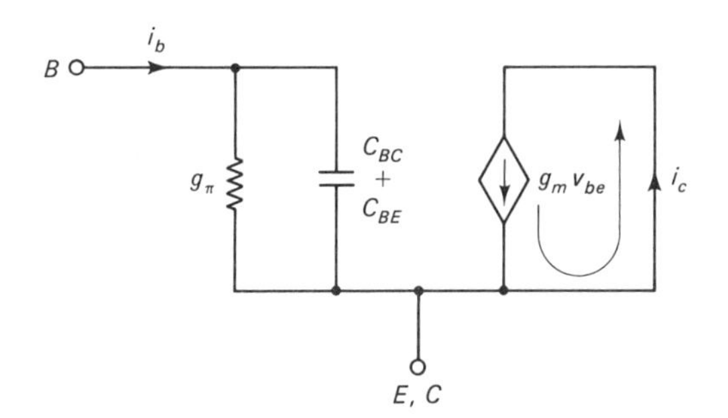

Figure 11.24 Hybrid-π equivalent circuit for a BJT in the active mode with the collector and emitter shorted together.

For high frequencies, such that $\omega(C_{BE} + C_{BC}) > g_\pi$, it follows that f_T, the frequency at which the gain reduces to unity, is

$$f_T = \frac{g_m}{2\pi(C_{BE} + C_{BC})} \tag{11.74}$$

There exists a relationship between this widely used figure of merit for high-frequency performance and the SPICE parameter TF, as we now show. For operation in the normal active mode, the total capacitance is likely to be dominated by the storage capacitance associated with electrons injected across the emitter–base junction and stored in the base (see Fig. 6.28). From (11.63) the relevant capacitance is

$$C_{BE} = \frac{q}{kT} \mathsf{TF\ IS} \exp\left(\frac{qV_{BE}}{kT}\right) \tag{11.75}$$

For the case of nearly horizontal I–V curves in the active region [i.e., VAF very large (see Fig. 11.16)], the expression for the collector current given in (11.46) simplifies such that (11.75) can be written as

$$C_{BE} = \frac{q}{kT} \mathsf{TF}\ I_C \tag{11.76}$$

Also, using the definition of g_m from (11.66) and (11.67), that is,

$$g_m = \left.\frac{\partial I_C}{\partial V_{BE}}\right|_{V_{CE}} \tag{11.77}$$

we note that differentiation of (11.46) gives

$$g_m = \frac{q}{kT} I_C \tag{11.78}$$

Finally, from (11.74), (11.76), and (11.78), it follows that

$$f_T = \frac{1}{2\pi \mathsf{TF}} \tag{11.79}$$

This equation is used in SPICE to compute f_T.

11.8 BJTS VERSUS FETS

At the end of this, the last of the device chapters in this book, it is appropriate to summarize briefly some of the differences and similarities of silicon bipolar and field-effect transistors.

Generally speaking, BJTs are preferable to FETs in applications involving ***high currents*** and ***high power signals***. This is due primarily to the differences in geometries of the "active" regions of the two types of transistors. To make a rough comparison, note that the surface area occupied by the gate of a MOSFET is far larger than the cross-sectional area offered to charge flow in the channel. This is because the gate length is much larger than the inversion layer depth. However, in a BJT the surface area occupied by the emitter is essentially the same as the cross-sectional area through which the main transistor charge flows. Thus, other things being equal, the MOSFET takes up more "real estate" than the BJT. As space is often at a premium in integrated circuits, BJTs prevail over MOSFETs for high-power applications.

An application where FETs offer a significant advantage over BJTs is in circuits that need a ***high input impedance***. Opamp input stages are a notable example (see Problem 10.6). High input impedance is readily provided by the insulated gate of MOSFETs, or by the reverse-biased gate–channel diode of JFETs and MESFETs. Contrarily, the base–emitter diode of BJTs is usually forward biased (saturation and active modes) and so offers a low impedance to incoming signals.

To compare transistor ***operating speeds***, consider the transit times of electrons moving from source to drain in an *n*-channel MOSFET, and electrons traversing the base in an *npn* BJT. In the former case, the limiting value of transit time is

$$\tau_t(\text{FET}) = \frac{L}{v_{\text{sat}}} \tag{11.80}$$

where L is the channel length and v_{sat} is the scatter-limited velocity. For a 1-μm gate Si MOSFET this gives a transit time of 10 ps. The corresponding cutoff frequency [see (9.41)] is thus about 16 GHz. For a BJT, an approximate expression for the common-emitter current gain is

$$\beta_F = \frac{2D_e\tau_e}{L_B^2} \tag{11.81}$$

where D_e is the electron diffusivity in the base, τ_e the electron minority carrier lifetime, and L_B the neutral basewidth. In deriving (11.81) from (11.37) it has been assumed that the emitter injection efficiency equals unity and that $L_B \ll L_e$, where L_e is the electron minority carrier diffusion length. Combining (11.81) and (11.62) gives for the transit time across the base,

$$\tau_t(\text{BJT}) = \frac{L_B^2}{2D_e} \tag{11.82}$$

Taking $D_e = 30\ \text{cm}^2/\text{s}$ (see Fig. 3.5) and $L = 0.15$ μm (typical of modern, high-speed BJTs), we find that the cutoff frequency is about 40 GHz. Although these calculations are admittedly rough, they do help to indicate why the fastest silicon logic presently available is a bipolar technology [emitter-coupled logic (***ECL***)].

The very fast nature of ECL makes it the prime candidate for applications such as large computers, where speed is of the essence. However, ECL has a major drawback in the form of its large ***static power dissipation*** (see Problem 11.13).

This places a practical limit on the packing density of transistors on a chip and precludes this technology's use in passively cooled VLSI circuits. The large power drain arises because there is always a branch of an ECL gate in which the transistors are on. The MOSFET technologies of NMOS and, in particular, CMOS, are considerably less power-hungry because of their employment of normally-off (enhancement mode) transistors.

From the foregoing comparisons it can be appreciated that each of the main types of silicon transistor, the BJT and the MOSFET, has its strengths and weaknesses. If it were possible to combine both types of transistor on the same chip, a very useful technology could result. For example, the low power dissipation and very large scale integration capabilities of CMOS could be employed in the signal processing part of the chip, and the high power-handling capabilities of bipolar could be utilized for driving large loads. Such a merged technology is presently under development. It is called ***BICMOS***.

11.9 CHAPTER SUMMARY

In this chapter detailed descriptions of the principles of operation of the bipolar junction transistor and of its electrical performance under both dc and ac conditions have been presented. The reader should now have an appreciation of the following topics:

1. ***Structure.*** The ***bipolar junction transistor*** comprises two very closely spaced *pn* junctions which share a common region called the ***base***. The base separates the ***emitter*** from the ***collector***. Invariably the emitter is more heavily doped than the base; this is to improve the ***emitter injection efficiency***. Usually, the base width is less than the minority carrier diffusion length; this is to improve the ***base transport factor***.

2. ***Modes of operation.*** (a) In the ***active mode*** of operation the emitter–base junction is forward biased and the collector–base junction is reverse biased. The collector current nearly equals the emitter current. In the ***common-emitter connection*** the small difference current, the base current, can be used to control the collector current; this is a manifestation of ***current gain*** and ***transistor action***. (b) In the ***saturation mode*** both junctions are forward biased, causing the net collector current to decrease below its active mode value. (c) In the ***cutoff mode*** both junctions are reverse biased, so all the terminal currents are very low.

3. ***Current gain.*** In the active mode of operation current gain increases with collector–base reverse bias due to ***basewidth modulation***. ***Avalanche breakdown*** in the collector–base junction or complete depletion of the base leading to ***punch-through*** set practical limits to the allowable amount of basewidth reduction. Current gain is adversely affected at low values of base–emitter voltage by the significance of recombination in the emitter–base depletion region, and at high values

of base–emitter voltage by high-level injection in the base. A ***built-in field*** in the base due to nonuniform doping can speed the passage of electrons in transit to the collector, thus improving the current gain.

4. *BJT modeling.* In the dc case the BJT can be modeled as two back-to-back diodes and a single current source between the collector and emitter terminals. This ***nonlinear hybrid-π equivalent circuit*** representation of the ***Ebers–Moll equations***, when augmented by the diode capacitances, is a useful equivalent circuit for ***large-signal ac analyses***. This model is implemented in SPICE and used in this chapter to demonstrate factors affecting the switching speed of BJTs. High-frequency operation is conveniently modeled by the ***small-signal, linear hybrid-π*** equivalent circuit, as demonstrated by the SPICE simulation of a ***single-stage BJT amplifier***.

5. *BJTs versus FETs.* Generally speaking, BJTs are superior in situations where the signal power levels are high, and FET circuits are superior in cases where low power dissipation or high transistor packing density are required.

11.10 REFERENCES

1. J. J. Ebers and J. L. Moll, "Large-signal behaviour of junction transistors," *Proceedings of the IRE,* vol. 42, pp. 1761–1772, 1954.
2. I. E. Getreu, *Modeling the Bipolar Transistor,* pp. 10–20, Elsevier Science Publishing Co., Inc., New York, 1976.
3. J. M. Early, "Effects of space charge widening in junction transistors," *Proceedings of the IRE*, vol. 40, pp. 1401–1406, 1952.
4. W. M. Webster, "On the variation of junction transistor current amplification factor with emitter current," *Proceedings of the IRE*, vol. 42, pp. 914–920, 1954.
5. R. A. Colclasser, *Microelectronics: Processing and Device Design,* p. 213, John Wiley & Sons, Inc., New York, 1980.

PROBLEMS

11.1. Consider a symmetrical n^+pn^+ transistor with uniform doping densities in all three regions of the device. Sketch energy band diagrams, showing approximate positions for the quasi-Fermi levels, for the transistor when operating in the following modes: active, cutoff, and saturation.

11.2. An epitaxial base transistor is fabricated by depositing silicon with acceptor impurity concentration $N_A = 2 \times 10^{16}\ \text{cm}^{-3}$ onto a substrate with $N_D = 10^{18}\ \text{cm}^{-3}$. The deposition technique of epitaxy is described in Section 12.2.12 and is of relevance here, as it affords a means of attaining a base doping density which is less than the collector doping density. A heavily doped emitter is diffused into the *p*-

type epitaxial layer. The width of the neutral base is 2 μm when no bias is applied to the transistor.

(a) For $V_{EB} = 0$, compute the base–collector voltage that will cause the base–collector depletion region to extend across the base and touch the emitter–base depletion region.

(b) This condition is called punchthrough. Why is it to be avoided?

11.3. A silicon *npn* transistor has an emitter that is uniformly doped with a donor concentration of 10^{19} cm^{-3}. The base is also uniformly doped and has an acceptor concentration of 10^{17} cm^{-3}. In both regions the impurities can be taken as being completely ionized. Assume that $D_h^E/L_h^E = D_e/L_e$; the collector–base junction is short-circuited, and $L_B(V_{CB} = 0)/L_e = 1$.

(a) Compute the emitter injection efficiency and the base transport factor. Use equations (11.13), (11.23), and (11.24) and consult Fig. 11.9.

(b) Compare the product of these two quantities with the value for the normal common-base current gain, as computed from (11.36).

11.4. α_F as computed from (11.36) is a device property. The actual common-base forward current gain (I_C/I_E), as would be measured for a transistor in a real circuit, is bias dependent. Explain how and why I_C/I_E and α_F would differ for:

(a) Operation in the saturation mode.

(b) Operation in the cut-off mode.

11.5. To obtain some numerical data on the actual current gain in the saturation mode, use SPICE as follows. Use a single transistor (type 172 of Example 11.1) in the common-base connection with a fixed emitter current of 100 μA. Sweep V_{CB} over the range -0.7 to $+0.7$ V. Use the PRINT instruction to obtain a list of I_C at each V_{CB}. Compute the actual common-base current gain I_C/I_E at each value of V_{CB}. Note that the turn-on voltage of the collector–base diode has to be exceeded before the current gain decreases significantly.

11.6. Given a bipolar junction transistor with unlabeled leads, one dc voltage power supply, and one ammeter, how would you determine whether the transistor was *npn* or *pnp*, and which lead was which?

11.7. The simplest version of the Ebers–Moll model for the bipolar transistor has three parameters: BF, BR, and IS. Describe how numerical values for these parameters can be obtained from I–V measurements on a real transistor.

11.8. In most modern BJTs, the electron diffusion length in the base is far greater than the base width. This condition gives the transistor a high current gain, since electrons injected from the emitter have a high probability of diffusing to the collector–base depletion region before they recombine.

In a transistor in which $L_e \gg L_B$, it is possible to derive expressions for the gains and terminal currents using a far simpler analysis than that given in Section 11.3.5. The steps in this simple analysis are outlined below.

(a) Show that if $L_e \gg L_B$, the general solution to (11.15) becomes

$$n_p(x) - n_{0p} = A + Bx \qquad (11P.1)$$

(In other words, the electron concentration varies linearly with position in the base, as illustrated in Fig. 11.10 for $L_B/L_e = 0.1$.)

(b) Show that in normal active operation with $V_{CB} = 0$ the boundary conditions on $n_p(x)$ at $x = 0$ and $x = L_B$ require that

$$n_p(x) - n_{0p} = n_{0p}(e^{qV_{BE}/kT} - 1)\left(1 - \frac{x}{L_B}\right) \quad (11P.2)$$

(c) Show that the collector current is given by

$$J_C = \frac{qD_e n_{0p}}{L_B} e^{qV_{BE}/kT} \quad (11P.3)$$

assuming that $V_{BE} \gtrsim 2kT/q$ and ignoring the reverse leakage current of the collector–base junction.

(d) Show that the total rate of recombination in the base is given by

$$\int_0^{L_B} \frac{n_p(x) - n_{0p}}{\tau_e} dx = \frac{n_{0p}L_B}{2\tau_e}(e^{qV_{BE}/kT} - 1) \quad (11P.4)$$

for normal active operation. The base current is therefore given by

$$J_B = q\frac{n_{0p}L_B}{2\tau_e}(e^{qV_{BE}/kT} - 1) \quad (11P.5)$$

(e) Confirm the result for J_B by using (11.23) and (11.24); take $V_{BC} = 0$ V and make appropriate expansions of the hyperbolic terms.

11.9. If the n region in a particular pn diode is called the emitter and the p region is called the base, derive from the ideal diode equation (i.e., for the case of no recombination in the depletion region and a very long base region) an expression for the emitter injection efficiency. How and why is this expression different from the corresponding expression for a npn transistor in which the base width L_B is much smaller than the electron diffusion length L_e? Take the transistor to have $V_{CB} = 0$.

11.10. Increasing the reverse bias on the collector–base junction reduces the width of the neutral base, leading to a finite conductance in the active mode (see Fig. 11.16) and an increase in current gain (see Section 11.5.1). To examine these effects, use SPICE and compare the I_C–V_{CE} characteristics of two transistors which are identical in all respects save the Early voltage. Specify a value of VAF for only one of the devices. To estimate VAF assume that the neutral basewidth L_B can be represented by a short Taylor series expansion, that is,

$$L_B(V_{BC}) = L_B(V_{BC} = 0) + V_{BC}\left.\frac{dL_B}{dV_{BC}}\right|_{V_{BC}=0} \quad (11P.6)$$

and that

$$\frac{L_B(V_{BC})}{L_B(0)} = 1 + \frac{V_{BC}}{\text{VAF}} \quad (11P.7)$$

From these two equations a definition for VAF follows as

$$\text{VAF} = \left[\frac{1}{L_B(0)}\left.\frac{dL_B}{dV_{BC}}\right|_{V_{BC}=0}\right]^{-1} \quad (11P.8)$$

Compute VAF using $L_B(0) = 0.5$ micron, $N_{A,\text{base}} = 10^{17}\ \text{cm}^{-3}$, $N_{D,\text{collector}} = 10^{16}\ \text{cm}^{-3}$. For the other SPICE parameters use BF = 354, BR = 0.4, IS = 5.33×10^{-17} A (i.e., the parameters for $Q171$ of Example 11.3).

11.11. Derive an expression for the voltage gain in a single-stage amplifier starting from the hybrid-π equivalent circuit of Fig. 11.23. Do not forget to include the collector load resistor R_L and other relevant components from Figure 11E.12. Take advantage of the fact that in the active mode, both C_μ ($\equiv C_{BC}$) and g_μ are very small.

Use the resulting equation for voltage gain, along with numerical values from the output listing of Example 11.6 for any hybrid-π parameters that are needed, to verify that the high-frequency voltage gain for the circuit of Fig. 11E.12 is as given in Table 11E.2.

11.12. To display time-dependent signals in SPICE visually, it is necessary to perform a transient analysis. Modify the input file given in Example 11.6 to do this for the circuit of Fig. 11E.12 and a sine-wave input signal of magnitude 10 mV and frequency 100 kHz. Plot both the input and output voltages and confirm that the phase difference between them is 180°.

11.13. Figure 11P.1 shows the basic switching circuit used in emitter-coupled logic (ECL). V_{IN} and V_{REF} are chosen so that when V_{IN} represents a logic high level Q_A is on and Q_R is off, and when V_{IN} is "low," Q_A is off and Q_R is on. Thus, current drawn from the power supply V_{EE} is switched between the two branches of the differential stage. When Q_A is on, V_{OUT2} is low and V_{OUT1} is high. When Q_A is off, V_{OUT2} is high and V_{OUT1} is low. By paralleling additional transistors with Q_A, a multiple-input NOR/OR gate, the basic ECL gate, can be realized.

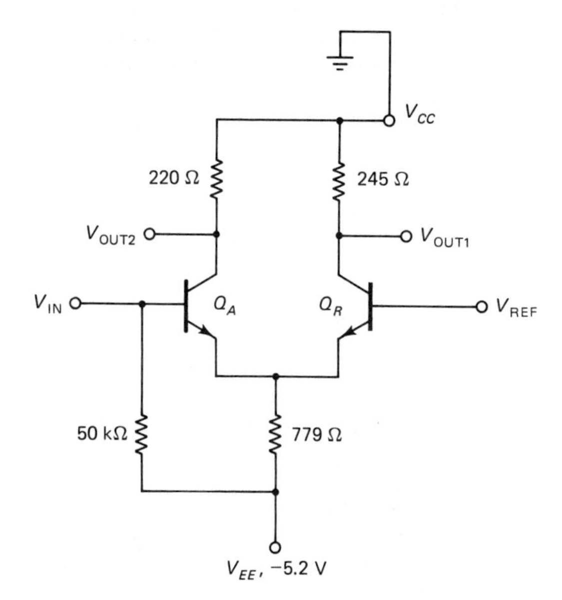

Figure 11P.1

The circuit components and bias levels are chosen so that the transistors never enter the saturation mode. This makes ECL very fast. The penalty for this high speed is a high static power dissipation, which arises because there is always current in one branch of the circuit. This contrasts with the situation in slower MOS logic families; for example, in CMOS logic there is no static power dissipation, and in NMOS there is static power dissipation only when the logic inputs are high.

In this problem the static power dissipation is to be computed via SPICE. Specifically, use the component values shown in Fig. 11P.1 (these are typical of the Fairchild 10K ECL family) and take $V_{\mathrm{REF}} = -1.3$ V, $V_{\mathrm{IN}}(\mathrm{high}) = -1.0$ V, and $V_{\mathrm{IN}}(\mathrm{low}) = -1.6$ V. Insert an ammeter in the circuit so that the product of the recorded current and $|V_{EE}|$ gives a good estimate of the static power dissipation. Perform the calculation for both high and low input levels. Also, by printing-out appropriate node voltages, confirm that the transistors never saturate. Model the transistors with the parameters of the $Q172$ device used in Example 11.2.

12 Semiconductor Device Processing for Integrated Circuits

12.1 INTRODUCTION

The volume of information published on the fabrication of semiconductor devices and integrated circuits is so enormous that only a brief outline of the subject can be given in this chapter. Readers interested in a more detailed treatment are referred to the excellent texts by Sze [1], Ghandhi [2], and Wolf and Tauber [3]. Emphasis here will be on silicon integrated circuit processing, for although the use of GaAs ICs is growing, silicon will be by far the most widely used material in microelectronics for the foreseeable future.

12.2 SILICON INTEGRATED CIRCUIT PROCESSING

Any silicon integrated circuit process consists of sequences of steps in which dopant atoms are introduced into selected areas of a silicon substrate, steps in which layers of silicon dioxide are grown on the substrate surface, steps in which thin films of other materials such as polycrystalline silicon, silicon nitride, or metals are deposited, and steps in which patterns are etched in these various layers. These steps are considered in order below. The preparation of the silicon wafers used as substrates is discussed in Section 2.2.

12.2.1 Solid-State Diffusion

Solid-state diffusion provides a relatively simple and inexpensive method of introducing dopant atoms into a silicon wafer. The process involves the motion of impurities through the silicon lattice in the solid phase. At first sight this appears

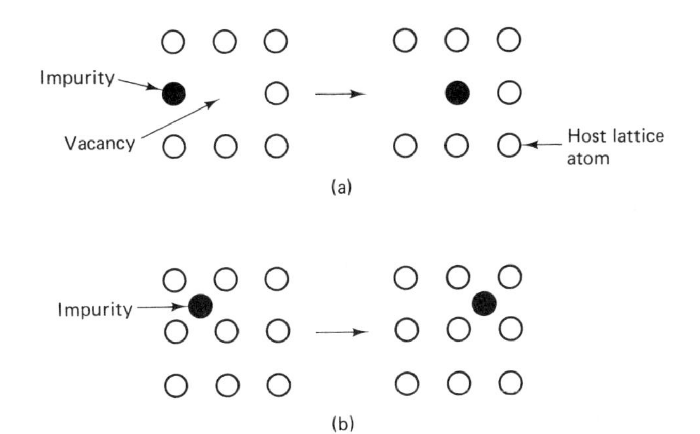

Figure 12.1 Diffusion of an impurity in silicon (a) via a vacant lattice site, and (b) as an interstitial.

improbable, but it must be remembered that the diamond lattice is a relatively loosely packed structure, and that at the temperatures usually employed in diffusion (1000°C is typical) atoms in the lattice have a substantial amount of thermal energy. On occasion, an atom can break the bonds with its neighbors and move into an adjacent empty region. Technically, the atom is said to move from a ***substitutional*** site to an ***interstitial*** site, creating a ***vacancy***. Once in an interstitial position, the atom may be able to move through the lattice quite readily (this is particularly true for transition metals such as gold, copper and iron, which are usually undesirable impurities in silicon). Alternatively, impurities may move through the lattice by exchanging places with an adjacent vacancy. Since there are relatively few vacancies in the crystal, this type of diffusion process is much slower than interstitial diffusion. It is widely but not universally accepted that the group III and group V impurities, such as boron, phosphorus, and arsenic, commonly used as n- and p-type dopants in silicon, diffuse by interaction with vacancies. These possible diffusion mechanisms are illustrated in Fig. 12.1.

The mathematics describing the diffusion of impurities in silicon is essentially the same as that governing the diffusion of electrons and holes (see Section 3.3). For an individual atom, diffusion is a random process, and it is impossible to predict the direction of motion. However, the combined random motion of large numbers of impurity atoms leads to a net flow of impurity from regions of high concentration to regions of lower concentration. ***Fick's law*** states that the flux F of the impurity (in atoms per unit area per unit time) is proportional to the concentration gradient; thus in one dimension,

$$F = -D\frac{\partial C}{\partial x} \tag{12.1}$$

where C is the impurity concentration and D is the diffusion coefficient. The diffusion coefficients of some common impurities in silicon are given in Fig. 12.2. It is worth noting that the diffusion coefficients of the common n- and p-type

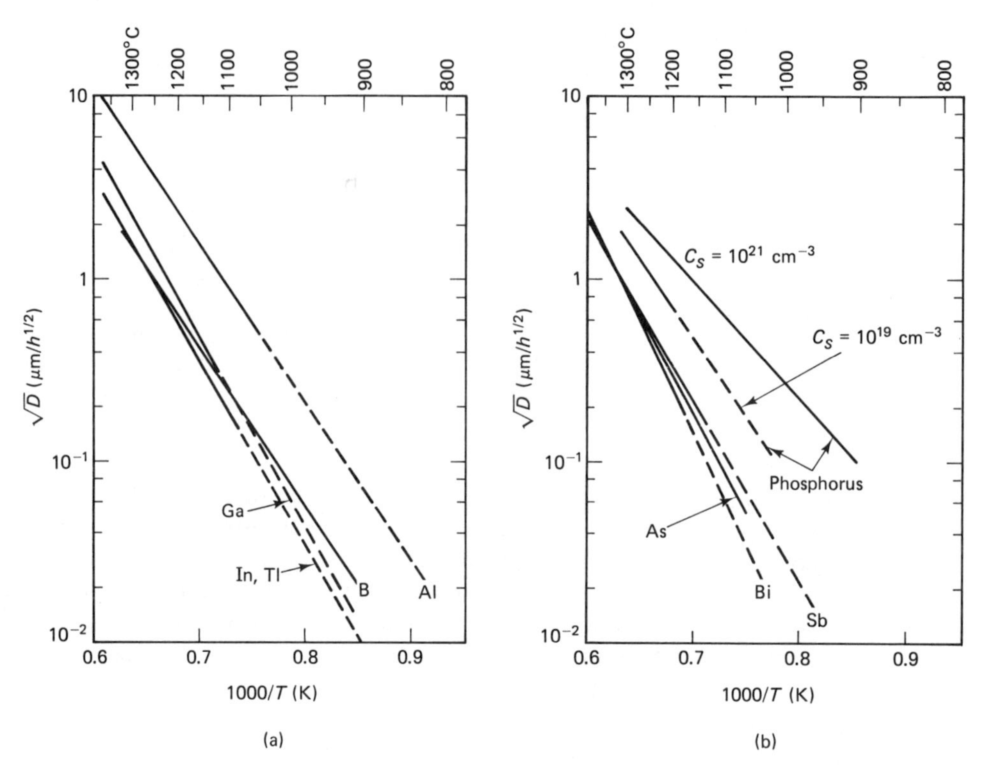

Figure 12.2 Diffusion coefficients for some common dopants in silicon: (a) acceptors; (b) donors. (From Grove [4]. Reprinted with permission of the publisher, John Wiley & Sons, Inc.)

dopants are very temperature sensitive, and that boron and phosphorus diffuse much more rapidly than arsenic and antimony.

Application of the continuity principle to (12.1) gives the "diffusion equation,"

$$\frac{\partial C}{\partial t} = \frac{\partial}{\partial x}\left(D \frac{\partial C}{\partial x}\right) \tag{12.2}$$

which governs how the dopant concentration profile changes with time. For low doping levels the diffusion coefficient is independent of the impurity concentration, and (12.2) becomes

$$\frac{\partial C}{\partial t} = D \frac{\partial^2 C}{\partial x^2} \tag{12.3}$$

For high doping levels, the diffusion coefficient tends to become a function of the impurity concentration, and (12.3) does not apply. However, since (12.3) has simple analytic solutions while (12.2) can in general be solved only by numerical methods, there is a tendency to use (12.3) even in situations in which it is not strictly applicable.

12.2.2 Diffusion Technology

The introduction of impurities into silicon by solid-state diffusion is usually done in two steps. In the first step, known as a ***predeposition*** or "predep," a shallow region with a high concentration of the impurity is introduced at the silicon surface, usually by exposing the wafer to a vapor containing the desired dopant at a temperature between 900 and 1000°C. A typical arrangement for carrying out phosphorus predepositions is illustrated in Fig. 12.3. The wafers are enclosed in a quartz furnace tube which is maintained at the required high temperature by a set of resistance heaters. To facilitate the insertion and removal of the wafers from the furnace, they are mounted on a slotted quartz carrier or "boat." Phosphorus oxychloride ($POCl_3$), which is a liquid at room temperature, is used as a phosphorus source. High-purity nitrogen is passed over the surface of the $POCl_3$, sweeping some of the vapor into the furnace tube. Oxygen is also added to the gas flow. At the high temperature in the tube the $POCl_3$ decomposes, leading to the formation of a glass layer containing Si, O, and P at the silicon surface. It is from this glass layer that the phosphorus actually enters the wafer.

At a given temperature, there is a certain maximum concentration of dopant known as the ***solid solubility*** which can be incorporated in the silicon lattice. The solid solubilities C_{SS} of boron, phosphorus, and arsenic in silicon are plotted as functions of temperature in Fig. 12.4. Predeposition furnaces are usually operated with a sufficient oversupply of dopant to allow the concentration C_S at the wafer surface to rise to C_{SS}; in other words, the silicon surface is saturated with dopant. Under these conditions, C_S is constant throughout the diffusion and (12.3) has the solution

$$C(x) = C_S \operatorname{erfc}\left(\frac{x}{\sqrt{D_1 t_1}}\right) \tag{12.4}$$

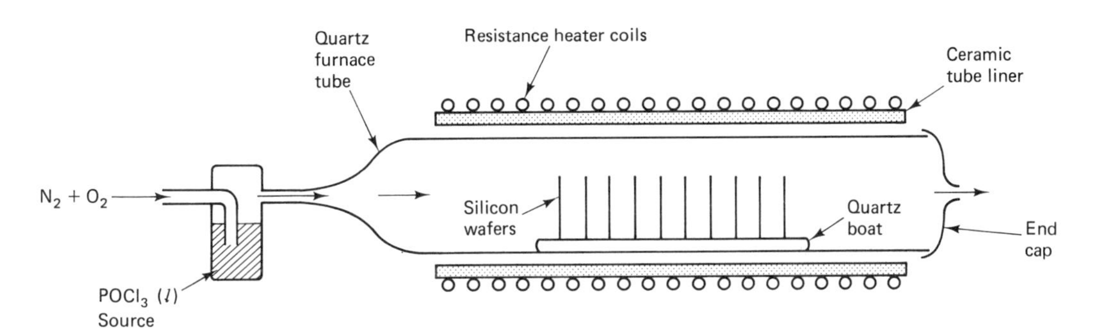

Figure 12.3 Schematic of quartz-tube furnace arrangement used for phosphorus diffusion.

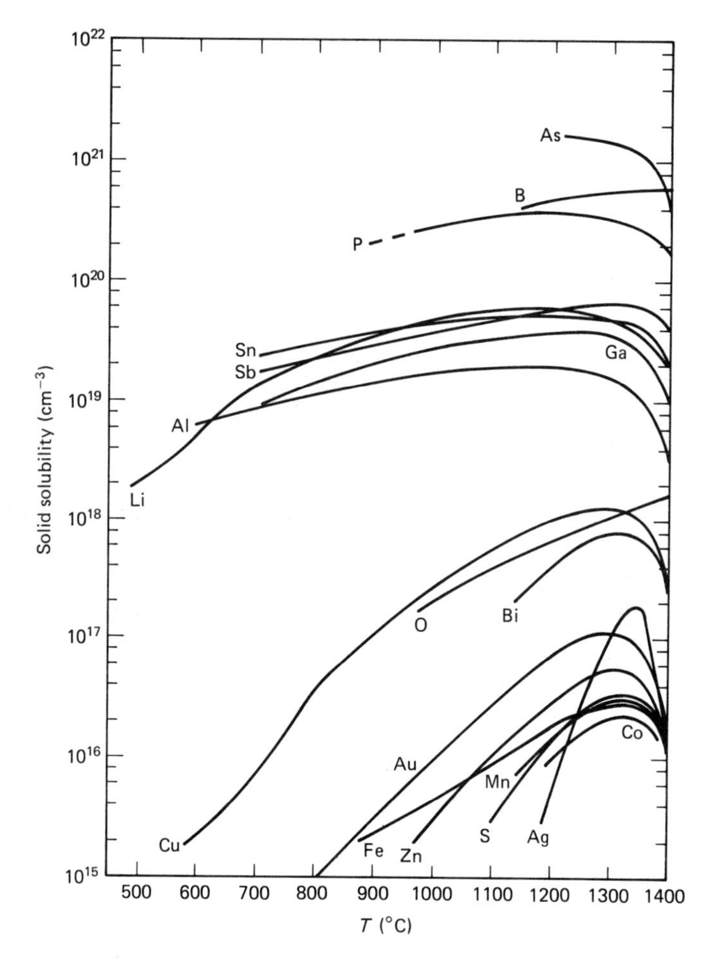

Figure 12.4 Solid solubilities in silicon. (From Grove [4]. Reprinted with permission of the publisher, John Wiley & Sons, Inc.)

where D_1 is the diffusion coefficient at the predeposition temperature and t_1 the time taken for this step. Here we have taken $x = 0$ at the silicon surface. Erfc (y) is the complementary error function:

$$\text{erfc}\,(y) = 1 - \text{erf}\,(y) = 1 - \frac{2}{\sqrt{\pi}} \int_0^y e^{-x^2}\,dx \qquad (12.5)$$

Although not available on most pocket calculators, the error function and complementary error function are widely tabulated. A graph of erfc (y) is given in Fig. 12.5.

An important parameter describing a predeposition is the total dose of dopant Q in atoms per unit area introduced during the diffusion. Integration of (12.4)

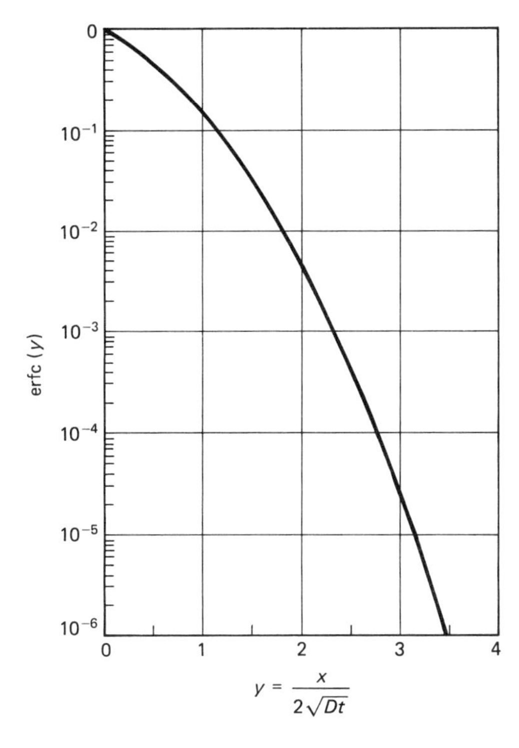

Figure 12.5 Complementary error function.

from the wafer surface to infinite depth gives

$$Q = \frac{2}{\sqrt{\pi}} C_S \sqrt{D_1 t_1} \tag{12.6}$$

Figure 12.6 shows the general shape of the doping profile obtained in a predeposition step, and how this doping profile changes with time. The main points

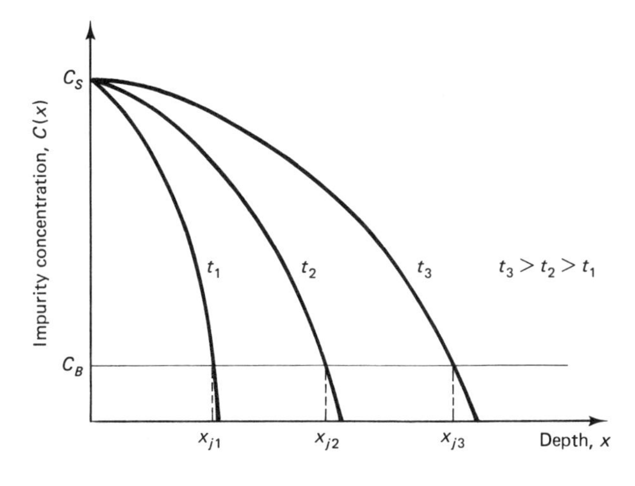

Figure 12.6 Evolution of dopant profile with time in a predeposition diffusion. C_B is the background impurity concentration in the substrate. The metallurgical junction x_j is defined as the location where the diffused impurity concentration equals C_B.

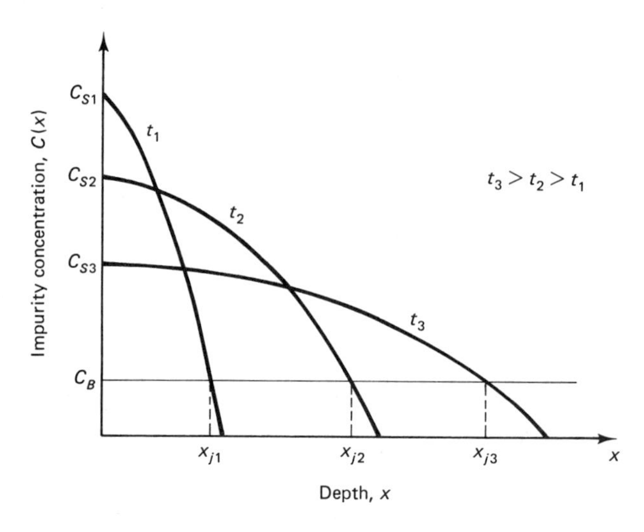

Figure 12.7 Evolution of dopant profile with time in a drive-in diffusion.

to note are that the surface concentration is constant, while the metallurgical junction moves deeper and the total dose of dopant added to the substrate increases as time passes.

As noted above, predeposition steps tend to produce a shallow, heavily doped layer at the wafer surface. It is frequently desirable to spread the dopant deeper into the semiconductor and lower the surface concentration. This is done using a second diffusion step, usually referred to as a ***drive-in***. Drive-ins are normally done at higher temperatures than predepositions, so the dopant moves into the substrate much more quickly. Furnaces used for drive-ins are much the same as those used for predepositions, but no additional dopant is added to the silicon as the diffusion proceeds. This causes the surface doping concentration to fall with time, as shown in Fig. 12.7. If the drive-in produces a doping profile that is substantially deeper than that left by the predeposition, then at the end of the drive-in the profile is given by

$$C(x) = \frac{Q}{\sqrt{\pi D_2 t_2}} e^{-x^2/4D_2t_2} \tag{12.7}$$

where Q is the amount of dopant introduced during the predeposition, D_2 the diffusion coefficient at the drive-in temperature, and t_2 the time taken for the drive-in. More precisely, (12.7) is a good approximation to the true $C(x)$ profile provided that $D_2t_2 \gtrsim 3D_1t_1$.

12.2.3 Nonideal Effects in Diffusion

For real diffusions carried out at high dopant concentration, (12.5) and (12.7) provide only an approximate description of the impurity profile produced in the silicon. For boron and arsenic the diffusion coefficient is enhanced in regions of

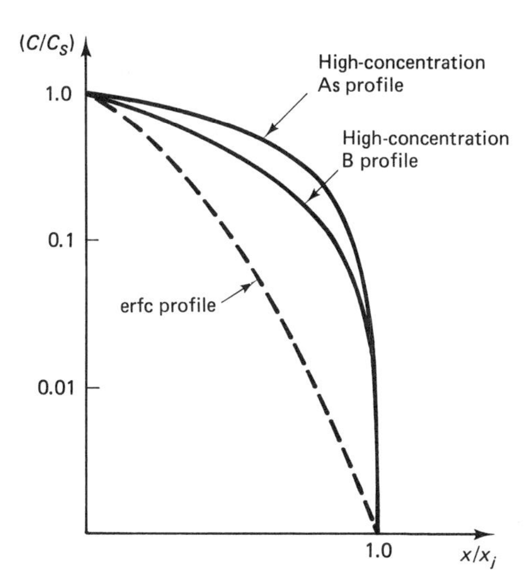

Figure 12.8 Boron and arsenic diffusion profiles at high dopant concentrations. (Adapted from Ghandhi [2], with permission of the publisher, John Wiley & Sons, Inc.)

high concentration, leading to dopant profiles that have a much more square-topped shape than the Gaussian or erfc profiles predicted by simple theory (see Fig. 12.8). Specifically, "high concentration" here refers to impurity concentrations greater than the intrinsic carrier concentration n_i at the diffusion temperature. High concentration phosphorus diffusions tend to produce a long "tail," as shown in Fig. 12.9. The existence of this tail is the reason two diffusion coefficients are specified for phosphorus in Fig. 12.2; the high concentration diffusion coef-

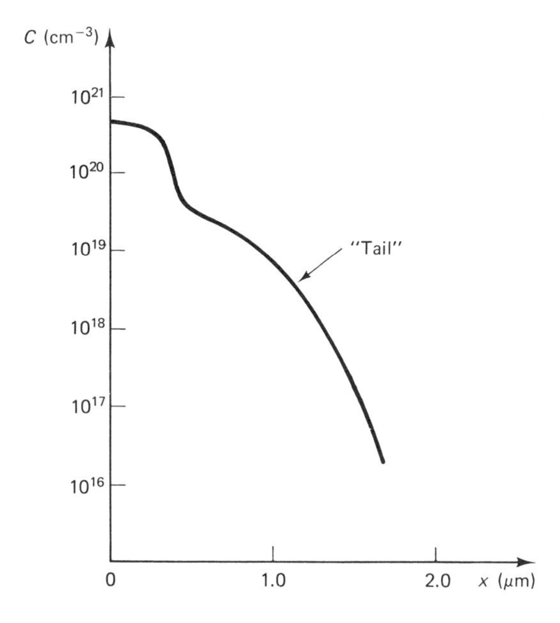

Figure 12.9 Doping profile obtained after a 1-hour phosphorus diffusion at 1000°C, showing characteristic "tail." (Adapted from Sze [5], with permission of the publisher, John Wiley & Sons, Inc.)

ficient is useful for estimating junction depths obtained in phosphorus predepositions. It should also be noted that although the maximum solid solubility of phosphorus in silicon is about 10^{21} cm^{-3}, the maximum electrically active concentration is only about 3×10^{20} cm^{-3}. At present it is not well understood why such a large fraction of the phosphorus should be electrically inactive.

12.2.4 Ion Implantation

Although solid-state diffusion is an inexpensive method of adding impurities to a silicon wafer, it does not offer a great deal of control over the amount of dopant introduced. In some instances, such as the adjustment of MOSFET threshold voltages (Section 7.4.7) or the formation of wells in CMOS ICs (Section 7.5), much more precise control is required. In this case it is necessary to resort to doping by ion implantation.

In ion implantation, a beam of impurity ions accelerated to energies ranging from tens to hundreds of keV is directed at the surface of a semiconductor wafer. At these high energies the dopant ions penetrate many atomic diameters into the target wafer. A schematic diagram of a conventional implanter is shown in Fig. 12.10. Dopant ions are typically generated by striking an arc discharge in a gas containing the desired dopant species; for example, phosphine (PH_3) is commonly used as a source of phosphorus ions. The ions are then accelerated to approximately 20 keV, and passed through a strong magnetic field. This magnetic field serves to separate ions of the desired dopant species from other impurities that may have been generated in the discharge, since the amount an ion deflects in passing through the magnetic field depends on its mass. (This principle is also used in the chemical analysis of substances in mass spectroscopy.)

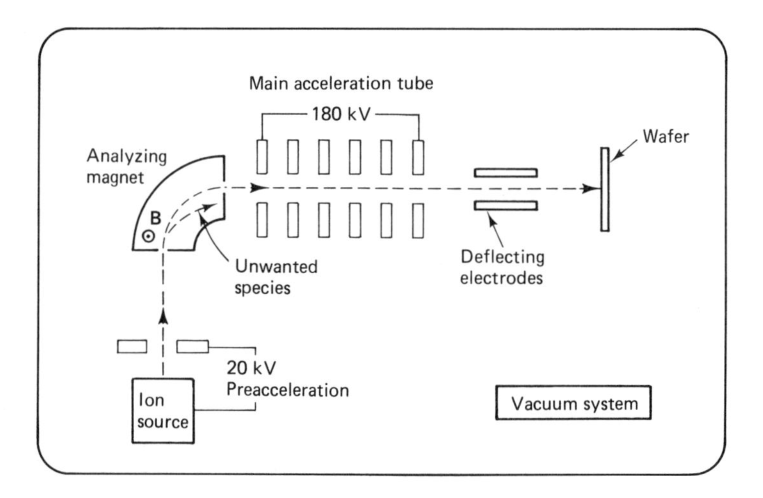

Figure 12.10 Schematic of ion implanter apparatus.

Following mass analysis, the ion beam is passed through additional accelerator stages, reaching a final energy that typically lies in the range 20 to 200 keV. The beam is then directed onto the surface of the target wafer. The amount of dopant added to the wafer can be determined to very high precision simply by monitoring the current in the ion beam.

Once an ion has penetrated the surface of the target wafer, it loses energy by collision both with electrons and with nuclei in the lattice. The latter type of collision disrupts the crystal structure of the target. The heavier an ion of a given energy, the more damage it produces.

The depth to which an implanted ion penetrates a target before coming to rest depends both on its energy and its mass. As a general rule, penetration depth increases with increasing ion energy and decreasing ion mass. The latter result is somewhat surprising, since more massive ions have greater momentum for a given energy. However, more massive ions also lose energy more quickly through collisions with the lattice. At the energies normally employed in IC fabrication, the profile of the implanted ions is roughly Gaussian and can be described reasonably well by the equation

$$C(x) = \frac{Q}{\sqrt{2\pi}\,\Delta R_p} e^{-(x-R_p)^2/\Delta R_p^2} \tag{12.8}$$

As usual, C is the concentration of the implanted impurity, while Q, R_p, and ΔR_p are parameters describing the implant. Q is simply the implant ***dose*** in ions per unit area, R_p is the ***range*** of the implant (a measure of how far the average ion penetrates into the target), and ΔR_p the ***straggle*** (a measure of the spread in the profile). Values of R_p and ΔR_p as functions of ion energy for the common dopants in silicon are given in Fig. 12.11.

As noted above, ion implantation leaves considerable damage in the target wafer. For high-dose implants of heavy ions, the surface of the wafer may even become amorphous. In general, this crystal damage will be detrimental to the performance of any completed device, and must be removed. In addition, implanted ions usually end their trajectories at interstitial sites in the silicon lattice. In order for these impurities to be activated—that is, to function electrically as dopants—they must be moved to substitutional sites. Amorphous layers can be recrystallized by ***annealing*** at temperatures as low as 600°C, but complete damage removal and dopant activation usually require heating the wafer to a temperature of 900°C or greater for approximately 30 minutes. Frequently, this annealing procedure can be combined with other high-temperature processing steps following the implant.

12.2.5 Thermal Oxidation

The microelectronics industry as we know it today would not exist if it were not for the special properties of silicon dioxide (SiO_2). High-quality SiO_2 layers are very chemically resistant, can be used for selective blocking of the diffusion of

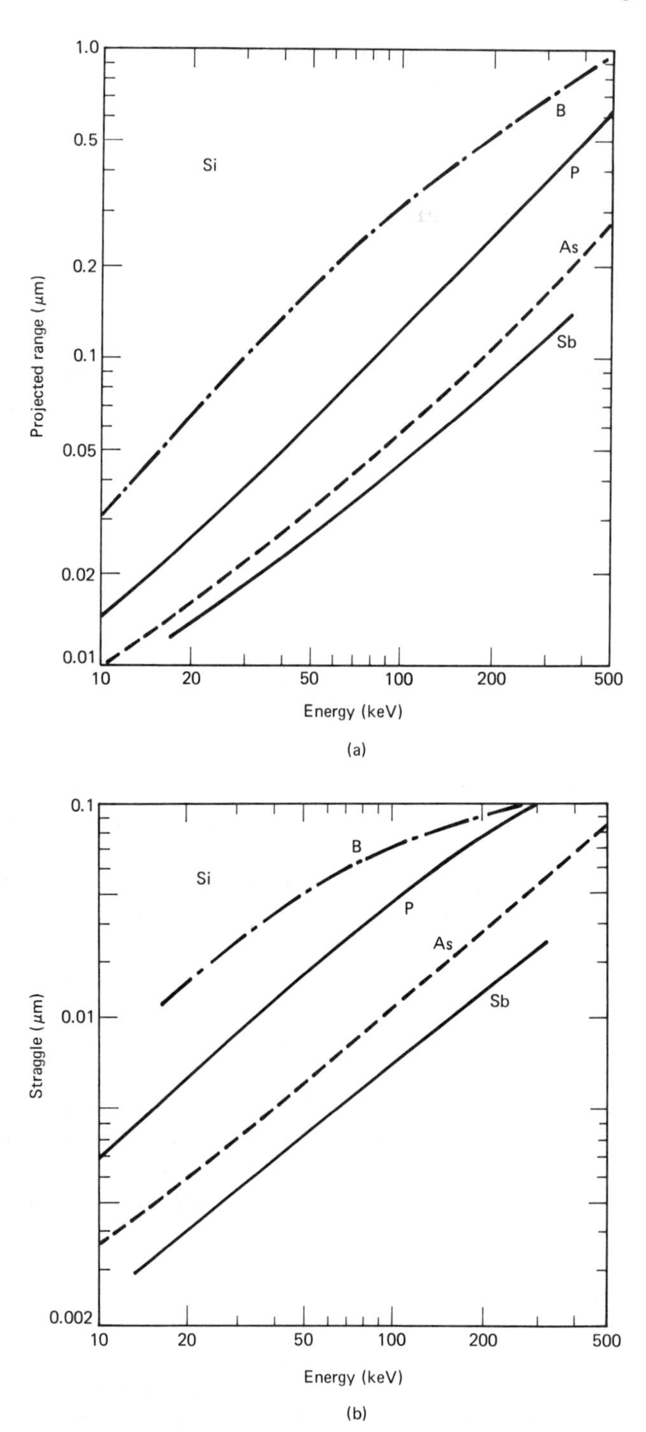

Figure 12.11 Projected range and straggle for common dopants implanted into silicon. (From Ghandhi [2]. Reprinted with permission of the publisher, John Wiley & Sons, Inc.)

dopant into silicon, and most importantly, to reduce the tendency of electrons to be trapped at the silicon surface. Without the latter property it would not be possible to make silicon MOSFETs. No other semiconductor forms an oxide that has such desirable features.

SiO_2 layers are easily grown on silicon by exposing the wafers to oxygen alone (so-called ***dry oxidation***) or to a mixture of oxygen and water vapor (***wet oxidation***) at temperatures comparable to those used in diffusion. Once again, quartz-tube furnaces are used to provide a clean environment in which the wafers can be held at high temperature. Figure 12.12 shows the rates of wet and dry oxide growth. Dry oxidation is a much slower process than wet oxidation, but produces oxides with considerably better electrical properties. For this reason, dry oxidation is used for critical tasks such as the growth of the gate oxide in most MOS processes.

12.2.6 Selective Oxidation

In most modern MOS processes, the silicon surface is oxidized in selected areas to produce regions of thick oxide that serve as bands of isolation between transistors (Section 7.5). This isolation scheme is known as LOCOS, for ***loc***al ***o***xidation of ***s***ilicon. LOCOS is made possible by the fact that silicon nitride oxidizes much more slowly than silicon itself. The actual LOCOS process is illustrated in Fig. 12.13. First, a ***pad oxide*** perhaps 100 nm thick is grown on the substrate. This oxide is necessary due to the extreme thermal expansion mismatch between silicon nitride and silicon. If it were not present, high stress would be placed on the wafers when they were heated to oxidation temperature, creating defects in the silicon crystal.

Following pad oxide growth, a layer of silicon nitride approximately 100 nm thick is deposited on the wafer surface. This layer is then patterned to leave silicon nitride only in regions where transistors are to be formed, and the wafer is oxidized. Typically, a 1-μm-thick layer of field oxide is grown in a wet ambient. Oxide only grows in the regions not coated with nitride, so after the oxidation step the wafer surface resembles Fig. 12.13b. After the nitride is removed, MOSFETs can be formed in the unoxidized regions. The thick oxide present in the field region provides electrical isolation between these transistors. Since oxidation proceeds laterally under the edge of the silicon nitride to some extent, the border between the field region and the active areas slopes gently, rather than forming a steep wall. This gradual transition region is often termed the ***bird's beak*** because of its characteristic beaklike shape.

12.2.7 Oxides for Diffusion Masking

The most commonly used *n*- and *p*-type dopants in silicon—phosphorus, boron, and arsenic—diffuse much more slowly in silicon dioxide than in silicon itself. This property allows a patterned layer of silicon dioxide to be used as a mask to

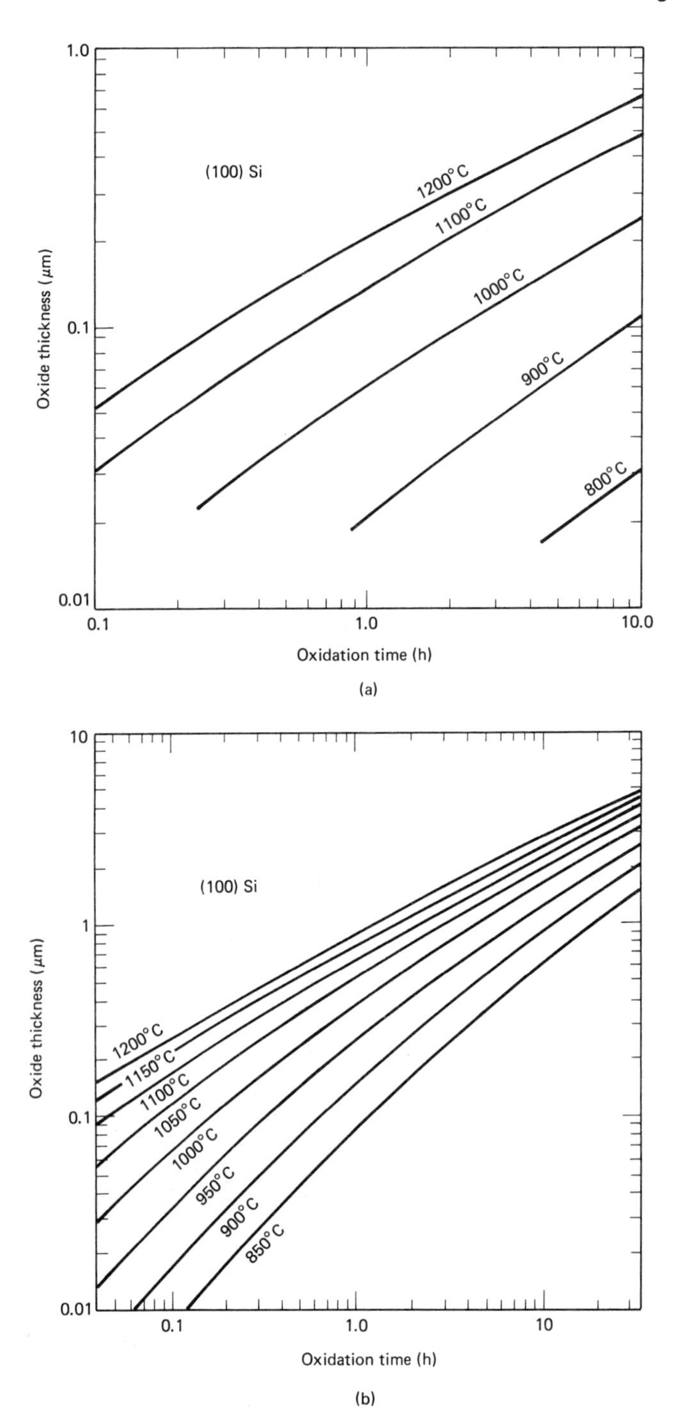

Figure 12.12 Oxidation rate for (100) silicon (a) in dry oxygen, and (b) in oxygen and water vapor. (From Ghandhi [2]. Reprinted with permission of the publisher, John Wiley & Sons, Inc.)

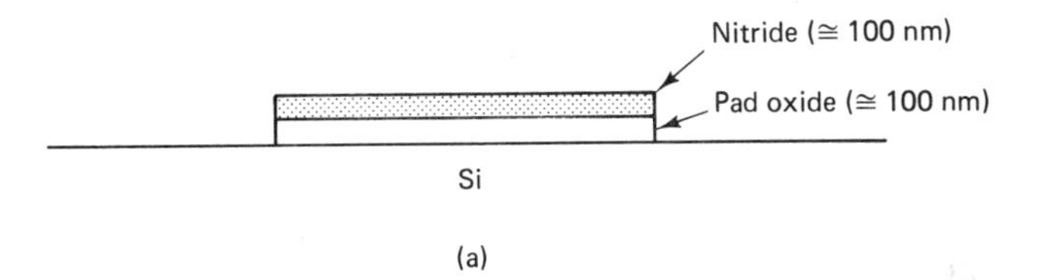

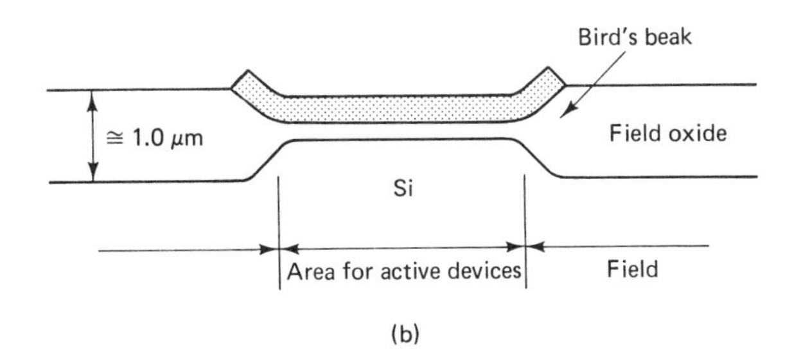

Figure 12.13 Local oxidation of silicon (LOCOS): (a) wafer with patterned nitride–pad oxide sandwich; (b) after completion of field oxidation.

define those regions on a wafer which an impurity is allowed to enter during a diffusion step, as illustrated in Fig. 12.14. Following a predeposition step, the top surface of the silicon dioxide mask is usually etched away to prevent dopant in this layer from diffusing through to the silicon surface in subsequent high-temperature steps. A 0.5-μm-thick layer of SiO_2 is sufficient to mask against a typical boron or phosphorus predeposition; frequently, a thinner layer of oxide will suffice.

Oxide layers can also be used to mask against ion implants. Once again, a 0.5-μm layer of SiO_2 is thick enough effectively to prevent boron, phosphorus, or arsenic ions of the energies usually encountered in ion implanters from reaching an underlying silicon substrate.

12.2.8 Film Deposition

Although the steps of oxidation, diffusion, and ion implantation are the most important in IC fabrication, production of a complete IC requires the deposition of films of silicon nitride, polycrystalline silicon (''polysilicon''), silicon dioxide, and aluminum as well. The silicon nitride is required in the LOCOS isolation

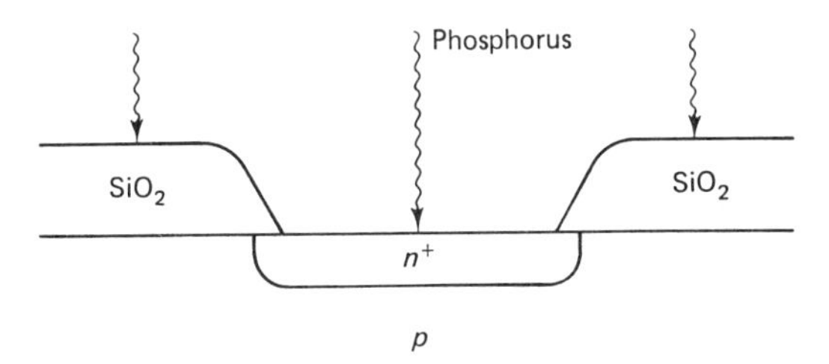

Figure 12.14 How a silicon dioxide layer acts as a diffusion mask.

scheme mentioned above, polysilicon is used almost universally to make gates in MOS ICs, silicon dioxide is used to provide electrical insulation between conducting layers, and aluminum is the metal most widely used to make interconnections (effectively, wiring) between transistors. Polysilicon, silicon dioxide, and silicon nitride are usually deposited by a process known as ***chemical vapor deposition*** (CVD). To take a specific example, polysilicon is normally deposited by heating the wafers to approximately 625°C and then exposing them to an atmosphere of silane (SiH_4). At this temperature, the silane decomposes according to the reaction

$$SiH_4(g) \longrightarrow Si(s) + 2H_2(g)$$

The silicon deposits on a silicon dioxide–coated substrate as a polycrystalline film. Silicon dioxide is commonly deposited at a temperature of 400°C using the reaction

$$SiH_4(g) + O_2(g) \longrightarrow SiO_2(s) + 2H_2(g)$$

Silicon nitride is deposited at 700°C from a mixture of dichlorosilane (SiH_2Cl_2) and ammonia (NH_3):

$$3SiH_2Cl_2(g) + 4NH_3(g) \longrightarrow Si_3N_4(s) + 6HCl(g) + 6H_2(g)$$

Several different techniques are available for depositing aluminum. One of the simplest and most common procedures is ***electron beam evaporation***. In this process, a group of wafers is placed facing a crucible containing very high purity aluminum in a high-vacuum system. The vacuum system is then evacuated to a pressure of roughly a billionth of an atmosphere. At this low pressure, residual gas molecules travel a distance of about 1 m between collisions. The aluminum is then heated using an electron beam until it first melts and then, at a temperature of roughly 1000°C, evaporates. The evaporated aluminum atoms cross the chamber and strike the wafers, leaving an aluminum coating. The basic components of an electron beam evaporation system are shown in Fig. 12.15.

12.2.9 Photolithography and Etching

At many stages in IC fabrication it is necessary to form patterns in layers of silicon dioxide or other materials. This is done using a technique known as ***photolithography***. There are several different approaches to photolithography in use today, but the simplest and most common is analogous to contact printing in ordinary photography. The wafer is coated with a complex photosensitive organic material known as ***photoresist***, pressed into contact with a glass plate known as a ***photomask*** carrying the pattern to be formed on the wafer, and then exposed to an intense ultraviolet light (see Fig. 12.16). The light reaches the wafer only through transparent regions in the photomask. Like photographic film, photoresists come in

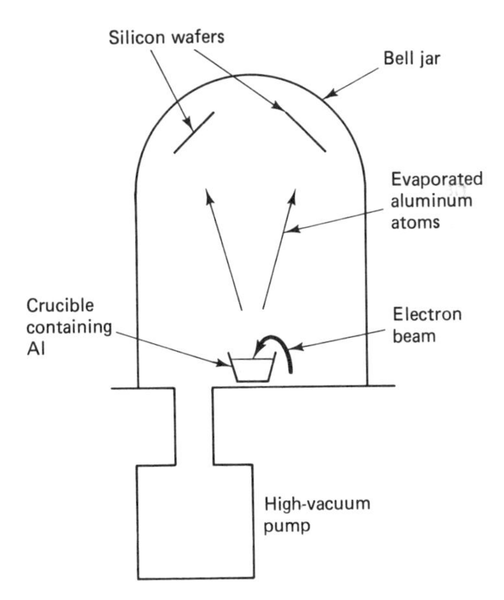

Figure 12.15 Schematic of an electron beam evaporation system, as used for coating silicon wafers with aluminum.

positive and ***negative*** varieties and must be developed after exposure. Exposure to light reduces the solubility of a negative photoresist in developer, so that after development the resist remains only in those areas that were exposed. Conversely, ultraviolet light enhances the solubility of a positive photoresist, so that the developer removes resist only from exposed regions. In both cases, the pattern carried on the photomask is transferred to the photoresist layer.

Once the photoresist has been patterned, it is used as a mask to etch an underlying layer selectively on the wafer. Naturally, the photoresist must be able to resist attack by the desired etchant. For example, to etch silicon dioxide the wafer can be immersed in a solution of hydrofluoric acid. This dissolves exposed silicon dioxide via the reaction

$$SiO_2(s) + 6HF(aq) \longrightarrow H_2SiF_6(aq) + 2H_2O(l)$$

Today it is more common to etch patterns by exposing a photoresist-coated wafer to a plasma containing reactive species such as atomic fluorine. Dry etching techniques of this kind allow better control over feature sizes than the simple method of etching in an aqueous solution, but the basic principle of removing material only in areas not protected by photoresist is the same.

It should be pointed out that photoresist layers can also be used to block ion implants at the energies normally used in IC fabrication. Patterned photoresist layers therefore provide a very convenient masking method for selectively doping a wafer by ion implantation.

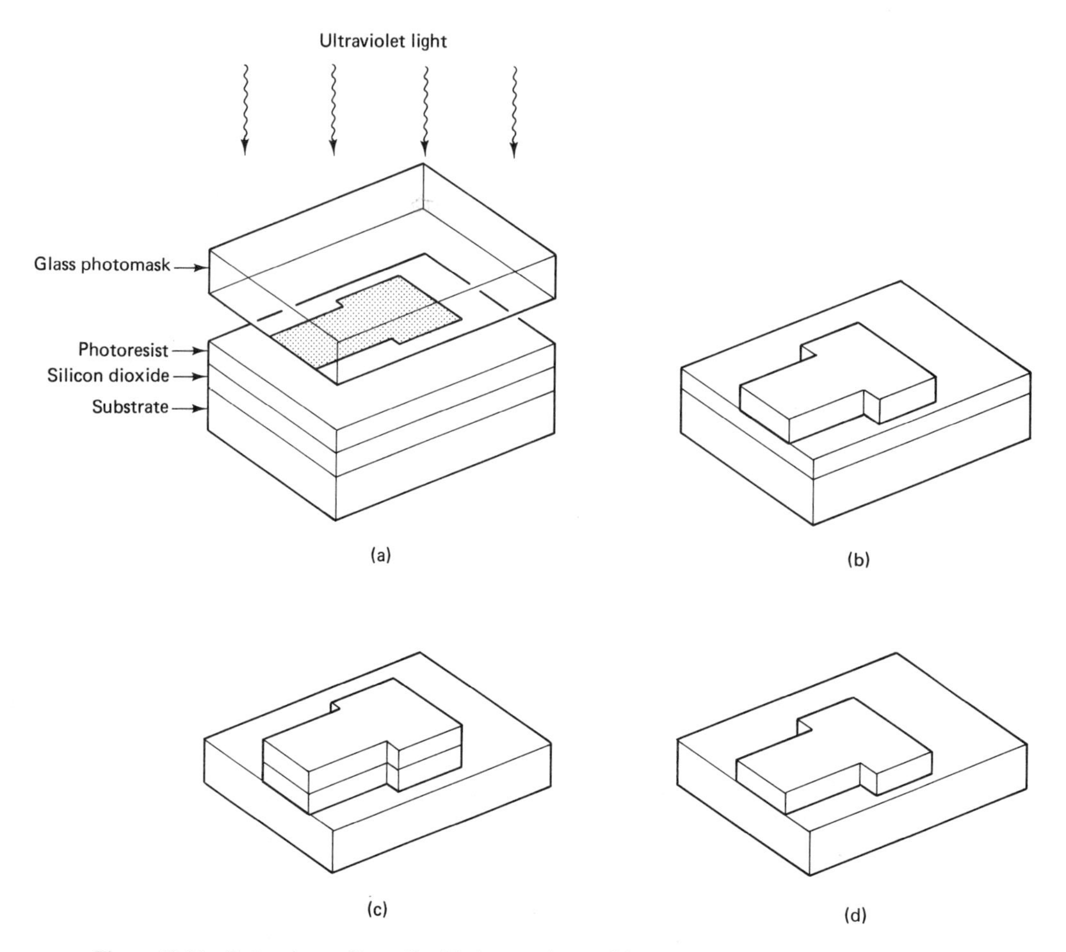

Figure 12.16 Patterning a silicon dioxide layer using positive photoresist photolithography. (a) Photoresist-coated wafer is pressed into contact with photomask and exposed to ultraviolet light. (b) Pattern on photomask is transferred to the photoresist during development. (c) Silicon dioxide is removed from those areas not protected by photoresist by etching in buffered hydrofluoric acid. (d) Photoresist is removed, leaving patterned silicon dioxide.

12.2.10 Ohmic Contacts

To connect a semiconductor device to an external circuit, it is necessary to make metal contacts to the semiconductor which can readily pass current in both directions. It is relatively simple to form such ohmic contacts to silicon. The most common technique is to deposit an aluminum film on the silicon surface, and then heat the wafer to a temperature of approximately 400°C. This procedure is believed to form a very shallow aluminum-doped p^+ region at the wafer surface which makes a good low-resistance contact to p-type silicon at any doping level. Aluminum also makes a good ohmic contact to heavily doped n-type silicon, presumably via the formation of a p^+n^+ junction through which electrons can readily tunnel (Section 6.5.3). However, aluminum contacts to lightly doped n-type silicon tend to be rectifying—that is, these contacts show Schottky-diode-like behavior (Chapter 8). For this reason, to contact a lightly doped n-type region, it is first necessary to form a heavily doped n^+ surface layer by diffusion or ion implantation.

A number of problems are encountered with aluminum contacts when junctions are very shallow or when the contact is forced to carry a high current density. Since these situations are common in modern VLSI circuits, a great deal of attention is presently being given to contact schemes that do not involve aluminum. An introductory discussion of some of these schemes is given in Ref. 1.

12.2.11 Typical CMOS Process

In this section we consider how the individual process steps described above can be put together to produce one of the most common products now manufactured by the microelectronics industry—a p-well CMOS IC. The process sequence needed to make a p-well CMOS inverter is outlined in Fig. 12.17.

The starting material for the CMOS process is a silicon wafer doped with phosphorus to a concentration of 10^{15} cm^{-3} or slightly higher. The first step in the process is the formation of the lightly doped p-type ***well*** in which the n-channel transistors are made, as shown in Fig. 12.17a. The threshold voltage of these transistors is critically dependent on the doping concentration at the surface of the well, necessitating the use of a boron ion implant to add the dopant needed to form the well. The first photomasking operation in the process is used to pattern a photoresist layer that defines the areas in which the boron is implanted, and hence in which the well is formed.

The source and drain regions of the n-channel transistor can be expected to extend to a depth of roughly 0.5 μm, with an additional depth allowance of approximately 1 μm required to accommodate the depletion regions under the source and drain. For this reason the p well must extend to a depth of several micrometers. From Fig. 12.11a it can be seen that even at the maximum energy of 200 keV available with a typical implanter, the range of boron ions in silicon is only 0.5 μm. For this reason, a long high-temperature drive-in must be done after the implant to diffuse the boron deep into the substrate. An implant dose on the order of 10^{12} cm^{-2} driven in for many hours at a temperature above 1100°C is normal for well formation.

To provide isolation between the MOSFETs in a CMOS IC, the threshold voltage of the parasitic transistors formed by interconnection lines running over the field region must be greater than the largest signals that will be generated in the circuit. High field threshold voltages are obtained by raising the surface dopant concentration in the field region using ion implantation and by growing a thick field oxide. This field oxide is grown using the LOCOS process described above;

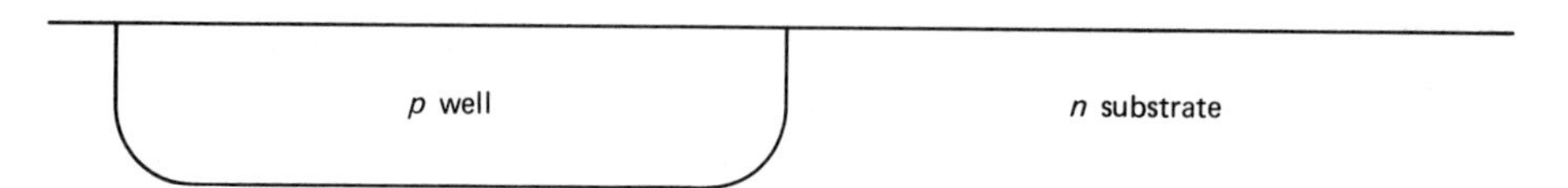

(a) After p-well implant and drive-in. Photomask #1 defines the location of the well.

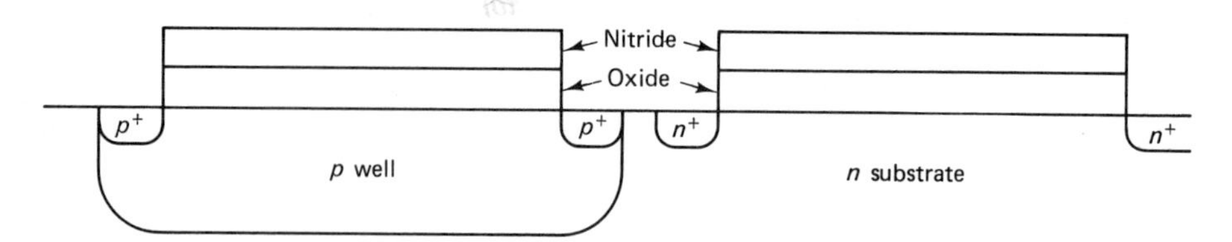

(b) After pad oxide growth, nitride deposition and patterning, and field threshold adjust implants.
Photomask #2 is used to pattern the nitride and pad oxide.
Photomask #3 defines the regions receiving the p^+ boron field threshold adjust implant.
Photomask #4 defines the regions receiving the n^+ arsenic field threshold adjust implant.

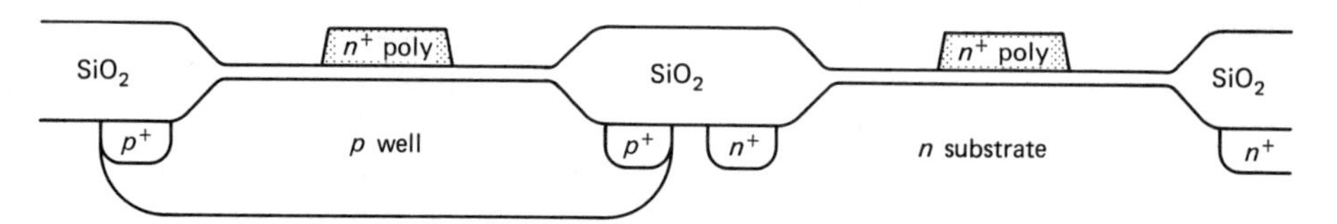

(c) After field oxidation, gate oxidation, and polysilicon gate deposition and patterning.
Photomask #5 is used to pattern the polysilicon gates.

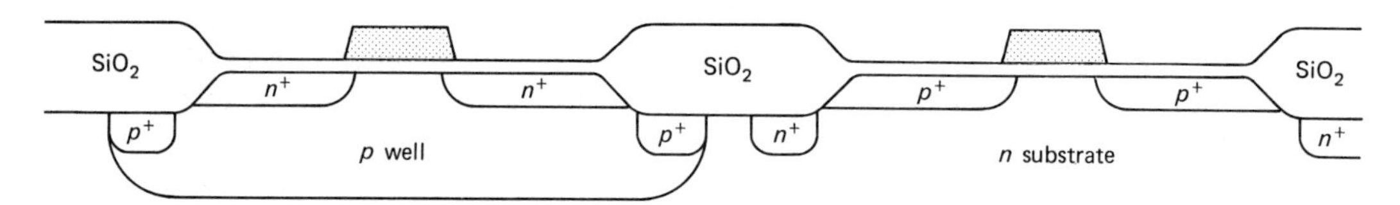

(d) After source/drain implants.
Photomask #6 defines the region receiving the n^+ arsenic implant.
Photomask #7 defines the region receiving the p^+ boron implant.

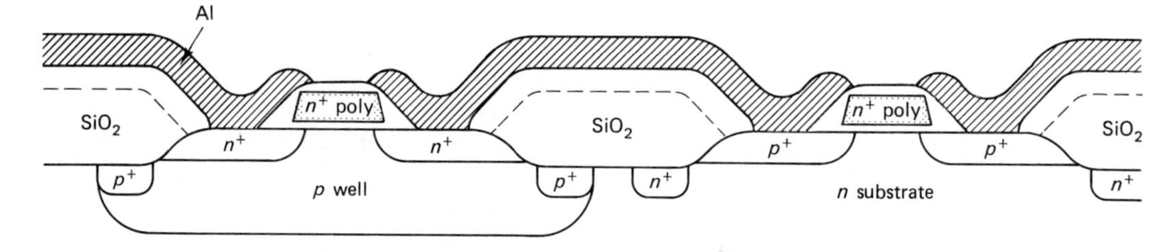

(e) After contact windows opened and aluminum metallization deposited and patterned.
Photomask #8 is used to open the contact windows.
Photomask #9 is used to pattern the aluminum.

Figure 12.17 Outline of a typical p-well CMOS process.

the silicon nitride oxidation mask required in the process is patterned in the second photomasking operation. After the nitride is patterned but before the field oxidation is carried out, boron is implanted into the p well and arsenic into the n-type substrate to raise the threshold voltage in these regions. The implant dose will be about 10^{13} cm^{-2}. These implants are referred to variously as ***guard implants***, ***field threshold adjust implants***, or ***channel stop implants***. The implants are in part blocked by the nitride layer required for the LOCOS step, so that the threshold adjust implant is self-aligned to the edge of the active area. However, photoresist masking must also be used to ensure that, for example, the boron implant does not reach the n-type regions of the wafer. The guard implants are defined using photoresist layers patterned in photomasking operations 3 and 4.

The field oxidation is done in a wet ambient to give a high oxidation rate. Even so, a 2-hour oxidation at 1100°C is required to produce a field oxide thickness of 1 μm. After the field oxidation, the nitride and pad oxide used in the LOCOS process are etched away and the gate oxide is grown. A gate oxide thickness of 50 nm is fairly typical today, but the tendency is to use thinner oxides to preserve long-channel electrical behavior as transistor dimensions shrink (see Section 7.6). The growth of the gate oxide is the most critical step in the entire CMOS process, since all oxide charges must be carefully controlled to give the completed transistors the correct threshold voltages. Although these threshold voltages are determined primarily by the doping level in the substrate and at the surface of the p well and by the gate oxide thickness, it is usually necessary to carry out a light active device threshold adjust implant to set the n- and p-channel thresholds to the desired values. If possible, a single "blanket" (unmasked) implant is used for this task. A dose of about 10^{11} cm^{-2} is typical.

Following the active device threshold adjust implant, the polysilicon layer that will form the transistor gates is deposited and doped n^+ by a phosphorus diffusion or implant. This layer is then patterned using the fifth photomasking operation in the process; the polysilicon is etched in a plasma containing atomic fluorine. After the polysilicon gates are defined, ion implantation is used to form the sources and drains of the two types of transistor. Like the field threshold adjust implants, the source and drain implants are self-aligned, since the polysilicon gate is thick enough to prevent the ions from reaching the underlying channel. Figure 12.18 illustrates the self-alignment. Apart from the fact that there

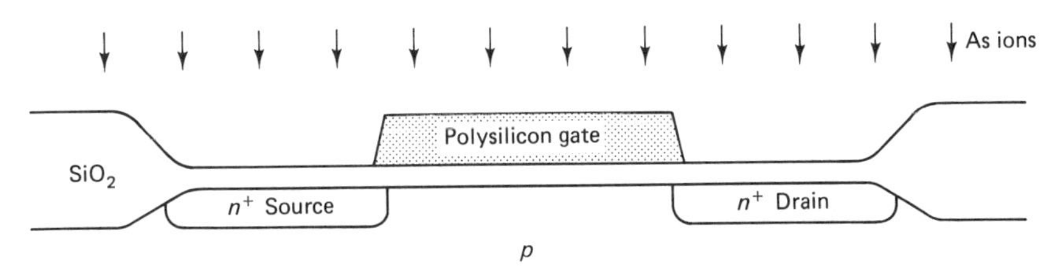

Figure 12.18 Principle of the self-aligned polysilicon gate process. The gate prevents the As ions from entering the channel region.

is some lateral scatter of the ions under the gate as they enter the substrate, and some lateral diffusion of dopant during the anneal that must follow the implant, the edge of each source and drain is perfectly aligned to the edge of the gate. This self-alignment feature of any polysilicon gate MOS process is critical in reducing the extrinsic gate–source and gate–drain overlap capacitances discussed in Section 7.7 and therefore obtaining the maximum possible circuit speed.

An important trade-off must be made in establishing the doping profiles in the sources and drains. On one hand, it is desirable that these regions be deep and heavily doped to minimize the parasitic resistance between the source and drain contacts and the edges of the channel. This resistance can result in significant RC delays in a completed integrated circuit. However, from Section 7.6.3 we find that shallow source–drain junctions are necessary to preserve acceptable electrical characteristics in transistors that are physically small. If nothing else, these considerations suggest that box-shaped doping profiles, such as those obtained with high-concentration arsenic and boron diffusions (Fig. 12.8), are preferable to those with deep "tails" like the high-concentration phosphorus profile (Fig. 12.9). For this reason, arsenic is the preferred dopant for forming n-channel source and drain regions. Arsenic also has a small projected range compared to phosphorus, a low diffusion coefficient, and a high solid solubility in silicon, all of which help in obtaining shallow but highly conductive sources and drains. Since boron is the only practically useful p-type dopant in silicon, one must simply accept the fact that its high implanted range and large diffusion coefficient will give relatively deep p^+ source–drain junctions. Naturally, the energy of the boron implant is kept as low as possible to minimize the junction depth.

It should be noted that very heavy implant doses are required to form source and drain regions of acceptably low resistance; doses in the range from 10^{15} to 10^{16} cm^{-2} are typical. The p-channel source–drain implant is generally kept light enough that the polysilicon gates of the p-channel devices retain their n-type doping. Both types of transistor therefore have n^+ polysilicon gates.

Once the source and drain implants have been carried out and the implant damage annealed, all the critical steps in the CMOS process are finished. To complete fabrication, a layer of silicon dioxide about 0.5 μm thick is deposited over the wafer to provide electrical isolation between the polysilicon gates and overlying metal lines. "Windows" are then etched open in this oxide to expose bare silicon wherever metal is to contact a source, drain, or gate. The eighth photomasking operation is used to define these windows. A 1-μm-thick layer of aluminum is next deposited over the wafer and patterned by etching in phosphoric acid (H_3PO_4) to define the interconnections in the circuit. The ninth photomasking operation in the process is used to pattern the aluminum. Finally, a glass layer is deposited over the wafer to provide protection from external contamination after processing is complete. The tenth photomasking and etching operation is used to remove this glass from the ***bonding pads***—large squares of aluminum that can be connected to the outside world by attaching very fine aluminum or gold wires in a process known as ***wirebonding***.

With all processing operations complete, the silicon wafer is sawn into individual ***chips*** or ***dice*** that can be mounted in the standard dual-in-line packages seen on circuit boards. Since chips of average complexity are typically 5 mm on a side, a single 100-mm-diameter wafer yields about 200 chips. Naturally, not all of these will be functional. In fact, if a chip is particularly complex and valuable, it may be possible to produce it profitably with a yield of 25% or even less.

12.2.12 Typical Silicon Bipolar Process

Integrated circuits first achieved widespread commercial acceptance in the mid-1960s. These early ICs were based on *npn* bipolar transistors with a structure closely resembling that of Fig. 12.19. Although this basic structure has been modified somewhat over the years to reduce parasitic capacitance and improve switching speed, present-day bipolar ICs contain very similar types of transistors.

The starting material for the device of Fig. 12.19 is a relatively heavily doped *p*-type substrate. A much more lightly doped *n*-type layer which will form the collector of the transistor is then grown on the substrate using a process known as ***epitaxy***. In general, epitaxy refers to the growth of a crystalline layer on a crystalline substrate. Epitaxial silicon layers are often grown by the decomposition of silane gas. The process is very similar to the deposition of polycrystalline silicon films discussed earlier, but must be carried out at a much higher temperature—typically, 1000°C—so that the deposited silicon atoms have enough thermal energy to rearrange themselves into a defect-free crystal structure as the film grows. It is also imperative that the substrate surface be clean on an atomic scale to allow the growing layer to seed properly to the underlying crystal.

To obtain the *n*-type doping needed in the collector, a small amount of phosphine is added to the silane flow during the epitaxial growth. To reduce the resistance of the collector, a heavy n^+ ***buried layer*** is formed by diffusing arsenic or antimony into the substrate prior to the epitaxial step. Arsenic or antimony are used as dopants in the buried layer to minimize diffusion of this layer during subsequent high-temperature steps.

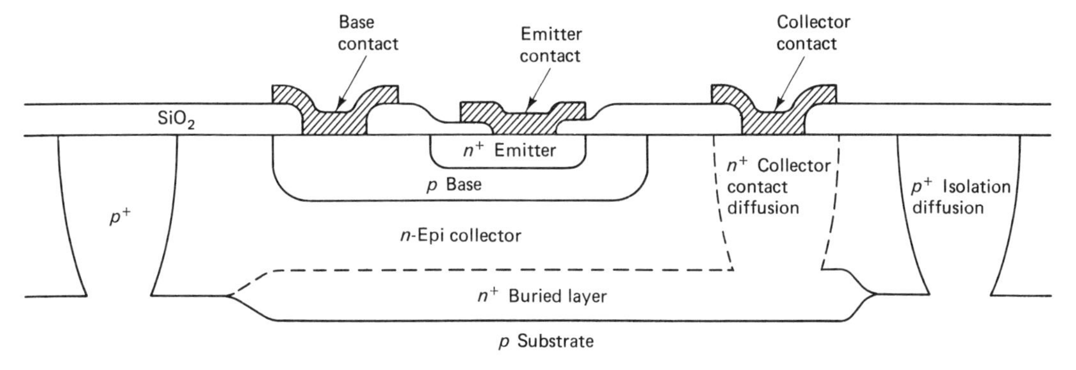

Figure 12.19 Cross section through a conventional silicon bipolar transistor.

Once the epitaxial collector has been grown, a long, high-temperature boron diffusion is used to create a p^+ curtain enclosing the transistor and isolating it from other devices. This p^+ isolation diffusion extends through the thickness of the epitaxial layer. A deep phosphorus diffusion is also done to form an n^+ plug reaching down to the buried layer. External contact to the collector is made through this n^+ plug to minimize resistance.

To complete fabrication of the transistor, the base is formed with a boron predeposition and drive-in, and then the emitter is formed with a phosphorus predeposition. The base–emitter metallurgical junction might typically lie at a depth of 1 μm, and the base–collector junction at a depth of 2 μm, giving a 1-μm active basewidth. To finish processing, contact windows are etched open through the oxide layers grown to serve as diffusion masks, and aluminum is deposited and patterned to form ohmic contacts to the base, emitter, and collector.

12.3 GaAs DIGITAL INTEGRATED CIRCUIT PROCESSING

Although silicon IC fabrication is still evolving rapidly in terms of minimum resolvable feature size and device geometry, most of the basic procedures used in silicon processing such as wafer cleaning, oxidation, and diffusion have been standardized across the IC industry. This simply reflects the fact that silicon ICs have been produced in enormous quantity for well over 20 years. In comparison, GaAs technologies are relatively immature. For this reason it is hard to specify a "representative" GaAs IC process. In this section a possible approach to producing GaAs digital ICs will be outlined, but it should be emphasized that other procedures are possible and that the technology is still in a developmental stage.

Figure 12.20 shows the cross section through a MESFET and a Schottky diode which might form part of a GaAs IC using *S*chottky *d*iode *F*ET *l*ogic (***SDFL***). The starting material for this circuit is a semi-insulating GaAs wafer of (100) orientation grown using the liquid encapsulated Czochralski (LEC) method discussed in Section 2.2.2. The semi-insulating substrate provides isolation between devices.

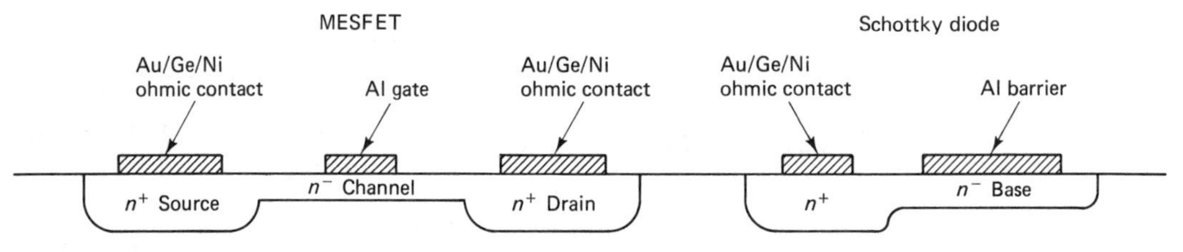

Figure 12.20 Cross section through a GaAs MESFET and Schottky barrier diode, as formed by ion implantation into a semi-insulating substrate.

All the doped regions in Fig. 12.20—the source and drain of the MESFET, the MESFET channel, and the n-type pocket in which the Schottky diode is located—are formed by ion implantation. Photoresist masks are used to define which regions of the wafer the implants penetrate. Different implant doses and energies and, in some cases, different species are used to create these different regions. Silicon is the most common choice of dopant. Silicon is an ***amphoteric*** impurity in GaAs—it can behave as either an n- or p-type dopant, depending on how it is introduced into the GaAs wafer and subsequently annealed. Under the implant and annealing conditions described here, silicon serves as an n-type dopant. A silicon dose of 10^{12} cm^{-2} at an energy of 100 keV is typically used for the channel, while far higher doses (10^{15} cm^{-2} or more) are needed to produce highly conductive drains and sources. Sulfur is sometimes used as a dopant for Schottky diode bases, since it diffuses very rapidly in GaAs, leaving a deep n-type pocket when annealed in the same manner as silicon. This allows the Schottky diode to have a wide depletion region and hence a low junction capacitance, leading to optimum high-speed performance.

To activate an implant in GaAs and remove the implant damage, an anneal at a temperature between 800 and 900°C is necessary. Here one of the main difficulties faced in all GaAs IC processing is encountered: when a GaAs sample is heated to a temperature above about 600°C the arsenic begins to evaporate and the material decomposes. For this reason, GaAs wafers are usually encapsulated with a thin layer of silicon nitride before annealing to prevent the loss of arsenic. In some processes, this nitride layer is added prior to implantation and remains on the wafer in all subsequent processing steps to provide surface protection. Alternatively, wafers may be annealed in an atmosphere saturated with arsenic vapor. A simple means of accomplishing this is ***proximity annealing***, in which the wafers are stacked in contact so that the arsenic lost from one sample can be replaced with that from another.

Once the implants have been completed and the implant damage annealed, ohmic contacts are made to the MESFET sources and drains and to the Schottky diode base regions. Ohmic contacts to GaAs are commonly formed by depositing an alloy of gold, germanium, and nickel and annealing for a very brief period at 400°C. Control of the anneal cycle is crucial to obtaining a smooth contact surface and low resistivity. Formation of the contact is believed to be due in part to the introduction of a germanium-doped n^+ layer at the wafer surface.

After the ohmic contact anneal, the metal layer that forms the barrier in the Schottky diode and the MESFET gate is deposited. Aluminum is a common barrier layer, although titanium overlaid with a layer of gold to provide high conductivity is also widely used. The speed advantages of GaAs MESFETs become apparent only at channel lengths of about 1 μm or less. A pattern transfer technique known as ***lift-off*** is commonly used to produce metal lines with such small dimensions. Lift-off is usually done using positive photoresists, which dissolve readily in solvents such as acetone. A typical lift-off procedure begins by coating the wafer with a 1-μm-thick layer of photoresist (Fig. 12.21). The wafer is next

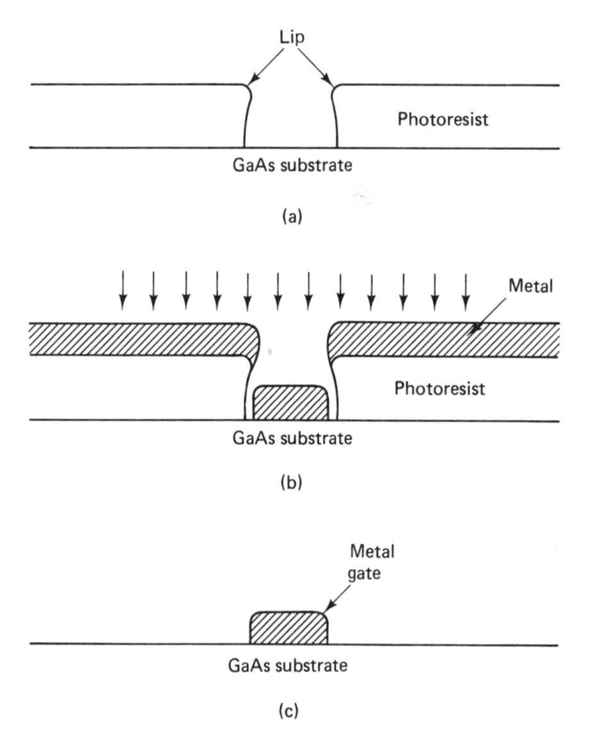

Figure 12.21 Typical lift-off process to form 1-μm gates for GaAs MESFETs. (a) Photoresist is treated so that lip forms after development. (b) When gate metal is deposited, the shadowing effect of the lip ensures a clean break between metal in contact with the substrate and metal overlying photoresist. (c) Photoresist is dissolved in organic solvent, and excess metal is lifted away.

immersed in an organic chemical (chlorobenzene) which reduces the solubility of the upper layer of resist in developer. The photoresist-coated wafer is then exposed and developed following conventional photolithographic procedures. Because of the special treatment received by the top layer of the photoresist, an overhanging "lip" is formed after development. The wafer is next coated with the metal used to form the Schottky gates in a vacuum evaporation system. The lip in the photoresist provides a shadowing effect during the evaporation, leaving a clean break between the metal actually contacting the GaAs surface and that overlying the resist. The wafer is then immersed in a solvent capable of dissolving the photoresist, at which time the excess metal simply floats away. Lift-off procedures of this kind can produce much smaller features in metal layers than could be obtained by standard pattern transfer techniques based on wet chemical etches.

In some GaAs IC processes, a single implant is used to form the source, drain, and channel, and precise control over the MESFET pinch-off voltage is obtained by etching away part of the channel region to form the recessed gate structure shown in Fig. 12.22.

Today it is relatively easy to produce silicon wafers that are essentially defect-free. This is not the case with GaAs, and most commercially available GaAs wafers have a high density of crystal defects. For this reason it is quite common to grow a high-quality epitaxial ***buffer layer*** a few micrometers thick on a GaAs substrate before devices are fabricated. This buffer layer can be grown

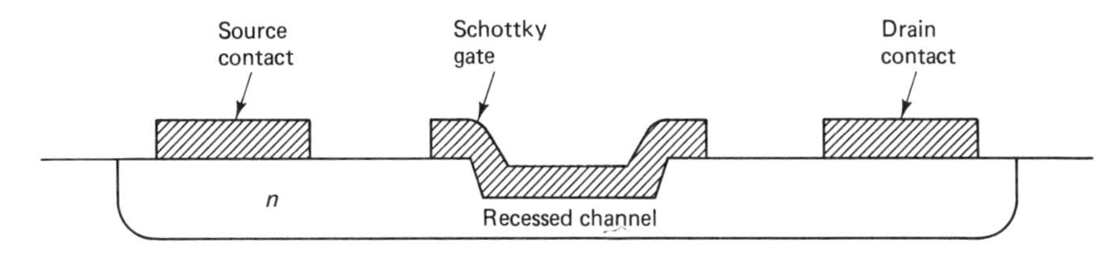

Figure 12.22 Recessed-gate MESFET.

by depositing GaAs from a mixture of arsine (AsH_3) and trimethyl gallium [$(CH_3)_3Ga$] gases. The overall deposition reaction is

$$AsH_3(g) + (CH_3)_3Ga(g) \longrightarrow GaAs(s) + 3CH_4(g)$$

As noted earlier, GaAs IC fabrication is still in a state of flux. One of the most important challenges faced at present is the development of processes capable of forming both enhancement and depletion MESFETs with reasonable yield. Enhancement MESFETs require very light channel doping, so that the work function of the gate metal alone is sufficient to deplete the channel completely even when no gate bias is applied. The doping profile must also be controlled with extreme precision, for the maximum forward bias that can be applied to turn the device on without giving substantial charge flow through the Schottky gate is only about 0.7 V. With this light doping it is essential that the source and drain regions be self-aligned to the edge of the gate to hold the parasitic resistances in the device to an acceptable level. Self-alignment requires that the gate be formed before the source and drain, in much the same fashion as the self-aligned polysilicon gate process used to make silicon MOS ICs. This, in turn, requires that the gate be capable of withstanding the temperatures encountered in annealing the source–drain implant. Gates made of refractory materials such as tungsten have been used successfully in this application.

12.4 CHAPTER SUMMARY

This chapter began with a brief examination of the basic procedures used in fabricating a silicon integrated circuit. These included the fundamental steps of ***thermal oxidation*** and doping by ***solid-state diffusion*** and ***ion implantation*** needed to make actual transistors, and the deposition of thin films such as aluminum (used to form contacts and interconnections) and silicon dioxide (used to provide isolation between conducting layers). The process of ***local oxidation of silicon*** (LOCOS), which is used extensively in both MOS and bipolar integrated circuits, was also mentioned. Attention then turned to how these basic steps could be combined to fabricate complete integrated circuits. A conventional CMOS process was considered in some detail, while bipolar and GaAs digital processes were treated more briefly.

12.5 REFERENCES

1. S. M. Sze, ed., *VLSI Technology,* 2nd ed., McGraw-Hill Book Company, New York, 1988.
2. S. K. Ghandhi, *VLSI Fabrication Principles: Silicon and Gallium Arsenide,* John Wiley & Sons, Inc., New York, 1983.
3. S. Wolf and R. N. Tauber, *Silicon Processing for the VLSI Era,* Lattice Press, Sunset Beach, CA, 1986.
4. A. S. Grove, *Physics and Technology of Semiconductor Devices,* John Wiley & Sons, Inc., New York, 1967.
5. S. M. Sze, *Semiconductor Devices: Physics and Technology,* p. 397, John Wiley & Sons, Inc., New York, 1985.

PROBLEMS

12.1. A *p*-well is to be formed for a CMOS process. The *n*-type substrate has a doping level of 2×10^{15} cm^{-3}. The hole concentration at the surface of the well is to be 10^{16} cm^{-3}, and the well–substrate metallurgical junction is to lie 5 μm below the silicon surface. The well is to be formed by a boron ion implant at 100 keV followed by a long drive-in at 1150°C. Specify the dose for the ion implant and the time required for the drive-in. Assume that all the implanted boron is concentrated at the silicon surface before the drive-in. Is this a good approximation from the viewpoint of finding the doping profile after the drive-in? Explain why.

12.2. Figure 7.33 shows the cross section through an inverter fabricated in a conventional NMOS process. It is important that the parasitic transistor formed by a polysilicon line crossing the field oxide never be turned on. If the field oxide thickness is 1.0 μm, what should the substrate doping be to make the field threshold voltage of this parasitic transistor equal to 10 V?

12.3. An oxide layer is grown on a silicon wafer in a two-step procedure. First the wafer is oxidized in dry oxygen at 1100°C for 60 minutes, and then a second oxidation step is carried out in a wet ambient for 30 minutes at 1000°C. Determine the final oxide thickness using the graphs of Fig. 12.12. (*Hint:* What oxidation time at 1000°C is required to grow an oxide thickness equal to that produced in 60 minutes in dry oxygen at 1100°C? What is the oxide thickness in a wet ambient at 1000°C at this time plus 30 minutes?)

12.4. The base region for a conventional silicon bipolar transistor is formed by a boron predeposition at 900°C for 30 minutes followed by a drive-in at 1050°C for 30 minutes. During the predeposition, the boron concentration at the silicon surface rises to the solid solubility value. The diffusion is done into an *n*-type substrate that will form the collector; the substrate doping is 10^{16} cm^{-3}. Compute the depth of the base–collector junction and the surface boron concentration at the end of the drive-in.

Appendix 1
SPICE Circuit Simulator

A1.1 INTRODUCTION

The main goal of this book has been the development of simple models for diodes and transistors. The term "model" here refers to equations specifying the current in a device and the charge stored in the device as functions of the biases applied at the terminals. Models of this kind can be used to simulate circuit performance.

The simulation of the electrical behavior of integrated circuits has become a very important task today, due both to the fact that it is extremely difficult, if not impossible, to build a discrete component equivalent of an integrated circuit, and because design of an IC by a trial-and-error approach involving the fabrication of a succession of prototypes is far too costly and time consuming to be feasible. For many electronics companies, the margin between financial success and failure is determined by the time required to take an integrated circuit from an original system-level specification to a manufacturable design.

The most popular circuit simulator in use at present is known as ***SPICE*** (the acronym stands for ***S***imulation ***P***rogram with ***I***ntegrated ***C***ircuit ***E***mphasis), which was developed at the University of California at Berkeley in the mid-1970s. The program and supporting documentation are available from the University of California for a nominal copying charge. At the time of writing, the current version is SPICE 2G.6. Although many major companies engaged in integrated circuit design have developed their own circuit simulators which in some circumstances may be more accurate and faster than SPICE, SPICE is still very widely used in industry, and is used almost exclusively in the academic community.

A1.2 MODIFIED NODAL ANALYSIS

To illustrate the operation of SPICE, and to show how SPICE employs a device model, we will consider some very simple examples. First, consider the circuit of Fig. A1.1. Analysis of this simple series combination of resistors is, of course, trivial, but it is instructive to follow the formal approach SPICE would take in this problem.

If we apply Kirchhoff's current law at each of the nodes of the circuit, we obtain the following set of equations:

node 1: $$i_1 - i_2 = 0 \tag{A1.1}$$

node 2: $$i_2 - i_3 = 0 \tag{A1.2}$$

Reexpressing i_2 and i_3 in terms of the voltage drops across each branch, we obtain

node 1: $$I_S - \frac{v_2}{R_A} = 0 \tag{A1.3}$$

node 2: $$\frac{v_2}{R_A} - \frac{v_3}{R_B} = 0 \tag{A1.4}$$

SPICE prefers to deal with node voltages rather than the voltage drop across each branch, and so rewrites (A1.3) and (A1.4) as

node 1: $$I_S + \frac{e_2 - e_1}{R_A} = 0 \tag{A1.5}$$

node 2: $$\frac{e_1 - e_2}{R_A} - \frac{e_2}{R_B} = 0 \tag{A1.6}$$

The node voltages e_1 and e_2 are measured with respect to a reference or ground node conveniently labeled as node 0. Needless to say, the choice of ground node in a circuit is arbitrary. Equations (A1.5) and (A1.6) are referred to as the ***nodal analysis*** (**NA**) equations. This set of linear equations can be solved to find the unknown node voltages e_1 and e_2.

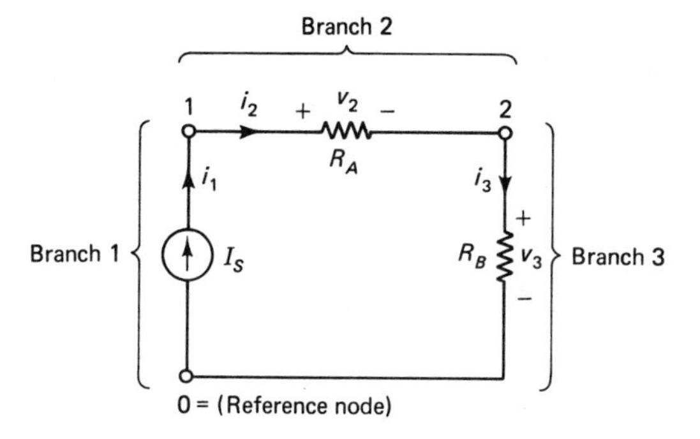

Figure A1.1 Simple circuit to illustrate the method of nodal analysis.

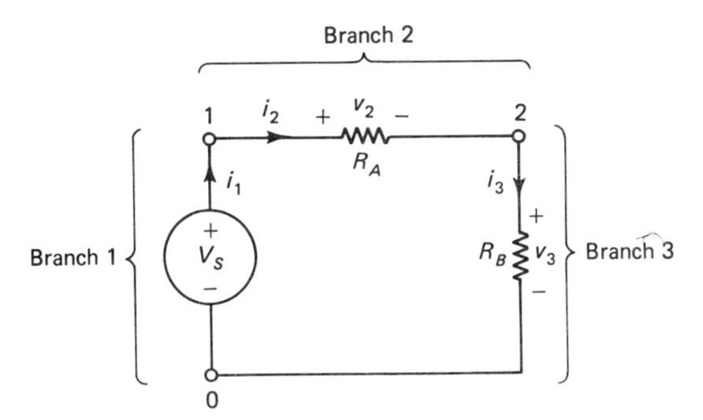

Figure A1.2 Simple circuit to illustrate the method of modified nodal analysis.

The circuit of Fig. A1.2, which contains an independent voltage source V_S, presents a slightly more difficult problem, since we cannot relate i_1 to V_S without knowledge of the currents in the other branches. In this case it is necessary to generate another equation by applying Kirchhoff's voltage law to relate V_S to other node voltages, producing the set of ***modified nodal analysis*** (**MNA**) equations listed below.

node 1: $$i_1 + \frac{e_2 - e_1}{R_A} = 0 \tag{A1.7}$$

node 2: $$\frac{e_1 - e_2}{R_A} - \frac{e_2}{R_B} = 0 \tag{A1.8}$$

$$e_1 = V_S \tag{A1.9}$$

Once again, linear algebra provides a variety of techniques that can be used to solve this system of linear equations for the unknowns e_1, e_2, and i_1.

A1.3 NONLINEAR CIRCUIT ELEMENTS

Naturally, any integrated circuit of practical importance will contain a variety of nonlinear devices as well as simple resistive elements. Such a circuit is shown in Fig. A1.3, in which one of the resistors of Fig. A1.2 has been replaced by a diode. Assuming that the polarity of the voltage source V_S is such as to place the diode

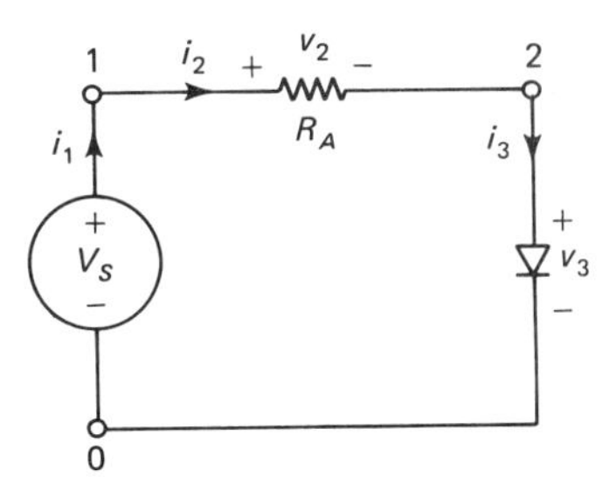

Figure A1.3 Simple circuit to illustrate the method of modified nodal analysis when nonlinear elements are present.

well into forward-bias operation, the steady-state current–voltage relation for the diode is

$$I = I_0 e^{qV/kT} \tag{A1.10}$$

where I_0 is the diode saturation current density. The **MNA** equations for Fig. A1.3 are therefore (A1.7), (A1.9), and

$$\frac{e_1 - e_2}{R_A} - I_0 e^{qe_2/kT} = 0 \tag{A1.11}$$

The **MNA** equations are now nonlinear, so cannot be solved by conventional linear algebra techniques. SPICE solves these equations using an iterative Newton–Raphson linearization procedure. Specifically, suppose that estimates e_1^k and e_2^k for the node voltages are known (here the superscript k will come to denote an iteration number). Applying a Taylor series expansion of the diode current I about some bias point V_0 and retaining only first-order terms gives

$$I \simeq I_0 e^{qV_0/kT} + \left.\frac{dI}{dV}\right|_{V=V_0} (V - V_0) \simeq I_0 e^{qV_0/kT} + \frac{q(V - V_0)}{kT} I_0 e^{qV_0/kT} \tag{A1.12}$$

With this linearization the **MNA** equations for Fig. A1.3 become

node 1:
$$i_1^{k+1} + \frac{e_2^{k+1} - e_1^{k+1}}{R_A} = 0 \tag{A1.13}$$

node 2:
$$\frac{e_1^{k+1} - e_2^{k+1}}{R_A} - I_0 e^{qe_2^k/kT} - \frac{q(e_2^{k+1} - e_2^k)}{kT} I_0 e^{qe_2^k/kT} = 0 \tag{A1.14}$$

$$e_1^{k+1} = V_S \tag{A1.15}$$

The linearized **MNA** equations (A1.13)–(A1.15) can be solved using standard matrix techniques to determine new approximations e_1^{k+1} and e_2^{k+1} to the node voltages. This procedure is repeated until the change in the "best approximation" values for these voltages between iterations is acceptably small.

A1.4 TRANSIENT ANALYSIS

Finally, we consider how SPICE carries out transient analysis. This will be done in reference to the circuit of Fig. A1.3, but we will now suppose that voltage source V_S changes with time.

In a transient situation, it is convenient to divide the current at each terminal of a device into two parts: a ***transport current***, which is simply the steady-state current that would exist under the given bias conditions, and a ***charging current***, which supplies the change in stored charge associated with the terminal in question. As discussed in Section 6.6, SPICE uses a ***quasi-static*** approach to computation of stored charge. Specifically, it is assumed that the stored charge is a function only of the present terminal voltages and does not depend on the previous biasing history of the device. This is, of course, an approximation, but it appears to work reasonably well for present-day integrated circuits and greatly simplifies analysis. When using quasi-static analysis, the transient response of a device can be determined completely by deriving expressions for the steady-state transport current and the steady-state stored charge as functions of applied bias. It is also very convenient if analytic expressions for the derivatives of the stored charge with respect to the terminal voltages are available. In an equivalent-circuit representation, these derivatives are viewed as small-signal capacitances associated with a device.

At moderate forward bias, the main component of stored charge in the diode is associated with electrons injected into the base. The stored charge Q is related to the applied bias V by

$$Q = Q_0 e^{qV/kT} \tag{A1.16}$$

where Q_0 is a constant. Adding this charging current to branch 3 of Fig. A1.3 results in the modification of **MNA** equation (A1.11) to the following:

$$\frac{e_1 - e_2}{R_A} - I_0 e^{qe_2/kT} - \frac{dQ}{dt} = 0 \tag{A1.17}$$

To determine the node voltages, it is therefore now necessary to solve a set of coupled ordinary differential equations (A1.7), (A1.9), and (A1.17). A variety of numerical methods is available for this task. SPICE proceeds as follows. Suppose that the node voltages $e_2(t_n)$ and $e_1(t_n)$ are known at some time t_n (here the n subscript will serve as a time-step index). The values of $e_2(t_{n+1})$ and $e_1(t_{n+1})$ at some time Δt later are then specified implicitly by

$$\begin{aligned} Q(e_2(t_{n+1})) - Q(e_2(t_n)) \\ = \frac{\Delta t}{2}\Bigg[\left(\frac{e_1(t_{n+1}) - e_2(t_{n+1})}{R_A} - I_0 e^{qe_2(t_{n+1})/kT}\right) \\ + \left(\frac{e_1(t_n) - e_2(t_n)}{R_A} - I_0 e^{qe_2(t_n)/kT}\right)\Bigg] \end{aligned} \tag{A1.18}$$

and (A1.7) and (A1.9). The right-hand side of (A1.18) represents the average of the charging currents at time t_{n+1} and at time t_n. This set of equations is then linearized and solved iteratively, following essentially the approach discussed above for the steady-state analysis of circuits containing nonlinear elements. The linearization of the steady-state transport current $I_0 e^{qV/kT}$ in the diode has already been discussed. The linearization of the expression for the stored charge Q is carried out in an analogous fashion. Let $e_2^k(t_{n+1})$ represent the "best guess" at iteration k for node voltage e_2 at time t_{n+1}; then

$$Q(e_2^{k+1}(t_{n+1})) = Q(e_2^k(t_{n+1})) + \left.\frac{\partial Q}{\partial V}\right|_{V=e_2^k(t_{n+1})} [e_2^{k+1}(t_{n+1}) - e_2^k(t_{n+1})]$$

$$= Q(e_2^k(t_{n+1})) + C(e_2^{k+1}(t_{n+1}) - e_2^k(t_{n+1})) \qquad \text{(A1.19)}$$

where C is the diode capacitance.

Naturally, to obtain an accurate estimate for $e_2(t_{n+1})$ and $e_1(t_{n+1})$ at the end of this iterative procedure, it is necessary that the time-step Δt be suitably small. Sophisticated procedures are followed by SPICE and by other circuit simulators to adjust the size of the time-step interval to preserve accuracy as the transient analysis proceeds.

A1.5 SUMMARY

In summary, the procedure followed by SPICE in the transient analysis of a nonlinear circuit is to obtain the solution for the node voltages at a sequence of discrete times. The solution at time t_{n+1} is generated from the solution at time t_n by assembling the set of **MNA** equations, including charging currents, and then solving these equations iteratively. The iterative procedure involves a Taylor series expansion of all nonlinear expressions for stored charge and transport current about a present "best guess" for the node voltages. This generates a set of linear equations that are solved to obtain an improved guess for these voltages. The iteration procedure continues until the node voltages at time t_{n+1} are known with sufficient accuracy, and then the next time point t_{n+2} is considered.

Although this discussion of the circuit analysis procedure followed by SPICE has been presented in the context of some very simple examples, the same basic principles are applied to more complex circuits. The only major extension required concerns the modeling of circuits containing devices with more than two terminals, such as BJTs and MOSFETs. Consider, for example, the transient analysis of a circuit containing a MOSFET. Each of the four terminals of the MOSFET has associated with it a stored charge which, in general, is a function of the biases applied to all the terminals; for example, the charge associated with the gate might be written as

$$Q'_G = Q'_G(V_D, V_G, V_S, V_B) \qquad \text{(A1.20)}$$

Equation (A1.20) is linearized in the form

$$Q_G'^{k+1} = Q_G'^{k} + \frac{\partial Q'_G}{\partial V_D}(V_D^{k+1} - V_D^k) + \frac{\partial Q'_G}{\partial V_G}(V_G^{k+1} - V_G^k)$$

$$+ \frac{\partial Q'_G}{\partial V_S}(V_S^{k+1} - V_S^k) + \frac{\partial Q'_G}{\partial V_B}(V_B^{k+1} - V_B^k)$$

$$= Q_G'^k + C'_{GD}(V_D^{k+1} - V_D^k) + C'_{GG}(V_G^{k+1} - V_G^k)$$
$$+ C'_{GS}(V_S^{k+1} - V_S^k) + C'_{GB}(V_B^{k+1} - V_B^k) \qquad \text{(A1.21)}$$

where, for example, $C'_{GS} = dQ'_G/dV_S$. Expressions for the stored charge associated with the terminals of the MOSFET and for the capacitances such as C'_{GS} are given in Section 7.7. It is worth noting here that there are 16 capacitance elements associated with the four-terminal MOSFET, and that these capacitance elements are in general nonreciprocal; for example, C'_{GS} is not necessarily equal to C'_{SG}.

As a final illustration of circuit analysis with SPICE, the **MNA** equations needed for transient analysis of the simple MOSFET inverter with a resistive load shown in Fig. A1.4 are given below.

$$i_1 - \frac{dQ'_G}{dt} = 0 \qquad \text{(A1.22)}$$

$$\frac{e_3 - e_2}{R_L} - \frac{dQ'_D}{dt} - I_D = 0 \qquad \text{(A1.23)}$$

$$i_4 - \frac{e_3 - e_2}{R_L} = 0 \qquad \text{(A1.24)}$$

$$V_{IN} = e_1 \qquad \text{(A1.25)}$$

$$V_{DD} = e_3 \qquad \text{(A1.26)}$$

In this simple circuit, the **MNA** equations reduce to a single equation in the one unknown node voltage, e_2:

$$\frac{V_{DD} - e_2}{R_L} - \frac{dQ'_D}{dt} - I_D = 0 \qquad \text{(A1.27)}$$

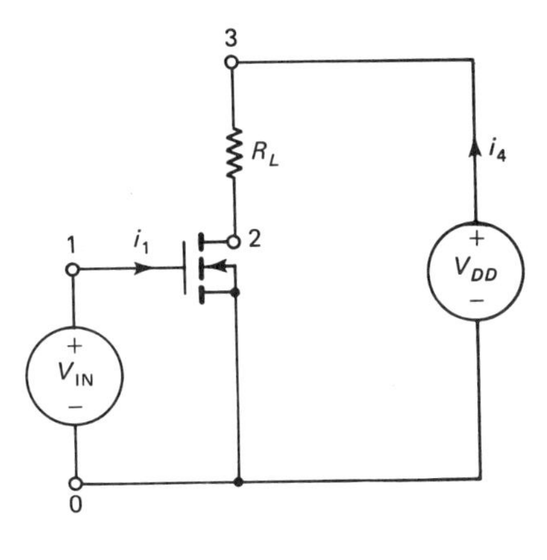

Figure A1.4 Simple circuit to illustrate the method of performing a transient analysis.

Here $Q'_D(V_D, V_G, V_S, V_B) = Q'_D(e_2, V_{\text{IN}}, 0, 0)$ and similarly $I_D(V_D, V_G, V_S, V_B) = I_D(e_2, V_{\text{IN}}, 0, 0)$. At iteration $k + 1$ at time t_{n+1}, the linearized equivalent of (A1.27) is

$$Q_D'^k(t_{n+1}) + \left.\frac{\partial Q'_D}{\partial V_D}\right|_k [e_2^{k+1}(t_{n+1}) - e_2^k(t_{n+1})]$$

$$= Q'_D(t_n) + \frac{\Delta t}{2}\left\{\frac{V_{DD} - e_2^{k+1}(t_{n+1})}{R_L} - I_D^k(t_{n+1})\right.$$

$$\left. - \left.\frac{\partial I_D}{\partial V_D}\right|_k [e_2^{k+1}(t_{n+1}) - e_2^k(t_{n+1})] + \frac{V_{DD} - e_2(t_n)}{R_L} - I_D(t_n)\right\} \qquad \text{(A1.28)}$$

Appendix 2
Properties of Silicon, Gallium Arsenide, and Silicon Dioxide[a]

Property	Unit	Si	GaAs	SiO_2
Atom or molecule density	cm^{-3}	5×10^{22}	2.2×10^{22}	2.3×10^{22}
Atomic or molecular mass	g/mol	28.09	144.63	60.08
Breakdown field	V/μm	30	40	~600
Crystal structure		Diamond	Zinc blende	Amorphous
Density	g/cm^3	2.33	5.32	2.27
Density of states effective mass	m^*/m_0			
Electrons		1.38[b]	0.0655	
Holes		0.946[b]	0.524	
Effective density of states in conduction band	cm^{-3}	4.07×10^{19}[b]	4.21×10^{17}	
Effective density of states in valence band	cm^{-3}	2.31×10^{19}[b]	9.51×10^{18}	
Electron affinity	eV	4.1	4.07	0.9
Energy bandgap	eV	1.12	1.42	8
Intrinsic carrier concentration	cm^{-3}	1.25×10^{10}	2.25×10^{6}	
Lattice constant	nm	0.543	0.565	
Melting point	K	1685	1510	1973
Minority carrier lifetime	s	$\sim 10^{-4}$[c]	$\sim 10^{-8}$[c]	

Property	Unit	Si	GaAs	SiO_2
Mobility	$cm^2\ V^{-1}\ s^{-1}$			
Electrons		1340^c	8500^c	
Holes		461^c	400^c	
Relative permittivity		11.9	12.9^d	3.9
Scatter-limited velocity	$cm\ s^{-1}$			
Electrons		$\sim 1 \times 10^7$	$\sim 1 \times 10^7$	
Holes		$\sim 8.4 \times 10^6$		

Sources:

W. E. Beadle, J. C. C. Tsai, and J. D. Plummer, eds., *Quick Reference Manual for Silicon Integrated Circuit Technology*, John Wiley & Sons, Inc., New York, 1985.

H. F. Wolf, *Semiconductors*, John Wiley & Sons, Inc., New York, 1971.

R. C. Weast, ed., *Handbook of Chemistry and Physics*, 60th ed., CRC Press, Inc., Boca Raton, Fla., 1981.

R. F. Pierret, *Advanced Semiconductor Fundamentals*, Vol. 6 in Modular Series on Solid-State Devices, Addison-Wesley Publishing Company, Inc., Reading, Mass., 1987.

J. S. Blakemore, "Semiconducting and other properties of GaAs," *Journal of Applied Physics*, vol. 53, pp. R123–R181, 1982.

R. Williams, "Photoemission of electrons into silicon dioxide," *Physical Review*, vol. 140, pp. A569–A575, 1965.

[a] At 300 K, unless otherwise stated.

[b] Proposed values (see Sections 4.4 and 4.4.1).

[c] Values for intrinsic material, [see (3.8), (3.9), and (2.33) for doping density dependence of mobility and minority carrier lifetime in silicon].

[d] 13-1 is also widely used, [see Table 9E.1].

Appendix 3
Physical Constants

Constant	Symbol	Value	
Avogadro's number	A	6.02×10^{23}	atoms/mole
Boltzmann's constant	k	8.62×10^{-5}	eV/K
Elementary charge	q	1.6×10^{-19}	C
Electron rest mass	m_0	9.1×10^{-31}	kg
Electron volt	eV	1.6×10^{-19}	J
Micron	μm	10^{-4}	cm
Permittivity of free space	ϵ_0	8.85×10^{-14}	F/cm
Planck's constant	h	4.14×10^{-15}	eV-s
Speed of light in vacuum	c	3×10^{10}	cm/s
Thermal energy at 300 K	kT	0.0259	eV
Thermal voltage at 300 K	kT/q	0.0259	V

Index

T